Elements of Mechatronic Systems

IEEE Press
445 Hoes Lane
Piscataway, NJ 08854

Elements of Mechatronic Systems

A Design Oriented Approach Using MATLAB®

Oguz A. Soysal
Frostburg State University
Maryland, USA

IEEE Press

WILEY

Published by John Wiley & Sons, Inc., Hoboken, New Jersey.

Library of Congress Cataloging-in-Publication Data
Names: Soysal, Oguz A., 1951– author
Title: Elements of mechatronic systems : A design oriented approach using MATLAB® / Oguz A. Soysal, Frostburg State University, Maryland, USA.
Description: First edition. | Hoboken, New Jersey : Wiley, [2026] | Includes bibliographical references and index.
Identifiers: LCCN 2026003969 (print) | LCCN 2026003970 (ebook) | ISBN 9781394339716 Hardback | ISBN 9781394339730 adobe pdf | ISBN 9781394339723 epub
Subjects: LCSH: Mechatronics–Design and construction | Mechatronics–Data processing | MATLAB | Computer simulation
Classification: LCC TJ163.12 .S69 2026 (print) | LCC TJ163.12 (ebook)
LC record available at https://lccn.loc.gov/2026003969
LC ebook record available at https://lccn.loc.gov/2026003970

Cover Design: Wiley
Cover Image: David/stock.adobe.com

Set in 9.5/12.5pt STIXTwoText by Straive, Pondicherry, India

To Hilkat, my lifelong companion and best friend.
This book would not exist without her motivation and support.

Contents

About the Author

Oguz A. Soysal, PhD, is Emeritus Professor of Electrical Engineering. He earned all his degrees from Istanbul Technical University and taught for 23 years at several universities in Turkey. During his academic career, he contributed as a visiting scholar to research projects at The Ohio State University in Columbus, Ohio, and the University of Toronto in Canada; he also served as a visiting professor at Bucknell University in Lewisburg, Pennsylvania. He served for 25 years at Frostburg State University in Maryland as a faculty member and department chair, leading several renewable energy projects. His teaching and research interests include energy systems, power electronics, control systems, and mechatronics. He has published more than 50 papers in leading journals and international conference proceedings and co-authored, with Hilkat S. Soysal, the textbook *Energy for Sustainable Society: From Resources to Users*, published by Wiley in 2020.

Preface

Mechatronics combines creativity and practical solutions to meet the demands of today's rapidly changing technology. This field blends mechanical systems, electronics, control theory, and computer science into one unified approach for designing and developing smart products and processes. This book is intended for advanced undergraduate students concentrating on electrical and mechanical engineering, providing foundational concepts and fostering innovative problem-solving within the interdisciplinary field of mechatronics.

This book is designed to meet the growing need for engineers who can work effectively across traditional disciplines. Contemporary industries including automotive, aerospace, robotics, and biomedical engineering require engineers to demonstrate practical proficiency across various fields.

The goal of this book is to bridge theoretical knowledge in diverse fields and practical applications. The material is designed to give students a thorough overview of key concepts and help them develop both analytical skills and hands-on understanding needed to tackle difficult engineering problems. By studying this book, readers will gain the skills to design, model, simulate, and build mechatronic systems that adapt responsively to environmental changes and user interactions.

Since the material is intended for advanced engineering students, readers are expected to have prior exposure to calculus, differential equations, linear algebra as well as fundamental principles of dynamics, electric circuits, magnetism, basic electronics, and programming concepts. This level of preparation enables a deeper exploration of the unique challenges and opportunities inherent in mechatronic system design.

While this book is primarily structured for undergraduate students, it is also suitable for early graduate students and practicing engineers seeking to advance or refresh their knowledge in this emerging field.

In standard electrical engineering programs, digital logic design, electric circuits, and microelectronics are typically delivered as distinct mandatory courses. Electric machinery, power electronics, signal processing, and control theory are usually offered as electives. The respective books provide a comprehensive and in-depth examination of theory and applications within each discipline. This book offers an integrated overview of key topics for mechatronic system analysis and design.

This book is organized into 13 chapters, with each chapter addressing a key component of the interdisciplinary field of mechatronic systems.

Chapter 1 outlines the scope of mechatronic systems and highlights their practical applications. Chapter 2 reviews the foundational concepts of classical mechanics and electrical circuits which engineering students typically encounter in separate introductory courses.

Chapter 3 introduces the key concepts of dynamic system modeling, system identification, and linearization techniques. It uses a unified approach to model basic mechanical, electrical, hydraulic, and thermal components, particularly in the context of system integration.

Chapter 4 examines sensors that convert physical quantities into electrical signals, while Chapter 5 focuses on signal conditioning techniques such as amplification, noise filtering, and modulation.

Chapter 6 covers the fundamentals of electromagnetic field theory, magnetic circuits, and static electromagnetic devices. Chapter 7 then introduces electromechanical energy conversion principles and their applications in transducers and actuators.

Chapter 8 examines rotating machines commonly used in mechatronic systems. Unlike traditional electric machine texts, it emphasizes reluctance motors, stepper motors, permanent-magnet DC motors, and fractional-horsepower induction motors because of their widespread use in mechatronics. It also covers permanent-magnet synchronous motors used in direct-drive systems with electronic converters.

Chapter 9 covers electronic converters, including rectifiers, DC choppers (DC-to-DC converters), and inverters. It also explains how variable frequency drives (VFDs) control AC motor speed in modern mechatronic systems by combining rectifier and inverter circuits.

Chapter 10 focuses on mechanisms for transmitting and converting rotary and translational motion. It includes practical examples and selection criteria. Analog and digital control systems are presented in separate chapters. Chapter 11 introduces fundamental control theory, progressing from open-loop motor dynamics to feedback control and proportional, integral, and derivative (PID) methods. It emphasizes position and velocity control systems commonly used in mechatronics and robotics. Chapter 12 focuses on digital control systems, including combinational and sequential logic circuits, programmable logic controllers (PLCs), microprocessors, and microcontrollers. It also covers digital signal conversion and conditioning techniques.

Chapter 13 integrates the preceding material and guides the reader through system-level design, simulation, prototyping, testing, as well as fault detection and diagnostics.

Each topic is approached from both theoretical and practical perspectives, with practical examples and illustrations to anchor concepts in tangible experience.

Teaching mechatronics involves more than simply conveying information; it includes engagement with the iterative design process and managing uncertainty. Therefore, this book is prepared considering the following pedagogical approach:

- **Conceptual Clarity**: Complex concepts are clarified using diagrams, analogies, and comparative analyses that connect new material to established knowledge.
- **Incremental Complexity**: Topics are introduced in manageable steps, building from simple models to more sophisticated systems, ensuring a solid grasp of each layer before advancing.
- **Interdisciplinary Connections**: Frequent cross-references are made between mechanical, electrical, and computational domains, highlighting both the unique contributions and the necessary compromises inherent in integrated design.
- **Real-World Relevance**: References to contemporary industry practice demonstrate the relevance and impact of mechatronic solutions in fields such as robotics, manufacturing automation, smart infrastructure, and medical devices.
- **Active Application**: Each chapter includes self-assessment quizzes, short answer questions, essay topics, numerical problems, and open-ended design problems that challenge students to synthesize knowledge, make informed decisions, and justify their approaches.

How to Use This Book

This book is designed to serve as either a required or reference book for an introductory course in mechatronics. Instructors will find the material adaptable to a range of course structures, from traditional lectures to project-based learning environments. Certain sections may be skipped depending on the students' background knowledge or assigned as homework.

To support both instructors and students, the book incorporates a variety of features aimed at enhancing comprehension and retention:

- **Clear Learning Objectives**: Each chapter begins with a concise summary of goals, providing a roadmap for focused study.
- **Worked Examples**: Step-by-step solutions illuminate problem-solving strategies and common pitfalls.
- **Practice Problems**: A diverse set of questions at the end of each chapter reinforces key concepts and encourages further exploration.
- **Software Tools**: Guidance is offered for simulation environments such as MATLAB/Simulink®, Multisim®, and Arduino, equipping students with the skills to model and test their designs virtually before physical implementation.
- **Supplementary Resources**: At the end of each chapter, references provide additional resources to expand readers' knowledge. An accompanying online portal provides additional content, including solved problems, code samples, and design examples.

Students should use this book as a dynamic learning tool instead of just static information. Every chapter encourages students to try out the examples, take on the challenges, and think about how various concepts lead to creative solutions. The companion website offers solutions to problems as well as additional teaching resources.

October 2025

Oguz A. Soysal
Frostburg, Maryland

Acknowledgments

This publication represents the synthesis of knowledge and expertise accumulated over more than 60 years as a hobbyist, student, engineer, and instructor. I deeply appreciate the excellent instructors, mentors, and colleagues at my alma mater, Istanbul Technical University (ITU), Turkey, especially Prof. M. Kemal Sarioglu and Prof. Yilmaz Tokad. I am thankful to my colleagues at Istanbul University (IU) and Karadeniz Technical University (KTU) in Turkey and Frostburg State University (FSU) in Maryland, USA, for their support in my courses and research endeavors. My professional perspective has been greatly shaped by the guidance and expertise of Prof. Stephen Sebo, Prof. Umit Ozguner, and Prof. Fusun Ozguner from Ohio State University; Prof. Adam Semlyen at University of Toronto; and Prof. Maurice Aburdene of Bucknell University. Many thanks to my students, whose questions and creativity have consistently driven my pursuit of clarity and excellence. I would like to express my gratitude to Mr. Evren Atli of Lockheed Martin for his recommendations regarding sensors and fault detection topics.

I wish to extend my sincere appreciation to Wiley's editorial team and technical reviewers for their valuable and constructive feedback.

I am truly grateful to Prof. Siddik Yarman and Mrs. Hilkat Soysal for their ongoing support and motivation throughout my work.

About the Companion Website

This book is accompanied by a companion website.

www.wiley.com/go/Soysal/emsdoamatlab

To access the website materials, click on the link under Related Resources on the book's page on wiley.com. You can also find this by searching www.wiley.com for the ISBN 9781394339716.

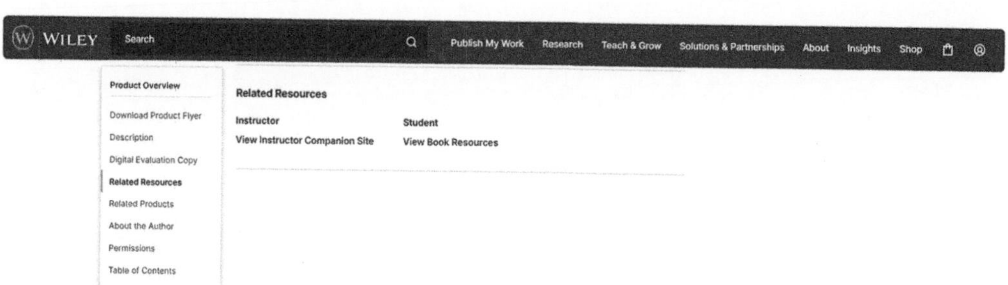

This website includes:

- Problems Solution Key
- Presentation Lecture Slides
- MATLAB Codes
- Questions Answer Key
- Problem solutions

1

Overview

OBJECTIVES

- Present an overview of mechatronics.
- Explain the anatomy and building blocks of mechatronic systems.
- Discuss mechatronic control strategies.
- Describe key applications of mechatronics.
- Outline the mechatronic system design process.

Mechatronic systems consist of mechanical parts controlled by electronic circuits to perform a particular task that involves motion. The term "mechatronics" combines the words "mechanics" and "electronics." First used in the late 1960s, the term rapidly became popular to denote a multidisciplinary field that deals with design and development of intelligent electromechanical products. Today, advanced mechatronics systems include analog and digital microchips, embedded microcontrollers, and sensor networks coordinated by computer software. The diagram in Figure 1.1 shows the flow from engineering disciplines to practical applications of mechatronics.

Primary engineering branches have evolved over time as specialized disciplines and established their own territory. Traditional engineering education focused on a particular field and provided limited exposure to other disciplines. Mechanical engineering has flourished with the industrial revolution and beyond to design and develop machinery and equipment for increasing needs in manufacturing, construction, agriculture, and transportation. The invention of electric machines, lightbulbs, gramophone, telephone, radio, and other electrical devices created a new field of expertise named electrical engineering. While early innovations in electrical and electronic devices were changing the lifestyle, electromechanical drives enabled more efficient and easily controlled industrial operations.

Mechatronics emerged as a multidisciplinary field of engineering in the second half of the twentieth century. Mechatronic systems have assumed an increasingly important role in our everyday life and all sectors of industry. Every aspect of our daily activities is affected by some type of mechatronic device or appliance (Billingsley, 2006; de Silva, 2005).

The complexity of advanced mechatronics devices changed the expectations of industrial companies for professionals employed to develop such products. The older model of electromechanical product design team included electrical and mechanical engineers to work on the parts of the system in their areas of expertise. A team of specialized elements needed to spend more time to achieve an optimal product design. In addition, for more efficient product development workflow, all aspects of the design need to be carried out simultaneously. Mechatronics engineers are typically

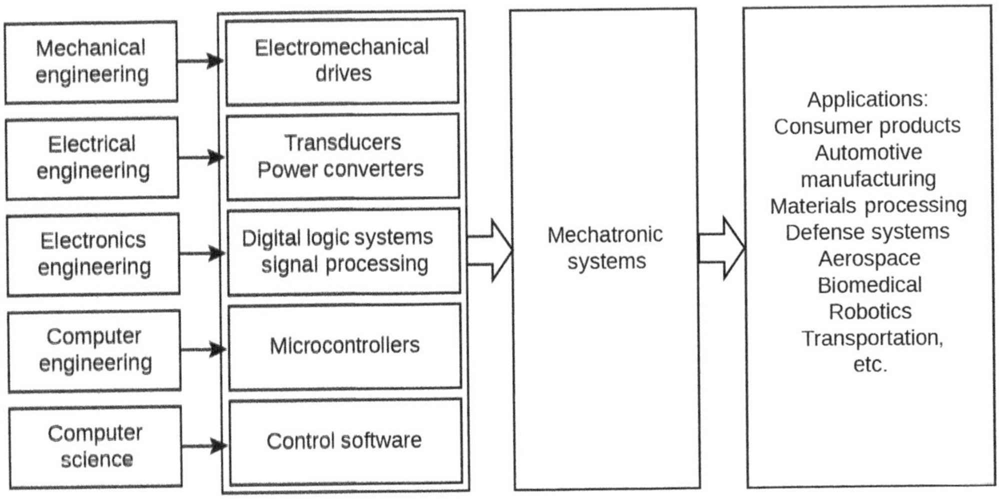

Figure 1.1 Engineering disciplines and topics that contribute to design of mechatronics applications

expected to design highly functional and adaptable products composed of electromechanical components, electronic converters, computer hardware, and software.

The goal of this text is to introduce the building blocks of mechatronic devices. The reader will acquire sufficient background to achieve relatively simple design works and understand more advanced topics in the field.

1.1 Evolution of Mechatronics

Initially, mechatronics technology emerged from electromechanical equipment used for industrial drives. In the nineteenth century, the invention of direct current (DC) motors was a breakthrough in the industry, allowing quiet, reliable, and more efficient operation of production lines. The first applications of electromechanical drives were based on servomechanisms that control either the speed or position of a mechanical device. The major challenge was, however, controlling the motor speed efficiently in a wide range because changing DC voltage required expensive additional motor–generator sets. This issue could not be resolved until power electronics devices became commercially available in the 1970s to make solid-state variable voltage converters.

Over time, electromechanical control systems have evolved to contain computer technology and programming to control more complex actions. Sequential process controls using electromagnetic relays and valves became popular in industrial automation. Programmable logic controllers (PLCs) simplified the design and installation of automated production lines.

During World War II (WWII), the defense industry developed adaptive control systems to track and intercept incoming missiles. While the defense sector has always supported progress in control systems, the space race that started following WWII in the Cold War era accelerated the development of intelligent mechanisms to be used in spacecraft.

In 1969, a senior engineer Tetsuro Mori combined the words "mechanism" and "electronic" to describe the industrial automation and robotics equipment that Yaskawa Electric Corp. was manufacturing (Brown, 2011). From 1971 to 1982, the company held the trademark rights of the term. In trademark application documents, the company explained the term as "technologies and developed products will be incorporating electronics more and more into mechanisms, intimately and organically, and making it impossible to tell where one ends and the other begins." Starting from 1982, the term has received broad public acceptance (Pannaga et al., 2013). The definition of mechatronics has evolved as microelectronics and computer technology continued to progress. Harashima, Tomizuka, and Fukada suggested one of the quoted definitions in 1996 as "the synergistic integration of mechanical engineering, with electronics and intelligent computer control in the design and manufacturing of industrial products and processes" (Bishop, 2002).

The Japan Society for the Promotion of Machine Industry (JSPMI) classified mechatronics products into four categories (Kyura & Oho, 1996):

Class I: Primarily mechanical products with electronics incorporated to enhance functionality. Examples include numerically controlled machine tools and variable speed drives in manufacturing machines.

Class II: Traditional mechanical systems with significantly updated internal devices incorporating electronics. The external user interfaces are unaltered. Examples include the modern sewing machine and automated manufacturing systems.

Class III: Systems that retain the functionality of the traditional mechanical system, but the internal mechanisms are replaced by electronics. An example is the digital watch.

Class IV: Products designed with mechanical and electronic technologies through synergistic integration. Examples include photocopiers, intelligent washers and dryers, rice cookers, and automatic ovens.

Today, mechatronics engineering societies operate in many countries to publish the latest developments and organize professional meetings. The Institute of Electrical and Electronics Engineers (IEEE) and the American Society of Mechanical Engineers (ASME) recognize mechatronics engineering as a multidisciplinary field of engineering by publishing the refereed journal: *IEEE/ASME Transactions on Mechatronics*.

Over the past five decades, advancements in semiconductor and computer technology enabled control of electromechanical systems by electronic converters and embedded microcontrollers. Hence, mechatronics engineering has become a multidisciplinary field of engineering combining mechanical, electrical, electronics, and computer engineering as well as computer science and information technology. Since the 1990s, mechatronics has found widespread applications in all sectors of industry and daily life because of the increased use of computational intelligence. In recent years, the automotive industry, consumer electronics, and various other industrial sectors have represented a substantial portion of the demand in the microelectronic chip market.

1.2 Anatomy of Mechatronic Systems

Mechatronic systems are a collection of subsystems interacting to perform a particular task that involves motion. Today, all mechanical devices include electronic components managed by some type of control strategy. The block diagram in Figure 1.2 shows the information flow in a typical mechatronic system.

The main components are power source, controller, sensors, actuators, electronic converters, and signal conversion-conditioning circuits (Alciatore & Histand, 2003). The mechanism to be controlled is often called plant or process. System components interact with each other by input and output signals. A signal is a voltage or current that carries some kind of information.

Sensors convert physical quantities to electric signals. In practice, the sensor outputs include some errors and electric noise, also called disturbance. Signal conditioning circuits filter out the

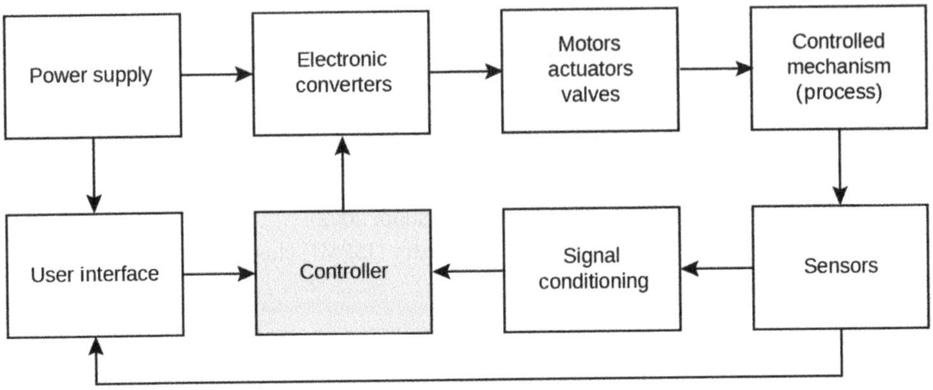

Figure 1.2 Mechatronic system architecture

noise. Conditioned signals are further processed by applying certain mathematical operations such as addition, subtraction, multiplication, and comparison. In addition, integral and derivative operations applied to the processed signals improve the system performance by increasing the accuracy and response speed (Cetinkunt, 2015).

Motors and actuators are electromechanical energy converters. Conventional motors convert electric energy into rotational motion. The actuator furnishes force to move a mechanism. In practice, both are electromagnetic mechanisms that convert electrical energy to mechanical energy.

The controller, the shaded block in Figure 1.2, manages the combination of conditioned and processed sensor signals and produces output signals applied to converters, which produce voltages and currents at sufficient power to drive motors, relays, valves, and actuators. Thin arrows in the diagram represent the flow of signals between elements. Arrows represent power supplied to electrical components.

Mechatronic system architecture may be analog or digital. In an analog control system, signals and all system variables are continuous, meaning that their values are known at any instant. Digital control systems operate with binary variables defined over fixed time intervals defined by a clock pulse. Therefore, analog-to-digital converters (ADCs) convert inputs received from analog sensors into binary numbers. Digital sensors generate binary numbers that can be applied to the processor directly.

Figure 1.3 outlines a typical digital control system. Digital and analog sensors collect information about the internal variables of the controlled mechanism. Analog signals are amplified and processed before being converted into digital form to align with the inputs required by the central processing unit (CPU). Digital signals are processed and transferred directly to the CPU. The CPU executes logic operations and calculations at each clock pulse and generates digital signals to manage system outputs. The CPU may be a simple digital logic circuit, an embedded microprocessor, or a comprehensive microcomputer.

Industrial system designers usually prefer PLCs for reliability, convenient installation, easy programming, and troubleshooting features. PLCs have been popular building blocks for assembly lines, machine control, and robotics.

Digital control has become a cornerstone of mechatronic system design, transforming how systems operate in both industrial and consumer domains. By integrating microprocessors, digital sensors, PLCs, and embedded control architectures, digital control enables precise, efficient, and adaptable management of electromechanical devices. Below are some of the most common applications for digital control within mechatronic systems.

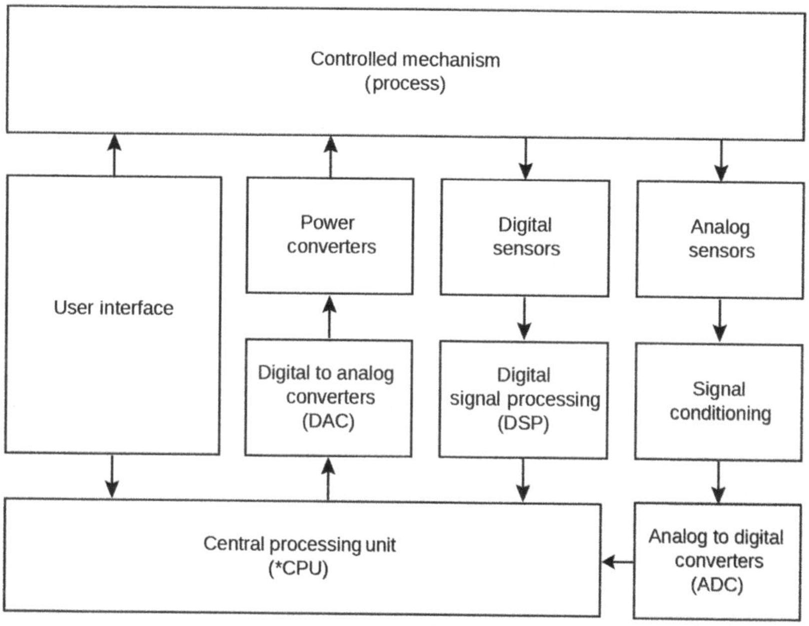

Figure 1.3 Digital control

From industrial automation to smart consumer products, digital control empowers mechatronic systems with greater intelligence, precision, and flexibility. By harnessing digital signals, embedded processing, and sophisticated software, these systems continue to revolutionize diverse sectors, improving both efficiency and quality of life.

Consumer electronics products, programmable appliances, electric cars, and some advanced tools employ embedded microcontroller chips. In more advanced systems, external PCs can perform more advanced control algorithms by executing a high-level computer code.

1.3 Building Blocks of Mechatronic Systems

A mechatronic system typically includes a combination of electrical and mechanical elements such as sensors, signal conditioning and processing circuits, actuators, electric motors, transmission elements, power converters, logic circuits, microprocessors, and controllers. Software modules coordinate digital control strategies and facilitate the user interface. The following building blocks are further explained in subsequent chapters.

1.3.1 Sensors

Sensors detect and convert physical quantities into signals that can be interpreted and processed by electronic components. In electronics, a signal is a voltage, current, or electromagnetic wave that carries information. Sensors are crucial for collecting data about the system's environment and its operational status. Temperature, pressure, position, velocity, and acceleration sensors are frequently used in mechatronics. Advanced mechatronic systems, including autonomous vehicles and robotics, employ sophisticated data acquisition technologies such as GPS (Global Positioning

System) and LIDAR (Light Detection and Ranging) for precise position detection and distance measurement.

1.3.2 Signal Conditioning Circuits

Sensors generate low-magnitude signals that need to be amplified for transmission throughout the mechatronic system and further processing. Filters eliminate the undesirable frequency components such as noise and interference to ensure the accuracy of the signals. ADCs and digital-to-analog converters (DACs) play an essential role in interfacing digital and analog components within a system.

1.3.3 Actuators

Actuators convert various forms of energy into linear or rotary motion. Electromechanical transducers and electric motors are crucial elements of mechatronic systems. Hydraulic cylinders and pneumatic actuators are often used alongside electromechanical elements in control architectures. Each type offers distinct advantages depending on the application, ranging from precision control to high power output.

1.3.4 Transmission Elements

Transmission elements, such as gears, belts, and shafts, transmit power from actuators to other parts of the system. These components must be precisely engineered to ensure smooth and efficient power transfer, which is vital for maintaining the system's performance.

1.3.5 Microcontrollers and Microprocessors

Embedded computer chips, microprocessors, and microcontrollers are the essential building blocks of digital systems. They execute complex algorithms and control signals to manage the system's operations. Most industrial systems employ PLCs to control assembly lines and robotic arms. These devices are responsible for processing input from sensors, making decisions, and controlling actuators.

1.3.6 Power Converters

Electronic power converters change the voltage and frequency of electricity to supply adequate power to various components of a mechatronic system. Power converters can provide either variable or regulated voltages. There are four essential types of converters to supply DC voltage DC source (DC/DC or DC chopper), DC from alternating current (AC) source (AC/DC or rectifier), AC from DC (DC/AC or inverter), and AC from AC (AC/AC or frequency converter).

DC choppers provide variable DC voltage from a fixed DC source like a battery. They are crucial elements of cordless power tools and larger variable speed DC motor drives.

Rectifiers provide DC power from AC utility to supply electronic circuits, DC actuators, and battery chargers. Switch-mode converters are high-efficiency, lightweight, and low-cost alternative to linear voltage regulators. High power AC-to-DC converters provide adjustable voltage to drive DC motors at variable speed by employing the pulse-width modulation (PWM) technique.

Inverters are the opposite of rectifiers; they provide AC power from DC sources like batteries and solar photovoltaic (PV) modules. Controlled rectifiers can be configured for bidirectional power flow. In hybrid and electric vehicles, efficiency is improved by utilizing AC/DC converter in a regenerative brake system, which charges the battery bank when the electric motor functions as a generator during deceleration.

AC-to-AC converters change the voltage and frequency of the AC power. They are particularly useful in residential-sized grid-tie wind turbines, which generate variable frequency and variable voltage electric power changing with wind speed.

1.3.7 Communication Modules

Communication modules enable data exchange between various parts of the mechatronic system. These can include wired communication, like Ethernet, or wireless communication, such as Bluetooth and Wi-Fi. Effective communication is essential for coordinated and synchronized operation of mechatronic systems.

1.3.8 Software

Software provides the platform for executing control algorithms, processing sensor data, and interfacing with users. Key aspects of mechatronic software include control algorithms, data processing, and user interface.

Control algorithms define the behavior of the system. These algorithms can range from simple proportional-integral-derivative (PID) controllers to advanced model predictive controllers (MPCs). The choice of algorithm depends on the complexity and demands of the application.

Data processing software analyzes the information gathered from sensors and makes appropriate decisions based on the system's objectives. This can involve filtering, statistical analysis, and machine learning techniques to improve system performance and adaptability.

User interface software allows operators to interact with the mechatronic system. This can include graphical user interfaces (GUIs), touch screens, and remote control applications. A well-designed user interface is crucial for ease of operation and monitoring.

1.4 Mechatronic Control Strategies

A mechatronic device or system can be operated manually or automatically. The simplest operation type is manual control. Automated control techniques are known as on–off, sequential, and feedback control.

1.4.1 Manual Control

The manual operation of a device is also called "open-loop control." In this control mode, a human operator observes the behavior of the system and then decides what action to take. Then, using the arms and feet, the operator takes a particular action to perform the task.

The eyes collect information about the state of the work being done; the brain processes the information and sends signals to the muscles to perform required actions.

Collecting information (observations) and decision-making do not require muscle power. The processed information lets the brain generate a signal traveling through the nerves to activate the muscles, which produce the powered (amplified) version of the output signal.

The term "open loop" indicates that such a system is unable to make decisions without the control of an operator to complete the intended task. In other words, the input is independent of the output; the human operator changes the input based on the observed output.

Driving a car without using cruise control is an example of open-loop control. The driver adjusts the speed using the gas pedal or brake at a pressure proportional to the desired acceleration. When the cruise control is turned on, the control system of the car takes control to maintain the speed at the value set by the driver.

1.4.2 On–Off Control

The simplest type of control is on–off control. The control system may simply turn a device on or off based on the state of a comparator. In such systems, the comparator output is binary, whether it is true or false. A thermostat-controlled heater is an example of on–off control. If the temperature falls below the desired level, the controller turns the heater on until the desired temperature is reached.

On–off control systems must have a certain error margin for the output quantity. If the system is set to turned on or off precisely at the same value of the output quantity, the system would oscillate. For example, an on–off temperature control system should turn the heater when the room temperature slightly exceeds the set value. As the room temperature increases, the switch should turn the heater off when the temperature slightly exceeds the set value. This feature is known as hysteresis.

1.4.3 Sequential Control

Sequential control systems are based on on–off control of different elements at certain time sequences. For example, a dishwasher can be programmed to turn the water inlet valve on until a certain amount of water is filled in the tank. Then, the heater is energized until the water temperature reaches a certain degree. In the meantime, the motor can spin the jets to wash dishes. When the washing sequence is completed, the spinner motor stops, and the pump motor is energized to empty the tank.

Sequential control of older appliances used to be achieved by a clock motor turning a cam. The cam would turn on and off several switches at different times to start or stop different elements of the system. As digital integrated circuits (ICs) became available, digital logic circuits replaced the mechanical controllers. In industrial production lines, PLCs were preferred. Today, sequential control is significantly simplified by microcontrollers.

Traffic lights are another example of sequential control. A traffic light can be programmed using logic flip-flops or binary counter chips. In more advanced traffic lights, a camera monitors the intersection using image-processing software. A microcomputer turns the traffic lights on or off sequentially.

In electronic sequence control, electromagnetic or solid-state relays must be used as an interface to energize the controlled elements according to the signals electronic circuits produce.

1.4.4 Feedback Control

A system may be designed such that sensors collect information from the output, electronic circuits process the signals, compare them with some reference inputs, and change the output to perform the desired task. Such control strategy is called "feedback," "closed-loop," or "automatic" control.

The role of the comparator is to compare the actual output quantities with the desired reference values provided by the external user. The difference between the reference values and the information collected from the plant outputs is often called "error signal."

An analog controller may contain operational amplifiers that magnify the error signal proportionally and perform analog integration and/or differentiation. A combination of mathematical operations in the controller is usually indicated by letters P, I, and D, which stand for proportional, integral, and derivative, respectively. For example, a PI controller combines proportional and integral, PID controller combines all three operations. Accurately designed PID control improves system performance by increasing accuracy and optimizing the response time.

Digital controllers may be designed with discrete logic gates, PLCs, or a microcontroller chip. Digital controllers are interfaced with actuators using transistors, relays, and solenoids. Mechanical actuators include mechanical elements, kinematic chains, cam mechanisms, gears, hydraulic, and pneumatic systems (Onwubolu, 2005).

1.5 Applications of Mechatronics Technology

Mechatronics technology originated in industrial automation systems. Automotive, semiconductor manufacturing, textile, aerospace, material processing, and office automation industries use mechatronic equipment and systems for more precise, cost-efficient, and faster production.

Digital control is extensively used in industrial automation systems. PLCs are the backbone of modern assembly lines, machine control, and robotic cells. These controllers offer reliable, programmable, and easily maintainable solutions for tasks such as coordinating robotic arms in manufacturing processes; managing conveyor belts and automated material handling systems; controlling welding, painting, and packaging machines; and enabling adaptive process control based on sensor feedback (Onwubolu, 2005).

Computer-controlled mechatronic systems play a critical role in the functionality of modern vehicles, electric cars, and autonomous systems. They are essential for the management of fuel injection systems, antilock braking system (ABS), and electronic stability control (ESC). In electric and hybrid vehicles, digital control applications are also integral to powertrain management and advanced driver-assistance systems (ADAS), including adaptive cruise control and automated parking features. All modern cars have computerized cruise control and driver assistance systems. Computer-controlled energy management has improved the efficiency, emission level, and reliability of transportation. The automotive industry has become so dependent on computer chips that between 2020 and 2023, the global computer chip shortage resulted in major crises in automobile production.

Modern consumer electronics heavily rely on embedded microcontrollers for digital control. Programmable washing machines and refrigerators, smart thermostats that optimize energy use based on user preferences and sensor data, audiovisual equipment with advanced signal processing functions, and home automation systems integrating lighting, security, and environmental controls are common examples.

Office automation equipment such as printers, plotters, scanners, and copiers are applications of mechatronics. Digital controllers manage the precise positioning and velocity of actuators and motors in systems that require high accuracy, such as computer numerical control (CNC) machines, 3D printers and additive manufacturing equipment, laser cutters, and engraving machines. Most power tools used in workshops and homes have electronic speed control and safety protection. CNC machines and 3D printers are based on precise position control. Digital cameras have fast micro machines for auto focus and shutter speed control. They also allow multiple image shooting with precise timing, image processing, data acquisition from image sensor, and data storage.

In the medical field, digital control ensures the safe and reliable operation of sophisticated equipment, like medical imaging devices (magnetic resonance imaging [MRI], computed tomography [CT] scanners), robotic surgery systems, and prosthetics and rehabilitation robots. Healthcare applications include laboratory test devices, patient monitoring and data logging instruments, and many other hospital equipment.

Robotic systems range from humanoids and autonomous mobile robots (like Mars Rovers) to industrial assembly and warehouse robots. Robots can perform most human functions automatically using a computer-supported controllers and advanced software. Advanced mobile robots have voice processing and vision capabilities.

Stationary robotic arms used in industrial assembly lines perform programmed actions more precisely and much faster than human workers. Warehouse robots perform sorting and packaging functions more effectively.

Mobile robots can effectively operate in hostile environments unsuitable for human workers. Radioactive debris after the Three Mile Island (TMI) nuclear accidents was cleaned by mobile robots. Mars rovers are popular examples of space exploration robots.

Mechatronics has abundant applications in our daily life. Most appliances, kitchen devices, and power tools used in homes involve some type of mechatronics application. Most of them have speed, temperature, and process controls. Any product publicized as "smart" is an example of mechatronics design. We may start the day with a smart coffee maker, which brews coffee at the strength we like at the time we previously programmed. Modern dishwashers are smart enough to adjust the washing cycle based on the type of dishes and glassware. A laundry washing machine can check the weight of the load and adjust the washing duration based on how dirty the laundry is.

1.6 Mechatronic System Design Process

As with any engineering design problem, mechatronic system design starts by identifying the end-user needs. Typically, customers express their needs in plain words. The design engineer transforms the user's needs into technical requirements and compiles a set of design specifications. The purpose of the engineering design process is to create a product that meets all customer requirements under certain limitations, either included in the list of requirements or implied by intended usage.

Brainstorming and generation of design ideas is the first step in a professional design process. Experience and looking for existing solutions may be helpful in the first steps of brainstorming, but they may limit creativity as well. The designer must be creative and utilize the fundamental knowledge of physics and underlying engineering to generate several viable solutions.

While starting by constructing a physical model or a functional prototype may be tempting, in the industrial design process, a trial-and-error approach is too costly and time-consuming. Design engineers must generate several alternative ideas and compare potential design concepts based on their feasibility, performance, cost, and other practical considerations. Using a mathematical model and computer simulations before building a physical prototype is the most cost-effective and time-saving approach.

A mathematical model consists of a symbolic diagram and a set of equations that describe the system's behavior. Mathematical modeling is a common ground for all engineering fields. It also facilitates efficient collaboration of engineers with different technical backgrounds who work together in an industrial design team.

Mechatronic systems include electronic circuits, electromagnetic devices, mechanical elements, several types of electric machines, and thermal, hydraulic, and pneumatic parts. While such parts employ principles and techniques of different engineering disciplines, fundamental laws of physics apply to their operations and interactions. In this chapter, we show analogies between electrical, mechanical, hydraulic, and thermal elements. Representation of all elements using their electrical equivalents allows the construction of a building a network model, which can be analyzed using powerful techniques that have been developed over more than a century.

The mathematical model, however, may not accurately reflect the performance of the final product. Its accuracy depends on many assumptions and approximations. Before completing the design work, engineers must build a functional prototype to perform physical tests and correct the practical issues that may arise. Such corrections may need modifications of the mathematical model and repetition of the calculations. Advanced simulation software with GUI is currently available to evaluate the operation and performance of complex mechatronic systems without writing and solving circuit equations. Therefore, using a mathematical model simplifies the mechatronic product design process and significantly reduces the cost and time needed to develop a full-size prototype.

The main intention of this book is to convey concepts of physics and engineering needed to design efficient, safe, and reliable mechatronic systems. Every chapter contains numerous solved problems and design examples to help readers understand practical applications and specific cases of the covered subjects.

1.7 Summary

This chapter provides an overview of mechatronics, a multidisciplinary field that integrates mechanical parts with electromagnetic devices, electronic circuits, and computers to perform tasks involving motion. The term "mechatronics" combines "mechanics" and "electronics" and has evolved significantly since its inception in the late 1960s, becoming vital in the design and development of intelligent electromechanical products. Modern mechatronic systems incorporate microcontrollers, sensors, and software, which enhance their functionality and adaptability in various applications, as illustrated in the diagrams presented in the chapter.

1) Mechatronics is a multidisciplinary engineering field that integrates mechanical parts, electronic circuits, and computer control to perform tasks involving motion. Originating in the late 1960s, it has evolved to include advanced microcontrollers, sensors, and software, enabling intelligent electromechanical products across various industries and daily-life applications.
2) Mechatronics emerged from electromechanical drives used in industrial automation, with the invention of DC motors in the nineteenth century, marking a significant advancement by enabling quiet and efficient production line operation. Challenges in motor speed control were overcome with the development of power electronics in the 1970s, leading to solid-state variable voltage converters. The integration of computer technology further advanced control mechanisms, popularizing PLCs in industrial settings. The term "mechatronics" was first coined by engineer Tetsuro Mori in 1969 and has since evolved to signify the synergistic integration of mechanical engineering, electronics, and intelligent computer control. The JSPMI classifies mechatronic products into four classes based on the degree of integration between mechanical and electronic components, ranging from primarily mechanical products enhanced by electronics to fully integrated systems designed synergistically.
3) A typical mechatronic system comprises interacting subsystems, including power sources, controllers, sensors, actuators, electronic converters, and signal conditioning circuits. Sensors detect physical quantities and convert them into electrical signals, which may be conditioned to reduce noise before processing. Actuators convert energy into mechanical motion, often via electromagnetic means. The controller processes signals and manages outputs to actuators through

electronic converters. Systems may be analog, with continuous signals, or digital, with discrete binary signals processed by microcontrollers or CPUs. Digital control systems rely on ADCs and DACs to interface between analog sensors and digital processors. PLCs are widely used in industrial automation for their reliability and ease of programming.

4) Key components of a mechatronic system include:
   - **Sensors**: Convert physical quantities such as temperature, pressure, position, velocity, and acceleration into electrical signals. Advanced systems may use GPS and LIDAR for precise detection.
   - **Signal Conditioning Circuits**: Amplify and filter sensor signals, removing noise and enabling accurate processing. ADCs and DACs facilitate communication between analog and digital components.
   - **Actuators**: Convert electrical or other energy forms into linear or rotary motion. Electromechanical, hydraulic, and pneumatic actuators each offer specific advantages based on application needs.
   - **Transmission Elements**: Gears, belts, and shafts transmit power from actuators to mechanical parts, ensuring smooth and efficient operation.
   - **Microcontrollers and Microprocessors**: Embedded chips execute control algorithms, processing sensor inputs, and managing actuators. PLCs are common in industrial applications.
   - **Power Converters**: Manage voltage and frequency to supply appropriate power. Types include DC/DC choppers, rectifiers (AC/DC), inverters (DC/AC), and AC/AC converters, each serving different functions such as variable speed motor drives or renewable energy integration.
   - **Communication Modules**: Facilitate wired or wireless data exchange within systems, enabling coordinated operation.
   - **Software**: Implements control algorithms (e.g., PID, MPC), processes sensor data, and provides user interfaces such as GUIs and remote control applications to ensure effective system management.
5) The primary control strategies implemented in mechatronic systems include the following:
   - **Manual (Open-Loop) Control**: Human operators observe system output and manually adjust inputs without automatic feedback. This method is simple but lacks precision and automation.
   - **On–Off Control**: The simplest automated control, switching devices fully on or off based on a threshold, with hysteresis to avoid oscillations. Common in thermostats.
   - **Sequential Control**: On–off control of multiple elements in a timed sequence, historically implemented by mechanical cams and now by digital logic circuits or microcontrollers. Examples include dishwashers and traffic lights.
   - **Feedback (Closed-Loop) Control**: Sensors continuously monitor output, and controllers compare actual values to desired references, adjusting outputs to minimize errors. Analog controllers use PID operations for improved accuracy and response. Digital controllers use microprocessors or PLCs.
6) Numerous industries derive significant advantages from the implementation of mechatronics solutions.
   - **Industrial Automation**: PLCs and robotic cells coordinate assembly lines, material handling, welding, painting, and packaging, improving precision and efficiency.

- **Automotive Industry**: Digital control manages fuel injection, ABS, ESC, powertrain management in electric/hybrid vehicles, and ADAS. The 2020–2023 global chip shortage highlighted the industry's reliance on mechatronics.
- **Consumer Electronics**: Embedded microcontrollers enable programmable appliances, smart thermostats, audiovisual equipment, and home automation systems.
- **Office Automation and Manufacturing**: Printers, scanners, CNC machines, 3D printers, laser cutters, and power tools rely on digital control for precision and safety.
- **Healthcare**: Medical imaging, robotic surgery, prosthetics, rehabilitation robots, and patient monitoring devices depend on mechatronic control for reliability and safety.
- **Robotics**: From industrial arms to autonomous mobile robots for hazardous environments and space exploration, mechatronics enables automation and advanced functionalities like voice and vision processing.
- **Everyday Life**: Smart home appliances such as coffee makers, dishwashers, and washing machines use mechatronics to optimize performance based on user preferences and sensor data.

7) Mechatronic system design process begins with identifying user needs and translating them into technical specifications. Creativity and fundamental engineering principles guide the generation of multiple design concepts. Mathematical modeling and computer simulations are essential for evaluating system behavior before building costly prototypes. These models use symbolic diagrams and equations to describe system dynamics, facilitating collaboration among multidisciplinary engineers. Prototypes validate the design and may lead to model refinement. Advanced simulation software further accelerates design by reducing reliance on physical prototypes.

This overview highlights mechatronics as a dynamic engineering discipline combining mechanical, electrical, electronic, and computer technologies to create intelligent systems. Its evolution, fundamental components, control strategies, diverse applications, and rigorous design processes underscore its critical role in modern technology and industry.

1.8 Check Your Understanding

1.8.1 Self-Assessment Quiz

1 A mechatronic system consists of

   a. Only electronic circuits secured in a mechanical structure.
   b. Mechanical parts tested using electronic instruments.
   c. Mechanical and electronic parts interfacing to perform a given task.
   d. An electronic converter that supplies a machine.

2 What is the primary goal of mechatronics?

   a. To create mechanical parts
   b. To integrate mechanical parts with electronic circuits and computers to perform tasks involving motion
   c. To design electrical circuits
   d. To develop software applications

3 How did the invention of DC motors impact industrial automation?
   a. It made production lines louder and less efficient.
   b. It allowed quiet, reliable, and more efficient operation of production lines.
   c. It had no significant impact.
   d. It increased the cost of production.

4 What are the key components of a mechatronic system?
   a. Only mechanical parts
   b. Only electronic circuits
   c. Power sources, controllers, sensors, actuators, and signal conditioning circuits
   d. Software modules only

5 How do sensors and actuators function within a mechatronic system?
   a. Sensors convert physical quantities to electric signals, and actuators convert energy into motion.
   b. Sensors convert electric signals to physical quantities, and actuators convert motion into energy.
   c. Sensors supply power to the system, and actuators detect physical quantities.
   d. Sensors and actuators have no specific function.

6 What are the differences between analog and digital mechatronic systems?
   a. Analog systems use continuous signals, while digital systems use binary variables defined over fixed time intervals.
   b. Analog systems use binary variables, while digital systems use continuous signals.
   c. Analog systems are more reliable than digital systems.
   d. There are no differences between analog and digital systems.

7 Can you explain the role of microcontrollers in mechatronic systems?
   a. Microcontrollers execute complex algorithms and control signals to manage the system's operations.
   b. Microcontrollers supply power to the system.
   c. Microcontrollers convert physical quantities to electric signals.
   d. Microcontrollers have no role in mechatronic systems.

8 What are some common applications of mechatronics in everyday life?
   a. Only industrial automation systems
   b. Consumer electronics, automotive systems, healthcare devices, and office automation equipment
   c. Only healthcare devices
   d. Only office automation equipment

9 How do feedback control systems enhance the performance of mechatronic devices?
   a. Feedback control systems do not enhance performance.
   b. Feedback control systems utilize sensors to gather output information, allowing for real-time adjustments to maintain desired performance levels.
   c. Feedback control systems supply power to the system.
   d. Feedback control systems convert physical quantities to electric signals.

10 What is the significance of mathematical modeling in the design process of mechatronic systems?
a. Mathematical modeling is not significant.
b. Mathematical modeling helps in creating a symbolic diagram and a set of equations that describe the system's behavior.
c. Mathematical modeling increases the cost and time needed to develop a prototype.
d. Mathematical modeling has no role in the design process.

11 How has the field of mechatronics evolved since its inception in the late 1960s?
a. It has remained the same.
b. It has evolved significantly, incorporating microcontrollers, sensors, and software to enhance functionality and adaptability.
c. It has become less important.
d. It has only focused on mechanical parts.

12 Which field below is not directly involved in mechatronic system design?
a. Mechanical engineering
b. Civil engineering
c. Electrical engineering
d. Computer engineering

13 Which fields of industry below use mechatronics applications?
a. Automotive
b. Biomedical
c. Materials processing
d. Manufacturing
e. All of the above

14 What is the purpose of sensors in mechatronic systems?
a. Convert electrical signals to physical quantities
b. Convert physical quantities to electric signals
c. Supply power to parts of the system
d. Sense whether the device is working or not

15 An open-loop system is
a. A group of automatically controlled analog elements
b. A computer-controlled mechanism
c. An uncontrolled system
d. A manually controlled mechanism by a human operator

16 Which element below is not involved in the operation of a closed-loop analog system?
a. Sensors
b. Comparator
c. Signal conditioning circuits
d. Human power

17 Which elements collect information from the output of a mechatronic system?
- **a.** Actuators
- **b.** Signal conditioning elements
- **c.** Sensors
- **d.** Motors

18 What is the purpose of analog-to-digital converters?
- **a.** Convert voltage to current
- **b.** Transform a digital signal to analog
- **c.** Change the frequency of a pulse train
- **d.** Convert an analog signal to a binary number

19 What is the purpose of digital-to-analog converters?
- **a.** Change the frequency of an AC signal
- **b.** Transform a digital signal to analog
- **c.** Change the binary code of a digital signal
- **d.** Convert an analog signal to a binary number

20 The central processing unit (CPU) of a digital control system performs
- **a.** Arithmetic operations
- **b.** Logic operations and calculations to control the system outputs
- **c.** Data collection and storage
- **d.** Energy management functions

21 In a mechatronic device, the purpose of electronic converters is to
- **a.** Convert DC to AC.
- **b.** Deliver voltage and current corresponding to the controller output signals at adequate power.
- **c.** Convert physical quantities to voltage and current.
- **d.** Supply power to the controller.

22 The first step in a design problem is to
- **a.** Construct a prototype and test its operation.
- **b.** Find a commercial device that performs similar functions.
- **c.** Brainstorm to generate design concepts.
- **d.** Identify customer needs and requirements.

23 A mathematical model consists of
- **a.** A CAD drawing.
- **b.** A symbolic diagram and a set of equations that describe the system's behavior.
- **c.** A set of algebraic equations to solve.
- **d.** A circuit diagram.

24 What is the purpose of a prototype?
- **a.** Use in product acceptance tests
- **b.** Serve as a product sample for marketing advertisements
- **c.** Test the operation and characteristics of the system before manufacturing
- **d.** Show that the designed product would attract customers

Answers:

1: c; 2: b; 3: b; 4: c; 5: a; 6: a; 7: b; 8: b; 9: b; 10: b; 11: b; 12: e; 13: b; 14: d; 15: d; 16: c; 17: d; 18: b; 19: b; 20: b; 21: d; 22: b; 23: a; 24: b.

1.8.2 Short Answer Questions

1 What are the key components of a mechatronic system?

2 What are the differences between analog and digital mechatronic systems?

3 Can you explain the role of microcontrollers in mechatronic systems?

4 What are some common applications of mechatronics in everyday life?

5 How do feedback control systems enhance the performance of mechatronic devices?

6 What is the significance of mathematical modeling in the design process of mechatronic systems?

7 What is the primary goal of mechatronics?

8 How did the invention of DC motors impact industrial automation?

9 How do sensors and actuators function within a mechatronic system?

1.8.3 Essay Questions

1 How has the field of mechatronics evolved since its inception in the late 1960s?

2 Discuss the impact of mechatronics on modern education. How has the integration of digital tools changed the way students learn, and teachers instruct?

3 Examine the role of artificial intelligence (AI) in mechatronics. What are the potential benefits and drawbacks of AI integration in various industries?

4 Evaluate the significance of mathematical modeling in the design process of mechatronic systems. How does it contribute to the development of efficient and reliable products?

5 Explore the applications of mechatronics in the automotive industry. How have advancements in mechatronics technology improved vehicle performance and safety?

6 Investigate the role of feedback control systems in enhancing the performance of mechatronic devices. What are the key principles and benefits of using feedback control?

7 Discuss the challenges and opportunities of integrating mechatronics in healthcare. How can mechatronic systems improve patient care and medical procedures?

8 Examine the impact of mechatronics on industrial automation. How have mechatronic systems transformed manufacturing processes and productivity?

9 Analyze the ethical implications of using mechatronics in defense systems. What are the potential risks and benefits of deploying mechatronics technology in military applications?

References

Alciatore, D. G. & Histand, M. B., 2003. *Introduction to Mechatronics and Measurement Systems.* Boston: McGraw Hill.

Billingsley, J., 2006. *Essentials of Mechatronics.* Hoboken, NJ: John Wiley & Sons.

Bishop, R. H., 2002. *The Mechatronics Handbook.* Boca Raton, London, New York, Washington DC: CRC Press.

Brown, A. S., 2011. *Mechatronics and the Role of Engineers.* [Online] Available at: https://www.asme.org/topics-resources/content/mechatronics-and-the-role-of-engineers [Accessed 15 8 2 024].

Cetinkunt, S., 2015. *Mechatronics with Experiments.* West Sussex, UK: John Wiley & Sons.

de Silva, C. W., 2005. *Mechatronics, An Integrated Approach.* Boca Raton, London, New York, Washington, DC: CRC Press.

Kyura, N. & Oho, J., 1996. Mechatronics – An Industrial Perspective. *IEEE/ASME Transactions on Mechatronics,* 1(1), pp. 10–15.

Onwubolu, G., 2005. *Mechatronics Principles and Applications.* Oxford: Elsevier.

Pannaga, N., Ganesh, N. & Gupta, R., 2013. Mechatronics - An Introduction to Mechatronics. *International Journal of Engineering Research & Technology (IJERT),* 8(2), pp. 129–134.

2

Basic Concepts

OBJECTIVES
• Present the essential principles required to understand the mechatronics topics discussed in subsequent chapters. • Provide a brief overview of physics concepts relevant to mechatronic system design, including classical mechanics and electric circuits. • Refresh the foundational physics and engineering knowledge needed for further study in mechatronics. • Guide readers to further explanations in recommended textbooks if concepts are difficult to understand.

This chapter presents the essential principles required to understand the mechatronics topics discussed in subsequent chapters. For further explanation, please refer to the textbooks listed at the end of this chapter.

Although mechatronic systems are composed of components created in different fields of engineering, the laws of classical physics apply to all of them. Each engineering discipline has developed laws, principles, and methods to study the behavior and interactions of components in its specific field. A mechatronic system engineer must always adhere to the fundamental principles of physics below to correctly apply engineering methods. The following is a brief overview of physics concepts that are relevant to mechatronic system design. If you find them difficult to understand, please refer to any college-level physics textbook (Giancoli, 2020; Ohanian & Market, 2006). A more comprehensive discussion of energy conversion and storage methods can be found in Soysal and Soysal (2020, Chapter 2, pp. 37–74).

2.1 Newton's Laws

Classical mechanics, also known as Newtonian mechanics, is founded on Newton's three laws of motion. These fundamental laws explain how forces affect the movement of objects (Giancoli, 2020).

Newton's first law of motion states that an object remains at rest or continues moving on a straight line unless a net force acts upon it. Consequently, an object's state of motion only changes when influenced by a force. As a result, when an object's velocity remains constant, it means that the total force acting on it is zero, as mathematically expressed below:

$$\sum \vec{F} = 0 \tag{2.1}$$

Newton's first law is simply a cause-and-effect rule, implying that to alter the velocity of an object, an external force is required.

Newton's second law of motion describes how force influences an object's motion. Isaac Newton experimentally discovered that an object's acceleration increases when a greater net force is applied but decreases as the object's mass grows.

Momentum is the product of an object's mass and its velocity. The change in momentum over time is directly related to the force exerted on the object. Thus, the force needed to move an object corresponds to how quickly its momentum changes, which is known as acceleration. The net force on an object equals the rate at which its momentum changes over time. Newton's second law can be expressed mathematically as follows:

$$\vec{F} = \frac{d}{dt}\left(m\vec{v}\right) = m\frac{d\vec{v}}{dt} = m\frac{d^2\vec{x}}{dt^2} = m\vec{a} \tag{2.2}$$

If the mass is constant, we can extract this term out of the derivative. Velocity is the derivative of displacement, and the derivative of velocity is acceleration, referred to as acceleration. Hence, the force acting on an object is proportional to its acceleration. The coefficient is the object's mass.

Newton's third law of motion states that "For every action, there is an equal and opposite reaction." This means forces come in pairs; when one object applies force to another, the second applies an equal and opposite force back.

Newton's Laws for Rotating Objects

Rotary motion is an important special case in mechatronic systems. A motor driving a mechanical load and a robotic arm rotating around its joint are examples of rotational mechatronic systems performing rotational motion. Rotation of a particle along the circumference of a circle is analyzed using rotational kinematics. Rotation of a solid object around an axis is the subject of rotational dynamics.

Consider a small object with mass m attached to a rigid arm with negligible mass that is rotating by an angle θ along the circumference of a circle of radius r. The length of the motion is $s = r\theta$, where θ is the rotational angle measured in radians. Angular velocity is the derivative of angular displacement.

$$\omega = \frac{d\theta}{dt} \tag{2.3}$$

Angular velocity is expressed in radians per second. In practice, however, rotation (or revolution) per minute (rpm) is common. Since one revolution is 2π radians, if the rotations per minute (rpm) value is N, the expression below gives the angular velocity in radians per second.

$$\omega = \frac{2\pi N}{60} \tag{2.4}$$

A tangential force F acting on an object rotating around the center of radius r performs the work below.

$$W = Fs = Fr\theta \tag{2.5}$$

The product of the tangential force and distance to the center of rotation is called torque.

$$\tau = F \cdot r \tag{2.6}$$

Consider an arm that is rotating around one of its ends with the effects of several tangential forces applied at different distances from the center. The total work done is

$$W_{total} = F_1 r_1 \theta + F_2 r_2 \theta + \cdots + F_n r_n \theta = \theta \cdot \sum_{1}^{n} \tau_k \tag{2.7}$$

For a body in rotational motion, Newton's first law becomes "in a system at rest, the sum of all torques relative to the same axis is zero."

For a mass m with dimensions negligible to the radius of rotation, the definition of angular momentum is similar to that of the linear momentum.

The rate of change of angular momentum over time is proportional to the total torque applied. In the torque equation below, the coefficient of the angular acceleration is defined as the moment of inertia for a point mass rotating around a circle.

$$\tau = r\frac{d}{dt}\left[mr\frac{d\theta}{dt}\right] = \left(mr^2\right)\frac{d^2\theta}{dt^2} = J\frac{d\omega}{dt} \tag{2.8}$$

For a point mass m rotating along a circle of radius r, the moment of inertia is mr^2. The moment of inertia of common 3D objects is given in physics textbooks (Giancoli, 2020) and engineering handbooks. Online sources are also available for moments of inertia of typical bodies (Engineering Toolbox, 2005).

If the moment of inertia is constant, the net total torque applied causes the rotating object to move at an angular acceleration proportional to the force. The constant of proportionality mr^2 is the "moment of inertia" of the point mass. The moment of inertia J depends on the mass and geometric shape of an object, and its standard unit is kg·m^2.

The equation below expresses Newton's second law for rotating objects, where J represents the moment of inertia of the solid object.

$$\tau = J\frac{d\omega}{dt} \tag{2.9}$$

Example 2.1 The winch in Figure 2.1 lifts a 50-kg block. The diameter of the pulley is 30 cm. At the start, the winch accelerates from stop to 60 rpm in 2 seconds. After it has reached the steady state, the pulley continues turning at 60 rpm constant speed (gravitational acceleration is 9.81).

a) Calculate the force and torque when the speed is constant (steady-state).
b) Find the initial force and torque as the block accelerates at the start.

Solution

a) At constant speed, the winch pulls the block with the force F_1 equal in magnitude to the gravity.

$$F_1 = mg = 490.5\ \text{N}$$

The torque on the pulley is the product of the radius and the tangential force.

$$\tau = \frac{D}{2}F_1 = \frac{0.3}{2} \times 490.5 = 73.575\ \text{N} \cdot \text{m}$$

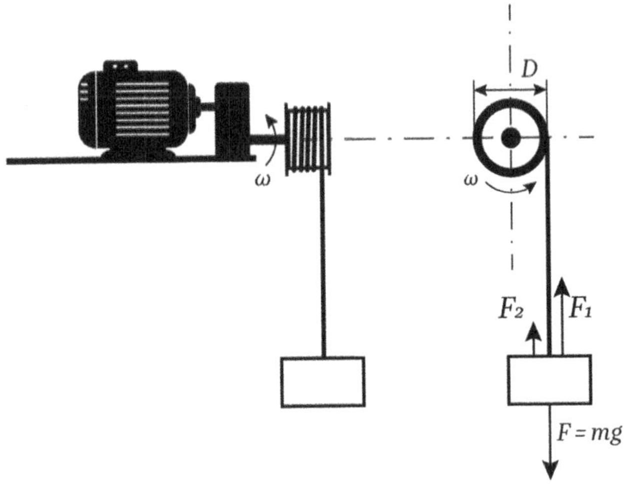

Figure 2.1 A winch lifting a block

b) To accelerate the block, the winch must apply an additional force equal to the product of mass and acceleration due to Newton's second law. First, let's calculate the linear velocity when the pulley rotates at 60 rpm.

$$v = \frac{N \cdot (\pi D)}{60} = \frac{60 \times \pi \times 0.3}{60} = 0.942 \text{ m/s}$$

Linear acceleration to reach the 0.942 m/s rotation speed from zero in 2 seconds is:

$$a = \frac{0.942}{2} = 0.471 \text{ m/s}^2$$

The force needed to accelerate the block:

$$F_2 = ma = 50 \times 0.471 = 23.55 \text{ N}$$

The block moves in a vertical direction. Therefore, the force due to the acceleration is added to the weight of the block. Thus, the force at the start is

$$F = F_1 + F_2 = 490.5 + 23.55 = 514.05$$

The starting torque is:

$$\tau = F\frac{D}{2} = 514.05 \times \frac{0.30}{2} = 77.1 \text{ N} \cdot \text{m} \therefore$$

2.2 Work, Energy, and Power

Work, energy, and power are essential concepts of classical physics. While these terms often appear in daily conversation, they have specific meanings and definitions in physics.

2.2.1 Work

Work is the measure of energy transfer that occurs when an object is moved over a distance by an external force. It is a scalar quantity and is calculated as the dot product of force and displacement. If the direction and magnitude of the force changes during the motion and the displacement is not a straight line, the incremental physical work (dw) done at any instant is equal to the dot product of the force vector $\left(\vec{F}\right)$ and the vector representing the differential displacement ($d\vec{s}$). Total work done when the object moves from a point a to b is calculated by integrating dw from a to b.

$$dw = \vec{F} \cdot d\vec{s} \Longrightarrow W = \int_a^b \vec{F} \cdot d\vec{s} \tag{2.10}$$

For linear displacement (translation) of an object resulting from a force with constant magnitude and angle, equation 2.10 can be expressed as the following algebraic equation:

$$W = L \mid \vec{F} \mid \cos\alpha \tag{2.11}$$

Here, L refers to total linear displacement, and α is the angle between force and displacement.

In the International System of Units (SI), force is measured in Newton (N), and distance is measured in meters (m). Therefore, the unit of work is Newton meters (Nm). Joule (J) is the alternative unit of work equal to Nm. The British or Imperial system (IS) measures work in foot-pounds force (ft·lb).

Work is positive when force and displacement are in the same direction and negative when force and displacement are in opposite directions. No work is done if there is no displacement or if the force is perpendicular to the displacement.

2.2.2 Energy

Energy is typically described as the ability to perform work. It takes many forms, including mechanical, electrical, thermal, and chemical energy. In mechatronic systems, we mostly concentrate on mechanical energy, specifically potential and kinetic energy. Meanwhile, in electrical and electronic circuits, components interact by means of electrical or magnetic energy.

The principles governing energy and the laws of physics concerning energy exchange remain consistent across all systems, including mechanical, electrical, thermal, chemical, and hybrid forms.

The standard unit of energy in SI is the same as work, either Newton Meter (Nm) or Joule (J). Joule and Nm are equivalent units; therefore, $1\,\text{J} = 1\,\text{Nm}$. In the British (Imperial) unit system, energy is measured in British Thermal Unit (Btu). One Btu is equal to 1,055 J.

2.2.2.1 Potential Energy

Potential energy is the energy stored due to the position or configuration of an object within a force field, such as gravitational, electrical, or magnetic fields, or because of elastic deformation. This energy represents the capacity of the object to perform work when its position or arrangement changes. For example, an object held at a certain height above the ground has gravitational potential energy, which can be converted into kinetic energy if the object is allowed to fall.

Gravitational potential energy is associated with the vertical position of a body. Suppose that an object is at a height h_1 measured relative to a certain reference elevation. If the object is released to fall with gravity, the only force that acts on it is the gravitational force. As the object moves from the position h_1 to h_2, the work done by the gravitational force is expressed as the product of the mass m and gravitational acceleration constant g (approximately 9.81). Therefore, the gravitational potential energy at the position h_1 relative to the reference position h_2 is obtained by multiplying the gravitational force by the difference of the two heights.

$$W_{p,grav} = mg(h_2 - h_1) \tag{2.12}$$

In common language, weight of an object is the gravitational force mg. In physics, if an object arrives at the same point after moving, its gravitational potential energy does not change. If it ascends to a higher level, it gains potential energy. In contrast, if it descends to a lower level, it loses potential energy. Since the gravitational force is always in the vertical direction, potential energy depends only on the difference in height between the initial and end positions, not the paths traveled.

In physics, a conservative force field is one in which the work done by the force on an object moving from one point to another depends only on the initial and final positions of the object, not on the specific path taken between them. This means that when an object moves in a closed loop within a conservative force field, the total work done by the force is zero. Gravitational and elastic force fields are common examples of conservative force fields.

Elastic potential energy is the energy stored in an object when it is deformed elastically. That means when its shape or dimensions are changed temporarily, it can return to its original form once the deforming force is removed. This type of energy is typically observed in objects like springs, rubber bands, and inflated balls, where the energy stored is proportional to the displacement from a reference position. The amount of elastic potential energy can be calculated using the formula involving a constant k (characteristic of the material or object) and the change in position, demonstrating how the energy depends on the extent of deformation.

$$W_{p,elastic} = k(x_1 - x_2) \tag{2.13}$$

2.2.2.2 Kinetic Energy

Kinetic energy is the energy an object possesses due to its motion. It depends on the mass of the object and the square of its velocity, meaning that as either the mass or speed increases, so does kinetic energy. This form of energy describes how moving objects can perform work because of their movement. Kinetic energy can be formulated using Newton's law of inertia:

$$F = ma = m\frac{dv}{dt} \tag{2.14}$$

The work performed as the object undergoes an infinitesimal displacement (dx) is given by

$$dW = F \cdot dx = m\frac{dv}{dt}(v \cdot dt) = m \cdot v \cdot dv \tag{2.15}$$

We can formulate kinetic energy W_k by integrating both sides of this equation:

$$W_k = \int_{v_1}^{v_2} m \cdot v \cdot dv = \frac{1}{2}m\left(v_1^2 - v_2^2\right) \tag{2.16}$$

To determine the kinetic energy of a rotating object, the moment of inertia is used in place of mass, and angular velocity replaces linear velocity in the calculation.

$$W_k = \frac{1}{2}J\left(\omega_1^2 - \omega_2^2\right) \tag{2.17}$$

An object gains kinetic energy as it accelerates and releases it when slowing down. At constant velocity, it retains kinetic energy but does not exchange it. To slow or stop a vehicle, kinetic energy is converted to another form: mechanical brakes turn it into heat via friction, while electric and hybrid vehicles convert it to electricity for storage or supply.

2.2.3 Mechanical Power

In physics, power is defined as the rate at which work is done, or energy is transferred over time. Mathematically, power (P) is expressed as the derivative of energy (W) with respect to time (t). This means that power measures how quickly energy is being used or converted from one form to another in a system.

$$p(t) = \frac{dw(t)}{dt} \Longrightarrow \; W = \int_{t_1}^{t_2} p(t)dt \tag{2.18}$$

Mechanical power refers to the amount of work a force accomplishes per unit of time. Since power is a scalar quantity, instantaneous power is calculated as the derivative of the dot product between force and displacement vectors. If the force remains constant throughout linear motion, power can be determined by taking the dot product of the force and velocity vectors, as shown in the following equation:

$$P = \frac{d}{dt}\left(\vec{F} \cdot \vec{x}\right) = \vec{F}\frac{d\vec{x}}{dt} = \vec{F} \cdot \vec{v} \tag{2.19}$$

Rotating machines, sensors, and actuators within mechatronic systems operate through rotational motion. As previously defined, torque serves as the rotational analog of force. By substituting angular velocity for linear velocity, the power associated with rotational motion is expressed as the product of torque and angular velocity ($P = \tau\omega$), where the angular velocity is in radians per second. In the metric system (SI), the unit of torque is newton meter ($N \cdot m$) and the unit of power is watt (W).

When power remains constant throughout a time interval t, it can be expressed in an algebraic form. Power's SI unit is the watt (W), which equals one joule per second. The primary distinction between energy and power lies in the time during which energy conversion happens, or when power is applied. In various engineering disciplines, it's common to describe energy using a unit of power multiplied with a unit of time. For example, 1 J = 1 W·s. In engineering, it is common to express power using metric prefixes such as microwatts, milliwatts, kilowatts, megawatts, and terawatts.

Horsepower (HP) is the imperial unit for power, with 1 HP equal to 550 ft lb/s or about 745.7 W. Metric HP equals 735.5 W and is based on lifting a 75 kg weight 1 m in 1 second. Electric motor ratings use 746 W per HP, expressing output in HP and input in watts. Mechatronic systems can range from milliwatts (10^{-3} W) to megawatts (10^6 W).

2.3 Conservation Principles

2.3.1 Conservation of Energy

Conservation of energy principle of classical physics is the unifying concept of all engineering fields. Also known as the first law of thermodynamics, the conservation of energy principle can be expressed in many ways, all leading to the same fundamental principle. The conservation of energy principle is often expressed as below:

> In an isolated system, the total energy exchanged between the elements remains constant.

In other words, if there is no energy exchange with the surroundings, the total energy of the system cannot increase or decrease. It can, however, be converted from one form to another, stored, or transferred from one place to another within the system.

According to the conservation of energy principle, there is no mechanism that can create energy without receiving an energy input from the external world. Similarly, no system can destroy energy but can release energy to its surroundings. Conservation of energy principle should not be confused with the term "energy conservation," which means reducing the consumption of energy by avoiding unnecessary and inefficient consumption. Terms like "renewable energy" and "sustainable energy" contradict the conservation of energy principle.

Although the conservation of energy principle indicates that energy cannot be produced nor destroyed, use of the terms "energy production," "energy generation," and "energy consumption" are common. Such terms technically imply that energy is converted from one form to another.

Efficiency

In any practical system, some part of the energy is converted to heat and dissipated to the surrounding space. If this heat is not recycled, it is lost energy. Due to the losses, the energy output of any physical system is smaller than the input energy. The ratio of the output to the input energy is efficiency.

$$\eta = \frac{W_{out}}{W_{in}} = \frac{W_{in} - W_{loss}}{W_{in}} = 1 - \frac{W_{loss}}{W_{in}} \tag{2.20}$$

In a mechatronic system, the losses mainly occur in friction between mechanical parts, heating of current-carrying wires, heating of electronic elements, and core of the electromagnetic structure. Since such losses are inevitable, the efficiency is always less than unity. In practice, it is common to express efficiency in percentage by multiplying equation 2.20 by 100.

2.3.2 Conservation of Mass

In a closed system where no mass enters or leaves, mass cannot be created or destroyed. When chemical reactions occur in such a system, the substances may change their composition, but the total mass remains unchanged. For instance, in a battery cell, the electrolyte's chemical makeup shifts during electricity production, yet its overall mass stays the same. The principle of mass conservation does not apply to nuclear reactions, since atomic properties change; however, these types of reactions do not take place in mechatronics.

In systems where mass is exchanged with the environment, the total mass must remain constant; that is, the rate of change of mass over time should be zero. Consider a scenario in which mass enters and leaves the system. If the system lacks mass-storage capability, the difference between incoming and outgoing mass is always zero at any given moment.

$$m_{in} - m_{out} = 0 \tag{2.21}$$

2.3.3 Conservation of Momentum

The product of the mass and the velocity of a body is called the linear momentum or just the momentum of the body. Since velocity is a vector quantity, so is momentum.

$$p = m\vec{v} = m\frac{d\vec{x}}{dt} \tag{2.22}$$

Momentum is a conserved quantity for a closed system, where the total mass is constant and no external force is acting on the individual elements. When objects of comparable sizes interact, the total momentum is zero.

$$\sum_k m_k \vec{v}_k = 0 \tag{2.23}$$

In mechatronic systems, interacting objects are at comparable mass and dimensions. Therefore, the principle of conservation of momentum applies to all mechatronic systems.

2.3.4 Conservation of Electric Charge

Electric charge is a property of atomic particles. A neutral atom contains the same number of electrons and protons. Electrons are negatively charged, and protons possess the same amount of positive charge. The unit of charge is coulomb (C).

The charge of a single electron, referred to as e, is known as the elementary charge with a magnitude of 1.602×10^{-19} C. Electric charge carried by any object is an integer multiple of e. If the total charge is zero, the object is neutral. In an isolated system where electrons are not exchanged with the surroundings, the electric charge is constant. Conservation of charge is the physical basis of basic laws of electric circuits and electrostatic systems.

Example 2.2 In an airport, a 30° inclined conveyor belt transports 100 kg packages to 3 m high platform at 0.5 m/s constant speed (see Figure 2.2). Assuming that the overall efficiency of the system is 85%, calculate the following:

a) Mechanical work done on each package.
b) Potential energy a package gains when it arrives at the top of the slope.
c) Minimum power rating of the motor driving the conveyor belt.

Solution

a) We calculate the work performed to move one package along the length of the conveyor belt using equation 2.11. The length of the conveyor belt is

$$L = \frac{3}{\sin(30^\circ)} = \frac{3}{0.5} = 6\,\text{m}$$

$$W = L(mg)\cos\alpha = 6 \times (100 \times 9.81)\cos(30^\circ) = 5{,}097.4\,\text{J}$$

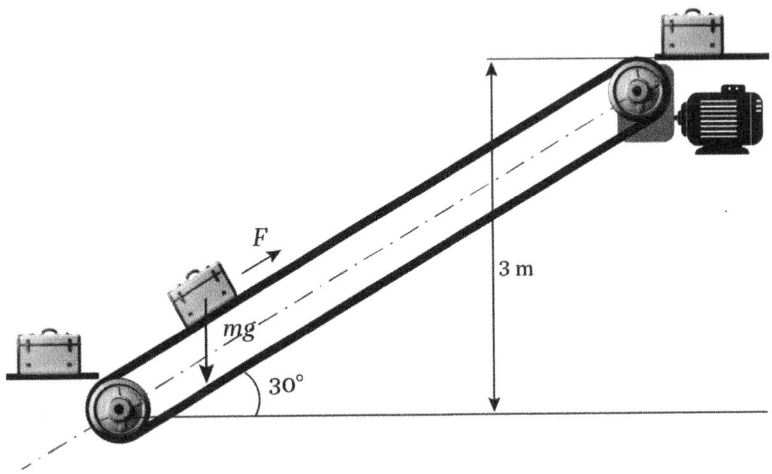

Figure 2.2 Inclined conveyor belt

b) As the conveyor belt transports the package to 3-m height, the package gains the potential energy: $W_p = mgh = 100 \times 9.81 \times 3 = 2{,}943$ J
c) Power is energy per unit time. The belt moves at 2 m/s velocity. The conveyor moves the package with a force equal to the component of the gravity in the direction of the motion.

$$F = mg\cos(30°) = 100 \times 9.81 \times 0.866 = 849.57\text{ N}$$

Mechanical power required to move the package in 1 second is

$$P = W_m = F \cdot v = 849.57 \times 0.5 = 424.8\text{ W}$$

$$P_e = \frac{424.8}{0.85} = 500\text{ W} \therefore$$

Since the overall efficiency is given as 85%, the electrical power rating of the motor must be at least 500 W.

Example 2.3 Forces on a Paraglider

In Figure 2.1, a motor-controlled winch helps a paraglider to launch in a flat area. The winch gives the initial velocity needed to take off. Once at full height, the pilot pulls a release cord, and the towline falls away.

During the free flight, three forces act on the paraglider. The wind flowing above and below the wing produces a lift (L) perpendicular to the airstream. This force allows the paraglider to rise and stay aloft. The resistance of the air against the surfaces of the wing and pilot causes a drag force (D), which opposes to the forward motion of the paraglider, slowing it down. The vertical vector W is the total weight of the paraglider including the equipment and the rider. At free glide, the sum of these three forces determines the vertical position, velocity, and acceleration of the paraglider *Invalid source specified* (Figure 2.3).

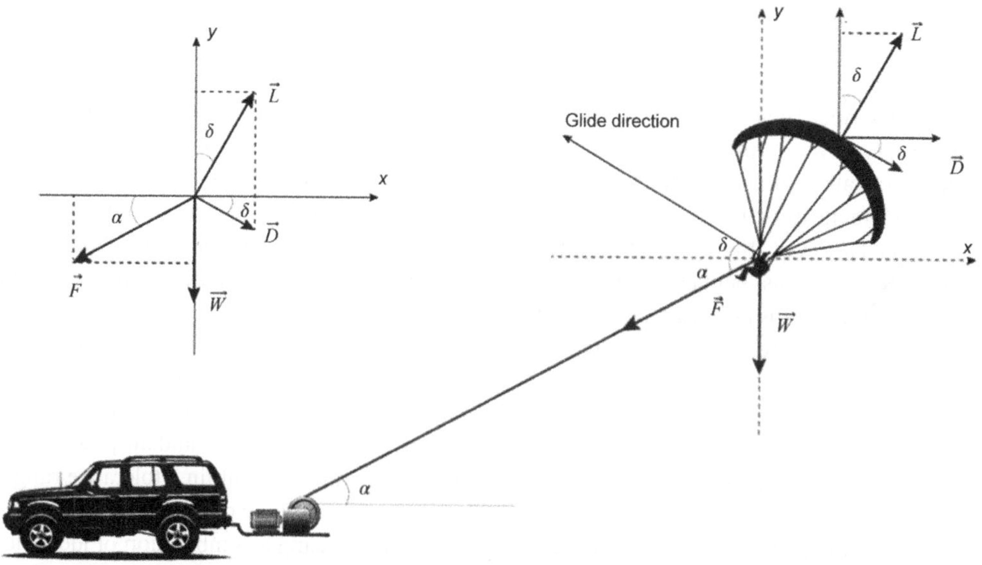

Figure 2.3 Forces acting on a towed paraglider

Problem

The weight of the paraglider with the pilot and equipment is 150 kg. At the moment free glide starts by releasing the towing cord, the angle that the aircraft makes with the horizontal ground, called the glide angle, is $\delta = 30°$. Find the lift and drag forces when the free glide starts.

Solution

At free glide, the towing rope has dropped, and the forces *L*, *D*, and *W* hold the paraglider aloft. According to Newton's first law, equilibrium occurs when the sum of all forces is zero. The vertical force pulling the paraglider and its pilot down is the gravity.

$$W = mg = 9.81 \times 150 = 1{,}471.5\ \mathrm{N}$$

$$W = F \cdot \sin\delta$$

The figure shows the free body diagram separately using the conventional coordinate axis relative to the horizontal ground. Since we have two unknowns, we can use a system of equations to solve them as shown below:

$$L \cdot \cos\delta - D \cdot \sin\delta = W$$

On the other hand, the drag to lift ratio is the tangent of the glide angle.

$$\frac{D}{L} = \tan\delta \Longrightarrow D = L \cdot \tan\delta$$

Thus, by substitution, we obtain

$$L = \frac{W}{\cos\delta - \tan\delta \cdot \sin\delta} = \frac{1{,}471.5}{\cos(30) - \tan(30) \cdot \sin(30)} = 2{,}548.7\ \mathrm{N}$$

$$D = 2{,}548.7 \times \tan(30) = 1{,}471.5\ \therefore$$

2.4 Electric Circuits

An electric circuit is a specific interconnection of circuit elements interacting with each other to perform prescribed functions. Voltage and current sources are active elements because they provide the necessary power to produce desired voltages and currents in the circuit. Resistors, capacitors, and inductors are common circuit elements that absorb power, thus referred to as passive elements. A circuit element must have at least two terminals to allow current flow caused by the voltage across them. However, three or more terminal elements are common in electronic circuits and electromechanical systems. Semiconductor devices such as various types of diodes, transistors, and integrated circuit chips are essential elements of electronic circuits.

Circuit elements are identified by their parameters and voltage–current (V–I) relationships observed on their terminals. To simplify the analysis and design of electric circuits, it is common to idealize circuit elements by describing their behavior with mathematical functions and assigning exact parameters, independent of external factors like temperature, humidity, and air pressure. In practice, however, circuit elements are available at standard parameters within a tolerance range. Furthermore, manufacturers guarantee rated specifications with current, voltage, power, and temperature limitations. Practical circuit elements may be modeled using a combination of ideal elements. A designer must verify the functioning of the circuit, considering the worst cases that might occur for commercially available components.

The terms below help identify the layout of an electric circuit in analysis and design problems (Alexander & Sadiku, 2013):

- **A branch** refers to a single circuit element or several elements connected in series that constitute a circuit.
- **A node** is the point where two or more branches join. In a practical circuit, any connection that can be accomplished by soldering various terminals together is a node. Sometimes, several components are connected using solid wire. An ideal junction wire without any inherent resistance, inductance, or capacitance is one node.
- **A loop** is any closed connection of branches. Suppose that we walk through circuit elements. If we arrive at the starting node without walking on any element twice, we have completed a loop.
- **A mesh** is a loop that does not contain other loops.

Circuit diagrams usually contain additional lines that do not correspond to any circuit element just to visually connect several elements. To avoid confusion in identifying the branches and nodes, it is convenient to use a graph representation. A graph diagram is an abstract representation of the circuit that contains lines corresponding to circuit elements and bold points representing the nodes.

Electricity is a form of energy associated with the existence and displacement of charged subatomic particles. Ancient Greek philosopher and mathematician Thales of Miletus, who lived between 624 and 546 BC, noticed that a piece of amber rubbed on fur attracted lightweight objects such as hair and feather. The word "electricity" comes from the Greek word "elektron" and the Latin word "electrum," both meaning amber.

Electric charge is a basic physical property of matter associated with subatomic particles. Electrons are elementary particles with negative charge, and protons have the same amount of positive charge. In a balanced atom, the total number of electrons equals the total number of protons; therefore, the net electric charge of a stable atom is zero.

Several physical effects can cause electron exchange between atoms, hence producing an excess negative or positive charge in an object. The standard unit of charge is Coulomb (C). Electric charge of an electron (-1.602×10^{-12} C) is the elementary charge. Total charge of an object is an integer multiple of electron charge with a positive or negative sign.

2.4.1 Voltage

Difference of potential energy between two points of the electric field is known as "potential difference," "tension." The unit for electric potential is volt (V), named after Italian scientist Alessandro Volta, who developed the first practical battery cell in 1801. Voltage is the common term for electric potential in respect to a certain reference. Since voltage is the physical work done per unit electric charge to move charged particles from one point to another, the term "electromotive force" (EMF) is alternatively used. In a mathematical sense, voltage is the rate of change of electrostatic energy per electric charge, as in the differential expression below:

$$v = \frac{dw}{dq} \tag{2.24}$$

2.4.2 Current

A balanced atom has the same number of negatively charged electrons and positively charged protons to balance each other. However, physical phenomena such as electromagnetic fields, heat, or light may increase the energy of electrons, forcing them to move from one atom to another. In solids, electrons are exchanged between atoms as their energy levels are increased. In liquids and gases, electrically charged molecules called ions can move with an electric or magnetic field effect.

Electric current is defined as the time rate of change of electric charge. In mathematical sense, electric current is the time derivative of the function that describes the variation of electric charge.

$$i = \frac{dq}{dt} \tag{2.25}$$

The standard unit of electric current is ampere (A). In mechatronic systems, lower current values are expressed in milliamps ($1\ \text{mA} = 10^{-3}\text{A}$) or microamps ($1\ \mu\text{A} = 10^{-6}\text{A}$).

2.4.3 Power and Energy

Voltage and current are essential variables that describe the state of all elements in an electrical system. Since both are defined in equations 2.24 and 2.25 based on electric charge, we can multiply and divide the time derivative of energy by differential electric charge dq and arrange the equation to express power in terms of voltage and current.

$$p(t) = \frac{dw}{dt} \cdot \frac{dq}{dq} = \frac{dw}{dq}\frac{dq}{dt} = v(t) \cdot i(t) \tag{2.26}$$

Electric power of an electrical device at any instant is the product of the voltage across its terminals and the current flowing through its terminals.

Energy processed in a circuit element is obtained by integrating electric power over a time interval.

$$W = \int_{t_1}^{t_2} p(t)dt \tag{2.27}$$

2.4.4 Reference Directions

In circuit analysis, the actual directions of voltages and currents are not relevant. However, before starting to write equations and solve the circuit, we must select reference directions. In all circuit problems in this book, we will first assign the voltage references arbitrarily and mark with a plus and minus sign. We will show the reference direction for the currents with an arrow entering the element from the positive terminal as shown in Figure 2.4. This selection is known as "passive notation" because if the product of the algebraic values obtained for voltage and current is positive, we will understand that the element is a passive element absorbing power. Otherwise, if the calculated power is negative, then the element is absorbing negative power, meaning that it is an active element delivering power.

Example 2.4 Figure 2.5 shows the voltages across and currents through the circuit elements. Find the unknown power P and give the appropriate reference directions.

Solution

In an isolated circuit, the algebraic sum of all powers must be zero. Using passive notation references, we can write the power equation.

$$-50 \times 1 + 30 \times 1 + 20 \times 3 + 30 \times 2 + P = 0 \Longrightarrow P = -100\,\text{W}$$

Negative power indicates that this element is a source supplying the circuit. Thus, the current must leave the element from the positive terminal.

2.4.5 Kirchhoff's Laws

The German scientist and mathematician Gustav R. Kirchhoff, among many other discoveries, coined two laws in 1845 that became the fundamental principles of circuit analysis. These concepts are known as Kirchhoff's circuit law (KCL).

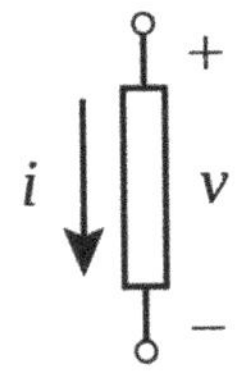

$P = vi > 0$ Passive element

$P = vi < 0$ Active element

Figure 2.4 Passive notation reference directions

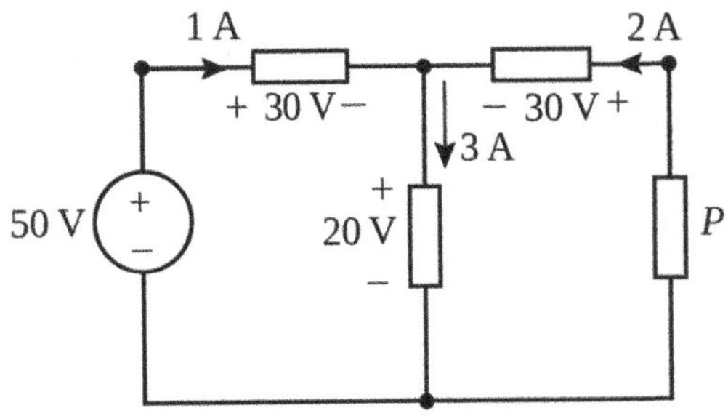

Figure 2.5 Circuit for Example 2.4

KCL: The algebraic sum of all currents entering a node is zero.

$$\sum_{k} i_k = 0 \tag{2.28}$$

KCL is a consequence of the conservation of electric charge principle. It applies to any electric circuit where the total electric field around the components and connecting wires is negligible. At high-frequency or voltage levels, KCL may not apply because of the electric charge flux leaving the elements. Microwave circuits and high-voltage power lines are examples of situations where KCL should be applied with caution. This limitation does not exist in circuits used in mechatronic systems.

Kirchhoff's voltage law (KVL): The algebraic sum of all voltages around a loop is zero.

$$\sum_{k} v_k = 0 \tag{2.29}$$

KVL is a consequence of the conservation of energy principle. In circuits where the magnetic field created by current flow through the loop is not negligible, KVL is an approximation. Such situation appears in high-frequency devices or high-current networks. KVL is applicable to mechatronic systems without any limitation.

2.4.6 Ohm's Law

A German physicist Georg Ohm experimentally discovered that the voltage across an electric wire is proportional to the current flowing through it. He named this parameter "resistance" since the wire resists the flow of electric current. The mathematical expression of Ohm's law is a simple linear equation where the current I is multiplied by the resistance R to obtain the voltage V across the wire. The unit of resistance is "ohm," named after Georg Ohm and represented by Ω. The inverse of the resistance is called conductance and is usually represented by the symbol G. The standard unit for conductance is Siemens (S). It is common to use Ω^{-1} as an alternative unit of conductance.

$$V = RI \longleftrightarrow I = \frac{V}{R} = GV \tag{2.30}$$

Ohm's law assumes that resistance does not depend on the current. In practice, the resistance of a wire changes with temperature as the current flows heats the wire. For example, the resistance of a lightbulb filament changes drastically as soon as current flows through it. But when the lightbulb reaches the operating temperature, the resistance remains constant. Semiconductor elements like diodes are examples of nonlinear resistances.

The resistance of a wire is proportional to its length ℓ and inversely proportional to its cross section A.

$$R = \frac{\rho \cdot \ell}{A} \tag{2.31}$$

In the above equation, the parameter ρ is the resistivity of the wire material. In practical calculations, it is sometimes more convenient to use the inverse of resistivity σ, named conductivity.

Metals such as copper, aluminum, and silver are used as conductor wires due to their low resistivity (in the order of 10^{-8}). Materials like glass, porcelain, ceramic, nylon, mica, polyester, and Teflon are used as insulators because of their extremely high resistivity in the order of 10^{16}. Online resources (Wikipedia, 2024) give extensive lists of electrical properties of materials. Resistivity and conductivity of materials frequently used in mechatronic devices are listed in Table 2.1.

Table 2.1 Resistivity and conductivity of selected materials

Material	Resistivity at 20°C (Ω·m)	Conductivity at 20°C (S/m)
Silver	1.59×10^{-8}	63.0×10^{6}
Copper	1.68×10^{-8}	59.6×10^{6}
Annealed (electrical) copper	1.72×10^{-8}	58.0×10^{6}
Aluminum	2.65×10^{-8}	37.7×10^{6}
Iron	9.70×10^{-8}	10.3×10^{6}
Amorphous carbon	0.5×10^{-3} to 0.8×10^{-3}	1.25×10^{3} to 2.00×10^{3}
Silicon	2.3×10^{3}	4.35×10^{-4}
Dry wood	10^{14} to 10^{16}	10^{-16} to 10^{-14}
Glass	10^{11} to 10^{15}	10^{-15} to 10^{-11}
Teflon	10^{23} to 10^{25}	10^{-25} to 10^{-23}

2.4.7 Circuit Elements

An electric circuit contains active and passive elements. Active elements provide power for the desired performance of the circuit. Passive elements either consume or exchange electric energy with other circuit elements. Elements used in basic electric circuits are described below. We will explain the structure and functions of semiconductor elements in more advanced chapters where signal conditioning, power electronics, or digital circuits are discussed.

2.4.7.1 Sources

Sources are active elements that supply power to an electric or electronic circuit. An ideal source is a conceptual device that can provide an arbitrary amount of energy. Ideal sources are categorized as voltage sources and current sources. Each category of sources may be independent or may depend on other voltages or currents in the circuit. Practical (or real) sources are equivalent to a combination of ideal elements, mostly an ideal source and a resistive circuit.

An ideal voltage source provides a known voltage that is not affected of the current it delivers. Therefore, an ideal voltage source can deliver infinite current without any change in voltage across its terminals. The voltage across the terminals of practical voltage sources such as battery cells and generators depends on the output current. Nonideal voltage sources are typically modeled with an ideal voltage source and a fictional internal resistance. Furthermore, the current output of a practical voltage source is limited.

An ideal current source is a device that provides a known fixed current to external elements connected to its terminals. The voltage across the terminals of an ideal current source is not a function of the output current. Even if the terminals of an ideal current source are connected to each other, in other words, short-circuited, the terminal voltage becomes zero while the output current remains the same. Practical examples of real current sources are battery chargers and some semiconductor devices that operate in the constant current region. While the output current of a physical current source is approximately constant, the voltage across its terminals depends on the external circuit conditions.

Dependent sources provide a voltage or current that is a function of a separate voltage or current in the circuit. The circuit model of a physical amplifier contains a voltage or current source that depends on the input voltage of the amplifier. Equivalent circuits of electric machines include a voltage source that depends on both the magnitude of the magnetic field and the rotation speed.

In practical mechatronic systems, two types of power sources are common. DC power sources supply circuit elements with a voltage or current that does not change in time. AC power sources provide sinusoidal voltage or current. Time variation of a sinusoidal quantity is in the form of a sine or cosine function. Characteristic parameters of an AC waveform are magnitude, frequency, and phase angle from a reference.

2.4.7.2 Resistor

Ohm's law defines the voltage–current relationship of a linear resistor. Resistors are passive elements that convert electrical power to heat. Power absorbed in a resistive element is

$$P = v \cdot i = (Ri) \cdot i = Ri^2 \tag{2.32}$$

Alternatively, power on a resistance can be found in terms of voltage and conductance.

$$P = \left(\frac{v}{R}\right) v = \frac{v^2}{R} = Gv^2 \tag{2.33}$$

Temperature rise due to the passage of current through a conductor is known as Joule heating, referring to British scientist James Prescott Joule. Useful practical applications of the Joule effect include electric heaters, incandescent light bulbs, electric fuses, and any device that produces heat by passing electric current through a resistance.

Resistors are essential components of electric and electronic circuits. Carbon film resistors are available in a wide range of values. Other types of resistors are made from metal wire, plate, or film. Commercial resistors are available at standard rated values and power limits. Common power ratings for resistors used in electronic circuits are 1/8, 1/4, 1/2, 1, and 2 W.

Electrical energy converted to heat dissipates to the surrounding environment. Overhead power lines, power cables, coils in electric machines, transformers, electronic circuits, and all kinds of electromechanical devices produce heat during their normal operation. If the electrical energy converted to heat does not serve a specific purpose, it is lost energy and reduces the efficiency of devices. Since all electrical and electronic devices are designed to operate below a certain temperature, special cooling elements such as heatsinks, fans, or circulated fluid may be used to maintain the temperature at an acceptable level.

Figure 2.6 shows the series and parallel connections of two resistors. In the series connection, the same current passes through each resistor. We obtain the equivalent resistor using Ohm's law to

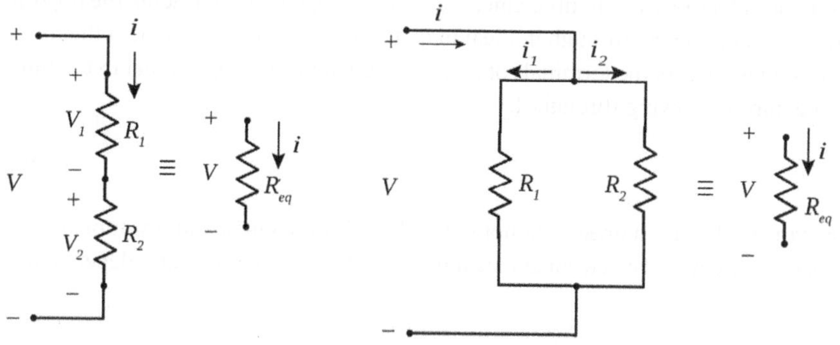

Figure 2.6 Series and parallel resistances

determine the voltage across each resistor and applying KVL to express the voltage across the terminals of the series connection.

$$V = (R_1 + R_2)I \tag{2.34}$$

To obtain the equivalent of several resistors connected in series, individual resistances are added.

$$R_{eq} = \sum_{k=1}^{n} R_k \tag{2.35}$$

Resistors connected in parallel have the same voltage across their terminals; we can obtain the equivalent resistor by using KCL.

$$I = \left(\frac{1}{R_1} + \frac{1}{R_2}\right)V = \frac{R_1 + R_2}{R_1 R_2}V \tag{2.36}$$

Therefore, the equivalent resistance of resistors connected in parallel is the sum of the inverses of all individual resistances. In the calculation of the equivalent of several resistors connected in parallel, it is more convenient to use the conductance than the resistance of individual elements. In the general case, if n resistors are connected in parallel, this result is generalized as:

$$R_{eq} = \sum_{k=1}^{n} \frac{1}{R_k}; G_{eq} = \sum_{k=1}^{n} G_k \tag{2.37}$$

Consequently, resistors in series add. Resistors in parallel are combined by adding the inverses of individual resistors. The equivalent conductance of parallel resistors is the sum of each individual conductance

2.4.7.3 Inductor

An inductor is a passive circuit element that has the ability to store energy in its magnetic field. Any conductor that carries electric current has inductive properties, but in order to enhance the magnetic effects, physical inductors are made by wrapping an insulated wire around a core that may be a nonmagnetic or a ferromagnetic material. Electric current flowing through the coil creates a magnetic field. As the magnetic field changes, an inductor stores or delivers energy with the circuit elements connected to its terminals. Due to the energy exchanging property, the electrical characteristic of an inductor is a function of the time variation rather than the magnitude of the electric current (Alexander & Sadiku, 2013).

Electrical property of inductors as a circuit element is inductance, which represents the opposition to the change of current flowing through it. Ideal inductor is a linear circuit element that does not have resistance. Voltage across the terminals of an ideal inductance is proportional to the time rate of change of the current flowing through it.

$$v = L\frac{di}{dt} \tag{2.38}$$

The unit of inductance is Henry, named in honor of the American scientist and inventor Joseph Henry (1797–1878), who discovered electromagnetic induction at the same time but independently from Faraday.

The derivative of the current implies that for a constant current (DC), the voltage across the terminals of an ideal inductor is zero. Hence, the ideal inductor acts as a short circuit for DC.

Inductors exchange magnetic energy with the circuit elements with which they interact. An inductor either absorbs or delivers energy depending on the voltage and current conditions.

Practical inductors have a broad range of applications in mechatronic systems. Transformers consist of two or more coupled inductors placed on a ferromagnetic core, typically electrical steel or ferrite. Inductors are an essential part of switching power converters. They limit the current load in AC systems without causing high-power loss like resistors. They are part of filters to block high-frequency components and harmonics of nonsinusoidal voltage and current waveforms.

Energy stored in an inductor is described by integrating the product of voltage and current.

$$W = \int_{t_1}^{t_2} \left(i \cdot L \frac{di}{dt} \right) dt = \frac{1}{2} L \left(i_2^2 - i_1^2 \right) \tag{2.39}$$

2.4.7.4 Capacitor

A capacitor consists of two conductors separated by an insulator. Consider two parallel conductor plates separated by air or another insulating material (also called dielectric). When we apply a constant voltage *V* across the terminals, an electric field proportional to the voltage occurs in the region between the plates. Combining equations 2.24 and 2.25, we obtain

$$V = \frac{d}{dq}(q \cdot E \cdot d) = E \cdot d \Rightarrow E = \frac{V}{d} \tag{2.40}$$

The insulator prevents the flow of a current through the terminals. However, an electric charge in an amount proportional to the current accumulates at any time.

$$Q = C \cdot V \tag{2.41}$$

The constant C is called capacitance because it represents the capacity of storing electric charge. The unit for capacitance is farad, named after Michael Faraday. The dimension of farad is coulomb per volt. One farad is a large quantity for conventional capacitors. In practice, pico- (10^{-12}), nano- (10^{-9}), and micro-farad (10^{-6}) are more common.

Now suppose that the applied voltage changes in time. The stored electric charge will also change in time.

$$q(t) = C \cdot v(t) \tag{2.42}$$

Electric current is the time derivative of the function that represents the electric charge. By taking the derivative of both sides of this equation, we obtain the voltage–current (*V–I*) relationship for the ideal capacitor.

$$i(t) = C \frac{dv(t)}{dt} \tag{2.43}$$

By multiplying both sides of this equation by *dt* and integrating, we obtain the expression of voltage in terms of current at an instant t_0.

$$v(t_0) = \frac{1}{C} \int_{-\infty}^{t_0} i(t) dt \tag{2.44}$$

If the initial voltage at time t_0 is V_0, then the voltage across the capacitor at time t_l would be

$$v(t) = \frac{1}{C}\int_{t_0}^{t_1} i(t)dt + V_0 \tag{2.45}$$

An ideal capacitor stores electric charge proportional to the voltage applied across its terminals. The voltage–current (V–I) characteristic is a differential equation, implying that the current is a function of the variation of the voltage, rather than its current value.

Energy stored in an ideal capacitor is calculated by integrating power in time.

$$W = \int_0^t C\frac{dv}{dt}\cdot v\cdot dt = C\int_0^t v\cdot dv \tag{2.46}$$

The integral yields

$$W = \frac{1}{2}Cv^2 \tag{2.47}$$

Capacitors connected in series and parallel can be combined to yield a single equivalent capacitance. Suppose n capacitors with different parameters are connected in parallel. Since they have the same voltage across their terminals, we can obtain the equivalent capacitor by writing equation 2.43 for each capacitor and using the KCL.

$$i(t) = (C_1 + C_2 + C_3 + \cdots + C_n)\frac{dv(t)}{dt} \tag{2.48}$$

Therefore, the equivalent capacitance of n capacitors connected in parallel is the sum of all capacitances.

$$C_{eq} = \sum_{k=1}^{n} C_k \tag{2.49}$$

Capacitors connected in series carry the same current. Therefore, we obtain the equivalent of n capacitors connected in series by writing equation 2.44 and applying the KVL.

$$v(t) = \left[\frac{1}{C_1} + \frac{1}{C_2} + \frac{1}{C_3} + \cdots + \frac{1}{C_n}\right]\int_{-\infty}^{t_1} i(t)dt \tag{2.50}$$

Hence, to obtain the equivalent of n capacitors connected in series, the inverse of individual capacitances is added.

$$C_{eq} = \sum_{k=1}^{n}\frac{1}{C_k} \tag{2.51}$$

Consequently, capacitances connected in parallel are added. The inverse of the equivalent capacitance equals the sum of the inverses of series-connected capacitances.

2.5 Summary

This chapter provides an overview of fundamental principles essential for understanding mechatronics, covering classical physics, electric circuits, and digital logic. It is intended to refresh the physics and engineering knowledge needed to understand the subsequent chapters. It presents a brief review of basic concepts of classical mechanics, electric circuits, and digital electronics. In the analysis or design of mechatronic systems, we frequently use a combination of the basic principles presented in this chapter.

1) Newton's first law states that an object remains at rest or in uniform motion unless acted upon by a net force, establishing the necessity of external forces to change velocity. The second law quantifies this by relating force to the rate of change of momentum, often simplified to force being proportional to acceleration with mass as the constant of proportionality. The third law highlights the symmetry of forces, indicating that forces always occur in equal and opposite pairs.
2) Rotational dynamics are also discussed, with angular velocity, torque, and moment of inertia introduced to analyze rotating systems such as motors and robotic arms. Newton's laws are extended to rotational motion, where torque plays a role analogous to force, and angular acceleration is proportional to the applied torque divided by the moment of inertia.
3) Work is defined as the product of force and displacement, or more generally, the integral of the force along the path of motion. Energy is described as the capacity to perform work, with potential energy linked to an object's position (gravitational or elastic) and kinetic energy related to motion. The chapter details the calculation of gravitational potential energy as the product of weight and height difference, and kinetic energy as proportional to mass and the square of velocity. Mechanical power is defined as the rate of energy transfer, expressed as the product of force and velocity in linear systems or torque and angular velocity in rotational systems.
4) The chapter outlines key conservation laws fundamental to engineering and mechatronics: conservation of energy, mass, momentum, and electric charge. It clarifies that energy cannot be created or destroyed but only transformed or transferred within an isolated system. Efficiency is introduced as the ratio of output to input energy, acknowledging inevitable losses such as heat dissipation. Conservation of mass applies to systems without mass exchange, and conservation of momentum governs interactions between bodies, especially relevant in systems with comparable masses. Conservation of electric charge underpins electric circuit laws.
5) Electric circuits consist of active elements (voltage and current sources) that supply power and passive elements (resistors, capacitors, inductors) that absorb or store energy. This chapter explains ideal and practical sources, emphasizing that real sources have limitations such as internal resistance or current constraints. Ohm's law relates voltage, current, and resistance linearly, with resistance depending on material properties, length, and cross-sectional area.
6) **Resistors** convert electrical energy to heat, governed by Ohm's law and power formulas. Series and parallel resistor combinations are explained with formulas for equivalent resistance. **Inductors** store energy in magnetic fields, with voltage proportional to the rate of change of current. The unit of inductance is Henry, and practical applications include transformers and filters. **Capacitors** store energy in electric fields between conductors separated by an insulator. Their voltage–current relationship involves the derivative of voltage with respect to time, and they combine in series and parallel with specific rules for equivalent capacitance.
7) KCL states that the sum of currents entering a node is zero, based on conservation of charge. KVL states that the sum of voltages around any closed loop is zero, reflecting conservation of energy. These laws form the foundation for circuit analysis and design.

2.6 Check Your Understanding

2.6.1 Self-Assessment Quiz

1 Newton's first law of motion states that an object remains at rest or in uniform motion unless acted upon by a net force.
True False

2 The unit of power in the international unit system (SI) is the Joule.
True False

3 An ideal capacitor stores electrical energy.
True False

4 The product of mass and velocity is known as momentum.
True False

5 Kirchhoff's voltage law states that the sum of currents entering a node is zero.
True False

6 What is the product of mass and velocity known as?
- **a.** Acceleration
- **b.** Force
- **c.** Momentum
- **d.** Displacement

7 Which unit describes power?
- **a.** Joule
- **b.** Volt
- **c.** Watt
- **d.** Ampere

8 What is the voltage across an ideal inductor proportional to?
- **a.** Current flowing through it
- **b.** Derivative of the current
- **c.** Flux density
- **d.** Magnetic field intensity

9 Which one of the following describes electric current?
- **a.** Energy of charged particles
- **b.** Time rate of change of electric charge
- **c.** Work done by moving an electric charge
- **d.** Product of voltage and power

10 What is the role of an ideal voltage source in an electric circuit?
- **a.** A prescribed voltage across its terminals
- **b.** A prescribed power to the connected circuit
- **c.** Constant energy
- **d.** Constant power

11 What is the unit of inductance?
  a. Farad
  b. Henry
  c. Joule
  d. Watt

12 Which law states that the sum of currents entering a node is zero?
  a. Ohm's law
  b. Kirchhoff's voltage law
  c. Kirchhoff's current law
  d. Newton's third law

13 What is the primary function of a capacitor in an electric circuit?
  a. To store electrical energy
  b. To dissipate electrical energy as heat
  c. To generate electric energy
  d. To act as a power source

14 Which of the following describes the relationship between voltage and current in an ideal resistor?
  a. Voltage is proportional to the square of the current.
  b. Voltage is inversely proportional to the current.
  c. Voltage is proportional to the current.
  d. Voltage is independent of the current.

15 What is the primary characteristic of an ideal voltage source?
  a. It provides a constant current.
  b. It provides a constant voltage.
  c. It provides constant power.
  d. It provides constant energy.

Answers:

1: True; 2: False; 3: True; 4: True; 5: False; 6: c; 7: c; 8: b; 9: b; 10: a; 11: b; 12: c; 13: a; 14: c; 15: b.

2.6.2 Short Answer Questions

1 Why is it important for a mechatronic system engineer to understand the fundamental principles of physics?

2 What is the difference between linear and angular momentum?

3 How is angular velocity calculated from rotations per minute (rpm)?

4 What is the relationship between torque and angular acceleration for a rotating object?

5 How does efficiency affect the energy output of a mechatronic system?

6 What is the significance of conservation of mass in battery cells?

7 How does the principle of conservation of momentum apply to mechatronic systems?

8 What is the unit of electric charge, and what is the charge of a single electron?

9 What is the difference between active and passive circuit elements?

10 How does the resistance of a wire depend on its length and cross-sectional area?

11 What is the function of an inductor in a mechatronic system?

12 How is the energy stored in a capacitor calculated?

13 What is the difference between series and parallel connections for capacitors?

14 What does Kirchhoff's voltage law (KVL) state?

15 How does a practical voltage source differ from an ideal voltage source?

16 What is the role of a resistor in an electric circuit?

17 Why are heatsinks and fans used in electrical and electronic devices?

18 What is the unit of inductance, and who is it named after?

19 How does a capacitor store electric charge?

20 What is the significance of reference directions in circuit analysis?

2.6.3 Essay Questions

1 Discuss the significance of Newton's laws of motion in the design and analysis of mechatronic systems. Provide examples of how these laws are applied in real-world mechatronic applications.

2 Explain the principles of rotational dynamics and their relevance to mechatronic systems. How do concepts like torque, angular velocity, and moment of inertia play a role in the functioning of devices such as motors and robotic arms?

3 Describe the concepts of work, energy, and power in classical physics. How are these concepts applied in the context of mechatronic systems, and what are the practical implications of energy conversion and efficiency in these systems?

4 Analyze the conservation principles of energy, mass, momentum, and electric charge. How do these principles underpin the operation and design of mechatronic systems, and what are the key considerations for engineers in ensuring these principles are adhered to?

5 Examine the role of electric circuits in mechatronic systems. Discuss the functions of active and passive circuit elements, and explain how Ohm's law, Kirchhoff's laws, and the behavior of resistors, capacitors, and inductors are fundamental to circuit analysis and design.

2.6.4 Problems

1 A robotic arm of mass 15 kg accelerates from rest to a velocity of 2 m/s in 4 seconds. Calculate the force required to achieve this acceleration.

2 A motor rotates a disk of radius 0.2 m and mass 5 kg. The disk accelerates from rest to 120 rpm in 3 seconds.

a. Find the angular acceleration.

b. Calculate the torque required, assuming the disk is a thin, solid disk.

3 An electric scooter of total mass 60 kg accelerates from rest to 15 km/h in 5 seconds.

a. Calculate the kinetic energy gained.

b. If the motor is 85% efficient, how much electrical energy was drawn from the battery?

4 A conveyor belt lifts 40 kg packages to a height of 2.5 m at a constant speed.

a. Calculate the gravitational potential energy gained by each package.

b. If the system is 80% efficient, what is the minimum power rating required for the motor if it lifts one package every 10 seconds?

5 A 500 kg elevator descends 20 m. The braking system recovers 90% of the potential energy as electricity. Calculate the energy recovered in joules and watt-hours.

6 Two carts, one of mass 2 kg moving at 3 m/s and another of mass 3 kg at rest, collide and stick together. Find their final speed after the collision.

7 A motor lifts a 75-kg load vertically at a speed of 0.8 m/s. Calculate the mechanical power output of the motor.

8 A hybrid car weighs 1,200 kg. Its electric brakes can convert 90% of the kinetic energy into electricity to be stored in the battery. Calculate the energy stored in Wh as the car slows down using the brakes from 70 to 30 mph.

9 An elevator has a total mass of 500 kg and is equipped with an electronic braking system that can return 95% of the mechanical energy back to the electric supply when the elevator goes down. Find how much energy is saved when the elevator goes down 30 m.

10 An electric car weighing 900 kg with its passengers travels 500 m on a 30° slope at constant speed. Assuming 90% overall efficiency, how much energy would its motor draw from the battery?

11 An electric scooter weighs 50 kg and has a 1 HP motor. In how much time would this scooter reach the speed of 20 km/h from the start carrying a 70-kg person?

12 A resistor of 10 Ω is connected across a 24-V battery.
 a. Calculate the current through the resistor.
 b. Find the power dissipated by the resistor.

13 Three resistors (5, 10, and 20 Ω) are connected in series and then in parallel. Calculate the equivalent resistance for both configurations.

14 A 100-μF capacitor is charged by a 12-V source. Calculate the energy stored in the capacitor.

15 An inductor of 0.5 H carries a current of 2 A. Calculate the energy stored in the inductor.

16 A car tows a boat from the road on the riverside. The angle between the rope and the direction of the boat's movement is 60°, and the car pulls the rope with a 600-N force. Determine the work performed as the boat is dragged 100 m.

17 A truck tows a car that weighs 1,200 kg on a horizontal road. The speed reaches 25 mph from the start in 20 seconds. Find the force on the towing rope.

18 A winch pulls a load that weighs 300 kg on a horizontal surface. The speed of the load reaches 2.5 m/s in 10 seconds from stand still. With how much force does the winch pull the load?

19 A winch pulls the load with 150 N to reach the speed from 0 to 2.5 m/s in 5 seconds. Calculate the work done by the winch.

20 A winch pulls a 100-kg load horizontally with a constant force of 250 N over a distance of 10 m. Calculate the work done by the winch.

21 A 30° inclined, 10-m long conveyor belt transports 50-kg packages in a warehouse. Calculate the mechanical energy required to transport each package.

22 Calculate the kinetic energy the 300-kg load gained when it has reached 2.5 m/s speed from standstill in 5 seconds.

23 Find the mechanical power needed to pull a 300-kg load when its speed reached 2.5 m/s from rest with an acceleration of $0.5\ m/s^2$.

References

Alexander, C. K. & Sadiku, M. N. O., 2013. *Fundamentals of Electric Circuits*. 5th ed. New York: McGraw-Hill.

Engineering Toolbox, 2005. Mass Moment of Inertia. [Online] Available at: https://www.engineeringtoolbox.com/moment-inertia-torque-d_913.html [Accessed 30 10 2024].

Giancoli, 2020. *College Physics*. New York: McGraw Hill.

Ohanian, H. C. & Market, J. T., 2006. *Physics for Engineers and Scientists*. New York, London: W. W. Norton & Company.

Soysal, O. A. & Soysal, H. S., 2020. *Energy for Sustainable Society - From Resources to Users*. Hoboken, NJ: Wiley.

Wikipedia, 2024. Electrical Resistivity and Conductivity. [Online] Available at: https://en.wikipedia.org/wiki/Electrical_resistivity_and_conductivity https://en.wikipedia.org/wiki/Electrical_resistivity_and_conductivity [Accessed 30 10 2024].

3

Dynamic System Modeling

OBJECTIVES

- Explain the transition from functional diagrams to mathematical models in engineering systems.
- Introduce mathematical modeling as a method for representing and analyzing the behavior of mechatronic systems, which integrate mechanical, electrical, and control components.
- Describe the process of system identification.
- Discuss different types of systems and their implications for modeling.
- Highlight the importance of linearization techniques for simplifying nonlinear systems.
- Introduce the concepts of through and across variables.
- Provide foundational modeling approaches for electrical, mechanical, hydraulic, pneumatic, and thermal elements in mechatronic systems.
- Present the basics of dynamic system response.
- Prepare the reader for analyzing and designing mechatronic systems using mathematical tools and simulation before physical implementation.

In earlier chapters, we used functional diagrams to show the interactions between system elements. These diagrams include blocks that briefly describe each element's purpose, with arrows showing the signal flow. This type of graphical representation is useful for visualization of the system operation but has limited use for analysis and design purposes.

Engineering systems are analyzed using mathematical models instead of verbal descriptions. A mathematical model is a set of equations that describe the behavior of a physical system. Mathematical modeling of mechatronic systems uses mathematical techniques to represent and analyze the behavior of systems that combine mechanical, electrical, and control components.

The mechanical subsystem consists of mechanical parts such as links, joints, gears, and hydraulic and pneumatic actuators. It often involves the study of dynamics, kinematics, hydraulics, and structural analysis. Electrical subsystems include sensors, electromechanical devices, motors, and electronic power converters. It involves circuit analysis, control theory, and signal processing. Control subsystems govern the behavior of the mechanical and electrical subsystems. It uses digital, analog, or hybrid architecture with components like operational amplifiers, logic circuits, and signal converters. Feedback control principles and algorithms are applied to examine their dynamic behavior, performance, and stability.

Mathematical modeling of mechatronic systems is an essential process that describes the complex interactions between mechanical, electrical, and computational components that define these systems. By creating a mathematical representation of mechatronic systems, engineers and researchers can analyze system dynamics, optimize performance, and simulate its behavior before physical implementation. This task typically encompasses several phases, which include defining the system, selecting appropriate modeling techniques, formulating mathematical equations, conducting simulations, and performing validation tests.

3.1 System Identification

The modeling process of a mechatronic system begins with a clear specification of the system's structure, the objectives of the model, physical constraints, and the desired level of accuracy. This initial step is critical for selecting the most appropriate modeling approach.

Typically, a technical drawing, functional block diagram, or circuit schematic outlines the system's architecture. These diagrams may be abstracted and simplified depending on the objectives of the modeling process. Structural symmetry can help reduce the number of unknowns. For steady-state performance analysis, algebraic equations can adequately describe the system's behavior. For transient analysis, differential equations are necessary.

The mathematical properties of the input–output variables and the interactions among the system components determine the appropriate type of model for accurate and realistic system representation.

3.1.1 Distributed and Lumped Parameter Systems

Systems with distributed variables like temperature in a heat exchanger, stress in a structure, fluid dynamics, or electromagnetic fields require partial differential equations (PDEs). Distributed parameter electrical systems are analyzed using electromagnetic field theory and wave propagation principles.

Numerical techniques are often employed to solve the PDEs used in distributed parameter systems. These techniques include the finite difference method (FDM), finite element method (FEM), and finite volume method (FVM).

If the distributed parameter representation is not necessary, complex systems can be simplified into manageable lumped parameter components where properties are averaged over a distinct volume. Examples include electric circuits with discrete resistors, capacitors, and inductors, as well as mechanical structures where forces are applied at specific points.

3.1.2 Linear and Nonlinear Systems

In mathematical modeling, linearity refers to a system's property where the output is directly proportional to the input. This indicates that the system adheres to superposition and homogeneity principles. Superposition means the response to multiple inputs is the sum of each input's individual response. Homogeneity indicates that scaling an input would scale the output by the same factor.

Homogeneity and superposition principles are combined in the following equation.

$$f(ax + by) = af(x) + bf(y) \tag{3.1}$$

Example 3.1 Demonstrate that Ohm's law defines the voltage–current (*V–I*) characteristic of a linear resistor.

Solution

The *V–I* characteristic of a linear resistor is defined by Ohm's law as $V = RI$. When the sum of the currents i_1 and i_2 flows through this resistor, the voltage across its terminals is

$$v = v(i_1) + v(i_2) = Ri_1 + Ri_2$$

When the currents change to $2i_1$ and $3i_{12}$, the voltage across the resistor becomes

$$v(2i_1) + v(3i_2) = 2Ri_1 + 3Ri_2$$

Then Ohm's law describes the voltage–current relationship of a linear resistor.

Equations derived from a linear system can sometimes be nonlinear. In a linear equation, coefficients are not functions of variables, and terms do not include operations like multiplication, division, exponential, or trigonometric functions of variables.

Example 3.2 Demonstrate that the power dissipated in a linear resistor is nonlinear.

Solution

Power dissipated in a resistor is

$$P = Ri^2$$

If the current doubles,

$$P(i^2) = R(2i)^2 = 4i^2 \neq 2P(i)$$

Suppose that the currents i_1 and i_2 flow through the resistor.

$$P(i_1 + i_2) = R(i_1 + i_2)^2 \neq Ri_1^2 + Ri_2^2$$

Therefore, the power equation is nonlinear.

Linearity simplifies the analysis and design of systems because it allows the use of powerful mathematical techniques such as the Laplace transform and Fourier analysis. The Laplace transform is a mathematical technique that converts differential equations into algebraic equations. Fourier analysis decomposes a signal into a function of its constituent frequencies. These techniques are particularly useful in the analysis and design of systems, as they enable engineers to solve dynamic equations using linear algebra methods.

However, most physical systems exhibit some degree of nonlinearity. Linearization is a method used to approximate a nonlinear system by a linear model around a specific operating point. This approximation enables the application of linear system analysis tools to study and design systems that would otherwise be difficult due to their inherent nonlinearity.

3.1.3 Time-Variant and Time-Invariant Systems

Parameters, characteristics, and behavior of time-variant mechatronic systems change over time. These changes can be due to factors such as wear and tear, environmental conditions, or operational variations. For instance, in a tape recorder, as the tape rewinds, the diameter and mass of the reel change, altering its moment of inertia. This variability complicates the modeling, analysis, and control of such systems, as engineers must account for these dynamic changes to ensure accurate predictions and efficient performance.

One approach to managing time-variant systems is to use adaptive control techniques, which adjust the control parameters in real time to accommodate the evolving system dynamics. Another method involves developing robust models that can account for a range of operating conditions and uncertainties.

Understanding and modeling time-variant systems is crucial for designing reliable and efficient mechatronic systems that can maintain optimal performance despite changing conditions. This knowledge is essential for advancing technologies in various fields, including robotics, automotive engineering, and aerospace.

In a time-invariant system, on the contrary, the characteristics and behavior do not change over time. In such systems, the output response to a given input is the same, regardless of when the input is applied. This property states that if an input signal is shifted in time, the output will be a corresponding time-shifted version of the original output. Time invariance greatly simplifies the analysis and design of systems, as engineers can rely on consistent system behavior over time. This predictability makes time-invariant systems particularly advantageous in applications where maintaining steady performance is crucial.

Time-invariant systems are often analyzed using methods like convolution, where the output is derived from the input and the system's impulse response. The predictability of time-invariant systems allows for efficient use of tools like the Laplace and Fourier transforms, which helps in frequency domain analysis and filter design. Engineers can leverage these tools to achieve desired system performance with high accuracy.

3.1.4 Continuous and Discrete in Time Systems

Continuous-time models represent systems with variables that are continuous over time, whereas discrete-time models are suitable for systems where variables are sampled at specific intervals.

Conventional servomechanisms used for position or speed control of electromechanical drives are typical systems continuous over time. Such systems employ linear electronic circuits for proportional, integral, and derivative (PID) control of the system outputs.

Examples of discrete-time systems in mechatronics include digital control systems, where controllers process input signals at discrete time intervals. Programmable logic controllers (PLCs) in industrial automation are another example, as they operate by executing control algorithms at specific sampling rates. Additionally, digital signal processing (DSP) systems used in robotics, such as sensor data processing and feedback control, rely on discrete-time models to perform operations like filtering, data compression, and error detection. These systems are fundamental in applications requiring precise timing and control, enabling mechatronic devices to function efficiently and effectively.

Continuous-time systems can be analyzed using differential equations: Laplace and Fourier transforms. Discrete-time systems are analyzed using difference equations and z-transform.

3.2 Limitations, Constraints, and Assumptions

Modeling mechatronic systems involves several typical limitations and constraints that engineers must consider for accurate and reliable representations. A significant constraint is the inherent nonlinearity and time-variant behaviors frequently observed in physical systems. These characteristics make it challenging to develop precise models that can predict system performance under varying conditions.

The principles of physics and engineering are applicable under ideal conditions. Physical systems can be approximated as ideal within certain limits. For example, Ohm's law assumes that the resistor is linear and time invariant. The resistance of practical resistors varies with temperature. When the power dissipation exceeds the rated value, the resistor's temperature rises, resulting in nonlinear behavior. The resistance of an incandescent lightbulb varies significantly in a short period when it is turned on due to the temperature rise.

Moore's law assumes that the deformation in the material is proportional to the applied force. This assumption is based on the elasticity theory, which holds true under small deformations and within the elastic limit of the material. However, when the material is subject to large forces or prolonged stress, plastic deformation can occur, leading to deviations from the linear behavior predicted by Moore's law.

Nonlinear systems can be approximated as linear for small variations in variables. Small-signal linearization is commonly used in modeling semiconductor components. The magnetic elements of transducers, actuators, and motors involve ferromagnetic materials, which display nonlinear properties. These components can be considered linear within a specific operational range.

Another key assumption is that the system components are isolated and do not interact in unpredictable ways. Interactions between components can introduce complexities that are not accounted for in simplified models. For example, in microelectronic circuits, parasitic effects such as capacitance and inductance between closely packed components can affect the overall performance and limit the scalability predicted by Moore's law.

Generally, environmental factors impose constraints on the modeling of mechatronic systems. Variations in temperature, humidity, and other environmental conditions can affect system components and their interactions, requiring models to account for these changes to ensure robust performance. It is crucial to explicitly define the environmental conditions under which the mathematical model accurately represents the physical system.

A major limitation in modeling is the availability of computational resources. High-fidelity models demand substantial computational capacity and time, which may not be feasible for real-time applications. Engineers must balance the level of model detail with the available computational capacity to attain a practical solution.

Additionally, the accuracy of input data plays a crucial role in modeling. Incomplete or noisy data can lead to errors and uncertainties in the model, which impact its reliability and effectiveness. Engineers must employ techniques such as data preprocessing, filtering, and validation to enhance data quality and model performance.

3.3 Linearization Techniques

Linearization techniques simplify the analysis and control of mechatronic systems by approximating a nonlinear system with a linear model around a specific point. This allows the use of ordinary differential equations (ODEs) and linear control theory. Piecewise linear approximation and small-signal modeling are common methods in mechatronic system analysis and design. Effective linearization techniques enable us to simplify complex mechatronic systems, thereby facilitating analysis, control design, and optimization.

3.3.1 Piecewise Linear Approximation

In this technique, the nonlinear system is divided into several linear segments, each valid over a specific range of operation. By switching between these linear models based on the system's state, engineers can approximate the overall nonlinear behavior with a series of linear models.

Piecewise linearization is a technique that enhances the analysis and control of mechatronic systems by dividing the nonlinear system into several linear segments, each valid over a specific operational range. This method provides a practical approach for approximating the system's overall nonlinear behavior with multiple linear models, making it more manageable for engineers to analyze and control complex systems. By switching between these linear models according to the system's state, engineers can effectively capture the dynamic behavior of the system across a broader range of conditions. This technique is particularly useful in applications where the system exhibits significant nonlinearity but can still be approximated using linear segments to simplify the design and optimization processes.

Example 3.3 Piecewise linear approximation can be effectively applied to model the magnetization curve of ferromagnetic materials used in electromagnetic transducers, actuators, and electric machines. The curve in Figure 3.1 depicts the relationship between the magnetic flux linkage λ and the magnetic field strength H within a solenoid core. Obtain an expression for the coil inductance if the solenoid operates in the linear region.

Solution

To apply piecewise linear approximation, the magnetization curve is divided into two linear segments, each representing a specific range of magnetic field strengths. By doing so, the nonlinear behavior of the core material's magnetization is approximated by two separate linear models. When the field intensity varies between 0 and 0.25 A/h, the transformer can be considered linear. The inductance is defined by the following equation:

$$L = \frac{d\lambda}{di} = \frac{\Delta\lambda}{\Delta i} = \frac{1.2}{0.25} = 4.8\,H.\therefore$$

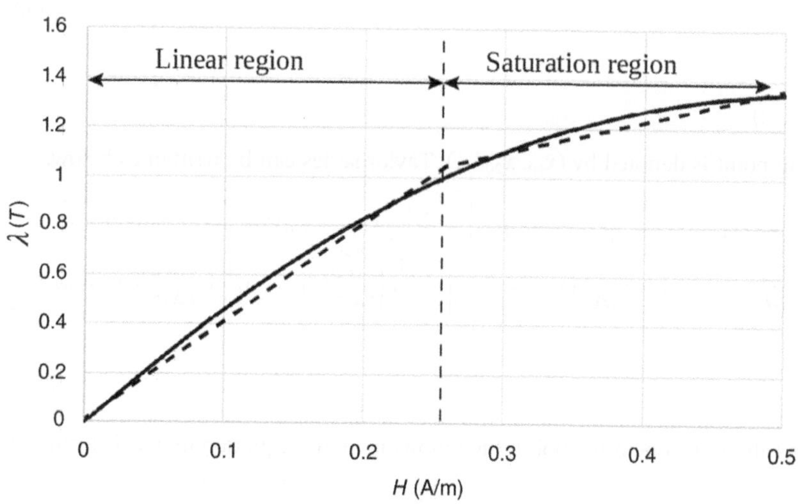

Figure 3.1 Magnetization curve of a solenoid core material

3.3.2 Small-Signal Linearization

This technique involves using a first-order Taylor series expansion of the nonlinear system dynamics around a steady-state point. By retaining only the linear terms, we can derive a linearized model that accurately describes the system's behavior for small deviations from the operating point.

Consider a system described by the nonlinear input–output relationship below where x is the input and y is the output variable.

$$y = f(x) \tag{3.2}$$

Suppose that $f(x)$ and its derivatives are continuous. The Taylor series of y about the operating point x_0 is

$$y = f(x) = f(x_0) + \frac{df}{dx}(x - x_0) + \frac{1}{2!}\frac{d^2f}{dx^2}(x - x_0)^2 + \cdots \tag{3.3}$$

If the variation $x - x_0$ from the operating point is small, higher-order terms in $x - x_0$ are negligible. Then, the Taylor series may be simplified.

$$y = y_0 + K(x - x_0) \tag{3.4}$$

where

$$y_0 = f(x_0)$$
$$K = \left.\frac{df}{dx}\right|_{x = x_0}$$

Equation 3.4 can be arranged as below:

$$y - y_0 = K(x - x_0) \tag{3.5}$$

This equation indicates that $y - y_0$ is proportional to the small variation $x - x_0$ (Ogata, 2002).

Linearization near an operating point may be generalized for a nonlinear system with multiple inputs. Consider the n-variable system below:

$$\begin{aligned} y_1 &= f_1(x_1, x_2, \ldots) \\ y_2 &= f_2(x_1, x_2, \ldots) \\ &\vdots \\ y_n &= f_n(x_1, x_2, \ldots) \end{aligned} \tag{3.6}$$

If a stable operating point is denoted by $(x_{10}, x_{20}, \ldots)$, Taylor series can be written as below:

$$\begin{aligned} y &= f_1(x_{10}, x_{20}, \ldots) + \left[\frac{\partial f_1}{\partial x_1}(x_1 - x_{10}) + \frac{\partial f_1}{\partial x_2}(x_2 - x_{20})\cdots\right] + \frac{1}{2!}\left[\frac{\partial^2 f_1}{\partial x_1{}^2}(x_1 - x_{10})^2 + \frac{\partial^2 f_1}{\partial x_1{}^2}(x_2 - x_{20})^2\cdots\right] \\ y_2 &= f_2(x_{10}, x_{20}, \ldots) + \left[\frac{\partial f_2}{\partial x_1}(x_1 - x_{10}) + \frac{\partial f_2}{\partial x_2}(x_2 - x_{20})\cdots\right] + \frac{1}{2!}\left[\frac{\partial^2 f_2}{\partial x_1{}^2}(x_1 - x_{10})^2 + \frac{\partial^2 f_2}{\partial x_1{}^2}(x_2 - x_{20})^2\cdots\right] \\ &\vdots \end{aligned} \tag{3.7}$$

For small variations about the operating point, the following matrix equation in matrix form can be obtained:

$$\mathbf{y} - \mathbf{y}_0 = \mathbf{A}(\mathbf{x} - \mathbf{x}_0) \tag{3.8}$$

In this equation, **x** and $\mathbf{x}_0$ are $n \times 1$ vectors. **A** is $n \times n$ matrix calculated as

$$\mathbf{A} = \begin{bmatrix} \frac{\partial f_1}{\partial x_1} & \cdots & \frac{\partial f_1}{\partial x_n} \\ \vdots & \ddots & \vdots \\ \frac{\partial f_n}{\partial x_1} & \cdots & \frac{\partial f_n}{\partial x_n} \end{bmatrix}_{\mathbf{x} = \mathbf{x_0}} \tag{3.9}$$

Example 3.4 The nonlinear characteristic given in Figure 3.1 can be approximated by the third-degree polynomial below:

$$\lambda = -0.47x\,H^3 - 4.56\,H^2 + 5.09\,H - 0.0021$$

Obtain the inductance for a small variation of H around the operating point 0.25 A/m.

Solution

For small variations around $H = 0.25$, the Taylor series approximation gives

$$[\lambda - \lambda(0.25)] = \left[\frac{d\lambda}{dH}\bigg|_{H=0.25}\right](H - 0.25) = \left(-0.47 \times 0.25^2 - 4.56 \times 0.25 + 5.09\,(H - 0.25)\right)$$

$$[\lambda - \lambda(0.25)] = 3.92(H - 0.25)$$

Therefore, the incremental inductance is

$$L = \frac{d(\lambda - \lambda_0)}{d(H - H_0)} = 3.92\,H$$

3.4 Through Variables and Across Variables

Through variables represent energy flow within a system, such as electrical current, mechanical force, and hydraulic fluid flow.

Across variables, on the other hand, represent the potential difference across a component within a system. For instance, voltage in electrical systems, displacement in mechanical systems, and pressure difference in hydraulic systems are all examples of across variables.

The correlation between through variables and across variables is crucial for the development of realistic and accurate models of mechatronic systems.

The core elements of mechatronic systems include electrical, mechanical, hydraulic, pneumatic, and thermal components. Chapter 2 covers the fundamental physics and engineering principles behind the behavior of these elements. Electrical components and circuit analysis techniques are essential for unified modeling of mechatronic systems. Analogies among electrical, mechanical, fluid dynamics, and thermal systems simplify the modeling of such complex systems with diverse elements and enable the use of advanced tools developed for electrical circuit analysis (Lewis, 1994).

3.5 Electrical Elements

Modeling of electrical components requires representing their behavior using dynamic equations, which can be formulated either in the time domain or in the s-domain by utilizing their Laplace transforms.

Table 3.1 Voltage–current equations of ideal circuit elements in *t*-, *s*-, and *jω*-domain

	Resistor	Inductor	Capacitor
t-domain	$v(t) = Ri(t)$	$v(t) = L\frac{di(t)}{dt}$	$i(t) = C\frac{dv(t)}{dt}$
	$i(t) = \frac{1}{R}V(t)$	$i(t) = \frac{1}{L}\int_0^{\tau} v(t)dt + i(0)$	$v(t) = \frac{1}{C}\int_0^{\tau} i(t)dt + v(0)$
s-domain	$V(s) = RI(s)$	$V(s) = sLI(s)$	$I(s) = sCV(s)$
	$I(s) = \frac{1}{R}V(s)$	$I(s) = \frac{1}{sL}V(s)$	$V(s) = \frac{1}{sC}I(s)$
jω-domain	$V(j\omega) = RI(j\omega)$	$V(j\omega) = j\omega I(j\omega)$	$I(j\omega) = j\omega V(j\omega)$
	$I(j\omega) = \frac{1}{R}V(j\omega)$	$I(j\omega) = \frac{1}{j\omega L}V(j\omega)$	$I(j\omega) = \frac{1}{sC}V(j\omega)$

Key electrical elements include resistors, capacitors, and inductors. Each component has specific characteristics that define its behavior in a circuit. Basic circuit components are often idealized as lumped, linear, and time-invariant elements. Nonideal components can be represented by equivalent circuits that combine ideal elements.

Ohm's law describes the linear correlation between the voltage across a resistor and the current flowing through it at any instant, regardless of how the current varies over time. Capacitors and inductors store energy in the form of an electric or magnetic field. The energy that has been previously stored influences the correlation between the voltage and the current at any given moment. Therefore, the time-domain relationship between voltage and current is defined by a derivative or integral.

The Laplace transform of the time-domain equations defines the *s*-domain characteristics of the circuit elements as algebraic expressions. The Laplace transform variable is a complex quantity defined as $s = \sigma + j\omega$. When the voltage and current are sinusoidal functions, in steady-state operation, the real part of s becomes zero, turning s to an imaginary variable that defines the $j\omega$ domain.

Table 3.1 presents a comparison of the voltage–current equations of ideal circuit elements across the time, Laplace transform, and frequency domains. It is important to observe that the equations governing energy storing elements in both the *s*-domain and *jω*-domain are analogous to Ohm's law.

Circuit components that change with motion and strain are often used in sensors to generate signals representing displacement. Common examples include strain gauges, variable capacitors, and moving coil magnetic devices. Voltage divider or bridge configurations incorporating these components generate a voltage or current that varies as a function of motion. Chapter 7 will provide an in-depth discussion on sensors constructed with motion-dependent circuit elements.

3.6 Mechanical Elements

The primary components of mechanical structures are masses, friction elements (dampers), and springs (Figure 3.2). These components can make either translational or rotational movement, or a combination of both, particularly in robotic systems. Additionally, mechanical linkages convert input forces and movements into predetermined output actions. These linkages are designed to

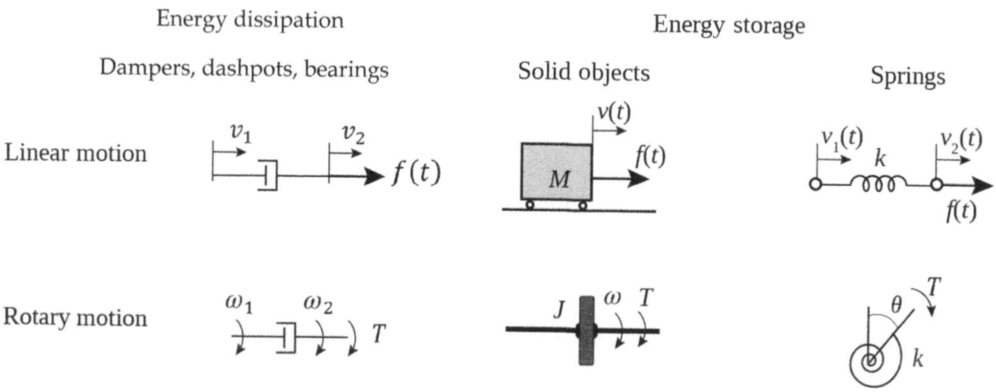

Figure 3.2 Basic mechanical elements

deliver precise motion profiles, ensuring that the mechatronic system operates with desired accuracy and efficiency.

The behavior of mechanical components is typically described by their force–motion characteristics. In mechatronic systems, masses and forces distributed over system elements can be treated as being concentrated at specific points. Consequently, Newton's laws of classical mechanics offer fundamental principles for modeling mechanical structures.

3.6.1 Linear Motion

Linear motion in mechanics refers to the movement of an object along a straight path or in a single direction. This type of motion involves translation where all parts of the object move in the same distance in a given time interval. Linear motion can be described using parameters such as velocity, acceleration, and displacement, and it follows Newton's laws of motion.

In linear motion, if the net force acting on an object is zero, the object remains at rest or continues to move at a constant velocity. According to Newton's law of inertia, the sum of all forces applied to an object is directly proportional to its acceleration, with the object's mass serving as the constant of proportionality.

$$\sum_k F_k = 0; k = 1, \ldots n \Leftrightarrow v = 0 \quad \text{or} \quad \text{constant} \tag{3.10}$$

Forces in a mechanical system are through variables, counterparts of currents in an electric circuit. Note that equation 3.10 is analogous to Kirchhoff's Current law (KCL).

Newton's second law of motion, or the law of inertia, states that the sum of all forces applied to an object is proportional to its acceleration. The proportionality constant is defined as the object's mass.

$$F = Ma = M\frac{dv}{dt} \tag{3.11}$$

In linear motion, a mass stores or delivers gravitational potential energy when its elevation changes in respect to a reference level. The kinetic energy stored in a mass is proportional to the square of its velocity.

3.6.2 Rotary Motion

Rotary motion, or rotational motion, refers to the movement of an object around a central axis. This type of motion is characterized by the angular displacement, angular velocity, and angular acceleration of the object. The fundamental principles that govern rotary motion are similar to those that govern linear motion, but with some key differences.

In rotary motion, an object will either remain at rest or rotate with a constant angular velocity if the net torque acting upon it is zero.

$$\sum_{k} T = 0; (k = 1, ...n) \Leftrightarrow \omega = 0 \quad \text{or} \quad \text{constant} \tag{3.12}$$

In rotary motion, the moment of inertia (J) plays a similar role to mass (m) in linear motion. The moment of inertia is a measure of an object's response to changes in its rotational motion. It depends on the mass distribution of the object relative to the axis of rotation. The rotational analog of Newton's second law is given by the equation:

$$T = J\alpha = J\frac{d\omega}{dt} \tag{3.13}$$

where T is the torque applied to the object, J is the moment of inertia, and α is the angular acceleration, the derivative of angular velocity ω.

In practical applications, rotary motion is found in numerous mechanical systems, such as gears, flywheels, and turbines. Understanding the dynamics of rotary motion is essential for designing and analyzing these systems.

3.6.3 Energy Dissipating Mechanical Elements

Friction between moving parts of a mechatronic system converts mechanical energy into heat, which is dissipated through natural or forced cooling methods. This dissipated heat represents energy losses that should be minimized to enhance system efficiency. Mechanical systems often incorporate controlled friction elements, including dampers, dashpots, and brakes. These components are essential for absorbing shocks, reducing vibrations, and regulating oscillations. The two main types of friction are viscous and dry friction. Viscous friction force is proportional to the velocity of motion and provides a resistance that is dependent on the fluid's viscosity within the damper. This type of friction is beneficial for stabilizing motion and reducing oscillations in dynamic systems.

Dry friction, on the other hand, is the resistance encountered when two solid surfaces slide against each other. This friction type is common in brakes, where it dissipates energy to stop moving parts.

Bearings, which are designed to minimize both types of friction, facilitate smooth and precise motion control by reducing the resistance between moving components. Dry friction is negligible in a bearing; hence, it can be considered solely viscous friction.

In linear motion, viscous friction generates a force that is directly proportional to the translational velocity. In rotational motion, viscous friction causes a torque that is proportional to the angular velocity.

$$\text{Linear motion} \qquad f(t) = \beta v(t) \Leftrightarrow v(t) = \frac{1}{\beta} f(t) \tag{3.14}$$

$$\text{Rotary motion} \qquad T(t) = \beta\omega(t) \Leftrightarrow \omega(t) = \frac{1}{\beta} T(t) \tag{3.15}$$

The force–velocity characteristic of a viscous friction shown in the equations above is analogous to the *V–I* characteristic of a resistance stated by Ohm's law in electric circuits. Viscous friction results in heat dissipation,

$$\text{Linear motion} \qquad P = f \cdot v = (\beta v)v = \beta v^2 \tag{3.16}$$

$$\text{Rotary motion} \qquad P = T \cdot \omega = (\beta \omega)\omega = \beta \omega^2 \tag{3.17}$$

Power dissipation on friction elements is consistent with the analogy to the electrical resistance. Viscous friction is the mechanical equivalent of resistors in electrical circuits.

3.6.4 Energy Storing Mechanical Elements

Masses of solid objects and springs are fundamental components of mechanical structures. These elements can store or transfer kinetic and potential energy. Maintaining energy balance is crucial when modeling mechanical systems.

3.6.4.1 Springs

Springs are used for storing and releasing energy, providing a restoring force, which makes them essential for applications requiring oscillatory motion or smoothing vibration. According to Hooke's law, the force exerted by a spring is directly proportional to the displacement from its equilibrium position in the linear region of its characteristic. If the force exceeds a certain value, the spring displays unrecoverable deformation. In a rotary spring, the torque is proportional to the displacement angle in the linear region. The equations below represent the external characteristics of linear and rotary springs in the linear region:

$$\text{Linear motion} \qquad f = k \cdot x = k\frac{dv}{dt} \tag{3.18}$$

$$\text{Rotary motion} \qquad T = k \cdot \theta = k\frac{d\omega}{dt} \tag{3.19}$$

where f is the force (T is the torque) exerted by the spring, x (or θ) is the displacement from the equilibrium position, and k is the spring constant. Linear and angular displacements are derivatives of the corresponding velocities, highlighting the spring's similarity to inductance in electric circuits.

Energy stored in a spring can be obtained by integrating instantaneous power.

$$\text{Linear motion} \qquad p(t) = \left(k\frac{dv(t)}{dt}\right)v(t) \Longrightarrow E = \int kv \cdot dv = \frac{1}{2}kv^2 \tag{3.20}$$

$$\text{Rotary motion} \qquad p(t) = \left(k\frac{d\omega(t)}{dt}\right)\omega(t) \Longrightarrow E = \int k\omega \cdot d\omega = \frac{1}{2}kv\omega^2 \tag{3.21}$$

3.6.4.2 Mass

Mass is another energy storage element, storing both kinetic and potential energy. For mechanical systems, force is a through variable while the velocity is an across variable. Newton's law of inertia as expressed in equation 3.11 is the mechanical analogous to the characteristic equation of the electrical capacitance.

$$F = M\frac{dv}{dt} \Leftrightarrow v = \frac{1}{M}\int Fdt$$
$$T = J\frac{d\omega}{dt} \Leftrightarrow \omega = \frac{1}{J}\int Tdt \qquad (3.22)$$

In rotational motion, the moment of inertia (J) replaces mass, and angular acceleration (ω) replaces linear acceleration. Note that the expressions in equation 3.22 are similar to the voltage–current characteristic of an electrical capacitance, which stores electrostatic field energy. Kinetic energy stored in a moving object is analogous to the electric field energy stored in capacitance.

Linear motion $$E_k = \frac{1}{2}Mv^2 \qquad (3.23)$$

Rotary motion $$E_k = \frac{1}{2}J\omega^2 \qquad (3.24)$$

Electrical capacitance $$E_k = \frac{1}{2}Cv_{12}^2 \qquad (3.25)$$

In equation 3.25, v_{12} is the voltage across the capacitance C. The last three equations align with the analogy between electrical capacitance, mass, and moment of inertia.

Example 3.5 Figure 3.3 illustrates a seismograph and its free-body diagram representing the forces acting on the moving disk suspended to the body of the device being a spring. The mass of the moving disk is M, and the spring constant is k. The needle fixed to the disk slides on a guide rail with viscous friction constant β. When an earthquake happens, the needle records the vibration on a rotating disk.

a) Derive the transfer function with the input as the vertical position of the base ($r(t)$) and the output as the vertical position of the disk $y(t)$.
b) Given $M = 0.01$ kg, $k = 5 \times 10^4$ N/m, $\beta = 30$ N.m.s, obtain the impulse response of the seismograph for the model parameters below.

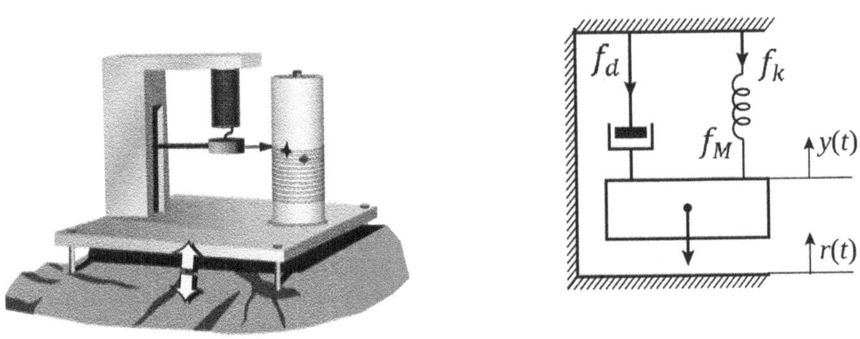

Figure 3.3 A seismograph and its simplified free-body diagram

Solution

a) Assume that both r and y are zero when the base is stationary (at equilibrium). When the ground is still, f_M equals the gravitational force balanced by the spring. When the base is moving due to a ground vibration, the mass M moves relative to a fixed reference by a distance $y(t)$. Assuming the spring is linear, the change in its length is proportional to the force applied across its ends. The damping element generates a force that is proportional to the velocity, due to viscous friction.

$$\begin{aligned}&\text{Spring force}: f_k = k[y(t) - r(t)]\\&\text{Damper force}: f_d = \beta\frac{d}{dt}[y(t) - r(t)]\\&\text{Force due to inertia}: f_M = M\frac{d^2}{dt^2}y(t)\end{aligned} \tag{3.26}$$

According to Newton's law of motion, the total sum of forces acting on a given mass must be equal to zero. Combining the three forces, the differential equation that describes the behavior of the seismograph motion is obtained as below:

$$\begin{aligned}&M\frac{d^2}{dt^2}y(t) + \beta\frac{d}{dt}[y(t) - r(t)] + k[y(t) - r(t)] = 0\\&M\frac{d^2y(t)}{dt^2} + \beta\frac{dy(t)}{dt} + ky(t) = \beta\frac{dr(t)}{dt} + kr(t)\end{aligned} \tag{3.27}$$

The Laplace transform of both sides with all initial conditions zero yields

$$\begin{aligned}&Ms^2Y(s) + \beta s(Y)s + kY(s) = \beta sR(s) + kR(s)\\&(Ms^2 + \beta s + k)Y(s) = (\beta s + k)R(s)\end{aligned} \tag{3.28}$$

Arranging the equation, we obtain

$$T(s) = \frac{Y(s)}{R(s)} = \frac{\beta s + k}{Ms^2 + \beta s + k} \tag{3.29}$$

b) For the given model parameters, the transfer function becomes

$$T(s) = \frac{30s + 5 \times 10^4}{0.01s^2 + 30s + 5 \times 10^4} \tag{3.30}$$

Unit-step response is obtained by multiplying the transfer function by the Laplace transform of the unit-step function $1/s$.

$$Y(s) = T(s)\frac{1}{s} \tag{3.31}$$

The time-domain unit-step response is determined by applying the inverse Laplace transform to $Y(s)$. The following MATLAB® script calculates and plots the unit-step response of the seismograph. The step response in Figure 3.4 demonstrates a 20% overshoot, leading to an undesirable error in the recorded vibration data. The optimal friction value to minimize the recording error can be determined using the MATLAB® script.

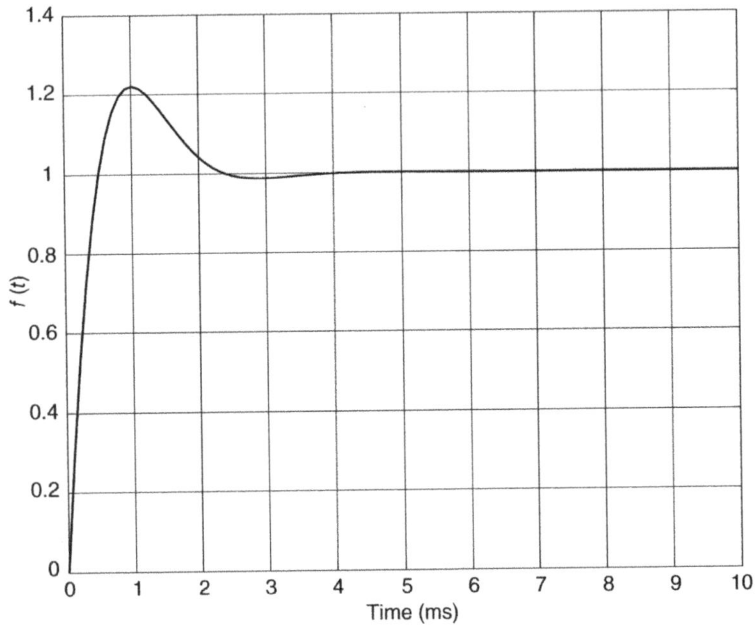

Figure 3.4 Unit-step response of the seismograph described in Example 3.5

MATLAB® Script for Example 3.5

```
% Define the symbolic variables
syms s t
% Define the Laplace transform
F = (1/s)*(30*s +5.e4) / (0.01*s^2 + 30*s + 5.e4);
% Compute the inverse Laplace transform
f = ilaplace(F, s, t);
% Convert the symbolic result to a MATLAB function for numerical
evaluation
f_numeric = matlabFunction(f);
% Define the time range for the plot
t_values = linspace(0, 0.01, 100); % Time from 0 to 10 ms
% Evaluate the function numerically
f_values = f_numeric(t_values);
% Plot the result
figure;
plot(1000*t_values, f_values, 'LineWidth', 1.5);
xlim([0 10])
grid on;
xlabel('Time (ms)');
ylabel('f(t)');
title('Unit Step Response of the Seismograph');
legend('f(t)');
```

3.7 Hydraulic and Pneumatic Elements

Hydraulic and pneumatic elements in a mechatronic system are fluid components that operate based on the principles of fluid mechanics to transmit power and control motion. Hydraulic systems use incompressible fluids like oil to generate large forces with high precision, making them suitable for heavy-duty applications. These systems include actuators, pumps, valves, and fluid reservoirs. Pneumatic systems, in contrast, employ compressed air for lighter, faster movements. They are integral in various industries, including manufacturing and aerospace, providing reliable and efficient control in automated processes.

A fluid can be a liquid or gas, which has the ability to flow and take the shape of their containers. Unlike solids, fluids do not have a fixed shape and can easily adapt to environmental conditions. The fundamental properties of fluids include viscosity, density, and compressibility, which determine their behavior and applications in various fields, such as hydraulics, pneumatics, and aerodynamics (Lewis, 1994).

Pressure and flow rate are external variables in devices that manage fluid dynamics. Pressure in fluid mechanics is defined as the force exerted by a fluid per unit area within the fluid or on a surface in contact with the fluid. It is a scalar quantity and is often measured in Pascal ($Pa = \mathrm{N/m^2}$) or pounds per square inch (psi). Pressure influences the behavior and movement of fluids in various applications. It can be exerted in all directions, contributing to the stability and functionality of systems involving fluids.

Flow rate is the volume or mass of fluid that passes through a given cross-sectional area per unit of time. Volume flow rate is more often used in liquid systems, while in gas flow systems, the mass flow rate is more commonly used. In a metric unit system, volume flow rate is typically expressed in cubic meters per second ($\mathrm{m^3/s}$), or liters per minute (L/min) in the standard international unit system (SI). In the imperial unit system, it is usually measured in gallons per minute (gpm). The mass flow rate is typically expressed in kg/s.

3.7.1 Hydraulic Power

Figure 3.5 (a) illustrates a hydraulic system powered by a pump filling a reservoir at a higher elevation. The force that moves the liquid in the pipe is the product of the pressure and the cross-sectional area of the pipe p_1A_1. The work done by the pump to force the fluid move the distance l is

$$W = p_1A_1l \tag{3.32}$$

The product of the area and length is the volume of liquid. The volume of liquid flowing in unit time is the volumetric flow rate. Power is the work done per unit time; therefore, it can be expressed in terms of the pressure and volumetric flow rate as below:

$$P = p_1q_v \tag{3.33}$$

3.7.2 Compressible and Incompressible Fluids

The conservation of mass principle is commonly used to analyze fluid flow by comparing flow conditions at different points within the fluid. Consider the pipe shown in Figure 3.5 (b). At one end of the pipe, the pressure is p_1, the area is A_1, and the velocity of flow is v_1. At the other end, the pressure

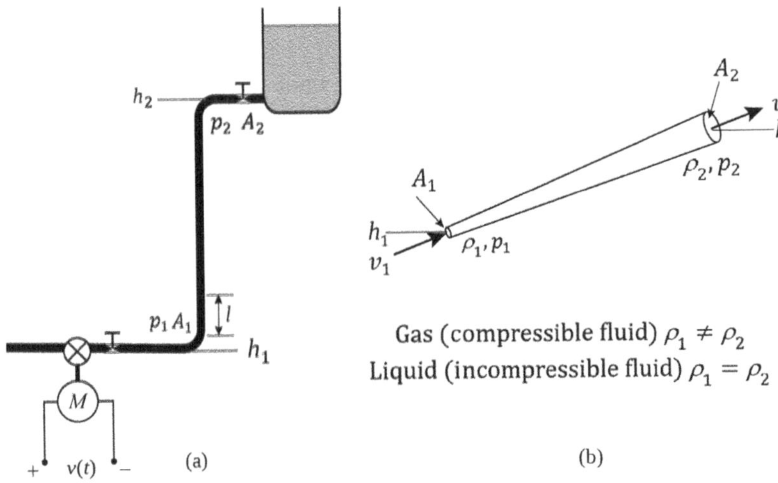

Figure 3.5 Fluid transport in pipelines (a: Illustration of a hydraulic system; b: Gas and liquid flow throuh a pipe with variable diameter)

is p_2, the area is A_2, and the velocity is v_2. If the fluid flowing in the pipe is gas, then the mass density (mass per unit volume) varies, being denoted as ρ_1 at one end and ρ_2 at the other end.

The mass flow rate entering and leaving the pipe must be constant since no mass is being stored inside the pipe. The mass continuity equation is expressed as

$$q_{m1} = q_{m2} \Rightarrow \rho_1 A_1 v_1 = \rho_2 A_2 v_2 \tag{3.34}$$

If the pipe carries a liquid, the mass density remains constant at all points. As a result, for liquids, the density term cancels out on both sides of equation 3.34, thereby simplifying the equation

$$A_1 v_1 = A_2 v_2 \Rightarrow q_{v1} = q_{v2} \tag{3.35}$$

This shows that the volume rate of flow q_v must be the same at both ends of the pipe.

3.7.3 Resistive Fluid Elements

A device or element within a fluid system that creates pressure decline is analogous to the electric resistance, which produces a voltage drop in the loop where the resistor is placed. Fluid resistors can be classified based on their design and the manner in which they impede the fluid flow. Some common types include:

- **Orifices**: Typically, sharp-edged openings that cause a sudden contraction of the fluid flow path, resulting in a drop in pressure and a change in velocity.
- **Valves**: Adjustable elements that can control the flow rate by varying the opening through which the fluid passes.
- **Porous Media**: Materials with many small passages that offer resistance to fluid flow due to friction and other factors.

The behavior of a fluid resistor is often analyzed using Bernoulli's equation, which relates the pressure, velocity, and elevation of a fluid at different points along its flow path. The Bernoulli's equation is based on the fundamental principle of conservation of energy and energy conversion within a fluid. For a steady, incompressible flow with no friction losses, Bernoulli's equation is expressed as

$$p_1 + \frac{1}{2}\rho v_1^2 + \rho g h_1 = p_2 + \frac{1}{2}\rho v_2^2 + \rho g h_2 \tag{3.36}$$

where

- p_1 and p_2 are the pressures at points 1 and 2.
- v_1 and v_2 are the velocities at points 1 and 2.
- h_1 and h_2 are the elevations at points 1 and 2.
- ρ is the fluid density.
- g is the acceleration due to gravity.

Fluid resistors often create nonlinear relationships because they introduce changes in velocity and pressure that do not scale linearly with flow rate (Lewis, 1994). However, linear expressions are desirable to simplify the solution of equations and establishing an analogy with linear electric resistances.

Some examples of fluid resistances widely used in engineering applications include:

- **Flow Measurement Devices**: Orifices and venturi meters are used to measure the flow rate in pipes by causing a pressure drop that can be correlated to the flow rate.
- **Control Systems**: Valves are used in fluid control systems to adjust and maintain the desired flow conditions.
- **Hydraulic Systems**: Fluid resistors help control the movement and pressure of hydraulic fluids in machinery and equipment.

In fluid dynamics, pressure is analogous to electric potential, while flow rate corresponds to electric current. Consequently, the parameter that correlates pressure and flow rate in a fluid system is comparable to an electrical resistor in circuits.

The Hagen–Poiseuille equation is a principle in fluid mechanics that explains the flow of a viscous fluid through a cylindrical pipe in a laminar motion, where the fluid moves in parallel layers without turbulence. The general form of this equation is

$$q = \frac{\Delta p \cdot \pi \cdot D^4}{128 \cdot \mu \cdot L} \tag{3.37}$$

where

- Q is the volumetric flow rate (m^3/s).
- Δp is the pressure difference across the length of the pipe (Pa).
- D is the diameter of the pipe (m).
- μ is the dynamic viscosity of the fluid ($Pa \cdot s$).
- L is the length of the pipe (m).

The fluid resistance, which defines the linear relationship between the pressure difference and volumetric flow rate, can be determined by rearranging equation 3.37.

$$R_f = \frac{\Delta p}{q} = \frac{128 \cdot \mu \cdot L}{\pi \cdot D^4} \tag{3.38}$$

Therefore, the pressure difference across the ends of a capillary tube is related to the volume rate of flow through the tube by

$$p_{21} = R_f q_v \tag{3.39}$$

We can derive the expression for the power dissipated in the fluid due to laminar flow by substituting this equation into equation 3.33. This expression is analogous to the equation describing electrical power dissipation in a resistor.

$$P = p_{21} q_v = R_f q_v^2 \tag{3.40}$$

The energy absorbed by a fluid flowing through a pipe is determined by integrating the dissipated power over a specific time interval.

3.7.4 Energy Storing Fluid Elements

Reservoirs, pressurized tanks, spring-loaded accumulators, and air-charged accumulators are commonly used for storing fluids. These components function similarly to electrical capacitors that store electrostatic energy in electrical circuits. Fluid capacitors are components used in various hydraulic and pneumatic systems.

3.7.4.1 Open Reservoir

An open reservoir is often used in a hydraulic system as a capacitor. Figure 3.6 (a) shows the general arrangement of such a capacitor. A volume rate of flow q_v entering the bottom of the tank increases the level of the fluid in the tank to h. This increased fluid level also increases the pressure p at the bottom of the tank.

The mass continuity equation applied to the tank gives us

$$q_v = A\frac{dh}{dt} \tag{3.41}$$

We can also see that the pressure at the bottom of the tank is equal to the weight of the fluid in the tank divided by the area. That is,

$$p = \frac{\rho g h A}{A} = \rho g h \Rightarrow \frac{dh}{dt} = \frac{1}{\rho g}\frac{dp}{dt} \tag{3.42}$$

By substituting equation 3.41 to eliminate dh/dt in the last equation, we have

$$\begin{aligned} q_v &= \tfrac{A}{\rho g}\tfrac{dp}{dt} \\ \tfrac{A}{\rho g} &= C_f \Longrightarrow q_v = C_f \tfrac{dp}{dt} \end{aligned} \tag{3.43}$$

The equation for fluid capacitance is mathematically analogous to the equation for an ideal electrical capacitor.

3.7.4.2 Accumulator

A spring-loaded accumulator shown in Figure 3.6 (b) is another form of fluid capacitor. In this type of accumulator, a spring provides the pressure increase instead of gravity. A volume flow rate q_v entering the tank's bottom raises the fluid height x, compressing the spring and increasing the tank's pressure p. The mass continuity equation applied to the tank gives

$$q_v = A\frac{dx}{dt} \tag{3.44}$$

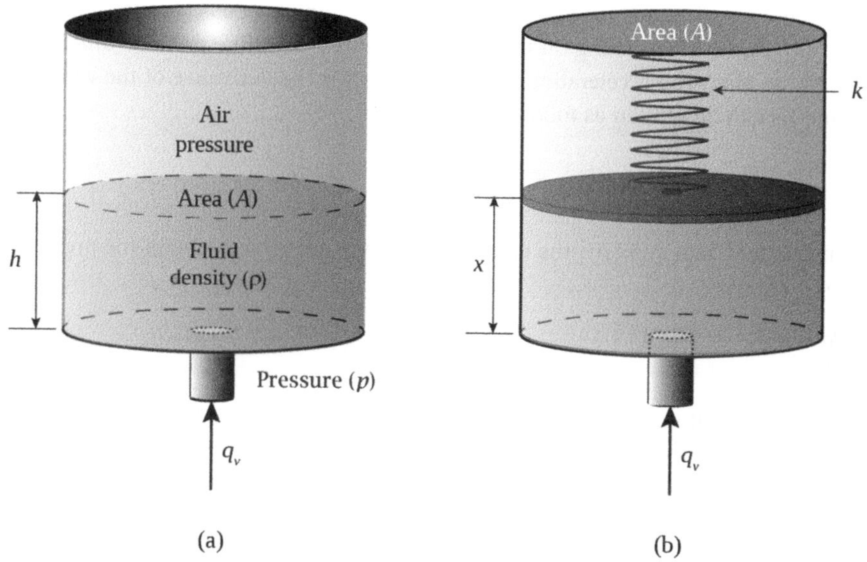

Figure 3.6 Fluid storage reservoirs (a: Open reservoir; b: Accumulator)

The pressure in the tank is equal to the force exerted by the spring on the fluid divided by the area. That is,

$$p = \frac{kx}{A} \tag{3.45}$$

where k is the spring constant. Equation 3.45 can be rearranged and then differentiated to obtain

$$\frac{dx}{dt} = \frac{A}{k}\frac{dp}{dt} \tag{3.46}$$

This equation can then be written as a fluid capacitor

$$q_v = \frac{A^2}{k}\frac{dp}{dt} \Rightarrow C_f = \frac{A^2}{k} \tag{3.47}$$

The energy delivered to a fluid capacitor in the time interval from $t = t_1$ to $t = t_2$ is found by integrating the power of the fluid flow.

$$\begin{aligned} E_b - E_a &= \int_{t_1}^{t_2} p_{12} q_v dt \\ q_v dt &= C_f dp \\ E_b - E_a &= C_f \int_{t_a}^{t_b} p dp = C_f \left[\frac{p^2}{2}\right]_{p_1}^{p_2} = \frac{1}{2} C_f \left(p_2^2 - p_1^2\right) \end{aligned} \tag{3.48}$$

3.7.4.3 Fluid Inertia

A moving fluid mass is comparable to a solid mass in motion because the fluid has inertia and requires a force to change its speed. Suppose that an incompressible ideal fluid (no viscosity) is flowing at a variable velocity through a pipe with no friction. Newton's laws of motion can be applied to the mass of fluid in the pipe.

If the cross-sectional internal area of the pipe is A, the length is l, and the density of the fluid is ρ, the fluid mass is ρAl. The net force acting on this mass is the area times the pressure difference between the ends of the pipe. The acceleration of the fluid mass is the derivative of the velocity. Newton's law of inertia can be written as follows:

$$A(p_2 - p_1) = \rho Al\frac{dv}{dt} \tag{3.49}$$

Since the volumetric flow rate $qv = Av$, the equation can be rearranged to obtain the pressure difference $p_{12} = p_2 - p_1$.

$$p_{12} = \frac{\rho l}{A}\frac{dq_v}{dt} \tag{3.50}$$

The volumetric flow rate in fluid systems is analogous to electric current and the pressure difference to voltage. Thus, this equation is comparable to the characteristic equation of an ideal electrical inductor:

$$\begin{aligned} V_L &= L\frac{di}{dt} \\ \frac{\rho l}{A} &= L_f \Rightarrow p_{12} = L_f\frac{dq_v}{dt} \end{aligned} \tag{3.51}$$

Mathematically, an ideal fluid inductor functions in the same manner as an ideal electrical inductor (Table 3.2). Energy delivered to a fluid inductor in the time interval from $t = t_1$ to $t = t_2$ is evaluated by integrating the fluid flow power.

$$\begin{aligned} E_1 - E_2 &= \int_{t_1}^{t_2} p_{12}q_v dt \\ p_{12} &= L_f\frac{dq_v}{dt} \\ E_2 - E_1 &= L_f\int_{t_1}^{t_2}\left(\frac{dq_v}{dt}\right)q_v dt = L_f\int_{q_{v1}}^{q_{v2}}(q_v)dq_v \\ E_2 - E_1 &= \frac{1}{2}L_f\left(q_{v2}^2 - q_{v1}^2\right) \end{aligned} \tag{3.52}$$

3.8 Thermal Elements

Temperature control systems include mechatronic elements designed to manage heat. But even more important task in mechatronic systems is to manage heat distribution to ensure proper operation of the electronic circuits within the specified temperature range.

Examples of mechatronic systems designed to control heat include HVAC (heating, ventilation, air conditioning) systems, which use a combination of sensors, actuators, and control algorithms to maintain comfortable indoor temperatures. Heat exchangers, boilers, condensers, and radiators work together to manage and control the flow of heat. Water heaters and heat pumps also fall under this category, utilizing precise controls to efficiently regulate water temperature for various household and industrial applications.

Table 3.2 Analogy between the mathematical models of mechatronics system components

		Mechanical Elements		
	Electrical Elements w	**Linear Motion**	**Rotary Motion**	**Hydraulic and Pneumatic Elements**
Through Variables	Current (i)	Force (f)	Torque (T)	Pressure (p)
Across Variables	Voltage (v)	Velocity (v)	Angular velocity (ω)	Flow rate (q)
Energy Dissipation	Resistor resistance (R) $v_{12} = Ri$ $i = Gv_{12}$ $G = \frac{1}{R}$ $P = Ri^2 = Gv_{12}^2$ 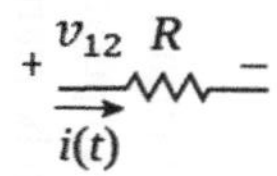	Linear damper viscous friction (β_v) $v_{12} = \alpha f$ $f = \beta v_{12}$ $\alpha = \frac{1}{\beta}$ $\Gamma = \beta v_{12}^2 = \alpha f^2$ 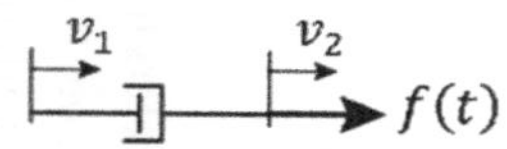	Rotary damper viscous friction (β_v) $\omega_{12} = \alpha T$ $T = \beta \omega_{12}$ $\alpha = \frac{1}{\beta}$ $P = \beta \omega_{12}^2 = \alpha T^2$ 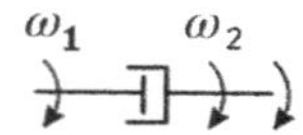	Capillar Tube Fluid Friction (R_f) $p_{12} = R_f q$ $q = G_f p_{12}$ $G_f = \frac{1}{R_f}$ $P = Rq^2 = G_f p_{12}^2$ 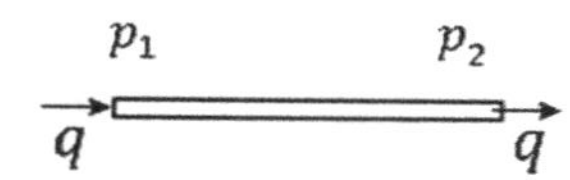
Capacitive Storage	Capacitor capacitance (C) $v_{12} = \frac{1}{C}\int i \cdot dt$ $i = C\frac{dv_{12}}{dt}$ $E = \frac{1}{2}Cv_{12}^2$ 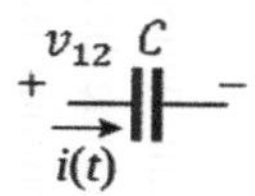	Solid object mass (M) $v_{12} = \frac{1}{M}\int f \cdot dt$ $f = M\frac{dv}{dt}$ $E = \frac{1}{2}Mv^2$ 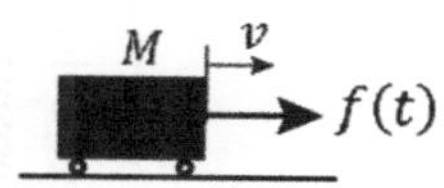	Rotational object moment of inertia (J) $\omega = \frac{1}{J}\int T \cdot dt$ $T = J\frac{d\omega}{dt}$ $E = \frac{1}{2}J\omega^2$ 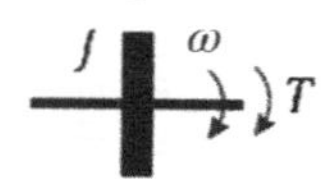	Fluid Reservoir Fluid Capacitance (C_f) $p(t) = \frac{1}{C_f}\int q \cdot dt$ $q(t) = C_f\frac{dp_{12}}{dt}$ $E = \frac{1}{2}C_f p_{12}^2$ 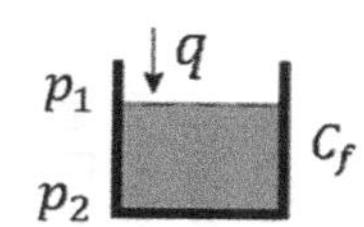
Inductive Storage	Inductor inductance (L) $v = L\frac{di}{dt}$ $i = \frac{1}{L}\int v \cdot dt$ $E = \frac{1}{2}Li^2$ 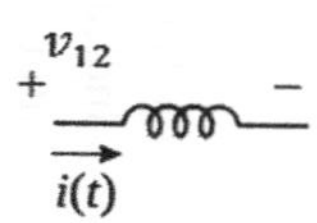	Spring constant (k) $v = \frac{1}{k}\frac{di}{dt}$ $f = k\int v \cdot dt$ $E = \frac{1}{2}kf^2$ 	Torsion spring constant (k) $\omega_{12} = \frac{1}{k}\frac{dT}{dt}$ $T = k\int \omega \cdot dt$ $E = \frac{1}{2}kT^2$ 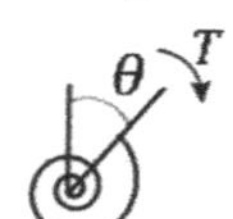	Tube, Pipe, Accumulator Fluid Inertia (L_f) $p_{12} = 1\frac{dq(t)}{dt}$ $q = \frac{1}{J}\int p \cdot dt$ $E = \frac{1}{2}Jq^2$ 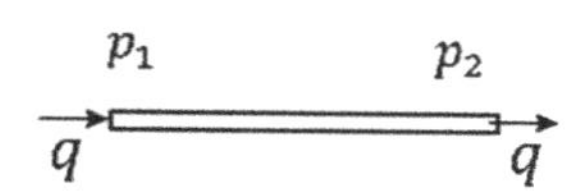

In automotive applications, engine cooling systems are critical to ensure optimal performance and prevent overheating. These systems integrate thermostats, fans, and radiators to manage heat dissipation effectively. Similarly, electronic devices such as computers and smartphones use fans, heat sinks, and thermal management software to prevent overheating and maintain performance.

Moreover, all elements of mechatronic systems dissipate heat due to losses when they operate. Thorough understanding of heat transfer process is essential for enhancing the performance and lifespan of motors, transformers, and electronic circuits in mechatronic systems.

Heat refers to the total energy of molecular motion in a substance, while temperature indicates how hot or cold an object is. Thermal energy, or heat, transfers from higher to lower temperatures.

Heat transfer between different regions of a system at different temperatures can occur through three primary mechanisms. Conduction is the direct transfer of thermal energy through solid materials. Convection refers to heat transfer through the movement of fluids, both liquids and gases. Radiation is the transfer of heat by electromagnetic waves, including the light spectrum from infrared to ultraviolet. As mechatronic systems are built with solid components that either touch or are in proximity to each other, conduction serves as the primary method of heat transfer.

Key components of thermal systems are:

- **Heat Sources**: Components or materials that generate heat, such as engines, electrical devices, and chemical reactions.
- **Heat Sinks**: Components or materials that absorb and dissipate heat, helping to maintain a stable temperature within the system. Examples include radiators, heat exchangers, and cooling fins.
- **Insulation**: Materials that reduce heat transfer between the inside and outside of a system, helping to maintain desired temperatures.
- **Thermal Control Devices**: Devices used to regulate and control the temperature within a thermal system, such as thermostats, temperature sensors, and thermal switches.

3.8.1 Thermal Resistance

The measure of a material's ability to resist the flow of heat refers to thermal resistance. Typical unit of thermal resistance is degrees Celsius per watt (°C/W). Lower thermal resistance indicates better heat conduction properties, essential for maintaining the performance and longevity of components in mechatronic systems. The rate of heat transfer q_h between two surfaces of area A is proportional to the area and the temperature gradient.

$$q_h = -k_c A \frac{d\theta}{dx} \tag{3.53}$$

The proportionality constant, denoted as kc, represents thermal conductivity. The negative sign indicates that heat flows from the direction of decreasing temperature. Assuming that the temperature distribution is homogeneous, the equation above can be rearranged to obtain the correlation between temperature difference and the heat transfer rate.

$$\theta_2 - \theta_1 = \theta_{21} = \frac{l}{k_c A} q_h \tag{3.54}$$

By defining the thermal resistance for conduction-type heat transfer, the relationship between the temperature difference and the heat transfer rate can be expressed similarly to Ohm's law.

Table 3.3 Common units in thermal systems

Symbol	Description	SI Units	Imperial Units
A	Cross-sectional area	m^2	ft^2
θ	Temperature	°C	°F
l	Material thickness of	m	Ft
q_h	Rate of heat flow	Joules/second (J/s)	BTU per hour (BTU/j)
k_c	Thermal conductivity	J/[s m^2 (°C/m)]	BTU/[h ft^2 (°F/ft)]

$$R_T = \frac{l}{k_c A} \Rightarrow \theta_{21} = R_T q_h \tag{3.55}$$

This equation suggests that the heat transfer rate is the through variable and temperature difference is the across variable in thermal systems. However, since heat is energy measured in Joules (or BTU), the rate of heat transfer has the dimension of power. The expression of power drawn from this equation is not analogous to the power dissipated on an electrical resistance.

$$P = q_h = \frac{1}{R_T}\theta_{21} \tag{3.56}$$

Table 3.3 summarizes the units commonly used in the analysis of thermal systems.

3.8.2 Thermal Capacity

Thermal capacity, also known as heat capacity, is the amount of heat required to change the temperature of a substance by 1°C. In the metric unit system (SI), the unit of thermal capacity is Joules per degree Celsius (J/°C). In the British unit system, the unit of thermal capacity is British Thermal Units per degree Fahrenheit (BTU/°F).

The amount of heat required to change the temperature of an object is

$$dq_h = mcd\theta \tag{3.57}$$

where the specific heat, or the heat required to raise the temperature of one unit mass by one degree, is denoted by c. Specific heat is a function of temperature; hence, this equation is nonlinear for large temperature variations. For small temperature variations, c can be assumed to be constant. By dividing both sides of the equation by dt, we obtain

$$\frac{dq_h}{dt} = mc\frac{d\theta}{dt} \tag{3.58}$$

Defining mc as thermal capacitance C_T and denoting the derivative of the heat transfer by Q_h, the last equation becomes similar to the characteristic equation of an electrical capacitance.

$$Q_h = C_T\frac{d\theta}{dt} \tag{3.59}$$

3.8.3 Thermal Inductance

Currently, there are no known thermal phenomena that store thermal energy based on the rate of change of the heat flow rate. As a result, it is not possible to draw analogies with electrical inductance.

3.9 Dynamic System Response

The analysis of dynamic systems involves examining the responses produced by their models to various inputs. Linear, time-invariant (LTI) systems are usually categorized based on the order of ODEs that describe their behavior. Most building blocks of mechatronic systems can be represented by first- and second-order systems. Interconnections of the system components and feedback loops result in higher-order systems.

First-, second-, and higher-order systems exhibit distinct response characteristics that are fundamental to control design and analysis. First-order systems are described by a single energy storage element, resulting in a characteristic exponential response. Second-order systems, often modeled with two storage elements, display behaviors such as overshoot, oscillation, and damping, which are critical in applications requiring precise control over motion or force. Higher-order systems encompass more complex dynamics and may require advanced analytical or computational methods to predict and optimize performance.

3.9.1 Impulse Response

The impulse response of a system describes how the output changes over time when the input is a unit impulse function. The unit impulse, often denoted as $\delta(t)$ and commonly referred to as the Dirac delta function, is a singularity function that goes to infinity at $t = 0$ and stays zero all other times as defined below:

$$\delta(t) = \begin{cases} 0 & \text{for} \quad -\infty < t < \infty \\ \infty & \text{for} \quad t = 0 \end{cases} \tag{3.60}$$

The unit impulse is an idealized mathematical function that does not occur in nature. However, real-world events such as lightning strikes or surge impulses resulting from instantaneous electrical discharges can be represented mathematically by a unit impulse function scaled appropriately. Many natural phenomena can be idealized as impulsive events. For example, the force exerted during a collision (such as a hammer striking a nail or a particle collision in quantum mechanics) can be represented as an impulse. In mechanics, the impulse–momentum theorem relates the change in momentum of an object to the integral of force over time; when the force is large and applied over a very short interval, the unit impulse function models this idealized force.

While it was first introduced as a mathematical abstraction, the unit impulse function has found many practical applications across a broad array of scientific and technological disciplines. One of the most important applications of the unit impulse function lies in the analysis of dynamic systems. In LTI systems, the impulse response is a central concept that describes how a system reacts to a brief, idealized input at a single point in time. In electrical and control engineering, the impulse function is used in system analysis, design, and testing.

The integral of the unit impulse function from negative to positive infinity is one.

$$\int_{-\infty}^{\infty} \delta(t)dt = 1 \tag{3.61}$$

Multiplying a signal by a unit impulse shifted in time by τ yields the value of the signal at that specific instant. This property is essential for quantization of sampled signals in digital control and communication systems.

$$y(t) \cdot \delta(t - \tau) = y(\tau) \tag{3.62}$$

Assume that an input signal $r(t)$ is applied to a system at the instant $\tau = 0$.

$$r(t) \cdot \delta(t) = r(0) \tag{3.63}$$

If the system's impulse response is denoted by $g(t)$, the time variation of the system's output can be described by the convolution integral as expressed below (Kuo, 1982):

$$y(t) = \int_0^t r(\tau) \cdot g(t-\tau)d\tau \tag{3.64}$$

The convolution integral is typically represented by an asterisk symbol. Considering the commutativity property of the convolution process, the response of a linear system with a single input and output can be expressed as below for any given input function:

$$y(t) = g(t) * r(t) \tag{3.65}$$

The unit impulse response defines the system's transfer characteristic since it describes its behavior for any input. The evaluation of the transfer characteristic plays an important role in the analysis and design of systems when the parameters and characteristics of the physical components are considered over time.

3.9.2 Transfer Function

In the LTI systems, the transfer function is a mathematical representation that relates the system's output to its input in the complex frequency (s) domain. The Laplace transform of the unit impulse function is unity.

$$\mathcal{L}\{\delta(t)\} = \int_0^\infty \delta(t)e^{(-st)}dt = 1 \tag{3.66}$$

This indicates that the unit impulse serves as the identity element in the s-domain. Consequently, this characteristic makes the impulse function a fundamental instrument for system analysis. Laplace transform changes the convolution operation into complex multiplication in the s-domain. Hence, the Laplace transform of both sides of equation 3.65 results in

$$Y(s) = G(s) \cdot R(s) \Longrightarrow G(s) = \frac{Y(s)}{R(s)} \tag{3.67}$$

The transfer function, commonly represented as $G(s)$ or $T(s)$, is defined as the ratio of the Laplace transform of the system's output to the Laplace transform of its input, under the assumption that all initial conditions are zero.

The transfer function is used to obtain the system's response to various inputs. Since the Laplace transform of the unit impulse function is unity, the transfer function effectively describes how the system responds to this fundamental input, serving as the identity element in the s-domain. It provides a powerful tool for analyzing and designing LTI systems by transforming complex time-domain relationships into simpler algebraic relationships in the frequency domain. Transfer function allows for a complete characterization of a system's input–output behavior and is especially useful when examining how changes in system parameters affect performance.

The time domain response of an LTI system is represented by ODEs. The following equation corresponds to the input–output relationship for an LTI system with a single input and a single output.

$$a_n \frac{d^n y(t)}{dt^n} + a_{n-1} \frac{d^{n-1} y(t)}{dt^{n-1}} + \cdots a_1 \frac{dy(t)}{dt} + a_0 y(t) = b_m \frac{d^m r(t)}{dt^m} + b_{m-1} \frac{d^{m-1} r(t)}{dt^{m-1}} + \cdots b_1 \frac{dr(t)}{dt} + b_0 r(t) \tag{3.68}$$

Assuming that all initial conditions are zero, the Laplace transform of both sides of this equation yields

$$a_n s^n Y(s) + a_{n-1} s^{n-1} Y(s) + \cdots a_1 s Y(s) + a_0 Y(s) = b_m s^m R(s) + b_{m-1} s^{m-1} R(s) + \cdots b_1 s R(s) + b_0 R(s) \tag{3.69}$$

The equation above can be arranged as a ratio of two polynomials:

$$T(s) = \frac{Y(s)}{R(s)} = \frac{b_m s^m + b_{m-1} s^{m-1} + \cdots b_1 s + b_0}{a_n s^n + a_{n-1} s^{n-1} + \cdots a_1 s + a_0} \tag{3.70}$$

Therefore, the transfer function of an LTI system with a single input and output is the ratio of the Laplace transform of the output to that of the input, assuming all initial conditions are zero. The denominator of the transfer function corresponds to the characteristic polynomial of the ODE defined in equation 3.68. Poles of a transfer function are defined as the roots of the characteristic equation, which is obtained by setting the denominator equal to zero. Conversely, zeros of the transfer function are determined by equating the numerator to zero. The locations of poles and zeros on the complex plane supply critical insights into the performance characteristics of LTI systems.

In practical systems, the power of the numerator is equal to or less than the power of the denominator.

$$m \leq n \tag{3.71}$$

A transfer function is considered "proper" if the degree of its numerator is less than or equal to the degree of its denominator. Systems in which the degree of the numerator is higher than that of the denominator are referred to as improper. Proper systems are common in control systems and signal processing due to their reliability and stability. In contrast, improper systems result in unbounded responses when they are subject to bounded inputs and are not realizable with physical components (Ogata, 2002).

Transfer function is a powerful tool in the analysis and design of LTI systems. They simplify complex differential equations into algebraic equations in the s-domain, making it easier to analyze system dynamics. Furthermore, transfer functions help in understanding the system's behavior and designing appropriate control strategies. However, it is limited to single-output systems, and it does not consider the initial conditions.

3.9.3 Block Diagrams

Block diagrams are a graphical representation of a system, illustrating the relationships between various components through blocks and connecting lines. Each block represents a distinct functional element, such as a process, a sensor, or a controller, and the connecting lines denote the signal flow between these elements. The primary purpose of a block diagram is to provide a clear and simplified overview of the system's structure and operation, making it easier to understand and analyze complex interactions within the system.

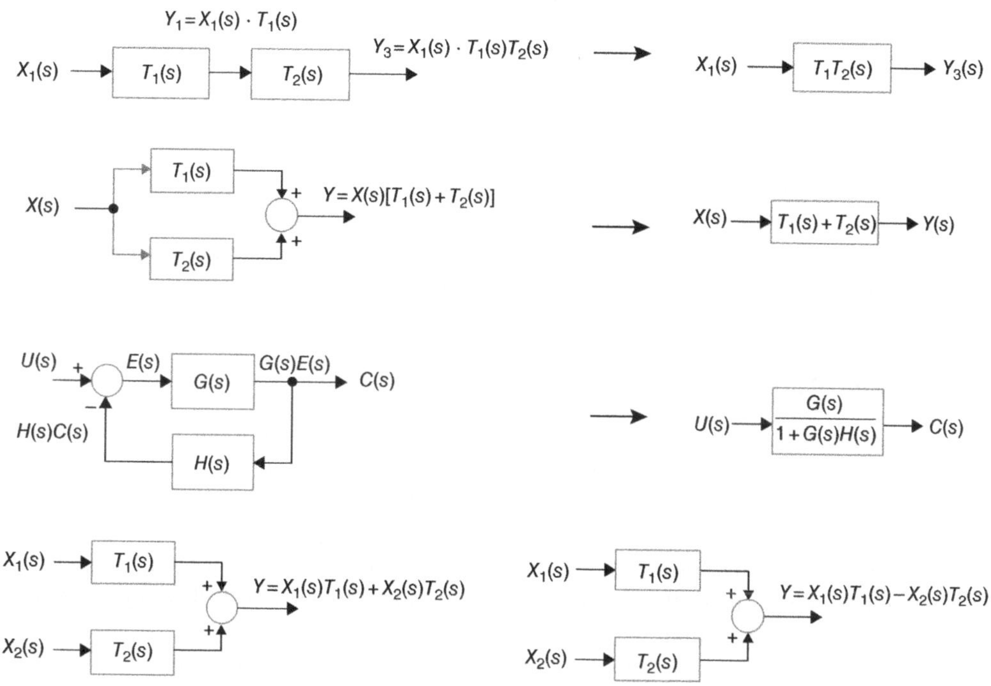

Figure 3.7 Block diagram algebra

In control systems engineering, block diagrams are used extensively to illustrate the flow of signals through different stages of the system, from inputs to outputs. These diagrams help engineers visualize how changes in one part of the system affect other parts, facilitating the design and troubleshooting of control strategies. By breaking down the system into smaller, more manageable components, block diagrams enable a systematic approach to problem-solving and performance optimization. Block diagrams are particularly useful when working with systems that can be described using mathematical models, such as transfer functions and state-space representations. By combining these models with block diagrams, engineers can create a comprehensive and intuitive representation of the system, which can be easily manipulated and analyzed using various analytical and computational tools. Figure 3.7 shows common algebraic operations for simplifying block diagrams. By using these, we can find the transfer function of complex systems with multiple feedback loops.

Block diagrams of feedback control systems do not include the initial conditions because they serve as a powerful tool for visualizing how different system components interact, particularly in the context of signal flow, transfer functions, and feedback loops.

When analyzing dynamic systems using the Laplace transform, initial conditions manifest as additional terms in the transformed equations. Specifically, the Laplace transform of a constant (such as an initial state) yields an impulse function scaled by the value of that constant. While mathematically important for determining the full solution to a differential equation, these initial condition terms do not alter the inherent transfer function of the system, which is what block diagrams are designed to illustrate.

Block diagrams are intended to simplify and standardize the study of signal flow and feedback. Including initial conditions would complicate the diagram, detracting from its clarity and utility as a tool for analyzing system structure and response to inputs.

The transfer function in a block diagram specifically characterizes how inputs are processed by the system. Initial conditions, which reflect the system's state at a particular moment, are transient and do not affect the input–output mapping that the transfer function represents.

In control design and tuning, engineers are usually concerned with the system's response to commanded inputs and feedback, rather than initial transients. Controllers are designed to regulate performance based on ongoing system behavior, not the state at startup.

3.9.4 Transfer Matrix

The superposition principle allows transfer functions to be applied to multi-input, multi-output LTI systems rather than just single input–output ones. A system with multiple inputs and outputs is referred to as a multivariable system. Each input–output pair can be represented by an ODE when other inputs are set to zero. The transfer function for each input–output pair is obtained by applying the Laplace transform to the corresponding ODE, with all other inputs held at zero. The superposition principle allows the total effect of all inputs on a specific output to be calculated by adding the individual effects from each input.

Suppose that the system has p inputs and q outputs. The following transfer function describes the relationship between the input i and output j while all other inputs are zero:

$$\begin{aligned} G_{i,j}(s) &= \frac{Y_i(s)}{R_j(s)}; \\ R_k &= 0 \quad \text{for} \quad k = 1\cdots p;\ k \neq j \end{aligned} \tag{3.72}$$

By applying the superposition principle, the Laplace transform of the output i when all inputs act together can be written as below (Kuo, 1982):

$$\begin{aligned} Y_i(s) &= G_{i1,}(s)R_1(s) + G_{i2,}(s)R_2(s)\cdots G_{ip,}(s)R_p(s) \\ Y_i(s) &= \textstyle\sum_{j=1}^{p} G_{i,j}(s)R_j(s)\ i = 1, 2, \ldots q \end{aligned} \tag{3.73}$$

It is convenient to represent the input–output relationship using the matrix notation.

$$\begin{aligned} &\mathbf{Y}(s) = \mathbf{G}(s)\cdot\mathbf{R}(s) \\ &Y(s) = \begin{bmatrix} Y_1(s) \\ Y_2(s) \\ \vdots \\ Y_q(s) \end{bmatrix}; R(s) = \begin{bmatrix} R_1(s) \\ R_2(s) \\ \vdots \\ R_p(s) \end{bmatrix}; G(s) = \begin{bmatrix} G_{11} & \cdots & G_{1p} \\ \vdots & \ddots & \vdots \\ G_{q1} & \cdots & G_{qp} \end{bmatrix} \end{aligned} \tag{3.74}$$

3.9.5 State-Space Analysis

State-space analysis is a mathematical method used to model and analyze systems through a minimal number of first-order differential equations. The set of variables, usually referred to as *state vector*, describe the state of a system in time domain.

State space refers to the n-dimensional space where each coordinate axis corresponds to a state variable $x_1, x_2, \dots, x_n$. Any state of the system can be depicted as a point within this state space. A state vector in state-space analysis comprises the state variables necessary to comprehensively describe the behavior of a given system. Once the state at $t = t_o$ is determined and the input $u(t)$ is specified for all $t \geq t_o$, the system state $x(t)$ is uniquely defined for any time (Ogata, 2002).

State-space analysis provides a framework for evaluating the dynamic behavior of a system that has multiple inputs and multiple outputs. The state variables are selected to describe the system's current condition, or current state. The input variables are known functions of time that drive the system, and the output variables reflect the system's response to the inputs. In contrast to the transfer function modeling, state-space models are not restricted to LTI systems with zero initial conditions.

A system's state-space model typically defines $\mathbf{x}(t)$ as an n-dimensional state vector, with $\mathbf{f}$ and $\mathbf{g}$ representing n functions of several variables.

$$\begin{aligned} \frac{d}{dt}\mathbf{x}(t) &= \mathbf{f}(\mathbf{x}, \mathbf{u}, t) \\ \mathbf{y}(t) &= \mathbf{g}(\mathbf{x}, \mathbf{u}, t) \end{aligned} \tag{3.75}$$

In the equation above, we define

$$\mathbf{x}(t) = \begin{bmatrix} x_1(t) \\ x_2(t) \\ \vdots \\ x_n(t) \end{bmatrix}; \mathbf{f}(\mathbf{x}, \mathbf{u}, t) = \begin{bmatrix} f_1(x_1, x_2, \cdots x_n, u_1, u_2, \dots u_n, t) \\ f_2(x_1, x_2, \cdots x_n, u_1, u_2, \dots u_n, t) \\ \vdots \\ f_n(x_1, x_2, \cdots x_n, u_1, u_2, \dots u_n, t) \end{bmatrix}$$

$$\mathbf{y}(t) = \begin{bmatrix} y_1(t) \\ y_2(t) \\ \vdots \\ y_n(t) \end{bmatrix}; \mathbf{g}(\mathbf{x}, \mathbf{u}, t) = \begin{bmatrix} g_1(x_1, x_2, \cdots x_n, u_1, u_2, \dots u_n, t) \\ g_2(x_1, x_2, \cdots x_n, u_1, u_2, \dots u_n, t) \\ \vdots \\ g_n(x_1, x_2, \cdots x_n, u_1, u_2, \dots u_n, t) \end{bmatrix}$$

State equations and output variables of LTI systems are usually expressed with matrices in the general form below:

$$\begin{aligned} \dot{\mathbf{x}}(t) &= \mathbf{A}\mathbf{x}(t) + \mathbf{B}\mathbf{u}(t) \\ \mathbf{y}(t) &= \mathbf{C}\mathbf{x}(t) + \mathbf{D}\mathbf{u}(t) \end{aligned} \tag{3.76}$$

In this equation, $\mathbf{x}(t)$ is the state vector, a column matrix with n rows. Generally, the derivative d/dt is simplified by placing a dot above the vector $\mathbf{x}$. The vector $\mathbf{u}(t)$ consists of p inputs, while the vector $\mathbf{y}(t)$ contains q outputs of the system. Therefore, $\mathbf{A}$ is the $n \times n$ state matrix, $\mathbf{B}$ is the $n \times n$ input matrix, $\mathbf{C}$ is the output matrix with q rows and n columns, and $\mathbf{D}$ is the feedthrough (or direct transmission) matrix with q rows and p columns.

State-space models are particularly useful in the study of complex systems that are difficult to analyze using transfer functions and transfer matrices. MATLAB® and similar mathematical analysis software provide efficient numerical methods to solve the set of differential equations in equation 3.75. Arranging the LTI system equations in a matrix format allows state-space analysis to utilize linear algebra techniques for solving and manipulating system equations. This method is commonly employed in control theory, signal processing, and various engineering fields.

Based on these definitions, the state-space model of the dynamic system is derived as follows:

$$\frac{d}{dt}\begin{bmatrix} x_1 \\ x_2 \\ x_3 \\ \vdots \\ x_{n-1} \\ x_n \end{bmatrix} = \begin{bmatrix} 0 & 1 & 0 & \cdots & 0 \\ 0 & 0 & 1 & \cdots & 0 \\ 0 & 0 & 0 & \cdots & 0 \\ \vdots & \vdots & \vdots & \cdots & \vdots \\ 0 & 0 & 0 & \cdots & 1 \\ -a_n & -a_{n-1} & -a_{n-2} & \cdots & -a_1 \end{bmatrix}\begin{bmatrix} x_1 \\ x_2 \\ x_3 \\ \vdots \\ x_{n-1} \\ x_n \end{bmatrix} + \begin{bmatrix} 0 \\ 0 \\ 0 \\ \vdots \\ 0 \\ 1 \end{bmatrix} u(t) \tag{3.77}$$

$$y = \begin{bmatrix} 1 & 0 & 0 & \cdots & 0 \end{bmatrix}\begin{bmatrix} x_1 \\ x_2 \\ x_3 \\ \vdots \\ x_n \end{bmatrix}$$

3.9.6 Derivation of Transfer Matrix from State-Space Model

If the state-space model of an LTI system is known, the transfer matrix can be obtained by matrix operations. The Laplace transform of both sides of the differential equations in equation 3.76 can be written in matrix form, assuming that all initial conditions are zero.

$$s\mathbf{X}(s) = \mathbf{AX}(s) + \mathbf{BU}(s) \tag{3.78}$$

By subtracting $\mathbf{AX}(s)$ from both sides and factorizing the left side, we obtain

$$[s\mathbf{I} - \mathbf{A}]\mathbf{X}(s) = \mathbf{BU}(s) \tag{3.79}$$

In this equation, s is multiplied by an $n \times n$ identity matrix to meet the dimension requirement for the matrix subtraction. The coefficient matrix of the state vector in the s-domain is always nonsingular; therefore, its inverse exists. $\mathbf{X}(s)$ can be obtained by multiplying both sides with the inverse of the $[s\mathbf{I} - \mathbf{A}]$ matrix from left.

$$\mathbf{X}(s) = [s\mathbf{I} - \mathbf{A}]^{-1}\mathbf{BU}(s) \tag{3.80}$$

Substituting $\mathbf{X}(s)$ in the algebraic output equation obtained

$$\mathbf{Y}(s) = \mathbf{C}[s\mathbf{I} - \mathbf{A}]^{-1}\mathbf{BU}(s) \tag{3.81}$$

The relationship between the outputs and inputs of the multivariable system in the s-domain is the transfer matrix.

$$\mathbf{T}(s) = \mathbf{C}[s\mathbf{I} - \mathbf{A}]^{-1}\mathbf{B} \tag{3.82}$$

3.9.7 State-Space Formulation of High-Order ODE

A higher-order ODE representing the dynamic behavior of an LTI dynamic system can be expressed as a set of first-order differential equations. Consider the following ODE:

$$\frac{d^n y(t)}{dt^n} + a_1 \frac{d^{n-1} y(t)}{dt^{n-1}} + \cdots + a_{n-1} \frac{dy(t)}{dt} + a_n y = u(t) \tag{3.83}$$

Let's define the set of state variables as below:

$$x_1 = y$$
$$x_2 = \frac{dy}{dt}$$
$$x_3 = \frac{d^2 y}{dt^2}$$
$$\vdots$$
$$x_n = \frac{d^{n-1} y}{dt^{n-1}}$$

Substituting these variables in equation 3.83, we obtain

$$\frac{dx_n}{dt} = -a_n x_1 - a_{n-1} x_2 - \cdots - a_1 x_n + u \tag{3.84}$$

Based on these definitions, the state-space model of the dynamic system is derived as follows:

$$\frac{d}{dt}\begin{bmatrix} x_1 \\ x_2 \\ x_3 \\ \vdots \\ x_{n-1} \\ x_n \end{bmatrix} = \begin{bmatrix} 0 & 1 & 0 & \cdots & 0 \\ 0 & 0 & 1 & \cdots & 0 \\ 0 & 0 & 0 & \cdots & 0 \\ \vdots & \vdots & \vdots & \cdots & \vdots \\ 0 & 0 & 0 & \cdots & 1 \\ -a_n & -a_{n-1} & -a_{n-2} & \cdots & -a_1 \end{bmatrix} \begin{bmatrix} x_1 \\ x_2 \\ x_3 \\ \vdots \\ x_{n-1} \\ x_n \end{bmatrix} + \begin{bmatrix} 0 \\ 0 \\ 0 \\ \vdots \\ 0 \\ 1 \end{bmatrix} u(t)$$

$$y = \begin{bmatrix} 1 & 0 & 0 & \cdots & 0 \end{bmatrix} \begin{bmatrix} x_1 \\ x_2 \\ x_3 \\ \vdots \\ x_n \end{bmatrix} \tag{3.85}$$

3.10 First-Order System

First-order systems contain a single energy storage element and are characterized by first-order differential equations. Examples include RC and RL circuits, moving mass, rotating flywheel, freely draining liquid tanks, and thermal systems.

The following differential equation describes the output of a first-order system where a is constant. In engineering systems, the reference input is usually represented by $r(t)$.

$$\frac{dx(t)}{dt} + ax(t) = r(t) \tag{3.86}$$

The solution of the linear differential equation consists of two parts: the natural response determined by the initial conditions and the forced response resulting from the input $r(t)$.

The natural response is obtained by solving the homogeneous equation where $(t) = 0$.

$$\frac{dx(t)}{dt} + ax = 0 \tag{3.87}$$

Rearranging this equation to place $x(t)$ on the right side and integrating both sides provides the resulting expression.

$$\int_{X_0}^{x} \frac{dx'(t)}{x'} = \int_{0}^{t} -at' \Longrightarrow \ln\frac{x}{X_0} = -at \tag{3.88}$$

By taking the exponent of both sides, and introducing and defining a new parameter $\tau = 1/a$, we can write

$$x_n(t) = X_0 e^{-\left(\frac{t}{\tau}\right)} \tag{3.89}$$

The particular solution due to the excitation effect refers to the steady-state response of the system after the natural response has faded. This solution is determined by aligning the steady-state behavior of the system with the mathematical function representing the excitation.

Step function is frequently utilized to evaluate engineering system performance. Applying a constant excitation such as a switch being activated, a constant force, torque, or any static physical effect that occurs suddenly at a specific moment can be described by a unit-step function scaled by a constant factor. Mathematical definition of a step function is

$$u(t) = \begin{cases} 0 & \text{for} \quad t < 0 \\ K & \text{for} \quad t \geq 0 \end{cases} \tag{3.90}$$

As time increases, the natural response decreases, and the state variable reaches its steady-state value X_s. Forced response, therefore, equals the steady-state value X_s.

$$\lim_{t \longrightarrow} \left[\frac{dx(t)}{dt}\right] = 0 \Rightarrow x_f = X_s = \frac{K}{a} \tag{3.91}$$

Consequently, the complete response is the sum of the natural response and the forced response due to the excitation.

$$x(t) = (X_0 - X_s)e^{-\left(\frac{t}{\tau}\right)} + X_s \tag{3.92}$$

This solution shows that the state variable changes exponentially from its initial value X_s toward the steady-state value X_s.

Due to its significance on system characteristics, τ is referred to as the *time constant* of a first-order system. The unit of the time constant is second. The time constant of a first-order system is a key parameter that characterizes the rate at which the system approaches its steady-state value. The unit of τ is second.

The tangent to the unit-step response at $t = 0$ intercepts the horizontal line that represents the steady-state value at τ.

If the initial and final states are known, the unit-step response of a first-order system may be sketched for a given time constant. Alternatively, the time constant of a system can be determined experimentally by plotting the unit-step response. When $t = \tau$, the value of the output can be determined from its initial and final values.

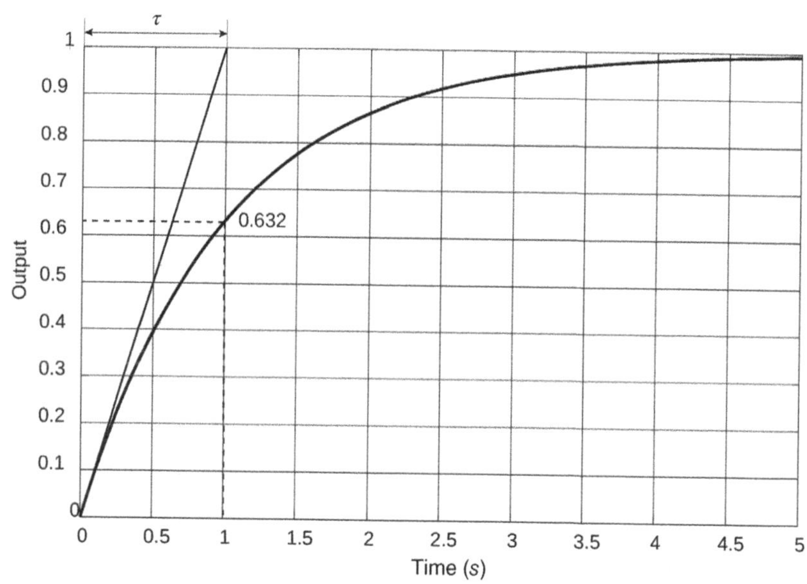

Figure 3.8 Unit-step response of a first-order system with $\tau = 1$

$$x(\tau) = X_s + \frac{X_0 - X_s}{e} \cong X_s + 0.368(X_0 - X_s) \tag{3.93}$$

Figure 3.8 illustrates the unit-step response of a first-order system, with a unity time constant and an initial condition of zero. The tangent at $t = 0$ intercepts the final value line at $t = 1$. According to the equation above, the output at $t = \tau$ is $1 - 0.368 = 0.632$.

Example 3.6 Flywheels serve as energy storage devices in mechatronic systems, retaining rotational kinetic energy and supplying it as needed to sustain steady rotational speed despite fluctuations in torque or motion. In analyzing the dynamic performance of a mechatronic system, it is necessary to know the total moment of inertia referenced to the motor shaft. In complex systems, calculating this parameter can be difficult, but it may be obtained through the test described below.

A flywheel is mounted to the shaft of a DC motor as illustrated in Figure 3.9. To determine the total moment of inertia on the shaft, the motor is powered, and the steady-state rotation speed is adjusted to 1,000 rpm. Then, the motor power is switched off at $t = 0$ and the motor speed is recorded via a data-logging tachometer. Figure 3.10 shows the plot of the recorded shaft speed. Given the viscous friction (damping) constant of the bearings is $0.3\ \text{N}\cdot\text{m}\cdot\text{s}$, determine the total moment of inertia referred to the motor shaft.

Solution

Dynamic equation of the flywheel after the motor power is switched off is represented by the first-order differential equation below:

$$J\frac{d\omega(t)}{dt} + \beta\omega(t) = 0 \Rightarrow \frac{d\omega(t)}{dt} + \frac{\beta}{J}\omega(t) = 0$$

The initial rotation speed is set to 1,000 rpm. Once the power has been switched off, the speed gradually decreases until it reaches zero. Therefore, $N_s = 0$. Because the angular velocity is

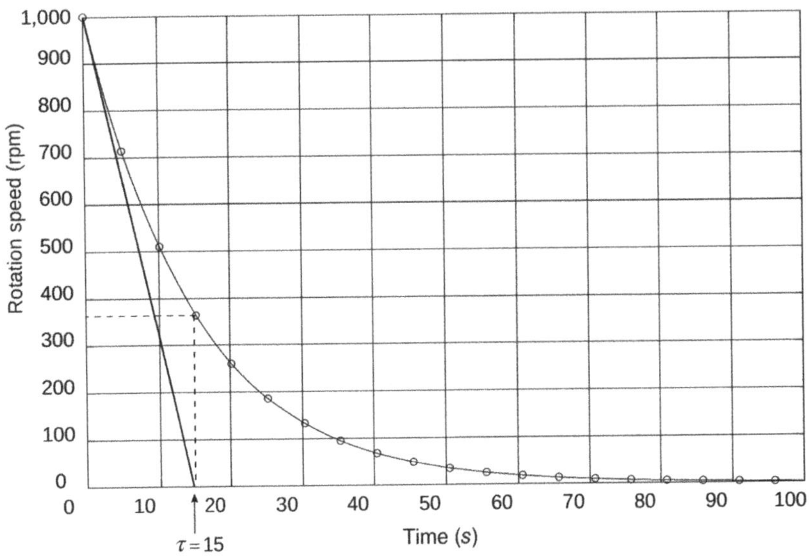

Figure 3.9 DC motor driving a flywheel

proportional to the rotation speed in rpm ($\omega = (2\pi/60)N$), the equation above remains unchanged if the rpm values are used.

The equation above shows that the time constant equals the moment of inertia divided by the damping constant ($\tau = J/\beta$). The tangent at $t = 0$ intercepts the time axis at $t = 15$ s, representing the time constant of the system. We find the total moment of inertia by multiplying the given damping constant.

$$J = \tau\beta = 15 \times 0.3 = 4.5\,\mathrm{N\,m^2} \therefore$$

Example 3.7 For verification the solution presented in Example 3.6, plot the shaft speed over time using the obtained moment of inertia using MATLAB®.

Solution

The MATLAB® code below plots the rotor shaft speed over time using the values provided and calculated in Example 3.6. The plot matches the motor speed recorded via a data-logging tachometer given in Figure 3.10.

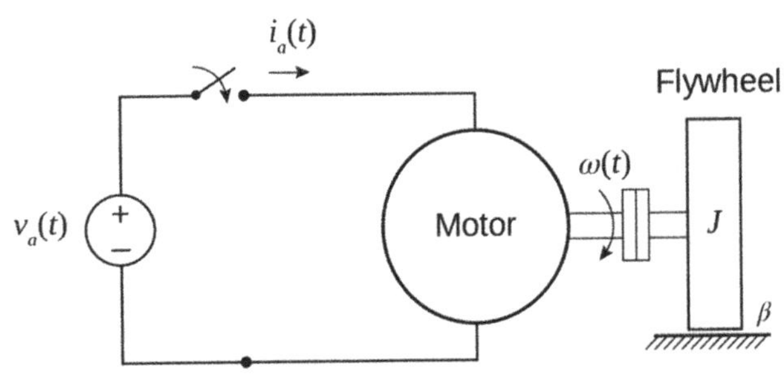

Figure 3.10 Speed of the rotor and flywheel assembly after motor power is turned off

```
% Parameters
J = 4.5; % Moment of inertia (kg·m²), example value
beta = 0.3; % Viscous friction constant (N·m·s), example value
omega0 = 1000 * 2 * pi / 60; % Initial speed in rad/s (1000 rpm)
% Time vector
t = linspace(0, 100, 1000); % Time from 0 to 100 seconds
% Shaft speed over time
omega = omega0 * exp(-beta/J * t);
% Convert rad/s to rpm for plotting
rpm = omega * 60 / (2 * pi);
% Plot
figure;
plot(t, rpm, 'LineWidth', 2);
xlabel('Time (s)');
ylabel('Shaft Speed (rpm)');
title('Flywheel Deceleration After Motor Power-Off');
grid on;
```

If the initial condition is zero, the response of a first-order system can be represented using the transfer function below derived from the Laplace transform of the dynamic equation in equation 3.86.

$$sX(s) + aX(s) = R(s) \tag{3.94}$$

In a linear system, the output is directly proportional to the state variable. Substituting the time constant $a = 1/\tau$ yields

$$T(s) = \frac{Y(s)}{R(s)} = K\frac{1}{s + \frac{1}{\tau}} = K\frac{\tau}{\tau s + 1} \tag{3.95}$$

This equation can be represented by a single block as shown in Figure 3.11 (a). When the initial condition is not zero, the Laplace transform of the differential equation becomes

$$sX(s) - X_0 + \frac{1}{\tau}X(s) = R(s) \tag{3.96}$$

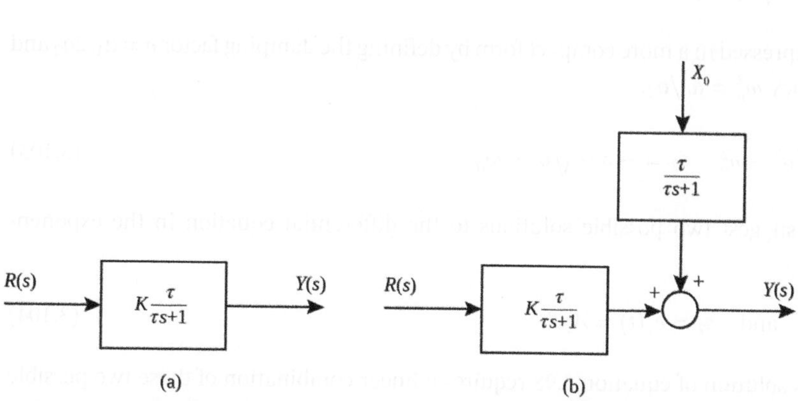

Figure 3.11 Block diagram representation of a first-order system (a: All initial conditions are zero; b: Initial conditions are non-zero)

By solving this equation for $X(s)$, we obtain

$$X(s) = \frac{\tau}{\tau s + 1}[R(s) + X_0] \tag{3.97}$$

A system with initial condition can be represented by a block diagram that features two distinct blocks as illustrated in Figure 3.11 (b). One block represents the transfer function, while the other reflects the effect of the initial condition on the output.

3.11 Second-Order Systems

Second-order LTI systems are characterized by the same order ODE of the following form, where $c(t)$ denotes the system output and $r(t)$ reference input:

$$a_2 \frac{d^2c(t)}{dt^2} + a_1 \frac{dc(t)}{dt} + a_0 c(t) = r(t) \tag{3.98}$$

Assuming all initial conditions are zero, applying the Laplace transform to this equation results in an algebraic equation in the s-domain.

$$\left(a_2 s^2 + a_1 s + a_0\right) C(s) = R(s) \tag{3.99}$$

The following transfer function is derived from this equation:

$$G(s) = \frac{C(s)}{R(s)} = \frac{1}{a_2 s^2 + a_1 s + a_0} \tag{3.100}$$

The poles of the transfer function correspond to the roots of the quadratic equation:

$$a_2 s^2 + a_1 s + a_0 = 0 \tag{3.101}$$

$$s_{1,2} = -\frac{a_1}{2a_2} \pm \sqrt{\left(\frac{a_1}{2a_2}\right)^2 - \frac{a_0}{a_2}} \tag{3.102}$$

The roots can be expressed in a more compact form by defining the damping factor $\sigma = a_1/2a_2$ and the resonant frequency $\omega_n^2 = a_0/a_2$.

$$s_1 = -\sigma + \sqrt{\sigma^2 - \omega_n^2} \quad s_2 = -\sigma - \sqrt{\sigma^2 - \omega_n^2} \tag{3.103}$$

The two roots in suggest two possible solutions to the differential equation in the exponential form:

$$c_1(t) = A_1 e^{s_1 t} \quad \text{and} \quad s_2 = c_1(t) = A_2 e^{s_2 t} \tag{3.104}$$

The homogeneous solution of equation 3.98 requires a linear combination of these two possible solutions where A_1 and A_2 must be determined from initial conditions.

a) Overdamped Response ($\sigma > \omega_n$)

In this case, the transfer function has two distinct negative real poles. Therefore, the natural response is the combination of two decaying exponents.

$$\begin{aligned} c(t) &= A_1 e^{s_1 t} + A_2 e^{s_2 t} \\ s_1 &= -\sigma + \sqrt{\sigma^2 - \omega_n^2} \end{aligned} \tag{3.105}$$

When ζ is significantly greater than unity, one of the two decaying exponentials decreases much faster than the other, so that the faster decaying exponential term (which corresponds to a smaller time constant) may be neglected. That is if $|s_2| \ll |s_1|$, then s_1 may be ignored for an approximate solution because the effect of s_1 on the response is much smaller than that of s_2, since the term involving it decays much faster than the term involving s_2. Once the faster decaying exponential term has disappeared, the response is like that of a first-order system.

b) Underdamped Response ($\sigma < \omega_n$)

A damping factor less than the natural frequency results in a negative discriminant of the quadratic equation. To calculate the poles, equation 3.103 can be written in the following form:

$$\begin{aligned} s_1 &= -\sigma + \sqrt{-\left(\omega_n^2 - \sigma^2\right)} = -\sigma + j\omega_d \\ s_2 &= -\sigma - \sqrt{-\left(\omega_n^2 - \sigma^2\right)} = -\sigma - j\omega_d \end{aligned} \tag{3.106}$$

In this equation, j denotes the unit imaginary number $\sqrt{-1}$ and ω_d is defined as damped oscillation frequency and can be calculated using the following equation:

$$\omega_d = \sqrt{\omega_n^2 - \sigma^2} \tag{3.107}$$

For complex conjugate poles, equation 3.111 becomes

$$c(t) = e^{-\sigma t}\left(A_1 e^{j\omega_d t} + A_2 e^{-j\omega_d t}\right) \tag{3.108}$$

The poles may also be expressed in terms of the damping ratio and damped natural frequency as follows:

$$\begin{aligned} &s_1 = -\sigma + j\omega_d; \quad s_2 = -\sigma - j\omega_d \\ &c(t) = e^{-\sigma}(A_1 e^{j\omega_d t} + A_2 e^{-j\omega_d t}) \end{aligned} \tag{3.109}$$

Euler identities allow expressing the natural response using trigonometric functions.

$$c(t) = e^{-\sigma t}[B_1 \cos(\omega_d t) + B_2 \sin(\omega_d t)] \tag{3.110}$$

The constant coefficients B_1 and B_2 are determined using the initial conditions.

When the transfer function is underdamped, the response to a disturbance or initial condition is characterized by oscillatory behavior that gradually diminishes over time. The presence of complex conjugate poles in the system's transfer function leads to sinusoidal components in the natural response, resulting in decaying oscillations about the steady-state value.

Key properties of the underdamped response include:

- **Oscillatory Motion**: The output exhibits sinusoidal oscillations, the frequency of which is determined by the damped natural frequency (ωd).
- **Exponential Envelope**: The amplitude of these oscillations decreases exponentially, dictated by the rate $e^{-\sigma t}$, where σ is the damping factor and ωn is the undamped natural frequency.

- **Overshoot and Ringing**: The response typically overshoots the steady-state value and may cross multiple times the final value before settling. This phenomenon is known as ringing.
- **Settling Time**: The time required for the oscillations to decay and the output to remain within a specified tolerance of the final value depends on both the damping ratio and the natural frequency.
- **Initial Conditions**: The exact trajectory of the response is influenced by the initial displacement and velocity, which set the coefficients for the cosine and sine terms in the solution.

Underdamped systems are common in mechatronics applications where some oscillation is tolerable or even desired, such as in suspension systems or tuned circuits. However, excessive underdamping can cause persistent oscillations and stability concerns, necessitating careful design and tuning.

c) Critically Damped Response ($\sigma = \omega_n$)

When the damping factor matches the natural frequency, the transfer function has a real negative double pole in the *s*-plane.

$$s_1 = s_2 = \omega_n \Rightarrow c(t) = (A_1 + tA_2)e^{\omega_n t} \tag{3.111}$$

The coefficients A_1 and A_2 are determined using the initial conditions.

A critically damped response is the boundary condition between overdamped and underdamped systems, providing the fastest return to steady state without exhibiting oscillations or overshoot. Such response is desirable in control applications where stability and speed are prioritized, such as in instrumentation where rapid, accurate positioning is essential or in servomechanisms designed to suppress transient vibrations.

d) Higher-Order Systems

Higher-order systems incorporate three or more energy storage elements. These systems exhibit complex transient and steady-state behaviors, with poles and zeros that affect response, stability, and disturbance sensitivity. Unlike the relatively straightforward responses seen in lower-order systems, higher-order responses exhibit a mixture of rapid and slow modes, multiple oscillation frequencies, and subtle interactions among system components. Advanced tools such as state-space modeling and frequency-domain methods are often required to thoroughly analyze these systems and develop effective controllers.

3.12 Summary

1) Mathematical modeling is essential for analyzing and designing mechatronic systems, which integrate mechanical, electrical, and control components. This chapter provides an in-depth discussion on various aspects of mathematical modeling, including system identification, types of systems, and analysis techniques.
2) The modeling process starts with defining the system's structure, objectives, and constraints. Technical drawings or block diagrams are used to outline the system's architecture, which can be simplified based on modeling objectives.
3) Distributed systems, such as temperature in heat exchangers, require PDEs for analysis. Numerical techniques like FDM and FEM are used to solve these PDEs. Lumped parameter systems simplify complex systems into manageable components.

4) Linear systems follow the principles of superposition and homogeneity, making them easier to analyze. Nonlinear systems, on the other hand, require linearization techniques for simplification.
5) Time-variant systems have parameters that change over time, complicating their analysis. Time-invariant systems, however, have consistent behavior over time, simplifying their analysis and design.
6) Continuous-time models are used for systems with variables that change continuously over time, while discrete-time models are suitable for systems where variables are sampled at specific intervals.
7) Deterministic systems produce consistent outputs, given the same initial conditions, whereas stochastic systems incorporate randomness, requiring probabilistic methods for analysis.
8) Modeling mechatronic systems involves several limitations, including nonlinearity, time-variant behaviors, and environmental factors. Computational resources and data accuracy are also significant constraints.
9) Linearization simplifies the analysis of nonlinear systems. Techniques like piecewise linear approximation and small-signal linearization are commonly used.
10) Various mathematical tools are used for analyzing engineering systems. The selection depends on the model type, desired accuracy, and assumptions made. This chapter focuses on lumped parameter, deterministic, and time-invariant systems.
11) State-space analysis models systems using first-order differential equations. It is particularly useful for complex systems with multiple inputs and outputs.
12) Modeling electrical components involves representing their behavior using dynamic equations in either the time-domain or the s-domain.
13) Mechanical structures include masses, friction elements, and springs. Their behavior is described by force–motion characteristics, which can be modeled using Newton's laws.
14) Hydraulic and pneumatic systems use incompressible fluids and compressed air, respectively, to transmit power and control motion. These systems are essential in various industries.
15) Temperature control is crucial in mechatronic systems to ensure proper operation. Thermal systems include heat sources, sinks, insulation, and control devices.

3.13 Check Your Understanding

3.13.1 Self-Assessment Quiz

1 Most physical systems exhibit nonlinearity when the variables vary over a wide range.
True False

2 A nonlinear system satisfies the properties of superposition and homogeneity.
True False

3 The transfer function is the ratio of the Laplace transform of the output variable to the Laplace transform of the input variable, with all initial conditions equal to zero.
True False

4 Transfer function cannot be used for a nonlinear system.
True False

5 To use transfer function method for analysis of a nonlinear system, the system must be linearized.

True False

6 A block representing a subsystem with transfer function $T_1(s)$ is directly feeding the input of a block with transfer function $T_2(s)$. Which one of the following is equivalent to these two blocks?

a. $T_1(s) \cdot T_2(s)$
b. $T_1(s) + T_2(s)$
c. $T_1(s)/(s)$
d. None of the above

7 In an electric circuit, the currents are

a. Through variables
b. Across variables
c. Either through or across depending on the element characteristics
d. None of the above

8 In an electric circuit, the voltages are

a. Through variables
b. Across variables
c. Either through or across depending on the element characteristics
d. None of the above

9 Which electrical variables are analogous to the forces in mechanical systems?

a. Voltages
b. Power
c. Currents
d. Charge

10 In a mechanical system, the translation of moving parts are

a. Through variables
b. Across variables
c. Either through or across depending on how they move
d. None of the above

11 In a hydraulic system, through variable is

a. Flow rate
b. Pressure
c. Temperature
d. Density

12 Under which conditions below are transfer functions defined? (Select all that apply)

a. The system is nonlinear time invariant; the initial conditions are zero.
b. The system is linear, time invariant; initial conditions are constant.
c. System is linear, time variant; the initial conditions are zero.
d. The system is linear, time invariant; initial conditions are zero.

13 What is the key difference between viscous and dry friction in mechanical systems?
  a. Dry friction is proportional to the velocity, whereas viscous friction is proportional to the force.
  b. Viscous friction is proportional to the velocity, whereas dry friction is constant regardless the velocity.
  c. Viscous friction is a function of the force, while dry friction is a function of the pressure.
  d. Viscous friction is constant regardless of the velocity, but dry friction changes with velocity.

14 What is the fundamental principle behind the conservation of mass in fluid dynamics, and how are they applied in hydraulic systems?
  a. Conservation of energy
  b. Conservation of charge
  c. Conservation of mass
  d. Conservation of momentum

15 What is the key property of incompressible fluids?
  a. Their density remains constant regardless of changes in pressure or temperature.
  b. Their volume flow rate changes with the pipe cross-sectional area.
  c. Their pressure remains constant when the pipe diameter changes.
  d. Their density changes with pressure.

16 What is the key characteristic of distributed parameter systems?
  a. Distributed parameter systems involve variables that are distributed over space.
  b. Distributed parameter systems can be analyzed using the Laplace transform.
  c. Parameters of a distributed parameter system are time variant.
  d. Parameters of a distributed parameter system are time invariant.

17 Which type of equations can be solved using the Laplace transform?
  a. Nonlinear differential equations
  b. Partial differential equations
  c. Linear algebraic equations
  d. Ordinary differential equations

18 Which mechanical element below is analogous to electrical resistance?
  a. Mass of a solid object
  b. Spring
  c. Friction
  d. Moment of inertia of a rotating object

19 What is the electrical analogous to a spring?
  a. Resistance
  b. Inductance
  c. Capacitance
  d. Voltage source

20 What is the electrical analogous to a fluid reservoir?
  a. Resistance
  b. Inductance

c. Capacitance
d. Voltage source

Answers:

1: True; 2: False; 3: True; 4: True; 5: True; 6: a; 7: a; 8: b; 9: c; 10: b; 11: a; 12: d; 13: b; 14: c; 15: a; 16: a; 17: d; 18: c; 19: b; 20: c.

3.13.2 Essay Questions

1 How do thermal systems manage heat distribution in mechatronic systems, and why is this important for their proper operation?

2 What role do hydraulic and pneumatic elements play in mechatronic systems, and how are they modeled?

3 Why is temperature control crucial in mechatronic systems, and what are some examples of thermal systems used for this purpose?

4 Discuss the differences between time-variant and time-invariant systems. Provide examples of each and explain the challenges associated with modeling and analyzing time-variant systems.

5 What are the differences between continuous-time and discrete-time models, and when is each type of model appropriate for mechatronic systems?

6 Describe the process of system identification for a mechatronic system. Explain how technical drawings, functional block diagrams, and circuit schematics are used to outline the system's architecture.

7 Discuss the similarities and differences between distributed parameter systems and lumped parameter systems. Provide examples of each and explain the mathematical techniques used to analyze them.

8 Explain the concept of linearization in the context of mechatronic systems. Discuss the piecewise linear approximation and small-signal linearization techniques and provide examples of their applications.

9 How does linearization simplify the analysis of nonlinear systems, and what are some common linearization techniques?

10 How does the piecewise linear approximation technique help in modeling nonlinear systems, and what are its limitations?

11 Explain the key steps involved in the system identification process for mechatronic systems.

12 How is the Laplace transform used to analyze the behavior of electrical components in mechatronic systems?

13 What is state-space analysis, and why is it particularly useful for complex systems with multiple inputs and outputs?

14 What are the main advantages of using state-space analysis for modeling complex systems with multiple inputs and outputs?

3.13.3 Problems

1 The following equation defines the input–output characteristic of a system. Obtain the small signal linear model of this system for $x = 0.8$.

$$y = 0.27x^3 - 1.44x^2 + 2.92x - 0.08$$

2 The magnetization curve of a ferromagnetic material utilized in the core of a solenoid can be approximated by the following equation:

$$\lambda = 1.1 + \frac{i - 0.8}{2i + 0.75}$$

a. Plot this magnetization curve using MATLAB®.
b. Apply piecewise linear approximation to obtain the inductance of the solenoid with less than 10% error.
c. Determine the limit of the linear region.

3 Using Symbolic Math toolbox of MATLAB®, compute the unit impulse response of the system modeled by the transfer function below:

$$\frac{C(s)}{R(s)} = \frac{s + 2}{s^2 + 4s + 5}$$

4 Figure 3.12 (a) is the circuit diagram of a band pass filter. Derive the transfer function of this circuit for the input V_{in} and output V_o.

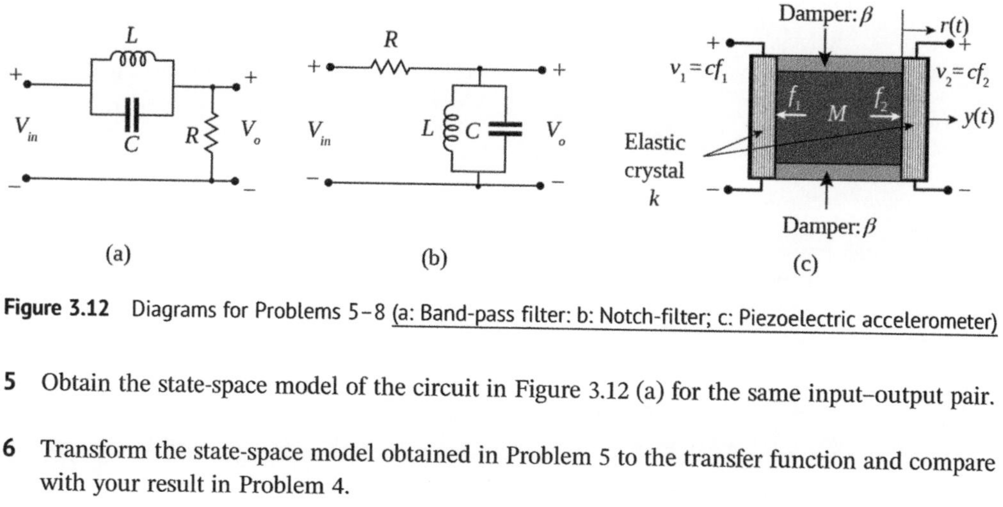

Figure 3.12 Diagrams for Problems 5–8 (a: Band-pass filter: b: Notch-filter; c: Piezoelectric accelerometer)

5 Obtain the state-space model of the circuit in Figure 3.12 (a) for the same input–output pair.

6 Transform the state-space model obtained in Problem 5 to the transfer function and compare with your result in Problem 4.

7 Repeat Problems 4–6 for the notch filter shown in Figure 3.12 (b).

8 Figure 3.12 (c) illustrates a piezoelectric crystal accelerometer. As the body moves horizontally, the mass M slides applying a force on the quartz crystals placed at each side. The elasticity constant of the crystal is k. The dampers mounted on the top and bottom of the moving mass have identical viscous friction constant β. Draw the free-body diagram of this device and obtain the transfer function for motion as input and the total voltage generated by the crystals connected in series as output.

9 Using the Hagen–Poiseuille equation, determine the fluid resistance of a cylindrical pipe with a given diameter, length, and fluid viscosity. Explain how the pressure difference across the ends of the pipe is related to the volume rate of flow through the tube.

References

Kuo, B. C., 1982. *Automatic Control Systems*. 4th ed. Englewood Cliffs, NJ: Prentice Hall.

Lewis, J. W., 1994. *Modeling Engineering Systems*. Eagle Rock, VA: HighText Publications, Inc.

Ogata, K., 2002. *Modern Control Engineering*. 4th ed. Upper Saddle River, NJ: Prentice Hall.

4

Sensors

OBJECTIVES
• Present an overview of the function and significance of sensors. • Describe the basic structure and operation of sensors. • Classify sensor types. • Explain common sensor technologies. • Discuss sensor integration with electronic circuits. • Highlight practical sensor applications. • Outline key features of sensors.

A sensor is a device or system that measures physical quantities and converts them into electrical signals. These signals may be interpreted, processed, or used for various purposes including process monitoring and control, experimental data collection, product testing, fault detection, alarm and warning generation, or surveillance. Output signal of a sensor in mechatronics and robotics, sensors provide information about the system outputs or to perceive and react to its surroundings.

The basic structure of a sensor comprises several key components working together to detect and convert physical events to a signal. A sensing element, which may be passive or active, responds directly to changes in the physical quantity being measured—such as temperature, light, pressure, or motion. This element's response induces a variation in some property of the sensor, such as resistance, capacitance, inductance, voltage, current, or magnetic flux. Then, a transduction mechanism converts this variation into an electrical signal, typically voltage or current, which reflects the magnitude of the measured quantity. In practical sensors, this primary signal has low power and may contain undesired effects termed as noise. The transducer's output is processed by amplifiers and filters to improve signal fidelity and reduce noise (Cetinkunt, 2015).

The resulting conditioned signal is then suitable for interpretation by electronic control units, data acquisition systems, or display modules. Figure 4.1 illustrates the integration of sensors within the system architecture. This modular configuration, which spans from the sensing element to signal output, constitutes the basic structure found in most sensors and serves a range of functions within mechatronic and robotic systems (de Silva, 2016).

Sensors may be grouped as analog and digital. Analog sensors generate electrical signals that change as a continuous function of the measured quantity. Digital sensors provide a binary output discrete in time.

Common types of sensors used in mechatronic and robotic systems measure physical quantities like position, motion, force, torque, pressure, flow rate, light, sound, humidity, and temperature. Advanced sensors utilize electronic circuits and embedded microprocessors to combine measured

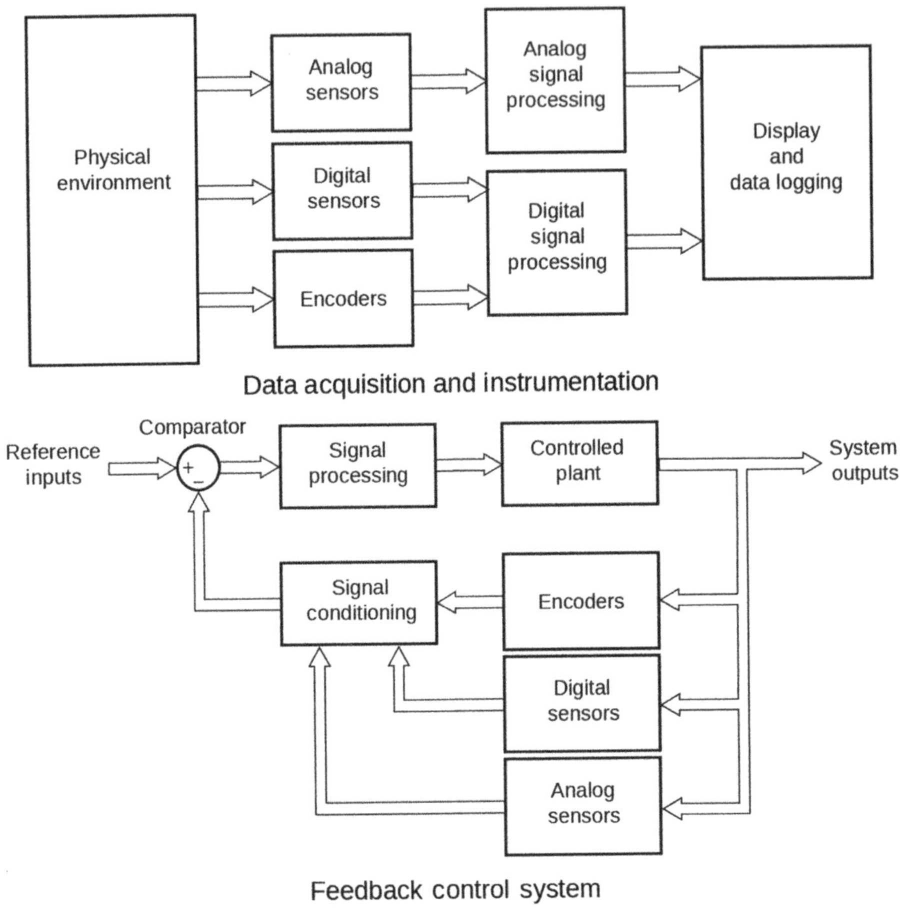

Figure 4.1 Integration of sensors in data acquisition and feedback control systems

quantities and apply mathematical operations to detect distances, orientation, proximity, vibration, stress, or chemical compositions.

Among advanced sensor technologies, GPS (Global Positioning System) and LIDAR (Light Detection and Ranging) are commonly employed in contemporary mechatronic and robotic systems (Pirc, 2025), including driver assistance features in automated vehicles, autonomous vehicles, and unmanned aerial vehicles (UAVs).

4.1 Sensor Types

Analog sensors employed in mechatronic and robotic systems can be broadly classified as passive or active, depending on how they interact with the physical quantity to be measured. Passive sensors rely solely on receiving and responding to external stimuli without requiring an additional energy source for their sensing operation. Analog sensors that function by detecting changes in resistance, inductance, or capacitance corresponding to the physical quantity are passive sensors. Examples include strain gauges, thermistors, and light-dependent resistors (LDRs), which simply change their resistance when exposed to mechanical deformation, variations in temperature, or light intensity. These sensors draw their output directly from the energy inherent in the measured phenomenon. To convert changes in physical quantities into electrical signals, passive sensors need to be part of an active circuit.

Active sensors, on the other hand, either generate their own energy like thermocouples or require an external power supply to operate. They emit some form of energy such as light, electricity, or sound into the environment and then detect the response after interacting with the target. Examples include photodiodes used in light detection, ultrasonic sensors for distance measurement, and certain types of proximity sensors. Some sensors operate based on voltage variations resulting from electromagnetic induction. By actively generating a signal and assessing its modification, active sensors can often achieve higher sensitivity or enable measurements that passive sensors cannot.

Digital sensors produce a binary output by switching on and off, generating a pulse train, or providing a coded binary number in response to factors such as motion, interruptions of a light beam by objects, or the reflection of audible sound, ultrasonic vibrations, or electromagnetic waves.

4.1.1 Resistive Sensors

Resistive sensors function by detecting changes in electrical resistance as a direct response to variations in a physical quantity, such as position, force, pressure, or temperature. When the measured parameter alters the geometry or material properties of the sensing element, the electrical resistance across specific terminals shifts accordingly. This principle allows resistive sensors to convert mechanical displacement or other physical effects into measurable changes in resistance, which can then be processed as an electrical signal.

A potentiometer is a typical resistive sensor whose output voltage depends on the wiper's position. Variable resistance sensors, such as strain gauges, thermistors, and LDRs, operate based on the resistance variations because of the changes in the physical quantity. These devices are valued for their simplicity, reliability, and direct translation of mechanical phenomena into electrical signals, making them indispensable in mechatronic and robotic applications for precise measurement and control.

4.1.2 Electromagnetic Sensors

Electromagnetic sensors operate based on the principle of induction to detect, measure, or interact with physical phenomena. These sensors are essential for noncontact measurement tasks and offer high precision in detecting parameters like displacement, position, velocity, and electrical quantities.

According to the principle of electromagnetic induction, when a conductor moves within a magnetic field or when the magnetic field surrounding a coil varies, an electromotive force (EMF) is induced. This fundamental property enables the development of sensors that can transduce mechanical motion into measurable electrical signals.

Common examples of passive electromagnetic sensors include inductive voltage dividers, which provide accurate voltage ratios for measurement and calibration purposes, and linear variable differential transformers (LVDTs), which utilize a movable ferromagnetic core within a set of windings to translate linear displacement into differential voltage outputs.

Examples of active electromagnetic sensors include dynamic microphones, which convert sound waves into electrical signals using electromagnetic induction, and electromagnetic flow meters used to measure the velocity of conductive fluids. Other commonly used devices are Hall-effect sensors, which detect the presence and strength of magnetic fields, and eddy current sensors that gauge displacement or material properties through the interaction of induced currents with nearby conductors. Even electric guitar pickups operate on electromagnetic principles, capturing string vibrations as variable magnetic flux to create unique audio signals.

These sensors are preferred for their robustness, reliability, and immunity to many forms of environmental interference. Their applications range from industrial automation and robotics to aerospace and geophysical instrumentation.

Given their ability to offer contactless measurements, electromagnetic sensors are also frequently integrated into systems requiring longevity or where physical linkage is impractical. In the sections that follow, we delve further into specific electromagnetic sensors such as the Inductive Voltage Divider and the LVDT, examining their construction, operation, and roles in precise measurement systems (Rizzoni & Kearns, 2007).

4.1.3 Capacitive Sensors

Capacitive sensors are transducers utilized for precise, contactless measurement of displacement, position, or proximity. Unlike electromagnetic sensors that rely on magnetic fields and induction, capacitive sensors are founded on the principle of capacitance, which is the ability of a system to store electrical charge between two conductive surfaces separated by a dielectric. The core of a capacitive sensor is a variable capacitor, whose capacitance changes in response to alterations in the physical configuration of its electrodes, such as their area of overlap or the distance between them (Alexander & Sadiku, 2013).

In operation, capacitive sensors exploit the relationship between physical displacement and the resulting change in capacitance. As one plate moves relative to the other, the overlap area or separation distance varies, causing a measurable shift in capacitance. This change is typically detected by integrating the variable capacitor into an AC bridge circuit; any deviation from the balanced state, induced by movement, produces an output signal proportional to the displacement. Owing to their high sensitivity, capacitive sensors are especially valuable in applications demanding non-contact, high-resolution measurements such as micro positioning, thickness gauging, and material characterization.

A capacitor is an electrical component that stores charge when a potential difference (EMF) is present across its terminals. A common type of capacitor formed by two parallel plates is illustrated below.

A capacitor is an electrical component that stores charge when a potential difference (EMF) is present across its terminals. A common type of capacitor formed by two parallel plates is illustrated below.

This example shows the plates being flat. In many capacitor types, the plates are parallel but formed by two concentric cylinders separated by an insulating dielectric. In terms of explaining how they work, treat them as being the same. The charge storage capacity is a function of the capacitance of the device, measured in Farads, and this may be determined from the equation:

$$C = \frac{A\varepsilon_o\varepsilon_r}{d}$$

The parameters of this expression are defined as:

- A: Area of the overlap between the two parallel plates in m^2
- $\varepsilon_o = 8.854 \times 10^{-12}$ F/m^2: permittivity of free space (vacuum)
- ε_r: Relative permittivity of the dielectric material (unitless)
- d: Distance between the plates in m

Because of the magnitude of ε_o, capacitance measured in Farad is extremely small and usually measured in micro Farads (mF) or pico Farads (pF).

If any or all of these parameters vary, then the capacitance changes. In the variable capacitor used in many capacitive sensors is the area of overlap of the plates since the capacitance is proportional to this area. Changing the area in direct proportion to displacement allows measurements of capacitance to be equated with displacement. In sensitive sensor systems, the variable capacitor is placed in an AC bridge such that any change in capacitance produces an imbalance and this is converted directly into a DC signal at the output.

4.1.4 Optoelectronic Sensors

Optoelectronic sensors are devices that detect and respond to changes in light by converting optical beams into electrical signals. These include various technologies such as photodiodes, phototransistors, and integrated sensor modules, each designed for specific uses like position sensing, object detection, or light intensity measurement. These sensors operate based on principles such as the photoelectric effect, where incident photons liberate charge carriers in a semiconductor material, resulting in a measurable current or voltage.

One common example is the optical encoder, which uses a light source and a photodetector to determine the position or speed of a rotating object. In industrial automation, optoelectronic sensors are frequently employed for noncontact detection of parts, counting, and safety interlock functions. They can also be found in environmental monitoring systems, where they measure ambient light levels or detect the presence of smoke or flame.

A significant advantage of optoelectronic sensors is their immunity to electromagnetic interference, making them ideal for use in environments where electrical noise might compromise the accuracy of conventional sensors. Additionally, their rapid response times and high sensitivity allow for precise detection, even of subtle changes in light conditions.

As technology advances, modern optoelectronic sensors are becoming more compact, energy-efficient, and capable of integrating with digital control systems, broadening their range of applications in science, industry, and everyday life.

Optoelectronic sensors utilize light sources (such as LEDs or laser diodes) and light-sensitive detectors (including photodiodes, phototransistors, or photomultiplier tubes). When a target object or environmental change alters the intensity, direction, or wavelength of the light reaching the detector, the sensor produces a corresponding electrical output. This enables precise measurement of properties such as position, displacement, presence, or the composition of materials.

Several common types of optoelectronic sensors include:

- **Photoelectric Sensors**: These detect the presence or absence of objects by monitoring the interruption or reflection of a light beam. They are widely used for object counting, position detection, and safety interlocks in industrial automation.
- **Photodiode and Phototransistor Sensors**: Sensitive to changes in light intensity, these devices are employed in applications ranging from ambient light detection and automatic brightness control to optical communication systems.
- **Fiber-Optic Sensors**: Utilizing optical fibers to guide light, these sensors can detect strain, temperature, or pressure changes along the fiber's length. They are ideal for environments where immunity to electromagnetic interference or remote sensing is critical.
- **Color and Spectral Sensors**: By analyzing the wavelength composition of incoming light, these sensors can determine the color or spectral properties of surfaces and materials, making them useful in quality control, sorting, and scientific instrumentation.

Optoelectronic sensors offer several advantages over traditional sensor technologies, including high sensitivity, noncontact measurement, and fast response times. Their versatility allows integration into a wide array of applications such as industrial automation, robotics, safety systems, environmental monitoring, and optical encoders for precise displacement measurement.

4.1.5 Piezoelectric Sensors

Piezoelectric sensors represent another important category of sensor technology, distinct from both optoelectronic and passive resistive components. These sensors utilize the piezoelectric effect—a phenomenon in which certain crystalline materials, such as quartz or lead zirconate titanate, generate an electric charge in response to applied mechanical stress. When subjected to forces like pressure, vibration, or acceleration, a piezoelectric element produces a voltage that is directly proportional to the magnitude and frequency of the mechanical input.

Piezoelectric sensors are exceptionally well suited for dynamic measurements, including vibration analysis, impact detection, and acoustic monitoring. Their inherent ability to convert minute mechanical variations into measurable electrical signals makes them invaluable in industrial monitoring, structural health diagnostics, and even consumer electronics such as microphones and touch-sensitive devices. These devices have characteristics including rapid response times, high sensitivity, and functionality in harsh environments without the need for external excitation power, as the sensed energy directly generates the signal.

However, piezoelectric sensors are primarily sensitive to changes rather than static or slowly varying forces, which limits their effectiveness for applications requiring measurement of steady-state conditions. Signal conditioning electronics, such as charge amplifiers or high-impedance buffers, are often used to process and interpret the low-current outputs generated by piezoelectric elements, ensuring accurate and reliable data acquisition for control systems or analysis.

4.2 Signal Generation and Interfacing

Resistive, capacitive, and inductive sensors are passive components. Signal generation circuits supply power and convert sensor-detected changes in displacement, pressure, temperature, or light intensity into electrical signals for display or control feedback.

The following paragraphs describe the use of voltage dividers and Wheatstone bridge circuits for converting variations in passive sensors into electric signals.

4.2.1 Voltage Dividers

Voltage dividers consist of two or more passive elements connected in series. When a voltage source is applied across the external terminals, a fraction of the applied source voltage is tapped across the connection nodes of the elements and a reference node. The general expression for the input–output relationship is obtained by applying Kirchhoff's voltage law (KVL) and Ohm's law. The voltage divider depicted in Figure 4.2 consists of n impedances. The input voltage source is applied between nodes labeled 0 and n, while the output signal is measured across node 0 and m.

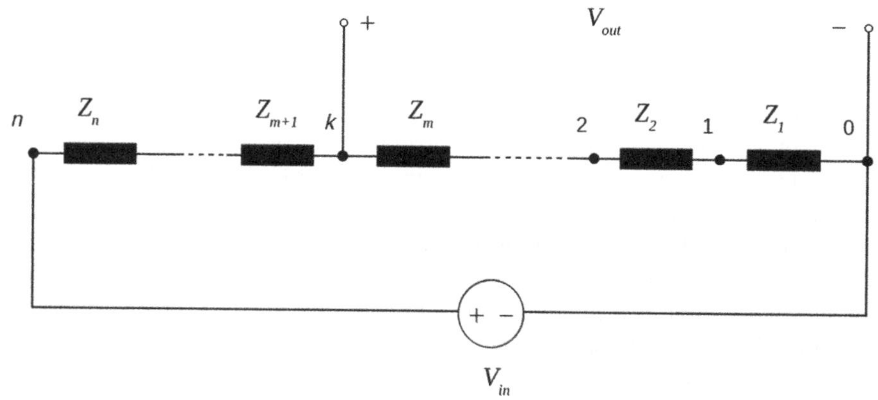

Figure 4.2 Multiple impedance voltage divider

$$V_{out} = \frac{\sum_{j=1}^{m} Z_j}{\sum_{k=1}^{n} Z_k} V_{in} \tag{4.1}$$

Voltage dividers consisting of two series impedances are more common in sensor-interfacing circuits. One of the impedances may be a sensor with variable impedance or pure *R*, *L*, *C* parameters depending on detected physical quantities. In a two-impedance divider, the expression of the output voltage is simplified to the following equation:

$$V_{out} = \frac{Z_2}{Z_1 + Z_2} V_{in} \tag{4.2}$$

When expanded to include three or more series impedances, voltage dividers become even more versatile, finding applications in a variety of sensor interfacing, signal conditioning, and circuit design tasks.

Multiple resistance voltage dividers are commonly employed used to generate precise intermediate voltages in a variety of analog-to-digital converter (ADC) and digital-to-analog converter (DAC), which are discussed in Chapter 9. Each tap along the divider provides a unique reference point for comparators or logic circuits, enabling fine-grained signal quantization or level detection.

Amplifiers and other analog circuits often require multiple voltage bias points for different circuit stages. A single voltage divider with several resistors can conveniently provide these, simplifying design and reducing component count.

When interfacing with sensor arrays (for example, temperature sensors, photodetectors, or strain gauges), a voltage divider with multiple variable impedances can enable simultaneous monitoring of several sensors. By measuring the voltage at various nodes, the system can infer the state of each sensor, making this approach valuable for cost-sensitive or space-constrained applications.

In resistive touchscreens or linear position sensors, a multi-point voltage divider enables the determination of touch location or slider position by measuring the voltage at the contact point, which corresponds to the ratio of impedances on either side of the touch.

Multi-impedance voltage dividers can be used to attenuate analog signals to several scaled levels. This is common in audio mixer circuits, sensor front ends, or wherever multiple output levels are required for parallel signal processing.

When interfacing devices operating at different logic voltages (for example, 5 and 3.3 V systems), a voltage divider with more than two resistors can create intermediate levels for safe communication or for driving multiple logic thresholds.

Multi-tap voltage dividers provide conveniently accessible test points for calibration, signal monitoring, or troubleshooting. In development and quality assurance, voltage dividers with selectable tap points can simulate various sensor output levels, allowing for controlled testing of sensor interfacing and signal processing circuits.

Interaction with load is one of the limitations of voltage dividers. Connecting a signal amplifier ADC or voltmeter with finite input impedance creates a parallel impedance with one leg of the divider, altering the voltage from its theoretical value. This "loading effect" becomes significant when the measurement device's input resistance is not much higher than the divider's impedances, leading to measurement errors. Especially in high-precision applications, this effect can lead to substantial discrepancies between the actual sensor output and the measured value. Compensation typically requires either amplifier buffers or recalculation of the voltage divider output. Voltage dividers are inherently unable to drive significant current without distorting the output voltage. This limits its use in circuits where the output needs to drive loads, relays, or other active devices without an intermediate buffer.

Another weakness of voltage dividers is the direct dependence of the output on the supply voltage. Any fluctuation or noise in the power source is directly reflected in the sensor signal, potentially introducing measurement errors unless a well-regulated supply is used.

Achieving desired sensitivity, accuracy, or output range typically requires manual calculation and selection of resistor values. In some cases, temperature compensation or trimming potentiometers may be needed for consistent long-term operation.

Voltage dividers are simple in design, but their limitations lead designers to explore other options for applications with greater requirements. Wheatstone bridges, for example, provide better sensitivity, linearity, and common-mode noise rejection, making them standard in precision strain measurement. Signal amplifiers or active conditioning circuits solve many of the divider's weaknesses, especially in environments with electrical noise or when interfacing with low-impedance loads.

4.2.2 Resistive Voltage Divider

A resistive voltage divider uses resistors connected in series across a voltage source. The voltage at the junction of the resistors represents a portion of the input voltage, calculated according to the proportion of their resistances. This is determined by substituting impedances with pure resistances in equation 4.1. This property makes the resistive voltage divider a useful configuration for signal conditioning, reference voltage creation, and sensor interfacing.

In sensor interfacing, one of the resistors is replaced with a variable resistance sensor element such as a strain gauge, thermistor, or LDR, or any two-terminal resistive sensor. The output voltage that reflects changes in the physical parameter being measured is usually fed to a signal conditioning circuit, an ADC, microcontroller, or measurement device for further processing.

While the concept is elementary, its implementation in sensor circuits comes with both distinct advantages and limitations.

Resistive voltage dividers are easy to design and implement. Their components are widely available, inexpensive standard resistors. The small number of components allows for compact circuit designs, which is especially beneficial in space-constrained applications. The voltage divider provides a real-time analog output that varies smoothly with the sensor's resistance, allowing for immediate monitoring of changes in the measured quantity. Furthermore, the conversion from resistance to voltage is inherent to the circuit, often eliminating the need for additional signal conditioning at the basic level. The basic voltage divider often serves as a first stage in more complex signal conditioning chains, feeding into operational amplifiers (op-amp), comparators, or ADCs for further processing.

Many resistive sensors (thermistors, LDRs) have a nonlinear relationship between resistance and the measured property. While the voltage divider itself is linear in resistance, the sensor's inherent nonlinearity means the voltage output is also nonlinear relative to the physical parameter. This can complicate calibration and may require additional signal processing or software correction.

The accuracy and stability of the voltage divider depend on the stability of the resistors. Standard resistors have temperature coefficients that can cause their values to drift under changing environmental conditions, leading to long-term or temperature-induced errors, unless high-quality precision resistors are used. Furthermore, voltage dividers do not provide differential output, which makes them susceptible to external interference, ground loops, and electromagnetic noise in electrically noisy environments.

Example 4.1 The output of the voltage divider circuit shown in Figure 4.3 is connected to a voltmeter with internal resistance of R_m. Find the values of the measured voltage for the following resistor values and determine the measurement error for each set of values.

a) $R_1 = 100\ \Omega$; $R_2 = 300\ \Omega$

b) $R_1 = 10\ \text{k}\Omega$; $R_2 = 30\ \text{k}\Omega$

c) $R_1 = 100\ \text{k}\Omega$; $R_2 = 300\ \text{k}\Omega$

Solution

When the meter is not connected (or $R_m = \infty$), the voltage across R_2 is

$$V_o^* = \frac{R_2}{R_1 + R_2} V_s$$

When the meter is connected, the branch of the voltage divider where the voltage is measured is found as the parallel equivalent of R_2 and R_m: $R_{eq} = \frac{R_2 R_m}{R_2 + R_m}$. Then, the measured voltage becomes

$$V_o = \frac{R_{eq}}{R_1 + R_{eq}} V_s$$

Figure 4.3 Resistive voltage divider supplying a voltmeter with finite input resistance

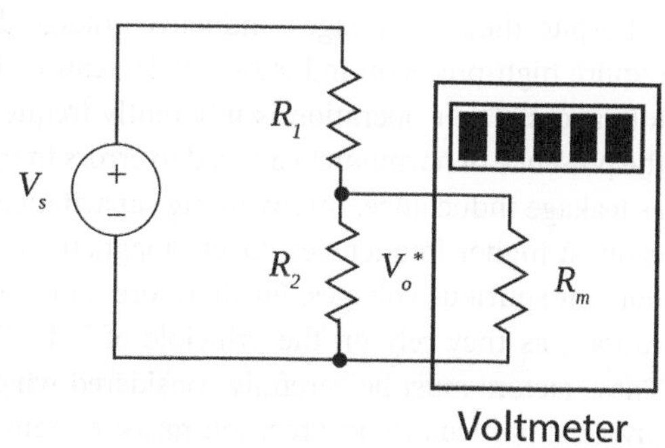

$$\%\varepsilon = \left(1 - \frac{V_o}{V_0^*}\right) \times 100$$

The calculations for the given resistance pairs are shown in the table below:

R_1 (Ω)	R_2 (Ω)	V_{R2}^*	R_{eq} (Ω)	V_{R2}	ε (%)
100	300	75	299.55	74.97	0.4
10,000	30,000	75	26,087	72.29	3.6
100,000	300,000	75	120,000	54.5	27

4.2.3 Inductive Voltage Dividers

Inductive voltage dividers are fundamental components used for scaling down voltage levels in AC circuits. Unlike resistive dividers, which rely solely on resistors, these dividers make use of the frequency-dependent reactance of inductors or capacitors to achieve specific voltage ratios. Variations in physical quantities identified by capacitive or inductive sensors may be converted into electrical signals via appropriate voltage divider configurations.

An inductive voltage divider consists of two or more inductors connected in series across an input voltage source. The output voltage is tapped across one of the inductors or a portion of the winding in the case of an autotransformer used for adjustable AC voltage in lab experiments. Because the voltage across each inductor is proportional to its inductance (assuming identical current through each series element), adjusting the inductance values allows precise control of the output voltage.

$$V_{out} = \frac{X_2}{X_1 + X_2} V_{in} = \frac{j\omega L_2}{j\omega L_1 + j\omega L_2} V_{in} = \frac{L_2}{L_1 + L_2} V_{in} \tag{4.3}$$

Inductive voltage dividers offer several notable advantages, making them valuable components in electrical and measurement systems, particularly for AC applications:

The output voltage from an inductive divider is directly proportional to the inductance values of the coils used. This means that by carefully selecting or adjusting the inductance, precise control over output voltage can be achieved, which is particularly advantageous in laboratory settings or calibration instruments. Because inductors ideally dissipate very little energy (except for minor core and copper losses), inductive voltage dividers are suitable for precision applications where heat generation and power loss must be minimized. Inductive dividers can be used in configurations designed to minimize loading of the measured circuit, making them attractive for sensitive measurement and instrumentation tasks.

Despite their advantages, inductive voltage dividers also have several limitations. First, they require high-precision inductors, which can be bulky, expensive, and sensitive to manufacturing tolerances. Their operation is inherently frequency-dependent; variations in signal frequency or the presence of harmonics can lead to errors in the output ratio. Additionally, parasitic effects such as leakage inductance, interwinding capacitance, and core losses can compromise accuracy, especially at higher frequencies. Electromagnetic interference from nearby equipment may introduce noise or spurious voltages. Furthermore, inductive dividers are generally unsuitable for DC applications, as they rely on the principle of inductive reactance, which vanishes at zero frequency. These factors must be carefully considered when designing circuits that utilize inductive voltage dividers, particularly in precision measurement or laboratory environments.

4.2.4 Capacitive Voltage Dividers

Capacitive voltage dividers use series-connected capacitors based on similar principles. The input–output relationship of a capacitive voltage divider is derived by substituting the capacitive reactances in equation 4.2.

$$V_{out} = \frac{1/j\omega C_2}{1/j\omega C_1 + 1/j\omega C_2} V_{in} = \frac{(j\omega C_1)(j\omega C_2)}{j\omega C_2(j\omega C_1 + j\omega C_2)} V_{in}$$
$$V_{out} = \frac{C_1}{C_1 + C_2} V_{in} \tag{4.4}$$

Capacitive dividers are often used in high-voltage measurements, oscilloscope probes, and impedance-matching circuits. Their inherent high input impedance minimizes circuit loading, making them well suited for sensitive measurement tasks or where isolation from the main circuit is required. Capacitive sensors may be used in a capacitive voltage divider to generate a signal corresponding to the measured variable.

Nevertheless, capacitive dividers present similar inherent limitations to the other types of voltage dividers, including the loading effect and weak differential signal output.

In addition, capacitive voltage dividers are susceptible to various parasitic effects, such as stray capacitance to ground, dielectric absorption, and leakage currents. These undesired effects can distort the intended divider ratio, especially at high frequencies or in circuits with significant electromagnetic interference. Parasitic capacitance and dielectric losses can also limit the accuracy and frequency range of the divider.

The accuracy of a capacitive voltage divider depends on the precise values of the capacitors used. Variations in capacitor tolerances, temperature coefficients, and aging can cause drift in the divider ratio over time. Environmental factors like humidity can also affect the dielectric material in capacitors, further impacting long-term stability and precision.

Because capacitors block DC and their reactance becomes very high at low frequencies, capacitive voltage dividers are unsuitable for DC or low-frequency AC applications. At such low frequencies, the output voltage may become negligible, rendering the divider ineffective for those types of measurements.

Capacitive voltage dividers used in high-voltage environments must be constructed with capacitors that can withstand the applied voltages without breakdown. Improper sizing or insulation can lead to catastrophic failure, posing safety hazards and risking damage to connected equipment.

4.2.5 Wheatstone Bridge

The Wheatstone bridge fundamentally consists of two voltage dividers arranged in parallel. It is a basic electrical circuit used to measure unknown resistances and is widely used in mechatronics and instrumentation. Originally invented by Samuel Hunter Christie in 1833 and popularized by Sir Charles Wheatstone, this bridge circuit has evolved into a key component of modern precision measurement systems. Its capability to detect small changes in resistance allows for use in various applications, particularly those requiring accuracy and sensitivity.

The Wheatstone bridge consists of two voltage dividers connected in parallel. In most circuit diagrams, the resistors are arranged in a diamond configuration as illustrated in Figure 4.4.

Output voltage corresponds to the potential difference between the central nodes of the voltage dividers. When no load current is drawn (open circuit), the same currents flow through each divider.

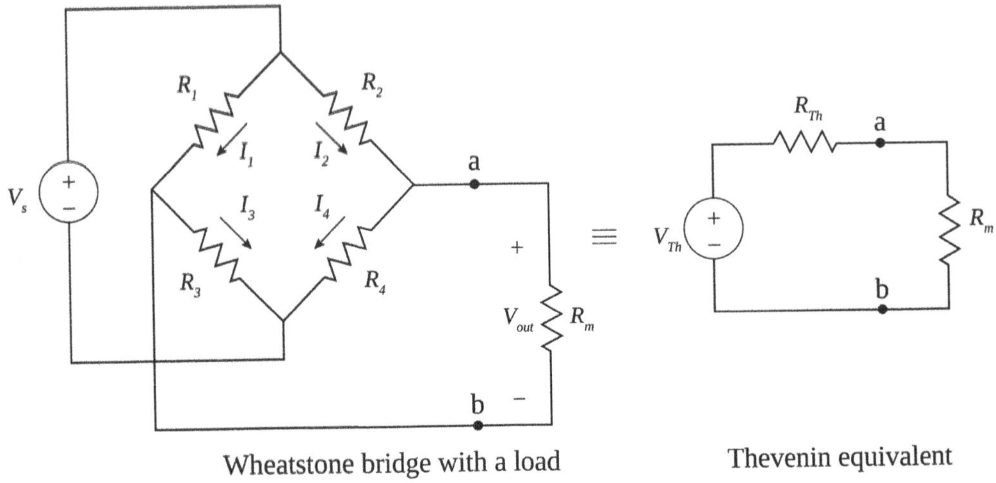

Figure 4.4 Wheatstone bridge configuration and its Thevenin equivalent circuit

$$I_1 = I_3 = \frac{V_S}{R_1 + R_3} \text{ and } I_2 = I_4 = \frac{V_S}{R_2 + R_4} \tag{4.5}$$

The bridge output voltage can be found as below by applying KVL.

$$V_{out} = V_S\left(\frac{R_3}{R_1 + R_3} - \frac{R_4}{R_2 + R_4}\right) \tag{4.6}$$

The bridge is balanced when the output voltage is zero. Solving the above equation for $V_{out} = 0$ yields the necessary and sufficient condition required for balancing in a Wheatstone bridge.

$$\frac{R_1}{R_3} = \frac{R_2}{R_4} \tag{4.7}$$

This relationship indicates that when the resistance ratios in each voltage divider matches the bridge is balanced, resulting in zero output voltage. The balance condition is independent of the supply voltage of the circuit. A variation in one or more resistance values can disturb this equilibrium, resulting in a voltage output that is proportional to the difference between the division ratios of the voltage dividers. Because of this property, the Wheatstone bridge is commonly used in instruments that detect changes in resistance.

Equation 4.6 is derived based on the assumption that no load current is drawn from the bridge. When the load resistance is finite, the output voltage will vary in the case of an unbalanced bridge. In that case, the voltage across the load resistance can be calculated by solving the complete circuit using one of the circuit analysis methods. Another alternative is using the Thevenin equivalent circuit of the Wheatstone bridge shown in Figure 4.4 (b).

Thevenin model consists of a voltage source corresponding to the no-load voltage given in equation 4.6 and the equivalent resistance seen from the output terminals when the voltage source is replaced with a short circuit as expressed below:

$$R_{Th} = \left(R_1 \| R_2\right) + \left(R_3 \| R_4\right) = \frac{R_1R_2}{R_1 + R_2} + \frac{R_3R_4}{R_3 + R_4} \tag{4.8}$$

By arranging this equation, we obtain the following expression for the Thevenin equivalent resistance.

$$R_{Th} = \frac{R_1R_2(R_3 + R_4) + R_3R_4(R_1 + R_2)}{(R_1 + R_2)(R_3 + R_4)} \tag{4.9}$$

Example 4.2 The resistor values of the Wheatstone bridge in Figure 4.4 are listed below, and the supply source provides 15 V DC.

a) Calculate the output voltage when no load resistance is connected.
b) Determine the Thevenin equivalent resistance.
c) Determine the voltage supplied to an amplifier with an input resistance $R_{in} = 500\ \Omega$.

$$R_1 = 10\ \text{k}\Omega, R_2 = 10\ \text{k}\Omega, R_3 = 15\ \text{k}\Omega, R_4 = 10\ \text{k}\Omega$$

Solution

a) Upon substituting the specified values into equation 4.6, we obtain the following result:

$$V_{out} = V_s\left(\frac{R_3}{R_1 + R_3} - \frac{R_4}{R_2 + R_4}\right) = 15 \times \left(\frac{15}{10 + 15} - \frac{10}{10 + 10}\right) = 1.5\ \text{V}$$

b) Thevenin equivalent resistance is found by using equation 4.9 with given values.

$$R_{Th} = \frac{R_1R_2}{R_1 + R_2} + \frac{R_3R_4}{R_3 + R_4} = \frac{10 \times 10}{10 + 10} + \frac{15 \times 10}{15 + 10} = 11\ \text{k}\Omega$$

c) The voltage supplied to the amplifier is the output voltage of the voltage divider Thevenin resistance and load resistance form and supplied by the open circuit voltage, which is calculated as 1.5 V in part a.

$$V_{input} = \frac{R_{in}}{R_{Th} + R_{in}} V_{out} = \frac{500}{11 \times 10^3 + 500} \times 1.5 = 0.043\ \text{V}$$

The output voltage of the Wheatstone bridge decreases because of the input resistance of the amplifier significantly lower than the bridge's Thevenin resistance. The voltage applied to the amplifier can be increased by using an amplifier with higher input resistance or by employing a buffer op-amp that has very high input resistance.

Wheatstone bridge is foundational in mechatronic systems for several reasons:

- **Strain Measurement and Force Sensing**: Wheatstone bridges are extensively used with strain gages, which are bonded to mechanical structures to detect strain, force, or pressure. As a structure deforms, the resistance of strain gage changes proportionally. Integrating strain gages into single, half, or full Wheatstone bridge configurations enables accurate measurement of mechanical deformation, often with high sensitivity and immunity to temperature variations. This is crucial in robotics, automated assembly, and precision manufacturing.
- **Load Cells and Weight Measurement**: Load cells, the primary transducers in weighing systems, frequently employ Wheatstone bridge circuits with strain gages. As a load is applied, the mechanical deformation of the cell causes a differential resistance change, producing a

corresponding voltage output. This allows for accurate mass measurement, which is vital in automated production lines, packaging, and quality control.

- **Robotic Feedback and Physical Sensing**: Robots use Wheatstone bridge strain gages in grippers or limbs to monitor contact forces, allowing precise and safe manipulation.
- **Position and Displacement Sensing**: Devices such as LVDTs and displacement sensors often use Wheatstone bridges to convert small mechanical movements into electrical signals. This is essential for accurate positioning and motion control in CNC machines, 3D printers, and automated inspection systems.

In electrical and electronic circuits, thermistors are commonly employed to prevent the overheating of semiconductor components. The resistance introduced by wiring and connections may impact the accuracy of test instruments. Utilizing a Wheatstone bridge configuration mitigates the influence of these extraneous resistances, as their effects are present in both arms of the bridge and can, therefore, be effectively balanced. This approach results in more stable and reliable measurements.

Instrumentation systems extensively utilize Wheatstone bridges to measure temperature, pressure, and fluid flow rate. Modified Wheatstone bridges, such as the Maxwell and Schering bridges that include capacitors, allow for the measurement of inductance and capacitance, respectively. These have important uses in the calibration and testing of electrical components and materials.

The versatility of the Wheatstone bridge has led to its widespread adoption in numerous applications in Weighing measurement systems from simple scales to massive weighbridges for trucks. Vehicle dynamometers, crash test instrumentation, and force sensors in steering and braking systems utilize Wheatstone bridge circuits for precise performance monitoring and safety analysis. Aircraft structures, bridges, and high-rise buildings are embedded with strain gage-based Wheatstone bridges to monitor stress, fatigue, and structural integrity in real time, offering early warning of potential failures. In medical devices like blood pressure transducers, respiratory flow meters, and biomedical force sensors, Wheatstone bridge circuits are used to deliver accurate physiological measurements, contributing to patient safety and effective diagnosis. Experimental physics, materials science, and engineering research frequently employ Wheatstone bridge circuits to investigate material properties, test prototypes, and validate design hypotheses under controlled conditions.

4.3 Practical Sensors

Common sensor types can be categorized based on the physical quantities they measure. These include sensors for detecting distance, motion, such as position, velocity, and acceleration sensors, as well as those designed for mechanical measurements, like force, torque, and strain sensors. Hydraulic sensors are also frequently used to measure parameters such as liquid level.

4.3.1 Distance and Detection

Distance sensors are essential tools for detecting and measuring the separation between objects or surfaces. They operate on various physical principles, such as the time delay of reflected light or sound waves, or by analyzing changes in electromagnetic fields. Common types include laser rangefinders, which emit a light beam to determine distance based on the return time of the reflection, and ultrasonic sensors, which transmit high-frequency sound waves and measure the echo. These devices are prevalent in automation, robotics, and safety systems, providing accurate and noncontact measurement for applications ranging from object detection to mapping environments.

Another notable distance measurement technology is LiDAR, which utilizes laser pulses to generate precise, high-resolution maps of surroundings. By emitting rapid bursts of laser light and measuring the time it takes for each pulse to reflect back from objects, LiDAR systems calculate distances with exceptional accuracy. This approach enables the creation of detailed 3D representations of environments, making LiDAR invaluable for applications such as autonomous vehicles, topographic mapping, and industrial automation. Unlike ultrasonic sensors, LiDAR can operate effectively over long ranges and delivers superior resolution, particularly in detecting small or distant features.

4.3.2 Proximity Sensors

Proximity sensors are designed to detect the presence of objects within a specified range without making physical contact. Unlike traditional contact switches, proximity sensors operate by emitting electromagnetic fields, light, or sound waves and monitoring the interaction with nearby objects. The most common types include inductive, capacitive, and infrared proximity sensors.

Inductive proximity sensors are particularly effective for detecting metallic objects. They generate an oscillating magnetic field and sense changes when a metal target enters the field, making them ideal for automation and machinery monitoring. Capacitive proximity sensors, on the other hand, can detect both metallic and nonmetallic objects by measuring changes in the capacitance caused by the presence of a target. These are frequently used for level detection in tanks or for sensing plastic, glass, or other materials.

Infrared proximity sensors use an infrared light source and detector to sense the reflection from nearby objects, commonly found in consumer electronics, touchless switches, and mobile devices. Ultrasonic proximity sensors, employing high-frequency sound waves, are well suited for applications requiring reliable detection over larger distances or through opaque materials.

Proximity sensors excel in environments where noncontact detection is crucial for safety, hygiene, or system longevity and are widely implemented in automated assembly lines, robotics, security systems, and touchless user interfaces. Their reliability, durability, and versatility make them indispensable components in modern sensing technology.

4.3.3 Position Sensors

Displacement, velocity, and acceleration are intrinsically linked by the fundamental principles of mechanics. Consequently, a sensor that measures any one of these quantities can be adapted, through electronic signal processing, to provide information about the other mechanical variables. Integrating velocity or acceleration signals by electronic or computational means produces displacement data, while differentiating displacement information through electronic circuits or digital algorithms results in velocity or acceleration values. The choice of sensor depends on the required measurement range, frequency response, sensitivity, and environmental conditions. By combining displacement, velocity, and acceleration sensors, complex dynamic behaviors can be accurately monitored and controlled within a wide variety of engineering systems.

4.3.3.1 Potentiometric Position Sensors

A potentiometer is an electrical component consisting of a resistor with a movable third contact, known as a wiper, that enables voltage adjustment based on the slider's position along the resistor.

Potentiometers are used in a wide range of electrical and electronic applications, available in various forms, sizes, and power ratings. For high-power applications, wire-wound potentiometers, also called rheostats, serve as adjustable voltage controls. In electronic devices, low-power carbon film potentiometers are utilized to adjust input signal levels or preset voltages on circuit boards. Most potentiometers have linear characteristics, while logarithmic potentiometers are commonly applied for volume control in audio equipment.

The basic operation of all potentiometer types is the same. One terminal serves as the common connection for both the input and output voltages. Adjusting the wiper's position on the fixed resistance element produces a portion of the input voltage between the wiper and the common terminal. In linear potentiometers, the output voltage changes proportionally with the wiper's position on the resistive element.

When the potentiometer wiper is properly linked to a mechanical system, any movement of the linked element results in the displacement of the wiper, thereby altering the output voltage. This change in voltage enables direct measurement of position or positional changes. As such, the potentiometer functions as an effective motion sensor for quantifying linear or angular displacement. A schematic representation of a potentiometer circuit is provided in Figure 4.5.

The output voltage is determined by the position of the wiper (C), which may be set at any point along the resistance between terminals A and B. For the general case, the output voltage is given by the voltage divider equation below:

$$V_{out} = \frac{R_{BC}}{R_{AB} + R_{BC}} V_{in} = \frac{R_{BC}}{R_{AB}} V_{in} = \frac{\ell_{BC}}{\ell_{AB}} V_{in} \tag{4.10}$$

In linear potentiometers, the resistance between the moving terminal and the common varies proportionally with rectilinear displacement. For rotary potentiometers, the resistive element typically spans an arc of 270°. As the wiper rotates, the output voltage changes in direct relation to the angular position within the total sweep. Some linear potentiometers use a worm gear-and-wheel assembly to convert linear motion into rotation, enabling precise voltage adjustment with multiple turns.

Potentiometric position sensors are widely used in mechatronic systems. This sensor detects the position based on the voltage division across the terminals of a linear or rotary potentiometer. As the wiper changes position, the resistance between the wiper and the ends of the resistive track varies proportionally, producing a measurable voltage output corresponding to the displacement.

Potentiometric position sensors are valued for their simplicity, reliability, and direct analog signal output. They offer high resolution over a wide measurement range and can be easily integrated into control systems for real-time feedback.

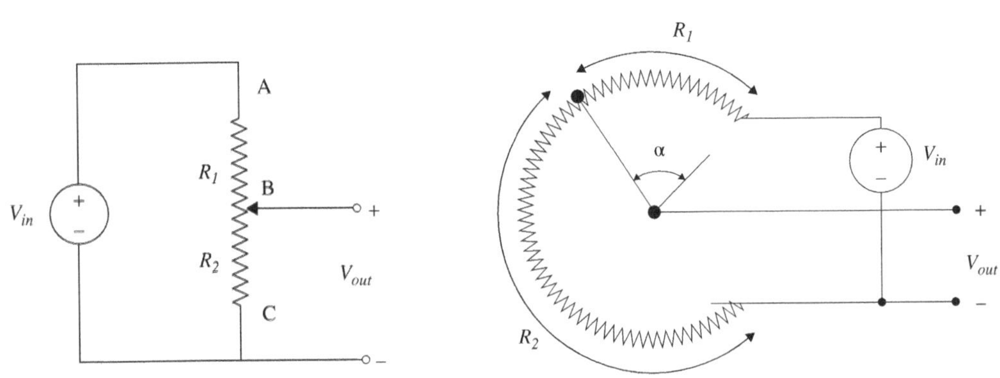

Figure 4.5 Circuit diagram of a linear and rotational potentiometer

Potentiometric sensors are a cost-effective choice in many applications such as industrial automation, robotics, and vehicle throttle control, where precise position feedback is crucial. However, a primary drawback of potentiometric position sensors is their susceptibility to dust accumulation and gradual wear over time.

4.3.3.2 Linear Variable Differential Transformer (LVDT)

The LVDT is a highly precise electromechanical transducer used for measuring linear displacement. Structurally, the LVDT consists of a cylindrical coil assembly and a movable soft iron core. The coil assembly includes one primary winding positioned between two identically wound secondary windings, which are series connected in opposite polarity as illustrated in Figure 4.6. The entire assembly is typically encapsulated for robustness and may be housed in a nonmagnetic casing to shield it from extraneous electromagnetic interference.

During operation, an AC is applied to the primary winding, generating an alternating magnetic field. The soft iron core, which can move linearly within the bore of the coil assembly, serves to channel the magnetic flux between the primary and secondary windings. When the core is centered, the magnetic coupling to both secondary windings is equal, resulting in zero differential output voltage between them due to their opposite phase connection.

As the core shifts away from the central position, the magnetic flux linkage becomes greater with one secondary winding and less with the other. This imbalance induces a differential voltage output that is directly proportional to the magnitude and direction of the core's displacement. The phase of the output signal indicates the direction of movement, while the amplitude reflects the extent of displacement. This principle allows LVDTs to provide highly accurate, frictionless, and continuous measurement of linear position.

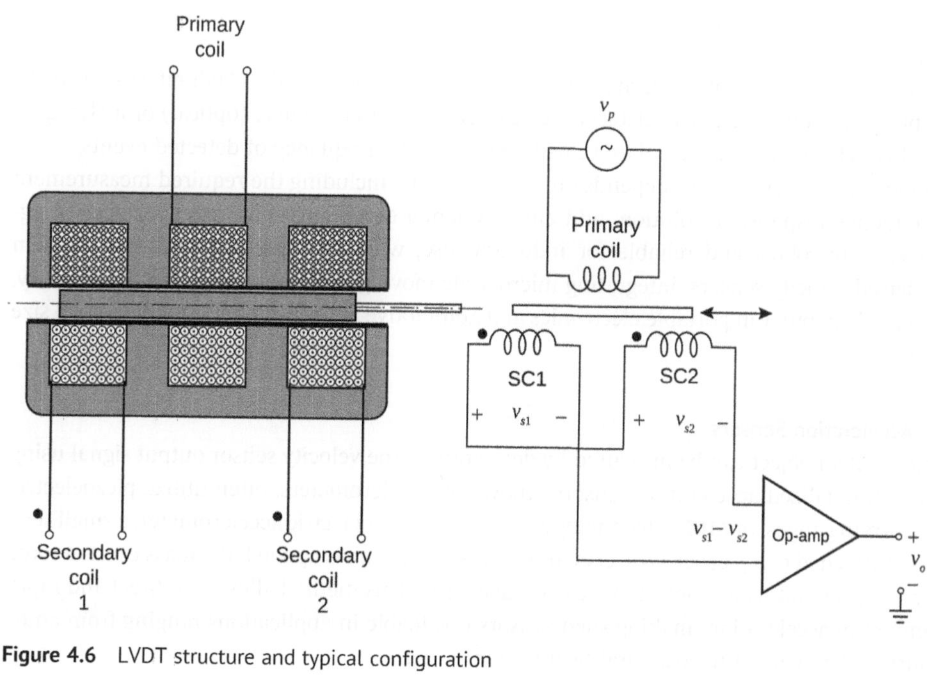

Figure 4.6 LVDT structure and typical configuration

The noncontact nature of the LVDT's operation ensures minimal wear, making it suitable for long-term applications in harsh environments. LVDTs are widely used in industrial automation, aerospace, and research instrumentation where precision and durability are paramount.

4.3.4 Motion Sensors

4.3.4.1 Velocity Sensors

Velocity sensors are essential elements for tracking movement and controlling speed in mechanical systems. These devices are engineered to provide a direct measurement of the speed at which an object or component is in motion—an essential parameter for feedback and control in numerous engineering applications.

Velocity sensors typically operate on the principle of electromagnetic induction. The fundamental operating principle of most velocity sensors is electromagnetic induction. In its classic form, a velocity transducer consists of a permanent magnet and a coil of wire. As the magnet moves relative to the coil (or vice versa), a voltage is induced across the coil that is proportional to the rate of change of magnetic flux linkage—directly correlating to the velocity of the moving element, as described by Faraday's law. This means that faster movement results in greater induced voltage, allowing for continuous- and real-time velocity monitoring.

One common type is the linear velocity transducer, often used in vibration analysis for rotating machinery, turbines, or structural health monitoring. Here, the sensor is affixed to a vibrating body, and as motion occurs, the relative shift between the internal magnet and the surrounding coil generates an AC voltage whose amplitude corresponds to the velocity of vibration. This is particularly valuable in predictive maintenance, where vibration velocity is a key indicator of mechanical health.

Rotational velocity sensors, also known as tachogenerators, operate on similar inductive principles but are typically designed to measure angular speed. As a shaft rotates, the changing magnetic field produces a voltage output proportional to the shaft's rotational velocity. These devices are commonly found in electric motor control, automotive speedometers, and industrial process automation.

Some velocity sensors employ alternative technologies, such as optical or Hall-effect sensors. In these implementations, the device detects the passage of a patterned disk (optical) or a changing magnetic field (Hall-effect) to calculate velocity based on the frequency of detected events.

The choice of a velocity sensor depends on several factors, including the required measurement range, frequency response, resolution, and environmental conditions. For instance, electromagnetic sensors are robust and reliable for industrial use, while microelectromechanical system (MEMS)-based velocity sensors, integrating micro-scale moving elements and electronic circuitry, are increasingly popular in portable electronics and automotive applications due to their small size and versatility.

4.3.4.2 Acceleration Sensors

Acceleration of an object can be measured by derivation of the velocity sensor output signal using an op-amp. Specialized acceleration sensors, known as accelerometers, often utilize piezoelectric materials, strain gauges, or MEMS technology. In a typical piezoelectric accelerometer, a small seismic mass is attached to a piezoelectric crystal. When acceleration is applied, the mass exerts a force on the crystal, generating a proportional electrical charge. This method allows for direct and rapid measurement of acceleration, making such sensors invaluable in applications ranging from automotive airbag deployment to structural health monitoring.

4.3.4.3 Thermistor

A thermistor is another type of resistive sensor, notable for its sensitivity to temperature. Constructed from semiconductor materials, a thermistor's resistance changes significantly with variations in temperature, making it an ideal component for temperature sensing and control applications. There are two main types: negative temperature coefficient (NTC) thermistors, where resistance decreases as temperature rises, and positive temperature coefficient (PTC) thermistors, in which resistance increases with temperature. Thermistors are commonly used in electronic circuits for temperature compensation, thermal protection, and precise temperature measurement due to their rapid and predictable response to thermal changes.

Thermistors are passive elements; therefore, they must be used in circuit configurations that include a voltage source to produce a temperature-dependent signal. A simple arrangement is to connect the thermistor in series with a known resistor to form a voltage divider supplied by a voltage source. A thermistor's resistance changes significantly with temperature, and using a simple voltage divider generates a signal that has low amplitude, is susceptible to noise, and is influenced by the supply voltage.

When the thermistor is placed in a Wheatstone bridge, the circuit magnifies the effect of even small resistance changes because the bridge output is proportional to the difference in potential between the two midpoints, allowing for the detection of very small temperature-induced resistance changes with high resolution. In addition, the signal amplitude is independent of the supply voltage.

Thermistors often have highly nonlinear resistance temperature characteristics. By carefully choosing the values of the other resistors in the bridge, one can linearize the response over a desired temperature range, making the relationship between temperature and output voltage more predictable and simplifying calibration and signal processing.

A Wheatstone bridge configured with a thermistor in one arm and fixed or temperature-compensating resistors in the other arms produces an output voltage as changes in temperature affect the thermistor's resistance. This voltage changes with temperature and allows for monitoring or control. The bridge generates a differential voltage that reflects the change in resistance of the thermistor. This differential signal can be used as input for amplifiers, ADCs, or other signal-conditioning circuits, facilitating further amplification and processing while providing resistance to common-mode noise.

By including a second thermistor (or another temperature-dependent resistor) in the opposing arm of the bridge, the system can be designed to compensate for ambient temperature fluctuations or drift, improving measurement stability. This is especially useful in environments where temperature gradients or thermal noise could otherwise distort the measurement.

4.3.4.4 Light-Dependent Resistor

An LDR, also known as a photoresistor, is a type of resistive sensor whose resistance changes as a function of the intensity of incident light. Typically made from semiconductor materials such as cadmium sulfide, an LDR exhibits high resistance in darkness and much lower resistance when exposed to light.

The operating principle of an LDR is based on photoconductivity. When photons of sufficient energy strike the surface of the sensor, they excite electrons into the conduction band, thereby increasing the number of charge carriers available. This results in a marked decrease in resistance as light levels increase. In practical terms, the LDR acts as a variable resistor that responds to ambient light intensity, making it invaluable in light-sensing circuits.

Common applications for LDRs include automatic lighting controls, such as street lamps that turn on at dusk and off at dawn, photographic exposure meters, alarm systems, and devices that require ambient light detection for adaptive responses. Their simplicity, reliability, and cost-effectiveness have led to widespread use in both everyday electronics and specialized industrial controls.

4.3.5 Force and Torque Sensors

Force and torque sensors are essential components in mechatronic systems, enabling accurate detection of mechanical interaction within a vast array of devices, from industrial equipment to biomedical instrumentation.

Force sensors are designed to quantify the amount of push or pull exerted on an object. They commonly employ principles such as strain-induced resistance change (as in strain gage-based sensors), piezoelectric effects, or variations in capacitance. When a force is applied, these sensors convert the mechanical deformation or displacement into an electrical signal that can be measured and analyzed.

Torque sensors, on the other hand, measure the twisting or rotational force applied to a shaft or other mechanical element. Like force sensors, they commonly utilize strain gages arranged in specific geometries to detect minute changes in resistance resulting from torsional strain. Piezoelectric and magnetoelastic principles are also sometimes employed, particularly in high-frequency or dynamic applications.

Typical force or torque sensors feature:

- **Sensing Element**: This is the component that physically responds to the force or torque, deforming slightly under load.
- **Transduction Mechanism**: Converts the physical deformation into an electrical signal, often through a resistive, capacitive, or piezoelectric effect.
- **Signal Conditioning**: The raw electrical output is typically very small and may require amplification and filtering for accurate measurement.

In practice, force and torque sensors are often integrated into bridge circuits for enhanced sensitivity and noise immunity, much like strain gages. Calibration and compensation techniques are crucial to ensure accuracy and to minimize the effects of temperature, hysteresis, and long-term drift.

Examples of applications include:

- Automated assembly systems, where precise force feedback enhances product quality and process reliability.
- Robotics, enabling delicate manipulation and tactile feedback.
- Medical devices, such as prosthetics and surgical instruments, where feedback on applied forces is critical for safety and performance.
- Vehicle testing and structural analysis, measuring dynamic loads and stresses in components.

Force and torque sensors bridge the gap between the mechanical and electrical realms, enabling real-time monitoring and intelligent control across countless fields of science and engineering.

4.3.6 Strain Gages

Strain is a fundamental concept in the mechanics of materials, describing how a material deforms in response to applied forces. It quantifies the relative change in shape or size of a body as it undergoes loading and is crucial for analyzing the behavior of structures, components, and materials under stress.

Strain is defined as the measure of deformation representing the displacement between particles in a material body relative to a reference length. It is a dimensionless quantity, typically expressed as a ratio or percentage.

Engineering strain (ε) is the ratio of the change in length ($\Delta\ell$) to the original length (ℓ_0) of a material under an applied force.

$$\varepsilon = \frac{\Delta\ell}{\ell_0} \tag{4.11}$$

This formula applies for axial (linear) deformation, such as when a rod is stretched or compressed. Strain provides insight into the deformation behavior of materials. A positive strain indicates elongation (tension), while a negative strain indicates contraction (compression). Shear strain describes the distortion of angles within the material.

Strain measurement is essential in structural engineering, materials science, aerospace, civil engineering, mechanical engineering, and many other fields. By monitoring strain, engineers can assess safety, predict failure, and design reliable structures.

Strain gage is a sensor commonly used to measure small deformations in a structure by detecting changes in electrical resistance as the object is strained. Fundamentally, it consists of a thin conductor arranged in a serpentine pattern and bonded securely to the surface under investigation. When the surface experiences stretching (tensile strain) or compression, the length and cross-sectional area of the conductor change, producing a corresponding change in its resistance.

Strain gages may be used in a voltage divider configuration as shown in Figure 4.7 (a) to generate signal proportional to the strain. However, the resistance change is typically quite small and often requires amplification or inclusion in a Wheatstone bridge circuit for precise measurements.

When a strain gage replaces one branch of a Wheatstone bridge as shown in Figure 4.7 (b), the bridge is balanced in the absence of strain on the gage. A strain disrupts the balance and results in an input signal to the amplifier proportional to the strain. It is obviously easier, more sensitive, and more accurate compared to the voltage divider strain gage circuit.

The half-bridge configuration depicted in Figure 4.7 (c) presents an advancement over the basic Wheatstone bridge. In this design, one branch consists of two strain gauges. Changes in output are influenced by variations in the resistance of both strain gages increasing and the other decreasing as the assembly moves in a specific direction. The fixed resistor arm provides a stable reference voltage, which is compared against the variable output of the strain gauge branch. The resulting difference between these two signals enhances the overall signal amplitude delivered to the amplifier.

The final improvement of the Wheatstone bridge circuits is to replace all fixed resistors as Figure 4.7 (d) shows with two tension gages and two compression gages. They are arranged such that when the linear assembly moves, the output voltage from one arm increases while the other arm decreases.

Strain gages are available in a wide range of specifications to suit different measurement scenarios. Key factors—gauge factor (GF), resistance, size, temperature range, fatigue life, construction material, and environmental resilience—should be carefully considered during selection. By understanding these specifications, engineers and technicians can ensure precise, reliable measurement of strain, guaranteeing both the efficacy of their designs and the safety of the structures they monitor.

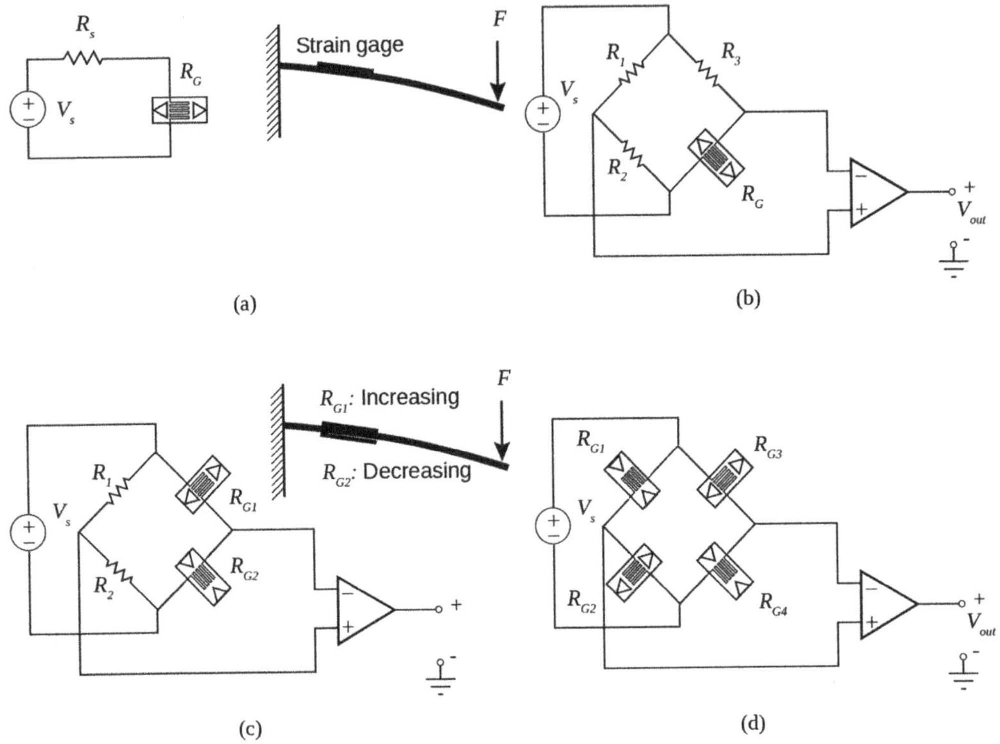

Figure 4.7 Strain gage circuit configurations. (a) Single strain gage; (b) One strain gage in resistive bridge; (c) Two strain gages in a bridge pasted on opposite sides of the beam, one is tension, the other compression gage; (d) Full bridge with four strain gages. RG1 and RG4 tension gages, RG2 and RG3 compression gages

Strain gages are specified by their nominal electrical resistance, which is essential for bridge circuits and signal amplification. Typical values 120, 350, and 1,000 Ω are common. Other resistances are available for special applications. Higher resistance gages draw less current, which reduces self-heating and is beneficial for battery-powered or low-power applications.

The GF is a measure of the sensitivity of the strain gage. It is defined as the ratio of the relative change in electrical resistance to the mechanical strain.

$$GF = \frac{\Delta R/R_o}{\Delta\ell/\ell_0}; \quad \varepsilon = \frac{\Delta\ell}{\ell_0} \Longrightarrow \Delta R = (R_o GF)\varepsilon \tag{4.12}$$

Accordingly, changes in gage resistance are directly proportional to changes in length. When a strain gage is attached to a material undergoing linear deformation, the resulting change in resistance provides an accurate measure of the applied force. For metallic strain gages, the GF is usually around 2.0. For semiconductor strain gages, it may range from 50 to 200 due to the piezoresistive effect. A higher GF means greater sensitivity to strain but may also result in increased susceptibility to temperature effects and noise (Dally et al., 1984).

Strain gages are invaluable in engineering applications where the detection of small deflections, stresses, or mechanical forces is required, such as in structural testing, civil engineering, aerospace, and mechanical systems. They can be applied to beams, bridges, or even machine components to monitor and analyze load-induced deformations, thus providing real-time feedback on the operational integrity and safety of structures.

Example 4.3 In the Wheatstone bridge shown in Figure 4.4 (a), all resistors are 120 Ω. The load resistance can be considered infinite.

1) Derive the output-to-supply voltage ratio as a function of the variation in R_3.
2) Calculate the output voltage for a 10-V supply when R_3 increases by 1%.

Solution

a) Equation 4.6 yields

$$\frac{\Delta V_{out}}{V_s} = \frac{R_3 + \Delta R_3}{R_1 + R_3 + \Delta R_3} - \frac{R_4}{R_2 + R_4}$$

a) Given $V_s = 10$ V and change in $R_3 = 1\%$

$$\frac{\Delta R_3}{R_3} = 0.01 \Longrightarrow \Delta R_1 = 0.01 \times 120 = 1.2\,\Omega$$

$$\frac{\Delta V_o}{V_s} = \frac{120 + 1.2}{120 + 1.2 + 120} - \frac{120}{120 + 120} = 0.0025$$

The output voltage is

$$\Delta V_o = 0.002488 \times 10 = 0.02488\text{ V} = 24.88\text{ mV} \therefore$$

4.4 Encoders

An encoder translates mechanical motion into electrical signals. Rotary encoders track angular movement for motors and servos, while linear encoders measure straight-line motion in actuators and CNC machines. An encoder provides real-time data on position, speed, and direction for feedback control systems like robotics, CNC machines, and motor drives.

Incremental encoders have a disk with evenly spaced slots or patterns and two output channels (A and B) placed 90° out of phase. As the shaft rotates, the disk interrupts light or magnetic fields, generating square-wave pulses. Relative position is detected by counting the pulses. Phase difference between A and B determines the direction. Incremental encoders lose position info when power is turned off and needs homing.

In absolute encoders, a disk with multiple concentric tracks, each representing a bit in a binary or Gray code, rotates with the shaft. Each shaft position corresponds to a unique digital code. Therefore, it can retain position even after power loss.

An encoder can detect motion by physical contact, light beam, or magnetic strips. Simple mechanical encoders with physical contact on conductive tracks are common in low-cost consumer electronics.

In an optical encoder, a light source such as an LED emits a light beam that passes through a slotted disk and triggers a photodiode. Optical encoders are fast and high resolution. They are common in robotics and CNC machines.

Magnetic encoders operate by detecting variations in magnetic fields. Their robust construction and durability make them particularly well suited for demanding environments, such as those encountered in automotive and outdoor applications.

4.5 Temperature Sensors

Temperature sensors are devices designed to detect and measure temperature changes in their environment, converting thermal energy into a readable electrical signal. They play a critical role in countless scientific, industrial, and everyday applications, providing real-time data for control, monitoring, and safety systems.

There are several types of temperature sensors, each with unique characteristics and suited to specific uses:

- **Thermistors**: These are temperature-sensitive resistors whose resistance varies significantly with temperature. They are widely used for their high sensitivity, fast response, and cost-effectiveness, especially in HVAC systems, digital thermometers, and battery protection circuits.
- **Resistance Temperature Detectors (RTDs)**: RTDs utilize materials like platinum that exhibit a predictable change in resistance with temperature. Renowned for their accuracy and stability, RTDs are commonly found in laboratory and industrial applications where precise temperature measurements are required.
- **Thermocouples**: These sensors consist of two dissimilar metal wires joined at one end. When exposed to a temperature gradient, they generate a voltage proportional to the temperature difference. Thermocouples operate over a wide range of temperatures and are robust, making them suitable for harsh environments such as engines, furnaces, and industrial processes.
- **Semiconductor Temperature Sensors**: IC temperature sensors output either analog voltage or digital signals corresponding to temperature. They are compact, linear, and well suited for electronic devices and embedded systems.

The selection of a temperature sensor depends on several factors, including required accuracy, response time, temperature range, environmental conditions, and budget constraints. Proper installation and calibration are essential to ensure reliable operation and minimize errors due to factors such as thermal lag, self-heating, or electrical interference.

Temperature sensors are indispensable in fields ranging from weather forecasting and climate control to process automation and biomedical monitoring. By providing precise and continuous temperature data, they help maintain system performance, enhance safety, and optimize energy usage.

4.6 Light Sensors

Light sensors, particularly LDRs and photodiodes, are crucial components in modern technology, enabling systems to detect and respond to varying light levels. They are used in a wide range of practical applications. For instance, light sensors are integral to automatic lighting control, where they help optimize energy usage by activating or deactivating lights based on ambient illumination. In consumer electronics, they enable adaptive brightness settings in smartphones, tablets, and TVs, ensuring comfortable viewing under different lighting conditions and enhancing battery life. Security and alarm systems also employ light sensors to detect unexpected changes in light, which may indicate unauthorized entry or movement. In industrial automation, light sensors contribute to quality control by monitoring the presence or uniformity of materials through changes in light transmission or reflection. Their application even extends to agriculture, where they assist in greenhouse automation by regulating artificial lighting, thereby optimizing plant growth.

- Automatic street lighting, where the circuit switches lamps on or off depending on ambient light levels.
- Light meters in photographic equipment.
- Brightness control in display screens and electronic billboards.
- Alarm systems that detect changes in lighting patterns.

The response time of an LDR to changes in light can vary—from milliseconds to seconds—depending on the material and construction, making them more suitable for applications where rapid detection is not critical. Temperature may also influence the resistance of an LDR, so compensation or calibration is sometimes necessary for precise measurements.

LDRs are typically incorporated into voltage divider circuits, where the changing resistance under different lighting conditions generates a varying voltage output. This output can then be measured directly or fed into amplifiers, comparators, or microcontrollers for further signal processing and control.

4.7 Integration of Sensors with Electric Circuits

Integrating sensors with electric circuits is a foundational step in designing systems that translate physical phenomena—such as light, temperature, or pressure—into usable electrical signals for measurement and control. The nature of this integration depends on both the characteristics of the sensor and the requirements of the application, including sensitivity, response speed, signal level, and immunity to environmental influences.

For resistive sensors such as LDRs, the most elementary approach involves configuring the sensor as part of a voltage divider circuit. Here, the changing resistance of the sensor due to physical stimulus results in a corresponding change in voltage across the sensor or its paired resistor. This voltage serves as the electrical representation of the sensed quantity and can be monitored with an ADC or further conditioned with amplifiers and comparators for digital processing.

Optoelectronic and other sensors often require additional signal conditioning to ensure that the output is within a suitable range and immune to noise. op-amps may be used to buffer, amplify, or filter the signal, while comparators can generate discrete logic-level outputs for digital control systems. For applications where noise immunity is critical, differential measurement techniques are favored.

A prominent method for differential signal measurement is Wheatstone bridge, especially useful with resistive sensors such as strain gauges. In this configuration, the sensor is incorporated into one or more arms of the bridge, and small changes in resistance lead to measurable voltage differences across the bridge output. When precisely balanced, the bridge nulls out baseline voltages, making it highly sensitive to minute changes induced by environmental factors or physical stress. This makes Wheatstone bridges ideal for precision applications requiring high resolution and accuracy.

The integration process may also include temperature compensation, calibration routines, and the use of reference elements to enhance stability and repeatability. As technology progresses, more sophisticated circuits—such as programmable gain amplifiers, integrated analog front-ends, and digital interfaces—are increasingly employed to interface sensors directly with microcontrollers and digital processing units, further enhancing the versatility and intelligence of modern sensing systems.

4.8 Key Features of Sensor Characteristics

4.8.1 Linearity

A sensor must produce reliable output that reflects the measured quantity accurately. Major sources of error occur due to effects such as nonlinear operation, saturation, hysteresis, delay, offset, dead band.

A sensor is linear if the output signal produced by the sensor has proportionality and additivity properties. Let be the measured quantity and be the corresponding sensor output signal.

Proportionality: $f(ax) = af(x)$

Additivity: $f(x_1) + f(x_2) = f(x_1 + x_2)$

Nonlinear characteristics of magnetic core or semiconductor elements used in the measurement and signal conditioning circuits may cause nonlinearity and hysteresis.

4.8.2 Loading Errors

Loading errors in measurement systems are errors introduced to the measurement of the variable due to the sensor and its associated signal processing circuit. There are two types of loading errors:

1) Mechanical loading error: Error due to the introduction of the sensor in the physical environment.
2) Electrical loading error: Error due to the current drawn by the measurement devices.

4.8.3 Static and Dynamic Response

The static input–output relationship of a sensor can be identified by changing the physical variable in known increments, then waiting long enough for the sensor output to reach its steady-state value before recording it, and repeating this process until the whole measurement range is covered. The result can be plotted to represent the static input–output characteristic of the particular sensor. If these nonideal input–output behaviors are repeatable, then a digital signal processor can incorporate the information into the sensor signal processing algorithm in order to extract the correct measurement despite the nonlinear behavior. Repeatability of the nonlinearities is the key requirement for accurate signal processing of the sensor signals. If the nonlinearities are known to be repeatable, then they can be compensated to obtain an accurate measurement.

4.8.4 Sensitivity

Sensitivity refers to the measure of how much the sensor's output signal changes in response to a change in the measured physical quantity. Mathematically, sensitivity is defined as the ratio of the change in output signal to the corresponding change in the input variable. High sensitivity implies that even a small variation in the input produces a significant output response, making the sensor particularly responsive and useful for detecting subtle changes in the measured environment.

4.8.5 Repeatability

Repeatability describes a sensor's ability to produce the same output when the identical input is applied multiple times under the same conditions. In other words, if a physical quantity is measured repeatedly by the same sensor, within the same environment and setup, a highly repeatable sensor will yield nearly identical results each time. High repeatability is essential for ensuring that measurements are consistent and reliable, especially in applications where precision is critical. Lack of repeatability leads to inconsistent data, making it difficult to trust the measurements or to calibrate the system accurately.

4.8.6 Noise and Interference

Sensors, being exposed to their environment and often connected to complex electronic circuits, are susceptible to noise and interference that can degrade the accuracy and quality of their measurements. Noise typically refers to random, unpredictable electrical signals that are superimposed on the desired sensor output, while interference refers to unwanted signals from external sources. Common sources of noise include thermal noise generated by electronic components, shot noise from current flow in semiconductors, and flicker noise that arises at low frequencies in electronic devices. Interference can be introduced by electromagnetic fields from nearby electrical equipment, radio-frequency signals, or even power line fluctuations. Mechanical vibrations, cross talk from adjacent circuits, and improper grounding can further contribute to measurement errors. Effectively mitigating noise and interference is essential to ensure the integrity and reliability of sensor data.

4.9 Calibration and Compensation

In general, a sensor needs to be calibrated to customize it for an application. In any control system applications such as an automated machine in an assembly line or a mobile equipment, which may involve hundreds of sensors, one of the first steps in implementing a control system is the sensor calibration. If the sensor exhibits drift in time, then it must also be calibrated periodically. Sensor calibration refers to adjustments in the sensor amplifier to compensate for the above variations so that the input (measured physical variable) and output (sensor output signal) relationship stays the same. The sensor calibration process involves adjustments to compensate for variations in gain, offset, saturation, hysteresis, dead band, and drift-in time.

Basic calibration techniques for sensors typically involve comparing the sensor's output with a known reference standard and making necessary adjustments to align the sensor readings with the true values. This process often starts with zero calibration, where the sensor is set to produce a baseline output when the input is at a known zero point. Next, span or gain calibration is performed by applying one or more known input values across the measurement range and adjusting the sensor response to ensure accuracy at these points. In some cases, a multi-point calibration may be used, where several reference values are measured to correct for nonlinearity or drift across the entire range. Regular calibration is crucial to compensate for changes due to aging, environmental conditions, or component variations, thereby maintaining measurement reliability and consistency.

Calibration may also include compensating for offset errors, adjusting for hysteresis or dead band, and correcting for drift over time. These adjustments can be performed manually using calibration screws or electronically through software algorithms in digital systems. Documentation of calibration settings and schedules helps ensure traceability and ongoing accuracy, especially in applications where precision is critical.

Another important consideration in sensor performance is the impact of environmental factors such as temperature, humidity, and pressure. Sensors may exhibit changes in output due to variations in ambient temperature or environmental conditions, which can introduce measurement errors or drift. To maintain reliable operation, it is often necessary to implement compensation techniques or select sensors with built-in environmental stability features. Proper housing and isolation can also help protect sensors from adverse external influences, ensuring consistent and accurate reading over time.

4.10 Design Examples

Example 4.4 Bathroom Scale
Design a digital bathroom scale with the following specifications.
Weigh range: 10–250 lb; Powered by 9 V battery; 4-digit display.
Available components:

Strain Gages

- **Nominal Resistance**: Common values include 120 Ω.
- **Resistance Rolerance**: Typically ±0.3% to ±0.5%
- **Gage Factor (GF)**: 2.0
- **Maximum Bridge Voltage**: 10 V

Digital Voltmeter Display

- 4-Digit red LED Display Panel
- **Measurement Range**: DC 0–500 mV
- **Supply Voltage**: DC 5 V/150 mA

Solution

Design Options:

1) One strain gage connected in series with a 120 Ω resistance to form a voltage divider.
2) One strain gage placed on one branch of a Wheatstone bridge, the other three resistors selected equal for the other branches.
3) A load cell with two strain gauges affixed to each side of a beam, the gauges connected in series to constitute one side of a Wheatstone bridge.
4) Two identical load cells as described in option 3 to form a full Wheatstone bridge.
5) Four identical load cells will be constructed as described in option 4, each equipped with two strain gauges positioned on opposite sides of the flexible beam, one on the top and one on the bottom.

Selection of the Design Concept:
Option 1 is the simplest and lowest cost. Wheatstone bridge options offer increasingly better sensitivity. For the maximum scale sensitivity, option 5 is selected. In addition, a potentiometer of 100 Ω will be included in the left side of the bridge for zero setting configured as shown in Figure 4.8.
Design Calculations:
All sensors have the same initial resistance of $R_o = 120\ \Omega$ and $GF = 20$. A change in the measured parameter results in an increase in ΔR in the value of two sensors and a decrease in the same amount in the value of the other two sensors. As Figure 4.8 shows, two load cells are arranged

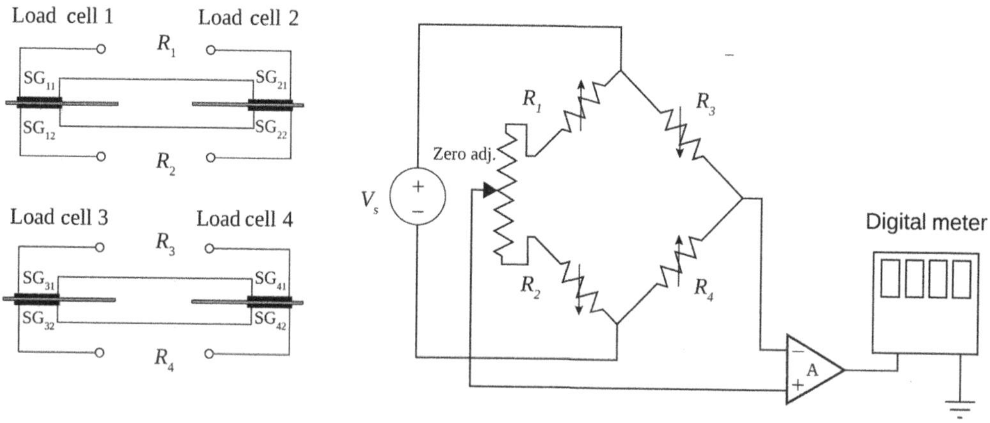

Figure 4.8 Wheatstone bridge configuration described in Example 4.4

so that the tension gauges are connected in series to form one branch of the bridge, while two compression gauges are connected in series to form the other branch.

The load cells will be placed above the corner.

The scale will be enclosed in a supporting frame, with each leg positioned to reliably bear the maximum load of 250 lb. When the weight is applied to the scale, the force is transmitted through the legs to the load cells, generating strain in the affixed gauges. This strain results in a change in resistances, which are detected and amplified by the Wheatstone bridge circuit, enabling precise measurement of the total force exerted on the legs. Since two upper strain gages and two lower strain gages are each connected in series, the combined resistance equals twice the resistance of a single gage.

$$R_1 = R_4 = 2(120 + \Delta R)$$
$$R_2 = R_3 = 2(120 - \Delta R)$$

The following equation gives the bridge output voltage:

$$v_o = V_s\left(\frac{R_4}{R_3 + R_4} - \frac{R_2}{R_1 + R_2}\right)$$

Substituting the values in this equation yields

$$v_o = V_s\left(\frac{2(120 + \Delta R)}{2(120 - \Delta R) + 2(120 + \Delta R)} - \frac{2(120 - \Delta R)}{2(120 + \Delta R) + 2(120 - \Delta R)}\right)$$

Simplified expression of the output voltage is

$$v_o = V_s\left(\frac{240 + 2\Delta R}{480} - \frac{240 - 2\Delta R}{480}\right) = V_s\frac{\Delta R}{120}$$

The GF is defined as the ratio between the change in resistance and the change in length. Therefore,

$$GF = \frac{\Delta R/R_o}{\Delta L/L_o}; \varepsilon = \frac{\Delta L}{L_o} \Longrightarrow \Delta R = (R_o GF)\varepsilon$$

Assuming that the beam deformation is proportional to the applied force, the output voltage of the Wheatstone bridge is also proportional to the force.

$$v_o = V_s(GF)\varepsilon = V_s KF$$

The voltmeter will be configured to operate within the 0–500 mV range, and the amplifier gain will be adjusted such that an applied force of 250 lb generates a 250 mV signal across the meter terminals.

4.11 Summary

This chapter provides an overview of sensors used in mechatronic and robotic systems. It covers sensor types, characteristics, practical examples, and integration with electronic circuits. The focus is on understanding sensor principles, their applications, and the factors influencing their performance.

- Sensors detect changes in physical quantities like temperature, light, pressure, or motion and transduce these changes into electrical signals, which are then processed for interpretation. Sensors can be analog, providing continuous output, or digital, providing discrete signals. Advanced sensors incorporate microprocessors for complex measurements and are used in systems such as GPS and LIDAR in autonomous vehicles and UAVs.
- Resistive sensors operate by changes in resistance corresponding to physical changes. A potentiometer, which includes a sliding wiper over a resistor, can measure linear or rotary displacement by outputting a voltage proportional to the wiper position. This principle is used in linear and rotary potentiometers. The Wheatstone bridge circuit enhances sensitivity by balancing voltage dividers and detecting small resistance changes, commonly used with strain gages. Half-bridge and full-bridge configurations improve measurement accuracy by compensating for tension and compression effects.
- Strain gages measure minute deformations by detecting resistance changes in a thin conductor bonded to a surface. They are widely used in structural testing and monitoring mechanical forces. Because resistance changes are small, Wheatstone bridges and amplification are employed for precision.
- Thermistors are temperature-sensitive resistors with resistance varying significantly with temperature. NTC thermistors decrease resistance with increasing temperature, while PTC thermistors increase resistance. They are used in temperature sensing and control applications.
- LDRs change resistance based on incident light intensity, exhibiting high resistance in darkness and lower resistance in light. They operate via photoconductivity and are used in automatic lighting, exposure meters, and alarm systems.
- These sensors exploit electromagnetic induction principles to measure displacement, velocity, or electrical quantities without contact. Examples include inductive voltage dividers and LVDTs, which use a movable ferromagnetic core to produce voltage changes proportional to displacement. They offer robustness and immunity to environmental interference.
- Capacitive sensors measure changes in capacitance caused by variations in plate area overlap, distance, or dielectric properties. They convert displacement into capacitance changes, which can be detected using AC bridge circuits to produce measurable signals.

- Optoelectronic sensors detect light changes and convert optical signals into electrical outputs. They include photodiodes, phototransistors, optical encoders, fiber-optic sensors, and color/spectral sensors. These sensors are valued for immunity to electromagnetic interference, fast response, and high sensitivity and find applications in object detection, position sensing, environmental monitoring, and industrial automation.
- Distance sensors measure separation between objects using principles like reflected light or sound waves. Examples include laser rangefinders, ultrasonic sensors, and LiDAR, which provides high-resolution 3D mapping for autonomous vehicles and industrial applications.
- Proximity sensors detect nearby objects without contact, using inductive, capacitive, infrared, or ultrasonic methods. Inductive sensors detect metallic objects via magnetic fields; capacitive sensors detect various materials by capacitance changes; infrared and ultrasonic sensors use light or sound reflection. These sensors are widely used in automation, robotics, and safety systems.
- Position sensors convert linear displacement into electrical signals using resistive, inductive, or capacitive effects. Velocity sensors use electromagnetic induction to measure rate of change of position, while accelerometers measure acceleration via piezoelectric or strain gauge effects. Integration and differentiation of signals enable comprehensive dynamic monitoring.
- Force sensors measure push or pull forces, often using strain-induced resistance changes, piezoelectric effects, or capacitance variations. Torque sensors detect twisting forces via strain gauges or other principles. These sensors include sensing elements, transducers, and signal conditioning to ensure accurate and noise-resistant measurements. Applications span automation, robotics, medical devices, and structural analysis.
- Temperature sensors convert thermal changes into electrical signals. Types include thermistors (resistive), RTDs, thermocouples (voltage generation from metal junctions), and semiconductor sensors (ICs). Selection depends on accuracy, range, response time, and environment. They are critical in industrial, scientific, and biomedical applications.
- Characteristic properties of practical sensors are linearity, accuracy, sensitivity, and repeatability.
- Sensors are susceptible to noise (random electrical signals) and interference. Mitigating these effects is essential for data integrity.
- Calibration adjusts sensor outputs to maintain consistent input–output relationships, compensating for gain, offset, saturation, hysteresis, dead band, and drift over time. It is vital in systems with many sensors and for long-term accuracy.

4.12 Check Your Understanding

4.12.1 Self-Assessment Quiz

1 What is the primary function of a sensor?
   - **a.** To generate electrical signals
   - **b.** To measure physical quantities and convert them into electrical signals
   - **c.** To control electronic circuits
   - **d.** To amplify electrical signals

2 In mechatronics, robotics, and control systems, the term sensor usually refers to
   - **a.** Devices or systems that measure current
   - **b.** Devices or systems that convert electrical quantities
   - **c.** Devices or systems that measure various physical quantities
   - **d.** Devices or systems that measure voltage

3 Data acquisition systems include
   a. Sensors
   b. Electromagnets
   c. Analog-to-digital converters
   d. Both a and c

4 Which type of sensor operates by changes in resistance corresponding to physical changes?
   a. Capacitive sensors
   b. Electromagnetic sensors
   c. Resistive sensors
   d. Optoelectronic sensors

5 Which actions below are involved in the operation of a sensor?
   a. Electric power is transmitted to a machine.
   b. The change of the measured quantity results in an equivalent change of a parameter or variable in the sensor.
   c. The change in the property of the sensor is converted into a low-power level electrical signal in the form of voltage or current.
   d. Both b and c are correct.

6 Which one below is not a fundamental property of sensors?
   a. Accuracy
   b. Voltage level
   c. Power
   d. Sensitivity

7 What is the main advantage of using a Wheatstone bridge with strain gages?
   a. It reduces the cost of the sensor.
   b. It enhances sensitivity by balancing voltage dividers and detecting small resistance changes.
   c. It increases the size of the sensor.
   d. It simplifies the sensor design.

8 Which sensor type is used for noncontact measurement of displacement, velocity, or electrical quantities?
   a. Resistive sensors
   b. Capacitive sensors
   c. Electromagnetic sensors
   d. Light-dependent resistors (LDRs)

9 What is the principle behind the operation of LDRs?
   a. Electromagnetic induction
   b. Photoconductivity
   c. Resistive changes
   d. Piezoelectric effect

10 A strain gage detects
a. Light
b. Deformation
c. Temperature
d. Position

Answers:

1: b; 2: c; 3: d; 4: c; 5: d; 6: d; 7: b; 8: b; 9: c; 10: b

4.12.2 Essay Questions

1 Explain the working principle of a potentiometer as a resistive sensor.

2 Describe the role of thermistors in temperature sensing and control applications.

3 What are the advantages of using optoelectronic sensors in industrial automation?

4 Discuss the importance of sensor integration with electronic circuits in mechatronic and robotic systems.

5 Compare and contrast the different types of proximity sensors and their applications.

4.12.3 Short Answer Problems

1 A strain gauge with a nominal resistance of 120 Ω is used in a Wheatstone bridge configuration. If the strain gauge experiences a 1% increase in resistance, calculate the output voltage of the bridge when supplied with a 10 V source.

2 A capacitive sensor has a plate area of 2 cm^2 and a separation distance of 1 mm. Calculate the capacitance if the dielectric constant of the material between the plates is 4. Assume the permittivity of free space is 8.85×10^{-12} F/m.

3 An inductive voltage divider consists of two inductors with inductances of 10 and 20 mH connected in series. If the input voltage is 15 V, what is the output voltage across the 10 mH inductor?

4 An optical encoder has 500 slots on its disk. If the encoder is rotating at 3,000 RPM, calculate the frequency of the output pulses.

5 A negative temperature coefficient (NTC) thermistor has a resistance of 10 kΩ at 25°C. If the temperature coefficient is −4% per °C, what will be the resistance at 35°C?

4.12.4 Design Problems

1 Design a temperature monitoring and control system for a laboratory incubator using thermistors and microcontrollers. Specify the sensor placement, circuit topology, signal conditioning,

and algorithm for maintaining the temperature within ±0.1°C. Justify your choices of sensor type, interface, and control logic.

2 An automated conveyor system requires detection of metallic and nonmetallic objects at various points along its length. Design a proximity sensing solution using a combination of inductive, capacitive, and optical sensors. Determine the optimal placements, interfacing circuits, and methods to distinguish between object materials in real time.

3 Propose a sensor integration scheme to provide real-time force feedback for a robotic arm manipulating delicate components. Select appropriate force/torque sensors, discuss their mounting and calibration, and outline the electronic circuits required to condition and transmit the signals to the robot's controller.

4 Design a distributed sensor network to continuously measure soil moisture, temperature, and pH over a large agricultural field. Specify sensor types, wireless communication protocols, power considerations, and the central data aggregation system. Present a plan for sensor calibration and long-term reliability.

5 You are assigned to develop a wireless system to monitor strain and vibration in a bridge using strain gauges and accelerometers. Specify the sensor types, signal conditioning, wireless modules, and power sources. Address issues of data integrity, synchronization, and energy efficiency for long-term deployment.

References

Alexander, C. K. & Sadiku, M. N. O., 2013. *Fundamentals of Electric Circuits*. 5th ed. New York: McGraw-Hill.

Cetinkunt, S., 2015. *Mechatronics with Experiments*. West Sussex: John Wiley & Sons.

Dally, J. W., Riley, W. F. & McConnell, K. G., 1984. *Instrumentation for Engineering Measurements*. 2nd ed. New York: John Wiley.

de Silva, C. W., 2016. *Sensors and Actuators - Control System Instrumentation*. 2nd ed. Boca Raton: CRC Press.

Pirc, T., 2025. Modern LIDAR Systems: Sensing for Every Industry. [Online] Available at: https://www.analog.com/en/signals/thought-leadership/modern-lidar-systems-sensing-for-every-industry.html [Accessed 20 7 2025].

Rizzoni, G. & Kearns, J., 2007. *Fundamentals of Electrical Engineering*. 6th ed. New York: McGraw-Hill.

5

Signal Conditioning

OBJECTIVES
• Describe the role of signal conditioning in mechatronic systems. • Explain types and representations of signals. • Explain signal conditioning functions. • Present fundamental signal conditioning circuits. • Discuss filters and their applications. • Outline modulation techniques.

A mechatronic system is a collection of elements interacting to perform the intended task. Elements that form a mechatronic system have diverse functions, structures, and characteristics including electronic sensors, digital logic integrated circuit (IC) chips, power electronics components, electromechanical actuators, and motors. Figure 5.1 illustrates the interaction of element groups in a mechatronic system.

Signal sources produce low-power signals, which are transmitted through wires, ultrasound, infrared, or electromagnetic waves for signal processing. In some situations, when the signal arrives at the input of a circuit that is intended to process the information, it may have been distorted or may contain errors and undesirable noise caused by environmental effects.

Signal conditioning circuits are essential components in mechatronic systems, serving to enhance the quality of signals produced by various sources. These circuits correct and improve the signals before they reach the processing stage. Key functions of signal conditioning include amplifying the signal's magnitude, reducing noise, and matching the impedances of interconnected electronic circuits to maximize overall gain. In feedback control systems, signal conditioning circuits can also add, subtract, or compare signals, as well as integrate or derive them to enhance system response, increasing response time and decreasing output error (Sedra et al., 2020).

5.1 Signals

In electronics, signals are a voltage, current, light, or electromagnetic wave that transmit information. Velocity, force, torque, distance, pressure, temperature, light, and liquid level are among the physical quantities of interest in mechatronic systems. A variety of sensors and transducers convert

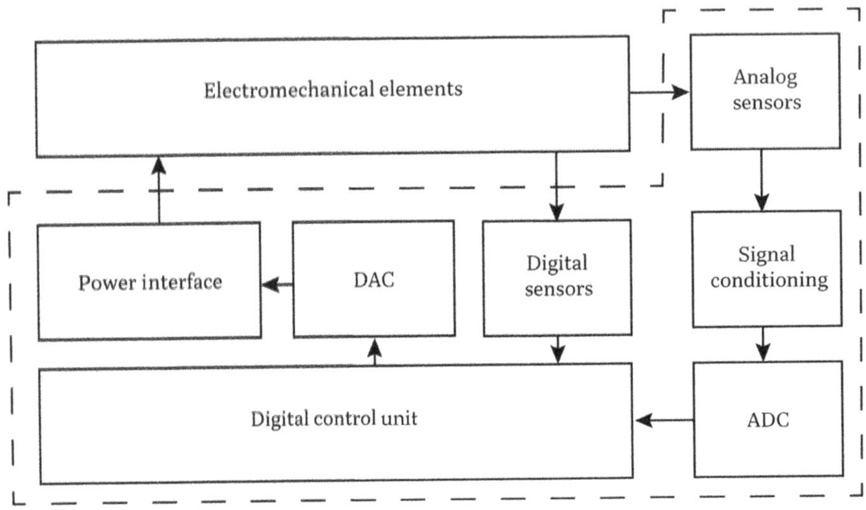

Figure 5.1 Interaction of groups of elements in a mechatronic system

such physical quantities into analog or digital signals. In the real world, variations of physical quantities are continuous; hence, most sensors produce analog signals, usually proportional to the physical value they represent.

An alternative way to represent signals is in the form of a sequence of numbers associated with the magnitude of the signal sampled at particular time steps. The resulting signal is called a digital signal. Digital signals are discontinuous in time, and their values remain constant over each sampling interval.

Some sensors used in mechatronic systems can produce digital signals directly. Examples include rotational and translational sensors that produce sequential pulses during the motion, optical sensors that detect the position of an object, and switching sensors that check a liquid level.

Time-domain representation of a signal means assigning a time function to represent the variation of a signal. A signal may be periodic or nonperiodic. A tachometer that converts the rotation speed of a shaft produces a periodic voltage if the velocity remains constant. If the motor accelerates or slows down, for instance, the tachometer signal is no longer periodic since both its amplitude and frequency change.

5.1.1 Periodic Signals

A periodic signal repeats its instantaneous values at a given period. Characteristic parameters of a periodic signal are its amplitude and frequency. Figure 5.2 (a) shows a periodic analog signal and its essential properties. A digitized version of the same signal is shown in Figure 5.2 (b). The mathematical definition of a periodic function is

$$\begin{aligned} f(t) &= f(t+T) \\ T &= \frac{1}{f} \end{aligned} \tag{5.1}$$

Sinusoidal signals are represented by a sine or cosine function in the form below:

$$\begin{aligned} v_s(t) &= \sin(\omega t + \varphi) \\ \omega &= 2\pi f \end{aligned} \tag{5.2}$$

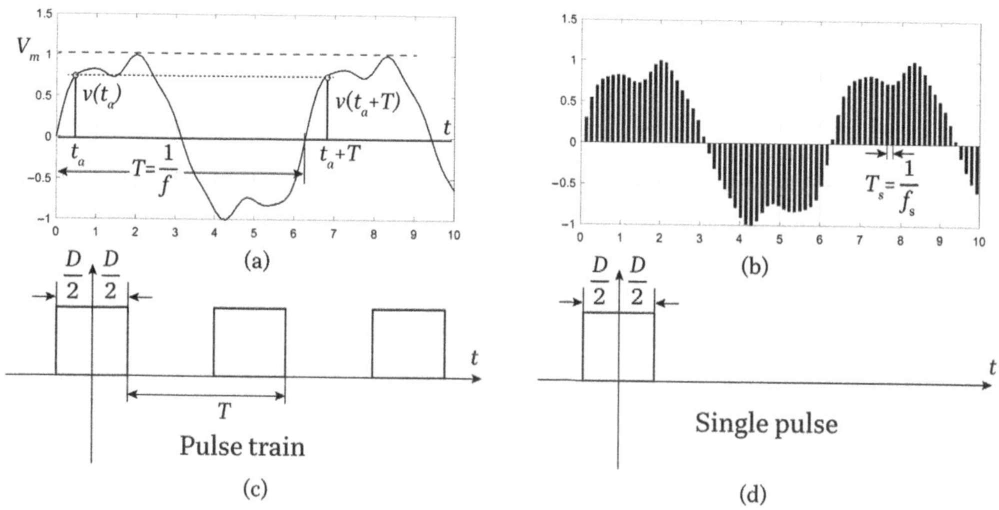

Figure 5.2 Periodic analog and digital signals (a: Periodic analog signal; b: Sampled signal; c: Periodic pulses; d: Aperiodic signal)

The angular frequency ω is measured in radians per second, and the phase shift φ is in radians.

Sensors could also produce a periodic signal that is not sinusoidal. Square, rectangular, triangular waveforms and pulse trains are common nonsinusoidal signals in mechatronic systems. Figure 5.2 (c) illustrates a pulse train. In mechatronic systems, timing circuits, optical sensors, and ICs that generate gate signals for switch-mode power converters typically produce pulse trains.

Fourier series is the essential mathematical method in the analysis and processing of nonsinusoidal periodic signals. Fourier series breaks down a periodic signal into a sum of the infinite number of sinusoidal functions. The general expression of Fourier series is shown below. In this equation, V_o is the average of one period, also known as the DC component. Suppose that the frequency of the periodic signal is f. The first sinusoidal term called fundamental component has the same frequency (fundamental frequency) as the periodic signal. Frequencies of the other components, called harmonics, are integer multiples of the fundamental frequency, that is, $\omega_n = n(2\pi f)$. V_n is the magnitude of the nth harmonic (Smith, 2003).

$$v_s(t) = V_o + \sum_{n=1}^{\infty} a_n \cos(n \cdot \omega t) + b_n \sin(n \cdot \omega t) = V_o + \sum_{n=1}^{\infty} V_n \cos(n \cdot \omega t + \theta_n) \tag{5.3}$$

Parameters of this equation are defined below.

$$V_0 = \frac{1}{T}\int_{-T/2}^{T/2} v_s(t)d(t)$$

$$a_n = \frac{2}{T}\int_{-T/2}^{T/2} v_s(t)\cos\left(n\frac{2\pi}{T}t\right)d(t)$$

$$b_n = \frac{2}{T}\int_{-T/2}^{T/2} v_s(t)\sin\left(n\frac{2\pi}{T}t\right)d(t)$$

$$V_n = \sqrt{a_n^2 + b_n^2};\theta_n = \tan^{-1}\left(\frac{-b_n}{a_n}\right)$$

It is often convenient to calculate Fourier series coefficients using Euler's formula.

$$\cos\omega t + j\sin\omega t = e^{j\omega t} \quad j = \sqrt{-1}$$
$$\cos\omega t = \frac{1}{2}\left[e^{-j\omega t} + e^{j\omega t}\right];\ \sin\omega t = \frac{j}{2}\left[e^{-j\omega t} - e^{j\omega t}\right]$$
$$v_s(t) = \sum_{n=-\infty}^{\infty} c_n e^{jn\omega t} \tag{5.4}$$
$$c_n = c_{-n} = \frac{1}{T}\int_{-T/2}^{T/2} v_s(t)e^{-jn\omega_0 t}dt$$

The integral equation above gives complex coefficients ranging from negative infinity to positive infinity (Chapra & Canale, 2006). Fourier series coefficients of waveforms encountered in mechatronic systems are listed in Appendix B, Table B.1.1.

Example 5.1 Consider the rectangular pulse train shown in Figure 5.2 (c). Given 5 V amplitude, 10 Hz frequency, and $D = 40$ ms pulse width, determine the Fourier series components.

Solution

By definition, the magnitude of a rectangular square wave is V_m during the pulse width D, and zero over the rest of the period. The mathematical definition is

$$\begin{cases} v(t) = V_m = 5\,\text{V for} -\dfrac{D}{2} \leq t \leq \dfrac{D}{2} \\ v(t) = 0 \qquad\qquad \text{otherwise} \end{cases}$$

The period of the 10-Hz periodic signal is

$$T = \frac{1}{f} = \frac{1}{10} = 0.1\ \text{s}$$

Angular frequency:

$$\omega = \frac{2\pi}{T} = 2\pi \times 10\ \text{rad/s}$$

The first term of the Fourier series corresponds to the DC component, the average of the periodic signal over one period.

$$V_0 = \frac{1}{T}\int_{-D/2}^{+D/2} V_m dt = V_m\frac{D}{T}$$

Components other than zero can be calculated using equation 5.3 or 5.4. First, let's write the integral expressions for a_n and b_n. It is more convenient to calculate the integral in terms of angular frequency.

$$a_n = \frac{2}{T}\int_{-D/2}^{D/2} V_m\cos\left(n\frac{2\pi}{T}t\right)d(t) = \frac{2V_m}{T}\left[\frac{T}{2\pi n}\sin\left(n\frac{2\pi}{T}t\right)\right]_{-D/2}^{D/2} = \frac{2V_m}{n\pi}\sin\left(n\pi\frac{D}{T}\right)$$

$$b_n = \frac{2}{T}\left[\int_{-D/2}^{D/2} V_m\sin(n\omega t)d(\omega t)\right] = 0$$

$$\frac{\pi D}{T} = \frac{\pi \times 0.4 \times 10}{10} = 0.4\pi$$

$$v_s(t) = 0.5 + \frac{10}{\pi}\left[\sin(0.4\pi t) + \frac{1}{2}\sin(0.8\pi t) + \frac{1}{3}\sin(1.2\pi t) + \cdots\right]$$

An alternative way is to calculate the series coefficients in complex form using equation 5.4.

$$c_n = \frac{1}{T}\int_{-D/2}^{D/2} V_m e^{-jn\omega t} d(\omega t) = \frac{V_m}{T}\frac{1}{jn\omega t}\left[e^{-jn\omega t}\right]_{-D/2}^{D/2} = \frac{V_m}{n\pi}\cdot\frac{e^{-j\frac{\pi D}{T}} - e^{j\frac{\pi D}{T}}}{2j}$$

$$c_n = V_m \frac{\sin(n\pi D)}{n\pi} \quad n = 0,\ \pm 1,\ \pm 2, \ldots, \infty$$

The ratio of the sine to its argument frequently appears in signal processing calculations, and it is defined as "sinc" function.

$$\mathrm{sinc}(x) = \frac{\sin(x)}{x} \quad x \neq 0$$

The sinc function simplifies the expression of the Fourier series coefficients of a pulse train.

$$c_n = V_m D \cdot \mathrm{sinc}(n\pi D) \quad n = 0,\ \pm 1,\ \pm 2, \ldots, \infty$$

The code below calculates the complex coefficients for the square pulse train and reconstructs the pulse train from calculated coefficients.

Example 5.2 Write a MATLAB® script to calculate the complex Fourier coefficients using the sinc function and reconstruct the original pulse train.

Solution

The MATLAB® code below calculates the complex Fourier coefficients using the sinc function and, for verification, reconstructs the original pulse trains. Note that the accuracy of the method depends on the number of coefficients as illustrated in Figure 5.3. The signal reconstructed with 50 harmonics in (a) is significantly more accurate than the signal reconstructed with 10 harmonics in (b).

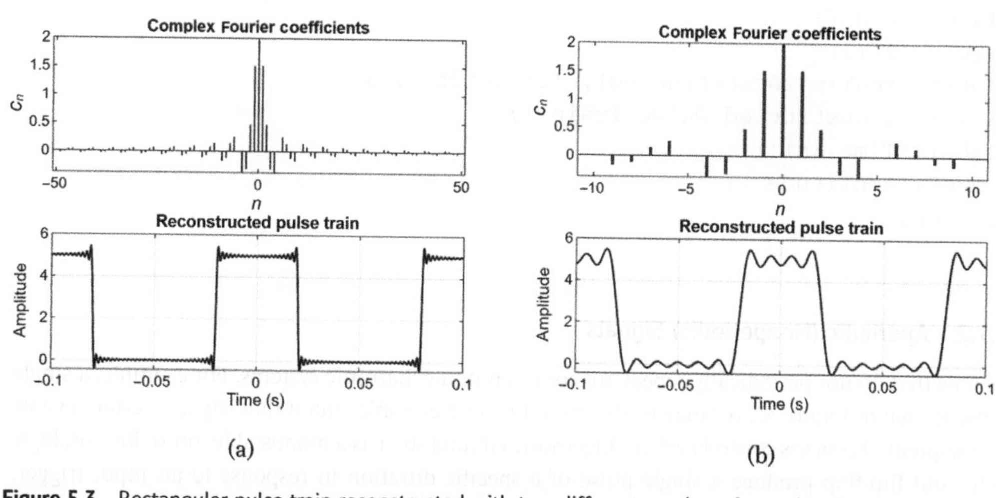

Figure 5.3 Rectangular pulse train reconstructed with two different number of complex Fourier series coefficients (a: 50 pairs; b: 10 pairs)

MATLAB® Script for Calculation of the Fourier Series Coefficients of a Square Pulse Train and Reconstruction of the Original Waveform

```
clear all
% Input parameters
T = 0.1;  % Period of the pulse train in seconds
A=5;      % Amplitude of the pulse train in volts
D = 0.04; % Pulse width in seconds
N = 50;   % Number of Fourier coefficients
% Time vector for reconstruction
t = linspace(-T, T, 1000);
% Compute Fourier coefficients
c_n = zeros(1,2*N+1);
for n = -N:N
  if n == 0
    c_n(n+N+1)=A*D/T; % Average value
  else
    c_n(n+N+1)=A*(D/T)*(sin(pi*n*D/T)/(pi*n*D/T));
  end
end
% Reconstruct the signal
x_reconstructed = zeros(size(t));
for n = -N:N
  x_reconstructed = x_reconstructed+c_n(n+N+1)*exp(1j*2*pi*n*t/T);
end
% Plot the original and reconstructed signals
figure;
subplot(2,1,1)
x_axis=-N:N;
bar(x_axis,c_n,0.1)
title('Fourier Series Coefficients of the Rectangular Pulse Train');
xlabel('n');
ylabel('c_n');
subplot(2,1,2)
plot(t, real(x_reconstructed), 'LineWidth', 1);
title('Reconstructed Pulse Train');
xlabel('Time (s)');
ylabel('Amplitude');
grid on;
```

5.1.2 Aperiodic (Nonperiodic) Signals

Signals that do not periodically repeat are common in mechatronic systems. For example, a single pulse shown in Figure 5.2 (d) may be the output of an electronic circuit that triggers certain events in a sequential process control system. Electronic circuits such as a monostable multivibrator, logic gate, and flip-flop produce a single pulse of a specific duration in response to an input trigger. A short-duration single rectangular pulse can represent an impulse, to simulate a car crash for design of safety features like airbags. Impulse function is also used to analyze the motion of robotic arms and legs to ensure smooth and precise movements. A lightning strike is an example of an extremely short-duration current pulse. Impulse function is used to analyze the impact of short voltage surges on electric devices and electronic circuits. In general, the impulse function is one of the test signals used to evaluate the performance of control systems.

Singularity functions, also known as switching functions, represent switching signals that arise in mechatronics. These are functions that are either discontinuous or have discontinuous derivatives. The three most widely used singularity functions in the analysis of mechatronic control systems are the *unit impulse*, the *unit step*, and the *unit ramp* functions.

a) The Unit Impulse Function

The unit impulse function, often denoted by the lower-case Greek letter delta (δ), is a singularity function referred to as the Dirac delta function. The unit impulse may be visualized as a very short-duration pulse of unit area. The Dirac delta function is often used to represent an idealized impulse, which is undefined or has an infinite amplitude at $t = 0$ and zero elsewhere, but its integral from negative infinity to positive infinity is one (Alexander & Sadiku, 2013). Expression of the unit impulse function is a conditional equation.

$$\begin{cases} t < 0\colon \delta(t) = 0 \\ t = 0\colon \delta(t) \to \infty \;\; \text{(undefined)} \\ t > 0\colon \delta(t) = 0 \end{cases}$$
$$\int_{0^-}^{0^+} \delta(t)dt = 1 \tag{5.5}$$

The unit impulse function is the key concept in digital signal processing (DSP) and digital control. In the conversion of an analog signal to digital, the sampling operation is mathematically equivalent to multiplying the continuous function by the unit impulse function applied at the sampling instant t_o.

$$\int_{t_o^-}^{t_o^+} f(t) \cdot \delta(t - t_o)dt = f(t_o) \tag{5.6}$$

This shows that when a function is integrated with the impulse function, we obtain the value of the function at the point where the impulse occurs. This is a highly useful property of the impulse function known as the *sampling* property.

b) The Unit Step Function

The unit step function is another important aperiodic function frequently used to analyze the performance of control systems. In practice, a step function represents a switch that turns on at the instant selected as the origin of the time axis and stays on. Similar to the unit impulse, the unit step is also defined as a conditional function.

$$\begin{cases} t \geq 0\colon u(t) = 1 \\ t < 0\colon u(t) = 0 \end{cases}$$
$$\frac{du(t)}{dt} = \delta(t) \tag{5.7}$$

Note that the unit impulse is the derivative of the unit step function.

c) The Unit Ramp Function

Integral of a unit step function is called the unit ramp function, because its time plot is similar to a ramp. This piecewise continuous function is used in signal processing and control systems to represent a signal that increases as a linear function of time. The derivative of the unit ramp function is the unit step, and the second derivative is the unit impulse.

$$\begin{cases} t < 0: r(t) = 0 \\ t \geq 0: r(t) = t \end{cases} \tag{5.8}$$

$$r(t) = \int_{-\infty}^{t} u(\tau)d\tau = t \cdot u(t)$$

Figure 5.4 illustrates the singularity functions frequently used in analysis and design of mechatronic systems.

5.1.2.1 Fourier Transform

Whereas the Fourier series represents periodic signals as a sum of sinusoids, the mathematical method known as "Fourier transform" converts an aperiodic function of time into the sum (integral) of the infinite number of sinusoids at frequencies spanning a wide range. A frequency spectrum is a visual representation of different constituent frequencies present in a signal, typically shown as a graph. The horizontal axis refers to the frequency and the vertical axis to the amplitude of each sinusoidal component. This visualization is useful in identifying dominant frequencies and interpreting the characteristics of the time-domain aperiodic signal.

An aperiodic function can be considered a periodic function with an infinite period. This assumption leads to a generalization of the Fourier series into an integral transformation. Given a function of time $f(t)$, the mathematical definition of its Fourier transform is

$$F(\omega) = \int_{-\infty}^{+\infty} f(t)e^{-j\omega t}dt \tag{5.9}$$

When the independent variable is time, the function representing a signal is in the time domain. If angular frequency is an independent variable, the function is in the frequency domain. Fourier transform results in a complex variable function in the frequency domain. Fourier transforms of common time domain functions are listed in Appendix A. In addition, software packages such as MATLAB® provide a variety of functions for frequency-domain analysis, including Fast Fourier Transform (FFT).

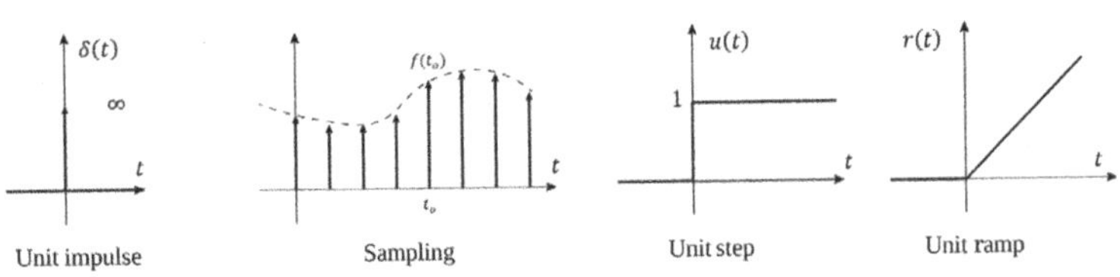

Figure 5.4 Switching functions and sampling

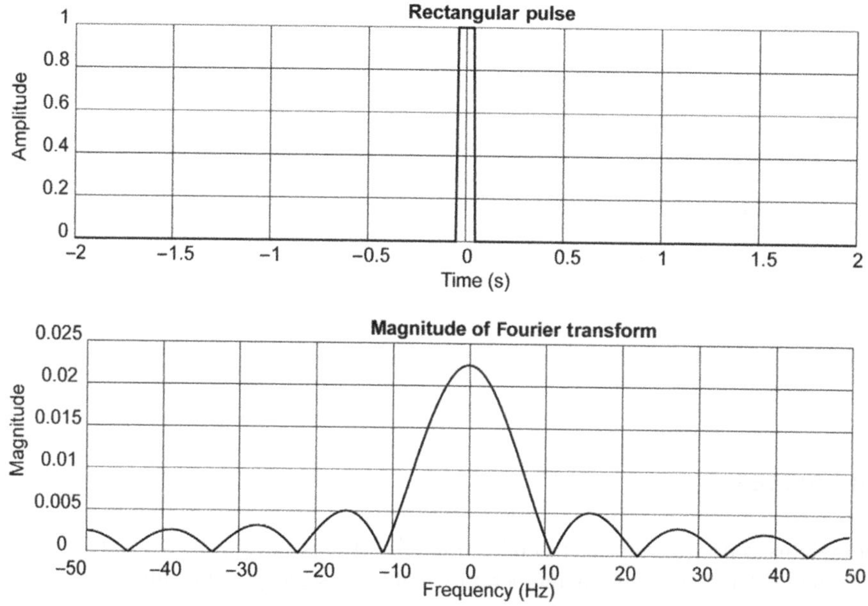

Figure 5.5 Frequency spectrum of a single rectangular pulse

Example 5.3 Compute the Fourier transform of a rectangular pulse with unit amplitude and 100 ms pulse width, and plot the frequency spectrum using a MATLAB® script.

Solution

The MATLAB® code below computes Fourier transform using the built-in fft() function based on the FFT algorithm. The single rectangular pulse of width T centered on the vertical axis and its frequency spectrum are plotted in Figure 5.5.

The MATLAB® Code to Compute Fourier Transform

```
% Define the parameters
T = 0.1; % Width of the rectangular pulse
t = -2:0.01:2; % Time vector
% Create the rectangular pulse
rect_pulse = double(abs(t) <= T/2);

% Compute the Fourier Transform using fft
N = length(t);
f = (-N/2:N/2-1)*(1/(t(2)-t(1)))/N; % Frequency vector
Rect_FT = fftshift(fft(rect_pulse));% fftshift centers the zero
frequency component

% Plot the rectangular pulse
figure;
subplot(2,1,1);
plot(t, rect_pulse);
```

```
title('Rectangular Pulse');
xlabel('Time (s)');
ylabel('Amplitude');
grid on;
% Plot the magnitude of the Fourier Transform
subplot(2,1,2);
plot(f, abs(Rect_FT)/N);
title('Magnitude of Fourier Transform');
xlabel('Frequency (Hz)');
ylabel('Magnitude');
grid on;
```

5.2 Signal Sources

In mechatronic systems, sensors, transducers, and various electronic circuits generate signals. Sensors are active elements that detect motion, temperature, light, moisture, pressure, and other physical events that occur in the environment. Transducers convert one form of energy into another. Microphones, speakers, thermocouples, and pressure sensors are typical examples. Transducers that convert any form of energy into electrical are signal sources. Some sensors are passive elements like resistors, capacitors, or inductors that change depending on physical variables. Variable resistors, capacitors, and inductors are commonly used as position sensors. A thermistor is a semiconductor that has a temperature-dependent resistor. Strain gages are resistors that change with a deformation caused by tension. LDRs have resistance that depends on light intensity. Such passive sensors require an electric circuit that contains voltage and current sources, electronic elements, or IC chips to produce an output voltage that corresponds to the observed event. Active sensors generate a voltage proportional to the observed event. For example, tachometers, used to measure rotational speed, are small generators that produce a voltage proportional to the rotation speed. A thermocouple, which consists of two separate metals welded together, produces a voltage proportional to temperature without needing an additional circuit. Advanced sensors include IC chips that perform functions such as conversion, linearization, preamplification, and impedance matching.

Feedback control systems contain electronic circuits and microcontrollers that provide control signals by processing the information received from sensors and transducers. Furthermore, advanced mechatronic systems receive signals from the eternal sources like data loggers, wireless sensor networks, and Bluetooth devices and provide external signals.

In the analysis, design, and simulation of mechatronic systems, complex signal sources that contain voltage and current sources can be represented by a simple equivalent circuit using Thevenin's and Norton's theorems. Thevenin's theorem states that a linear, active, and two-terminal circuit can be replaced by an equivalent circuit consisting of a voltage source V_{Th} connected in series with a resistance R. The Thevenin voltage V_{Th} is equal to the open-circuit voltage of the active circuit V_{oc}. The equivalent resistor R is the resistance seen at the terminals when the independent sources are turned off (Alexander & Sadiku, 2013). Norton's theorem, which is a similar concept, states that a current source I_N in parallel with an equivalent resistor R can represent a linear active circuit with two terminals.

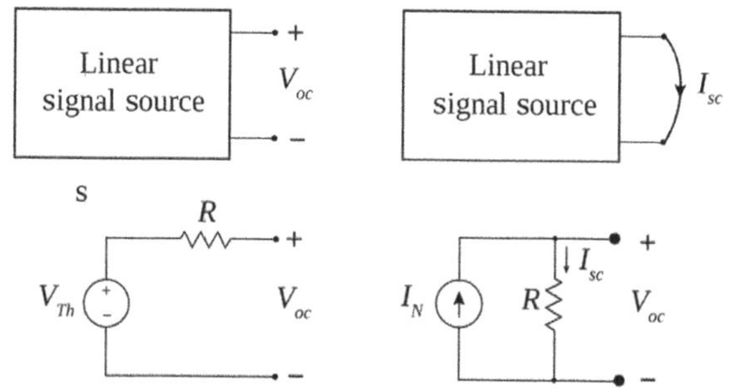

Figure 5.6 Thevenin and Norton equivalent circuits

Thevenin and Norton equivalent circuits illustrated in Figure 5.6 are interrelated by the equations below. They have the same impedance or resistance. The voltage and current sources are interrelated.

$$V_{Th} = V_{oc}; I_N = I_{sc}; R = \frac{V_{oc}}{I_{sc}}$$

$$V_{TH} = I_N R \Longleftrightarrow I_N = \frac{V_{TH}}{R} \tag{5.10}$$

Thevenin's and Norton's theorems apply to sources that produce sinusoidal signals by replacing the equivalent resistor with an impedance. They can be used for periodic nonsinusoidal sources too by expressing the output waveform as a sum of sinusoidal components (harmonics), each with its own amplitude and phase, then calculating the Thevenin equivalent circuit for each harmonic component separately. Thevenin equivalents for all the harmonics are combined using the superposition principle. Since the circuit is linear, the total response can be obtained by summing the responses due to each individual harmonic.

Example 5.4 A variable resistor (potentiometer) can serve as a rotational position sensor. Using the potentiometer as a voltage divider connected to the terminals of a reference voltage source, we can produce a voltage proportional to the rotation angle, as illustrated in Figure 5.7. Given the reference voltage of 10 V and a 10 kΩ potentiometer, obtain the Thevenin equivalent circuit of this resistive rotation sensor. The total rotation of the potentiometer is 270°.

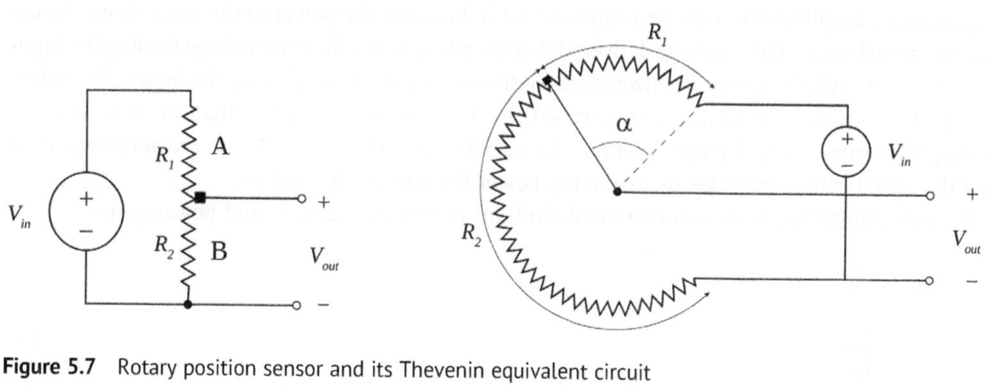

Figure 5.7 Rotary position sensor and its Thevenin equivalent circuit

Solution

The potentiometer works as a voltage divider. Therefore, the open-circuit voltage is

$$V_s = \frac{R_2}{R_1 + R_2} V_r = 10\frac{R_2}{10} = \frac{\theta}{270}$$

Thevenin equivalent resistance is the resistance seen at the terminals when the independent sources are turned off. Turning off a voltage source means the voltage across its terminals is zero. Therefore, the reference voltage source should be replaced by a short circuit (not an open circuit). With a short circuit in the place of the voltage source, the equivalent resistor seen at the terminals is the parallel equivalent of R_1 and R_2.

$$R_s = \frac{R_1 \cdot R_2}{R_1 + R_2} = \frac{\theta(270-\theta)}{270} = \theta - \frac{\theta^2}{270}$$

We see that the open-circuit voltage of the rotary sensor is directly proportional to the rotational angle. The internal equivalent resistance is a nonlinear function of the rotation angle. If the circuit connected to this sensor has sufficiently high input resistance, then the magnitude of the signal represents the rotation angle.

5.3 Signal Amplification

Practical sensors used in mechatronic systems produce power signals as low as that of milli- or microwatt levels and typical output voltage in the millivolt (mV) or even microvolt (μV) range. Such signals are usually too small to be accurately processed. A signal amplifier increases the magnitude of the signal to a level suitable for the input of analog or digital circuits that accomplish various tasks.

Analog signal amplification must preserve the characteristics of the input signal. That is, the amplifier should not change the information the signal contains or add new information. Ideally, the output signal of the amplifier must have exactly the same time variation as the input signal (Sedra et al., 2020). Any distortion on the input signal due to nonlinearity or clipping is undesirable.

The output voltage of an ideal signal amplifier is proportional to the input signal. The ratio of the output voltage amplitude to the input voltage amplitude is the voltage gain. Similarly, the ratio of the amplitudes of output and input currents is the current gain.

Obviously, power gain is the product of voltage and current gains. An important feature that distinguishes an amplifier from a transformer is that it increases the power of the input signal. In the case of a transformer, if the voltage delivered to the load is greater than the voltage feeding the input side (the primary), the current is decreased by the same ratio. Considering the losses, the power delivered to the load (from the secondary side of the transformer) is less than or at most equal to the power supplied by the signal source. An amplifier provides the load with greater power than that the signal source provides by extracting power from its supply source.

The expressions below summarize the definitions of voltage, current, and power gains.

$$\begin{aligned} A_v &= \frac{V_o}{V_{in}} \\ A_i &= \frac{I_o}{I_{in}} \\ A_p &= \frac{p_o}{p_{in}} = \frac{v_o i_o}{v_{in} i_{in}} = A_v A_i \end{aligned} \tag{5.11}$$

Amplifier gains are ratios of quantities with the same dimension. Hence, they will be expressed as either dimensionless numbers or, for emphasis, V/V for the voltage gain, A/A for the current gain, and W/W for the power gain.

Alternatively, for a number of reasons, an amplifier's gain is expressed in a logarithmic measure denoted as decibel (dB). If A represents either the voltage or current gain, its value in dB is

$$A(\text{dB}) = 20 \log |A| \tag{5.12}$$

Since power is related to voltage (or current) squared, the expression of power gain in decibels is

$$A_p(\text{dB}) = 10 \log |A_p| \tag{5.13}$$

Note that in the expression of the voltage and current gains, absolute values are used. If there is a 180° phase difference between input and output signals, they have opposite signs. Such a situation does not imply that the amplifier is attenuating the signal. If an amplifier attenuates the input signal, voltage gain is negative. For example, −20 dB means that the amplifier is attenuating the input signal by a factor of 10 (i.e., $Av = 0.1$ V/V). Since the power gain is always positive, the absolute value sign is not needed (Sedra et al., 2020).

Signal amplifiers are two-port circuits. The conventional amplifier symbol is in a triangular shape to indicate the direction of the signal flow. Another representation is a rectangular box, which shows the input and output impedances and the amplifier gain. The common ground of the input and output ports may be combined.

Input and output impedances are crucial parameters of amplifiers. An ideal amplifier has an infinite input impedance and zero output impedance. In practice, the inductive and capacitive components of the input and output impedances are small, so that in most design problems, we can consider only the input and output resistances to simplify the calculations (Dally et al., 1984).

Signal amplifiers are designed to present extremely high input resistance such that the input current they draw from the signal source is negligible. The output resistance must be as small as possible to reduce the voltage drop due to the output current.

Example 5.5 Designers may need to connect several signal amplifiers in tandem (cascade) to amplify very small signals. In that case, the input and output resistances impact the overall amplification. Consider the tandem connection of two amplifiers in Figure 5.8. Find the overall gain defined as the ratio of the voltage across the load to the voltage generated by the signal source.

Solution

The internal resistance of the signal source forms a voltage divider with the input resistance of Amplifier 1, the output resistance of Amplifier 1 forms a voltage divider with the input resistance of Amplifier 2, and the output resistance of Amplifier 2 forms a voltage divider with the load resistance.

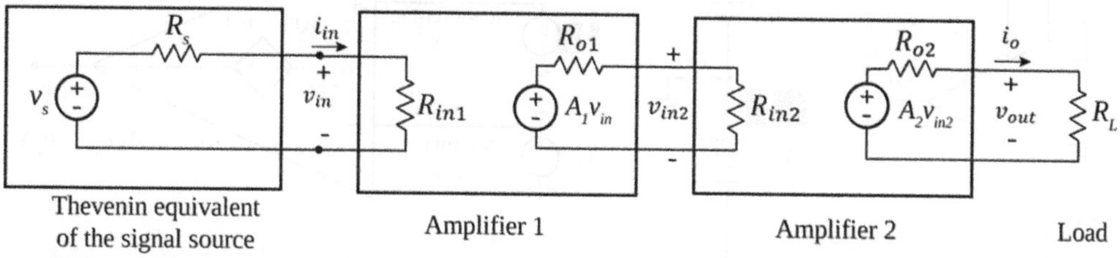

Figure 5.8 Cascade amplifiers

$$G = \frac{v_{out}}{v_s} = \left[\frac{R_{in1}}{R_s + R_{in1}} \cdot \frac{R_{in2}}{R_{o1} + R_{in2}} \cdot \frac{R_L}{R_{o2} + R_L} A_1 \cdot A_2\right] \tag{5.14}$$

Equation 5.14 clearly shows the importance of impedance matching. The input resistance of a signal amplifier must be as high as possible, and the output resistance must be as small as possible. Voltage dividers formed by input and output resistances of tandem-connected circuits cause attenuation, reducing the overall gain that would be the product of the two gains if the amplifiers were ideal.

5.4 Operational Amplifiers

Operational amplifiers, shortly op-amps, are important components of analog signal conditioning, instrumentation, and control circuits in mechatronic systems. An op-amp is a microelectronic circuit composed of 15–20 transistors, several resistors, and capacitors packaged typically in 8- or 14-pin IC chips.

Most op-amps require symmetrical positive and negative supply voltages, while single power supply op-amps are also available. All input and output voltages are defined with respect to ground, but an op-amp chip does not have a separate ground terminal. Ground is the common node of the two supply voltages. Electronic circuit diagrams usually show op-amps as three-terminal circuit elements. The additional terminals for power supply and compensation are usually not explicitly displayed in circuit diagrams.

In Figure 5.9, terminals 1 and 2 are signal inputs. The terminal marked with a + sign is noninverting input, and the other marked with a—sign is the inverting input. Ideal op-amp does not draw any current from the signal sources; thus, its input impedance is infinity. The gain of the ideal op-amp is infinite. The output impedance of the ideal op-amp is zero. Practical op-amp chips have a finite but extremely high gain and draw negligible current from the signal sources.

The basic properties of an ideal op-amp listed below are the basis of all electronic circuit analysis and synthesis works involving op-amps.

$$\begin{aligned} i_1 &= i_2 = 0 \\ v_1 &= v_2 \\ A &= \infty \\ v_o &= A(v_1 - v_2) \end{aligned} \tag{5.15}$$

Practical op-amps present characteristics close to ideal op-amp properties; therefore, assuming that a commercial op-amp is ideal does not cause a significant error. Op-amp chips available in

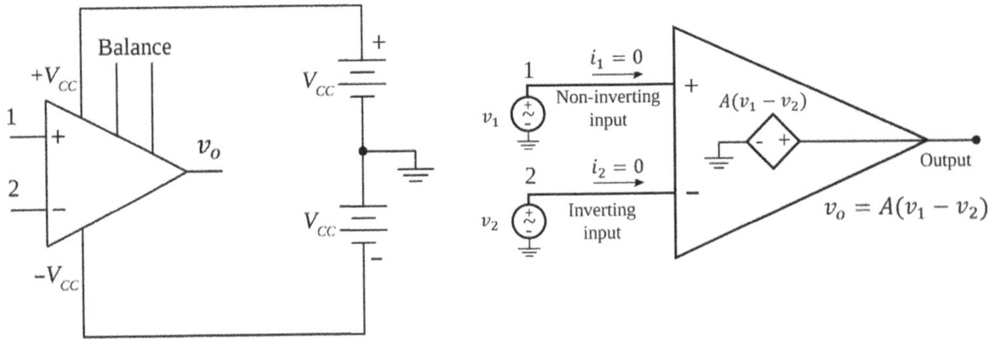

Figure 5.9 Op-amp terminals

8-terminal standard package or two op-amps in a 14-pin package have become versatile and inexpensive electronic circuit elements. Op-amps enable the design and implementation of a wide range of analog signal conditioning and processing circuits. Signal amplifiers, active filters, analog integration and differentiation, signal comparison, and instrumentation amplifiers are among the most common applications.

Op-amps are direct-coupled or DC amplifiers that can amplify DC as well as AC signals. The ideal op-amp amplifies signals of any frequency with equal gain, therefore providing infinite bandwidth. This property is especially desirable in mechatronic systems where temperature, motor speed, acceleration, and liquid-level variables change slowly. By adding a few external components and applying feedback, it can be used in a broad range of applications. Figure 5.10 shows fundamental signal processing circuits.

5.4.1 Comparator

Because the gain of an op-amp is infinite, it amplifies the smallest difference between its inputs up to its positive or negative supply voltage. An op-amp without feedback as shown in Figure 5.9 can operate as a comparator. If the noninverting (positive) input is connected to the signal source and

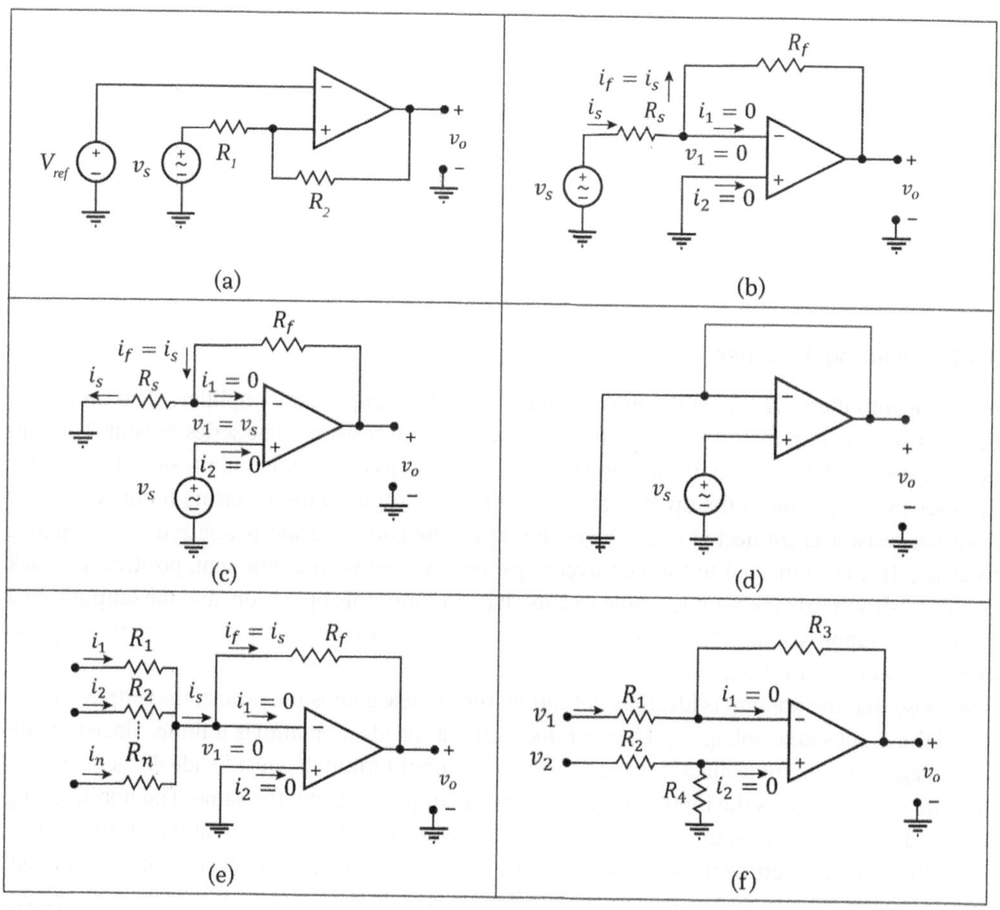

Figure 5.10 Common op-amp signal amplifiers (a: Comparator; b: Inverting amplifier; c: Non-inverting amplifier; d: Voltage follower (buffer); e: Summing amplifier; f: Differential amplifier)

the other input to a reference voltage, the op-amp compares the signal level with the reference. When the signal level is higher than the reference, the op-amp output jumps to the positive supply voltage. When the signal is below the reference, the output changes to the negative supply voltage. This type of operation is referred to as bistable mode, meaning that the output can be in one of the saturation states defined by the positive and negative supply voltages. An inverting comparator can be made by connecting the signal source to the inverting input and the reference voltage to the noninverting input. In this case, the output will go to the negative saturation state when the signal level is above the reference.

Although an op-amp does not need any external components to operate as a comparator, it has a weakness. If the signal contains noise or varies very slowly, like in a temperature control system, the op-amp may oscillate when the signal is close to the reference voltage, switching its output back and forth between the two saturation states. Applying positive feedback as illustrated in Figure 5.10 (a) resolves this problem by adding hysteresis to the transfer characteristic (Sedra et al., 2020). The resistors R_1 and R_2 form a voltage divider that defines the feedback fraction.

Hysteresis allows the comparator to change its state to upper or lower saturation levels at two distinct values. As the signal reaches the upper trip point V_{UT}, the op-amp output turns to positive saturation; when the signal reaches the lower trip point V_{LT}, the op-amp output goes to negative saturation. Trip points are calculated by multiplying the supply voltage V_{CC} with the feedback fraction.

$$V_{UT} = \frac{R_1}{R_1 + R_2} V_{CC}$$

$$V_{LT} = -\frac{R_1}{R_1 + R_2} V_{CC}$$

Comparators are utilized in feedback control systems and ADCs. Dedicated voltage comparators (LM311, LM393, LM339) are widely available and perform much better than a standard op-amp comparator.

5.4.2 Inverting Amplifier

Inverting amplifiers are one of the most frequently used op-amp signal amplifiers. The configuration of inverting amplifier in Figure 5.10 (b) has two external resistors. The series resistor R_s is connected between the signal source and the inverting input marked with the minus sign. The resistor R_f closes the loop around the op-amp by connecting the output to the inverting input. When the feedback resistor is connected to the inverting input, the configuration is referred to as negative feedback. If it is connected to the noninverting input marked with a plus sign, positive feedback occurs. Positive feedback is undesirable because it causes unstable operation, and the output signal reaches the supply voltage level. In the inverting amplifier configuration, the noninverting input is directly connected to the ground.

Suppose that the op-amp is ideal. By definition, the voltage gain is the ratio of the output voltage v_o and the input signal voltage v_s. The open-loop gain of the ideal op-amp is infinite. Since the output voltage is finite, the voltage between the op-amp input terminals must be ideally zero. Therefore, the voltages across the input terminals of the op-amp are exactly the same. The noninverting input is directly connected to the ground, while the inverting terminal is also at the ground potential. Although the inverting input is not physically connected to the ground, it is virtually grounded.

$$v_2 - v_1 = 0 \Rightarrow v_1 = v_2 = 0 \tag{5.16}$$

Ohm's law gives the current flowing through the series resistor.

$$i_s = \frac{v_s - 0}{R_s} = \frac{v_s}{R_s} \tag{5.17}$$

Applying the Kirchhoff's current law to the node where the two resistors are connected, we can write $i_s = i_f + i_1$.

Because the input currents of the ideal op-amp are zero (equation 5.15), the current coming from the signal source must directly flow through the feedback resistor. The Kirchhoff's voltage law applied to the feedback loop gives

$$v_{o+} + R_f \frac{v_s}{R_s} = 0 \tag{5.18}$$

By arranging the equation, we find the gain

$$A = \frac{v_o}{v_s} = -\frac{R_f}{R_s} \tag{5.19}$$

The negative sign indicates that the inverting amplifier reverses the input signal or shifts its phase angle by 180°. This equation shows that the gain of an inverting amplifier depends only on the external resistor values. In practice, the gain of an inverting amplifier can be easily changed or adjusted by changing the resistor values. Furthermore, using precision resistors, the gain can be set to a desired value accurately. Temperature dependency of practical resistors is much smaller compared to semiconductor elements. Therefore, designing an inverting amplifier using op-amp is a simple, reliable, efficient, and accurate solution for signal amplification.

Now let's consider a practical op-amp instead of ideal. Since the open-loop gain A of a physical op-amp is finite, the voltage across the inverting input and ground is not zero.

$$i_s = \frac{v_s - \left(-\frac{v_o}{A}\right)}{R_s} = \frac{v_s + \frac{v_o}{A}}{R_s} \tag{5.20}$$

The input impedance of practical op-amp chips is extremely high such that the current entering the op-amp is negligible. Therefore, the signal current flows through the feedback resistor.

$$v_o = -\frac{v_o}{A} - i_s R_f = -\frac{v_o}{A} - \frac{v_s + \frac{v_o}{A}}{R_s} R_f \tag{5.21}$$

Thus, the closed-loop gain of an inverting amplifier made using a nonideal op-amp becomes (Sedra et al., 2020):

$$G' = \frac{v_o}{v_s} = -\frac{R_f}{R_s} \frac{A}{\left[1 + A + \frac{R_f}{R_s}\right]} \tag{5.22}$$

Note that as the open-loop gain A approaches infinity, the closed-loop gain approaches the ideal value of $-R_f/R_s$. Typically, the open-loop gain of a practical op-amp ranges from about 10,000 (80 dB) to over 1,000,000 (120 dB) at low frequencies. However, this gain decreases as the frequency of the input signal increases due to internal compensation and bandwidth limitations of the op-amp.

The input resistance of an inverting amplifier is equal to the series resistance R_s. For an ideal op-amp, the output resistance is zero.

Example 5.6 A thermocouple produces 40 μV/°C temperature difference. Design an inverting amplifier to increase the voltage of this thermocouple to 10 V when it measures 100°C. First assume that the op-amp is ideal, then find the error if the LM741 op-amp with 200,000 open-loop gain is utilized.

Solution

At 100°C, the thermocouple produces

$$V_s = 100 \times 40 \times 10^{-6} = 4 \times 10^{-3} = 4\,\text{mV}$$

The inverting amplifier must increase 4–10 V. Therefore, the gain should be

$$A = -\frac{10}{4 \times 10^{-3}} = 2{,}500$$

Using equation 5.19, we find the ratio of the feedback resistor to the series resistor as

$$\frac{R_f}{R_s} = 2{,}500$$

Since there are two unknowns, the number of options is infinite. To avoid extremely high feedback resistance, let's select a small series resistance like 10 Ω. The corresponding feedback resistance is 25 kΩ.

The open-loop gain of LM741 is given as 200,000. If the LM741 chips were used, according to equation 5.22, the actual gain would be

$$G' = -\frac{25 \times 10^3}{10}\,\frac{200{,}000}{1 + 200{,}000 + 2{,}500} = -2{,}469$$

The error is

$$\left(1 - \frac{2{,}469}{2{,}500}\right) \times 100 = 1.24\%$$

5.4.3 Noninverting Amplifier

Figure 5.10 (c) shows the noninverting amplifier configuration. The signal source is connected between the ground and the noninverting input of the op-amp. The feedback loop is from the output to the inverting input where the series resistor is also connected. The other terminal of the series resistor is grounded. The same principles as the inverting amplifier apply to the noninverting amplifier. We will again assume that the op-amp is ideal, thus having infinite gain and infinite input impedance. Because the op-amp input currents are zero at both input terminals, the inverting and noninverting inputs are virtually short-circuited. Since the signal source is directly connected to the noninverting input, the voltage across the inverting input and ground is equal to the signal voltage.

$$i_f = i_s = \frac{0 - v_s}{R_s} \tag{5.23}$$

Applying the KVL to the feedback loop, we can write

$$v_s - R_f v_s \frac{-v_s}{R_s} - v_o = 0 \tag{5.24}$$

$$\frac{v_s}{v_o} = 1 + \frac{R_f}{R_s} \tag{5.25}$$

The closed-loop gain of the noninverting amplifier is positive; hence, it does not change the polarity of the input signal. If R_f/R_s is much bigger than unity, the gain is approximately equal to the ratio of feedback and series resistances, similar to the inverting amplifier but with a positive sign.

Because no current flows into the noninverting input of the op-amp, the input impedance of the closed-loop amplifier is ideally infinite. This is the major advantage of the noninverting amplifier. The output resistance is zero as for the inverting amplifier.

One of the important applications of noninverting amplifiers is the voltage follower shown in Figure 5.10 (d). Since the feedback and series resistors are zero, the amplifier gain is unity. Therefore, the output signal follows exactly the input signal. The advantage of this amplifier is that its input resistance is infinite. This configuration is also known as a buffer circuit.

5.4.4 Summing Amplifier

Figure 5.10 (e) illustrates the op-amp circuit known as a summing amplifier. Its analysis is similar to that of the inverting amplifier. Since $i_s = i_1 + i_2 + \cdots i_n$, the output voltage is

$$v_o = -R_f\left[\frac{v_1}{R_1} + \frac{v_2}{R_2} + \frac{v_3}{R_3} + \cdots\frac{v_n}{R_n}\right] \tag{5.26}$$

The summing amplifier adds up the input signals, each multiplied by the ratio of feedback resistance to the individual resistor connected in series to each signal source. Note that the summing amplifier inverts the polarity of the signal source. This is usually not an issue since the terminals of the signal sources can be reversed.

Example 5.7 A bathroom scale designed for a 200-kg maximum weight has four load cells (force sensors) mounted under its corners. Each sensor generates a 10 mV/kg output voltage. Design an op-amp circuit to produce a total output voltage of 10 V when the maximum weight is on the scale.

Solution

To add up the output voltage of four sensors, we need to use a summing amplifier. Even if the load distribution is uneven, the total sensor voltage for the maximum weight will be

$$V_s = 0.01 \times 200 = 2\,\text{V}$$

The op-amp should amplify the total voltage to 10 V. Therefore, the gain should be

$$A = \frac{V_o}{V_s} = \frac{10}{2} = 5$$

We can use identical resistors for each sensor. We can ignore the negative sign in equation 5.26 for this problem since the output terminals of the sensors can be reversed. The ratio of the feedback resistor to each one of the series resistors is $R_f/R_s = 5$. If we select the feedback resistor 10 kΩ, the series resistors connected to each sensor should be 2 kΩ.

To verify the design, we can simulate the summing amplifier using NI Multisim™. Figure 5.11 shows the schematic of the circuit. Four interactive DC voltage sources represent the sensors. The voltmeters connected to the input and output of the op-amp demonstrate that the actual gain is

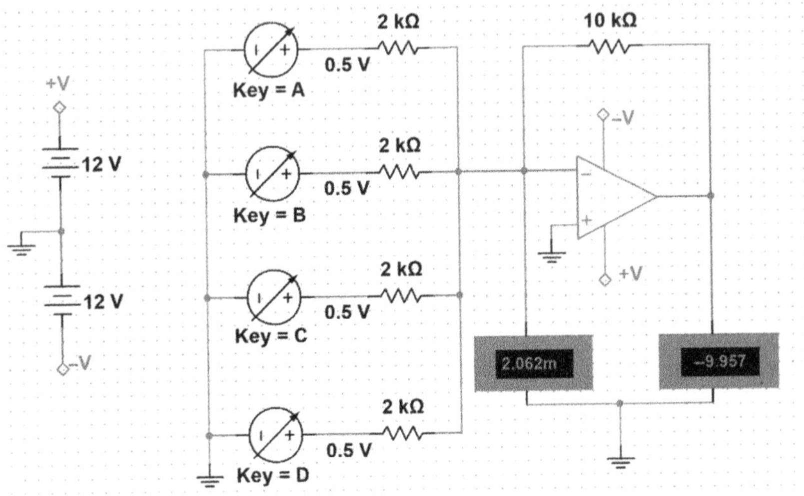

Figure 5.11 Schematic to simulate the circuit designed in Example 5.6 using NI Multisim™ software

slightly different from the calculated value of 5 because the op-amp used in this simulation is a nonideal commercial component.

5.4.5 Differential Amplifier

Differential amplifiers have many applications in mechatronic systems, especially for feedback control to compare the output signal with the reference input. The differential amplifier is basically the combination of an inverting and noninverting amplifier. We can analyze the differential amplifier by applying the superposition principle.

In Figure 5.10 (f), signal v_1 is connected to the inverting input through a series resistance, exactly the same way as the inverting amplifier. The voltage divider applies to the noninverting input a fraction of the signal v_2. Ideally, the gain of inverting and noninverting parts of the circuit must be equal.

$$\frac{R_3}{R_1} = \frac{R_2}{R_1 + R_2}\left(1 + \frac{R_3}{R_2}\right) \Longrightarrow \frac{R_4}{R_2} = \frac{R_3}{R_1} \tag{5.27}$$

If this condition is satisfied, the differential gain of the amplifier becomes

$$A_d = \frac{v_o}{v_2 - v_1} = \frac{R_3}{R_1} \tag{5.28}$$

While differential amplifiers are widely used in mechatronic systems for feedback control, they have certain weaknesses. A key limitation is their low input resistance, which can lead to loading effects and signal attenuation when interfacing with high-impedance sources. Additionally, any mismatch in the resistors used in the amplifier can lead to an imbalanced gain in the inverting and noninverting parts of the circuit, resulting in inaccurate signal amplification. Furthermore, their sensitivity to common-mode signals is another concern, as any common-mode noise present in the input signals can be partially amplified if the common-mode rejection ratio (CMRR) is not sufficiently high. This can introduce unwanted noise and distort the desired differential signal, impacting the overall performance and reliability of the amplifier in practical applications.

5.4.6 Instrumentation Amplifier

Instrumentation amplifiers are high-gain, DC-coupled differential amplifiers with high input impedance, low output impedance, and high CMRR.

CMRR is a vital specification of differential and instrumentation amplifiers. It measures the ability of the amplifier to reject common-mode signals, which are signals that appear simultaneously and in-phase on both input terminals. Ideally, the differential amplifier should amplify only the difference between the input signals (the differential mode), while completely rejecting any signals that are common to both inputs (the common mode).

A high CMRR indicates that the amplifier effectively rejects common-mode signals, thereby minimizing unwanted noise and interference that could distort the desired signal. This is particularly important in applications where the differential signal is small and susceptible to external noise, such as in medical instrumentation, sensor signal conditioning, and data acquisition systems (Horowitz & Hill, 1989).

CMRR is often expressed in decibels (dB). A higher CMRR value signifies better performance of the amplifier in rejecting common-mode noise. For example, a CMRR of 100 dB means that the common-mode signal is attenuated by a factor of 100,000 compared to the differential signal.

Figure 5.12 shows the standard instrumentation amplifier configuration with three op-amps (Rizzoni & Kearns, 2007). The op-amps on the left are voltage followers with high input resistances. They supply input signals to the main op-amp, which amplifies their difference. It can be demonstrated that the closed-loop gain of the instrument amplifier depicted in the figure is

$$G = \frac{v_o}{v_1 - v_2} = \frac{R_F}{R}\left(1 + \frac{2R_2}{R_1}\right) \tag{5.29}$$

To ensure the best possible match between the resistors, the whole circuit is often packaged in a single IC. Commercial instrumentation amplifier ICs like AD624 and INA848 available from several manufacturers contain this configuration.

5.4.7 Integrator

One of the widely used applications of op-amps is the integrator. Including a capacitor in the inverting amplifier circuit changes the transfer characteristic into a time function. In the circuit shown in Figure 5.13 (a), the feedback resistor of the inverting amplifier is replaced with a capacitor. Since the

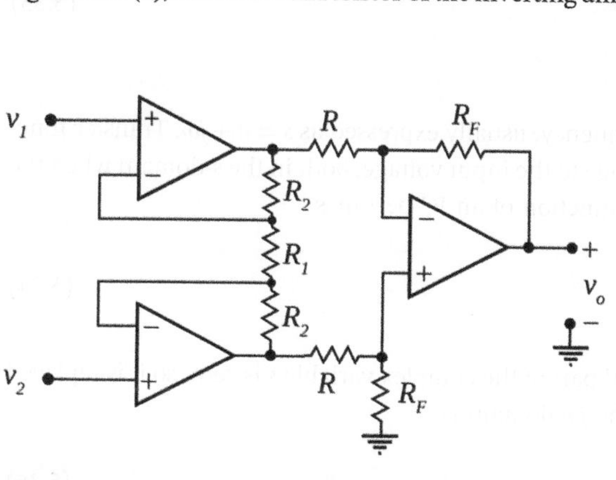

Figure 5.12 Basic instrument amplifier configuration

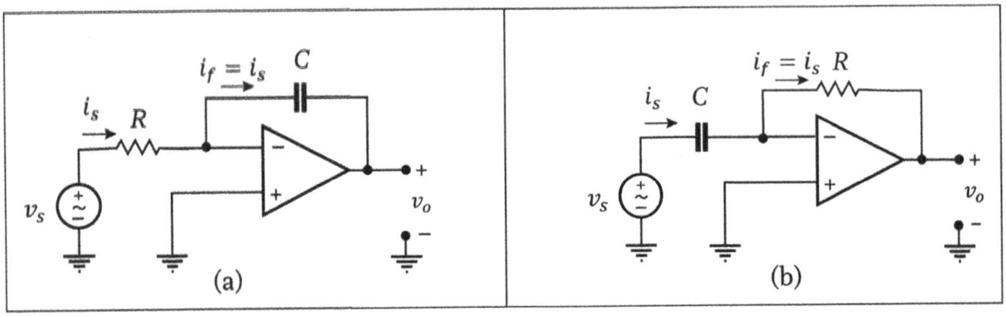

Figure 5.13 Integrator and differentiator circuits (a: Integrator; b: Differentiator)

op-amp does not draw any current from the input terminals and applying the virtual grounding concept, $i_f = i_s$. Applying Ohm's law and *V–I* relationship of the capacitor, we can write

$$\frac{v_s}{R} = -C\frac{dv_o}{dt} \tag{5.30}$$

This equation can be arranged as below:

$$dv_o = -\frac{1}{RC}v_s dt \tag{5.31}$$

By integrating both sides, we obtain

$$v_o(t) = -\frac{1}{RC}\int_{\tau=-\infty}^{t} v_s(\tau)d\tau \tag{5.32}$$

We see that the output voltage is proportional to the integral of the input signal.

It is more convenient to write circuit equations by taking the Laplace transform of both sides. Consider a continuous-time domain function *f*(*t*). The Laplace transform technique transforms the derivative and integral operations in the time domain into algebraic expressions in the s-domain.

$$\begin{aligned} \mathcal{L}\left\{\frac{df(t)}{dt}\right\} &= sF(s) + f(0) \\ \mathcal{L}\left\{\int f(t)dt\right\} &= \frac{F(s)}{s} \end{aligned} \tag{5.33}$$

In the equations above, s is a complex frequency, usually expressed as $s = \sigma + j\omega$. Transfer function is defined as the ratio of the output voltage to the input voltage, both in the s-domain when the initial condition is zero. Thus, the transfer function of an IC becomes

$$\frac{V_o(s)}{V_s(s)} = -\frac{1}{sRC} \tag{5.34}$$

In sinusoidal steady-state analysis, the real part of the complex variable s is zero, so it is replaced by $j\omega$. Therefore, the transfer function in the $j\omega$ domain is

$$\frac{V_o(j\omega)}{V_s(j\omega)} = -\frac{1}{j\omega RC} \tag{5.35}$$

The magnitude of the transfer function is

$$\left|\frac{V_o(j\omega)}{V_s(j\omega)}\right| = \frac{1}{\omega RC} \tag{5.36}$$

And the phase angle is $\phi = +90°$.

Op-amp integrators are very useful in signal processing, feedback control, and analog computers. For example, it is easier to measure the velocity of a vehicle from the rotational speed and the radius of the wheels than from the distance. The distance traveled can be measured by integrating the signal produced by the velocity sensor.

Another common application of op-amp integrator is to obtain a triangular waveform by integrating a square wave, which is easier to generate using bistable circuits like a multivibrator or flip-flop. For example, an integrator can transform a rectangular wave or a rectangular pulse train into a triangular wave. A microprocessor can generate a pulse train with a simple code. An op-amp integrator can transform the pulse train into a sawtooth waveform.

Example 5.8 Suppose that the input signal of the integrator shown in Figure 5.13 (a) is the square wave defined below.

$$\begin{cases} 0 < t < 0.5\text{ ms: } v_s = 5\text{ V} \\ 0.55\text{ ms} < t < 1\text{ ms: } v_s = -5\text{ V} \end{cases}$$

Given $R = 1\text{ k}\Omega$ and $C = 0.5\ \mu\text{F}$, find the time variation and amplitude of the output signal assuming that $v_o(0) = 0$.

Solution

According to equation 5.32,

$$0 < t < 0.5\text{ ms: } v_o(t) = -\frac{1}{RC}\int_{\tau=0}^{t} 5d\tau + v_o(0) = -\frac{1}{10^3 \times 0.5 \times 10^{-6}}5t = -(10^4)t$$

At $t = 0.5$ ms, $v_o = -10{,}000 \times 0.5 \times 10^{-3} = -5$ V

$$0.5 < t < 1\text{ ms: } v_o(t) = -\frac{1}{RC}\int_{\tau=0.5}^{t} -5d\tau + v_o(5) = \frac{1}{10^3 \times 0.5 \times 10^{-6}}5t - 5 = (10^4)t - 5$$

The op-amp output is a triangular wave changing between 0 and −5 V. If this signal is shifted to +5 V by adding 2.5 V DC voltage, the output signal would be a symmetrical triangular waveform.

5.4.8 Differentiator

In the circuit shown in Figure 5.13 (b), the capacitor is connected in series between the signal source and the inverting input of the op-amp, and the resistor is in the feedback loop. We can write the input–output relationship by applying the same principles.

$$v_o(t) = C\frac{dv_s(t)}{dt} \tag{5.37}$$

$$v_o(t) = -RC\frac{dv_s(t)}{dt} \tag{5.38}$$

Because this configuration takes the derivative of the input voltage, it is known as a differentiator. Clearly, the output voltage is the derivative of the input signal amplified by a negative gain.

Transfer function of the differentiator in the s-domain is

$$\frac{V_o(s)}{V_s(s)} = -sRC \tag{5.39}$$

In the frequency domain, by replacing s with $j\omega$ domain, the transfer function of a differentiator becomes

$$\frac{V_o(j\omega)}{V_s(j\omega)} = -j\omega RC \tag{5.40}$$

The magnitude of the transfer function is

$$\left|\frac{V_o(j\omega)}{V_s(j\omega)}\right| = \omega RC \tag{5.41}$$

And the phase angle is $\phi = -90^{\circ}$.

5.5 Filters

Signals in mechatronic systems typically vary from DC to low-frequency range. However, external electromagnetic fields may induce noise on sensor signals, resulting in measurement errors. Moreover, switch-mode converters and turning on or off power devices may cause voltage surges, leading to high-frequency oscillations on sensor wiring and connections. In the conversion of analog signals to digital, high-frequency components must be filtered to avoid aliasing. Signal filters are essential components in mechatronic systems to enhance accuracy and quality by removing undesired disturbances on the signals.

Electronic filters are two-port circuits intended to reduce the undesired frequency components present in their input signal while keeping constant or enhancing the desired frequency components.

The frequency response of a filter describes how the filter affects the amplitude and phase of different frequency components of a signal. Frequency response is often represented as a function $H(f)$ or $H(\omega)$, where f is the frequency in Hz and ω is the angular frequency in radians per second. The frequency response shows how the filter attenuates or amplifies the harmonic components of the input signal in a wide frequency range.

The transfer function of an LTI system, denoted by $H(s)$, is a function of the complex frequency $s = \sigma + j\omega$, where σ is the real part and $j\omega$ is the imaginary part. It describes the system's output in the frequency domain for a given input. The frequency response refers to the evaluation of the transfer function within a specified frequency range. To find a filter circuit's transfer function, write the circuit equations using voltage–current relationships in the s-domain below.

$$\begin{aligned} &\text{Resistor: } V_R = R \cdot I_R(s) \\ &\text{Inductor: } V_L = sL \cdot I_L(s) \\ &\text{Capacitor: } I_C(s) = \frac{V_C(s)}{sC} \end{aligned} \tag{5.42}$$

The frequency response $H(j\omega)$ is then obtained by substituting s with $j\omega$ in the transfer function $H(s)$.

$$H(j\omega) = H(s)|_{s=j\omega} = \frac{V_o(j\omega)}{V_{in}(j\omega)} = |H(j\omega)|\angle\phi \quad \omega = 2\pi f \tag{5.43}$$

Since the frequency range of interest spans from DC to several megahertz levels, usually, the frequency axis is in the logarithmic scale. The magnitude of the transfer function is the ratio of two voltages. Therefore, the magnitude in dB is calculated similarly to the voltage gain as below:

$$|H(j\omega)|_{dB} = 20\log_{10}|H(j\omega)| \tag{5.44}$$

This graphical representation is known as Bode plot or diagram. The frequency range where the input frequency is not attenuated compared to the output is passband, and the frequency range where the input signal begins to be attenuated is the stop band. Cutoff frequency is the transition boundary between the passband and stop band, which is when the output voltage drops to $1/\sqrt{2}$(or 0.707) times the input voltage. On the decibel scale, this value is −3 dB. Since power is proportional to the square of the voltage, at the cutoff frequency, the signal power decreases to one half of its level over the passband.

The roll-off rate is a crucial parameter in filter design, defining how sharply the filter attenuates frequencies beyond the cutoff point. For example, a first-order filter exhibits a roll-off rate of 20 dB/decade, which means the signal's amplitude reduces by 20 dB for every 10-fold increase in frequency beyond the cutoff.

In more sophisticated applications, such as the Butterworth and Chebyshev filters, higher-order configurations are preferred due to their steeper roll-off rates. A second-order filter, for instance, doubles the roll-off rate to 40 dB/decade. This enhanced performance is achieved by incorporating additional reactive components, such as inductors and capacitors, in analog filters, or additional delay elements in digital filters.

Designers often choose even higher-order filters, such as third-order or fourth-order filters, which offer even sharper attenuation and better control over the frequency response. This is particularly beneficial in applications requiring precise signal conditioning, where unwanted frequencies must be effectively suppressed to maintain signal integrity.

Filters are grouped as low-pass, high-pass, band-pass, and band-reject based on their frequency response. Table 5.1 shows the circuit diagrams and ideal transfer characteristics of common active and passive filters.

Filters can be designed in various circuit configurations. The filters shown in Table 5.1 are examples of basic circuits, and they cannot provide the ideal characteristics illustrated in the third column. Higher-order filters can be implemented to improve the transfer characteristics. A thorough filter design is beyond the scope of this text. Comprehensive filter design techniques and tables can be found in several specialized books on analog circuit design (Zumbahlen, 2008).

5.5.1 Passive Filters

Passive filters are two-port RLC circuits. Such circuits have frequency-dependent transfer characteristics that allow the Fourier components with the desired frequencies to pass to the output and block the others. The circuit elements are selected to achieve the required frequency response. RC filters are preferred for voltage waveform conditioning in low-frequency applications, as inductors can be too large for microelectronic circuits. In high-frequency switching power circuits, ferrite core inductors have become common to filter current harmonics (Zumbahlen, 2008).

Table 5.1 Basic filter configurations and their frequency responses

	Passive Filters	Active Filters	Ideal Transfer Characteristic
Low pass filter	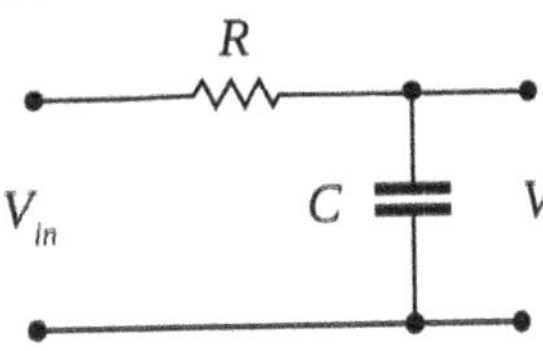	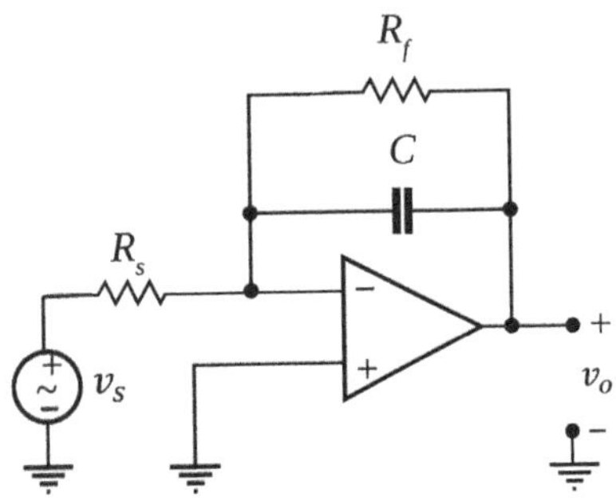	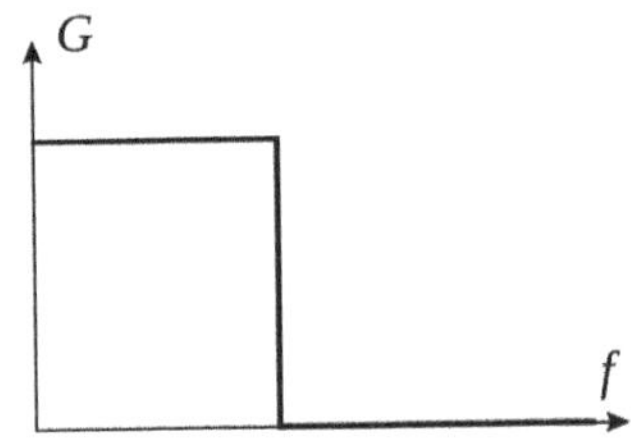
High-pass filter	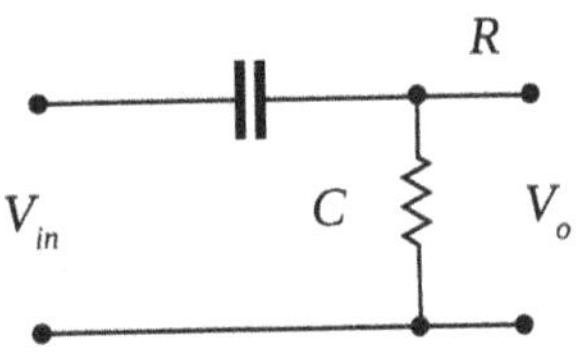	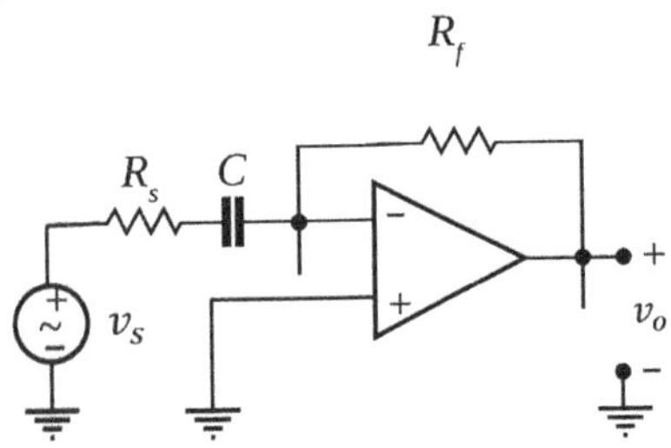	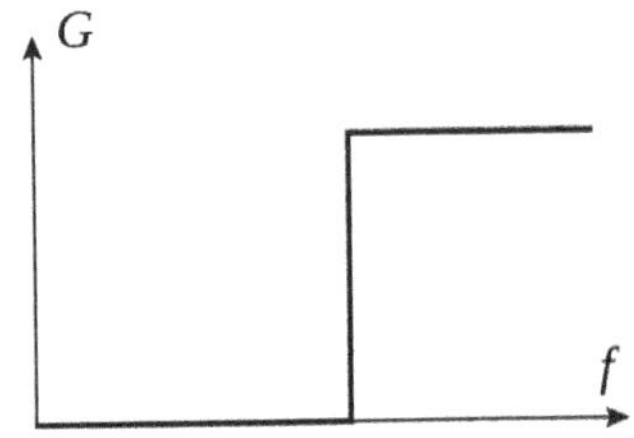
Band-pass filter	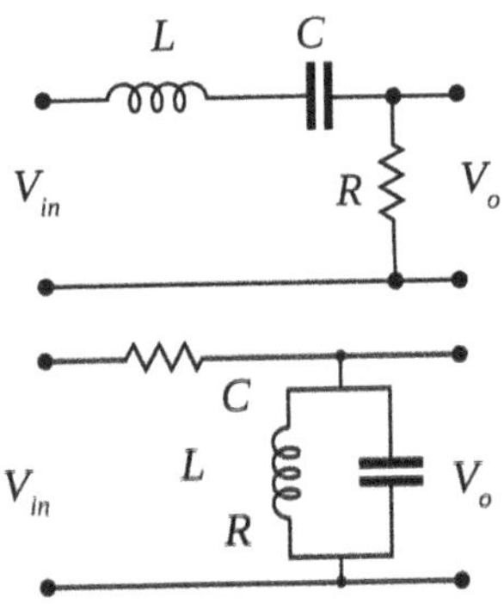	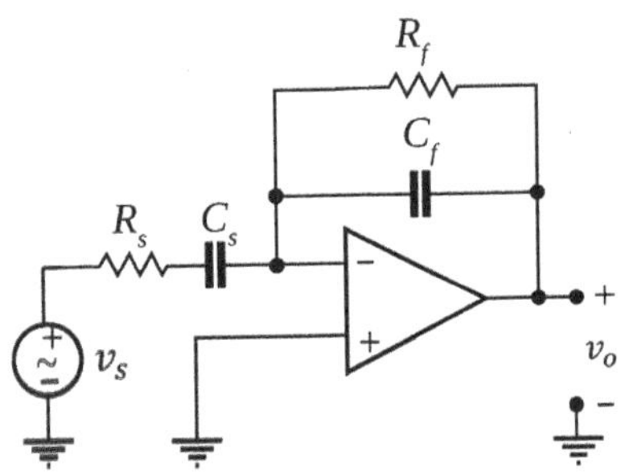	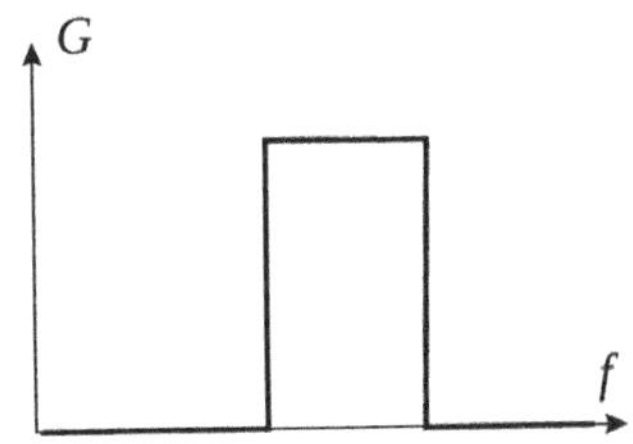
Band-reject filters	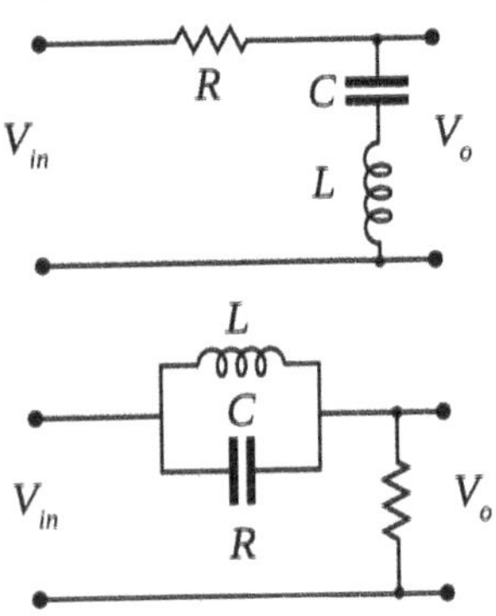	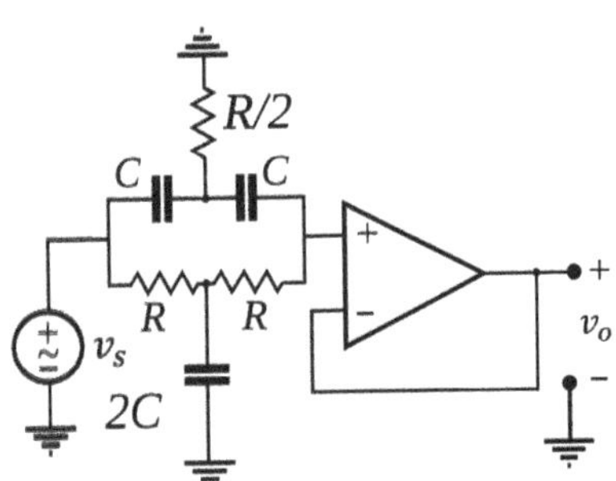	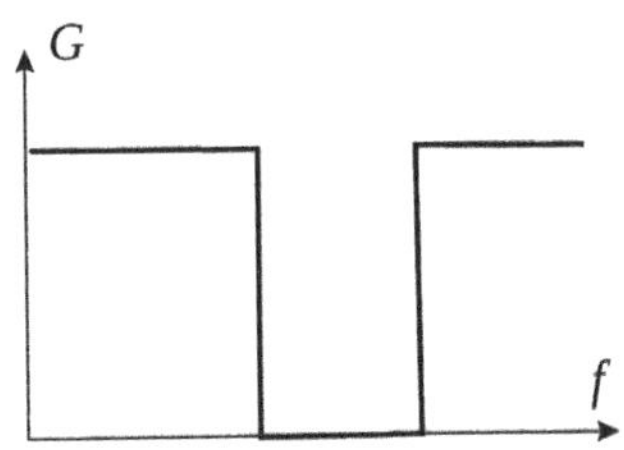

In signal processing, the "order" of a filter is a measure of its complexity. It refers to the highest power of the filter's transfer function. In other words, it is the number of reactive components (inductors and capacitors) in an analog filter. A first-order filter has a single reactive component, typically capacitor or delay, while a second-order filter would have two and so on.

Roll-off is a measure of how sharply a filter attenuates signals beyond its cutoff frequency. It is a critical parameter that determines the effectiveness of a filter in separating the desired signals from unwanted frequencies. The roll-off rate is expressed in decibels per decade (dB/decade), where each decade represents a 10-fold increase in frequency. For example, a filter with a roll-off rate of −20 dB/decade attenuates the signal by 20 dB for every 10-fold increase in frequency beyond the cutoff point. The steepness of the roll-off is influenced by the order of the filter; higher-order filters exhibit a steeper roll-off, providing better rejection of unwanted frequencies.

5.5.2 Active Filters

Active filters are essentially signal amplifiers designed for a given frequency response using feedback elements. Practical active filters are inverting amplifiers that contain RC circuits similar to the integrator and differentiator. Active band-pass filter does not require an inductor like passive band-pass filter. Inserting a capacitor in the series impedance and connecting a capacitor in parallel to the feedback resistor is sufficient to create two cutoff frequencies. In order to obtain the transfer function of an active filter, the resistances in the gain function of the inverting amplifier are replaced with impedances (Sedra et al., 2020).

Three basic active filter configurations are displayed in the third column of Table 5.1. These circuits are similar to the inverting amplifier. The difference is R_f and R_s are replaced by RC circuits. Transfer function of the active filters can be obtained by replacing the gain of an inverting amplifier R_f and R_s with the equivalent impedances of the corresponding RC circuits.

5.5.3 Comparison of Passive and Active Filters

While similar frequency responses can be obtained by using either active or passive filters, each type has benefits and drawbacks.

Passive filters contain only resistors, capacitors, and inductors, and they don't require a power supply. These properties make them more reliable in certain applications. They tend to be more cost-effective and energy efficient, as they don't require active components like transistors or op-amps. Passive filters usually introduce less noise into the signal path, which can be advantageous in low-noise applications. Passive filters can handle higher voltage and current levels without the need for additional components to manage heat dissipation.

Active filters involve op-amps or specialized microelectronic circuits. They can achieve more complex signal processing operations and may be programmable. Active filters can amplify signals, providing gain control. This makes them particularly useful in situations where signal strength is weak and needs boosting. They offer more flexibility in terms of tuning frequency and Q-factor (quality factor). You can easily adjust these parameters without needing to change components. Active filters generally provide better performance in terms of selectivity and stability compared to passive filters. Another practical benefit of active filters is that they don't require inductors, which can be bulky and prone to losses at higher frequencies. This makes active filters more compact and easier to implement in ICs. Furthermore, active filters can maintain better signal integrity over a wider frequency range, due to their ability to handle impedance matching more effectively.

In summary, active filters are preferred for applications requiring high performance, gain control, and flexibility in design, while passive filters are suitable for simpler, cost-effective, and high-power applications.

5.5.4 Low-Pass Filters

Low-pass filters transfer lower-frequency components of the signal below a certain cutoff frequency to output but present higher transfer impedance to higher-frequency components. The cutoff frequency is the boundary where the filter starts to reduce the amplitude of higher-frequency signals. The effectiveness of this attenuation depends on the design and order of the filter.

Low-pass filters are ideal for applications where it is necessary to remove high-frequency noise from a signal or to smooth out a signal in analog-to-digital conversion processes.

In practical applications, low-pass filters are used in various domains, including audio engineering, telecommunications, and control systems. In audio equipment, they help in reducing unwanted high-frequency noise, thus enhancing the quality of sound. In telecommunications, low-pass filters are employed to limit the bandwidth of signals, ensuring that transmissions do not interfere with each other. In control systems, low-pass filters play a crucial role in eliminating high-frequency noise from sensor signals, enabling accurate data processing and control.

Low-pass filter attenuates the signal beyond the cutoff frequency, which is a critical factor in determining the filter's performance. For instance, a first-order low-pass filter has a roll-off rate of 20 dB/decade, whereas higher-order filters have steeper roll-off rates, making them more effective in attenuating unwanted frequencies.

Low-pass filters are essential in mechatronics for enhancing signal quality and ensuring stable system performance. Low-pass filters eliminate high-frequency noise from sensor signals, ensuring accurate data is sent to controllers and processors. This is crucial for precision in robotic systems and automation. They smooth out the pulse-width modulation (PWM) signals used in controlling motor speed, resulting in more stable and precise motor operation.

In machinery monitoring, low-pass filters help isolate low-frequency vibrations that indicate specific faults or operational conditions, improving the accuracy of diagnostics.

Low-pass filters are used to filter out high-frequency components present in the output voltage of switch-mode power supply circuits to prevent electromagnetic interference on other circuits in the vicinity. They are also used to eliminate harmonics in the supply current, enhancing the power quality. In control systems, low-pass filters are integrated into control loops to filter out high-frequency noise from control signals, enhancing the stability and responsiveness of the control system.

5.5.4.1 Passive Low-Pass Filter

Consider the passive low-pass filter shown in Figure 5.14. Using the KVL and Ohm's law, the output voltage can be expressed directly in the s-domain as

$$
\begin{aligned}
V_o(s) &= \frac{1}{sC} \cdot \frac{V_{in}(s)}{R + \dfrac{1}{sC}} = \frac{1}{1 + sRC} V_{in}(s) \\
G(j\omega) &= \frac{V_o(j\omega)}{V_{in}(j\omega)} = \frac{1}{1 + j\omega RC} \\
|G(j\omega)| &= \frac{1}{\sqrt{1 + (\omega RC)^2}} \quad \phi = -\tan^{-1}(\omega RC)
\end{aligned}
\tag{5.45}
$$

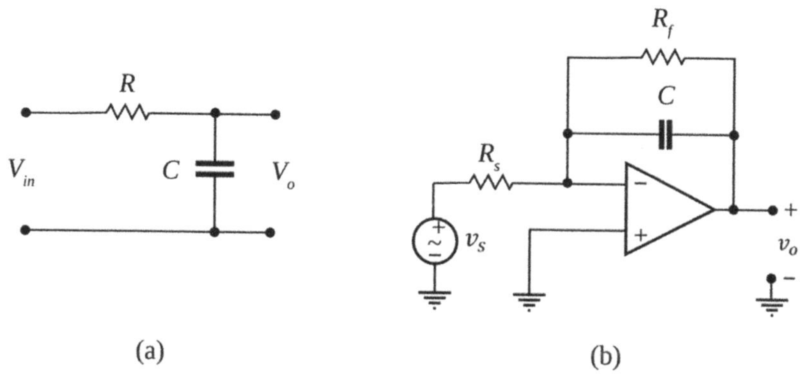

Figure 5.14 Passive and active low-pass filters (a: Passive; b: Active)

At low frequency, the magnitude is approximately 1, or 0 in dB. At very high frequencies, the denominator of the magnitude approaches ωRC because 1 is negligible. Thus, the magnitude approaches zero, or $-\infty$ in dB. When $\omega = 1/RC$, the magnitude is $1/\sqrt{2}$ (or 0.707), which results in $20\log_{10}(0.707) = -3$dB. This special value is named "cutoff" or "corner" frequency. At frequencies higher than the cutoff frequency, the slope of the magnitude plot in dB on the logarithmic horizontal scale is approximately -20 dB/decade.

At low frequencies, the phase angle is approximately 0 but becomes $-45°$ at the cutoff frequency. As the frequency increases further, the phase angle reaches asymptotically $-90°$.

Example 5.9 Design a first-order low-pass passive filter with 1 kHz cutoff frequency using MATLAB®.

Solution

The cutoff frequency corresponds to the angular frequency for which $\omega RC = 1$. Therefore,

$$RC = \frac{1}{2\pi f_0}$$

The problem involves two unknowns, the values of one resistor and one capacitor. In a practical design problem, it is usual to select one of the components arbitrarily and calculate the other. Commercially available capacitors are more limited than the resistors. Let's select 0.1 μF capacitor and calculate the resistance to match this capacitance. If the cutoff frequency is not critical, the closest standard resistor may be used. Any resistance can be created by series and parallel combinations of resistors, or a potentiometer may be used to accurately tune the cutoff frequency.

MATLAB® Code to Calculate the Resistance Needed for the Desired Cutoff Frequency Using a 0.1 μF Capacitor and Then to Plot the Magnitude and Phase Responses of the Filter

```
% Define the parameters
C = 0.1e-6; % Resistance in ohms (1 kOhm)
f_c = 1e3; % Cutoff frequency in Hz (1 kHz)
R = 1/(2*pi*C*f_c); % Resistance in Ohms
```

```
% Display the calculated Resistance
fprintf('Calculated Resistance: %.2e Ohms\n', R);
% Frequency range for plotting
f = logspace(1, 5, 1000); % Frequency range from 10 Hz to 100 kHz
omega = 2 * pi * f; % Angular frequency
% Transfer function of the low-pass filter
H = 1 ./ (1 + 1j*omega*R*C);
% Plot the magnitude response
figure;
subplot(2,1,1)
semilogx(f, 20*log10(abs(H)));
xlabel('Frequency (Hz)');
ylabel('Magnitude (dB)');
title('Magnitude Response of the Passive Low-Pass Filter');
grid on;
% Plot the phase response
subplot(2,1,2)
semilogx(f, angle(H) * 180/pi);
xlabel('Frequency (Hz)');
ylabel('Phase (degrees)');
title('Phase Response of the Passive Low-Pass Filter');
grid on;
```

The MATLAB® script above calculates the resistance value as 1.59 kΩ for the given values and generates Bode plots pasted in Figure 5.15. Control Toolbox provides the special function "bode()" to generate both the magnitude and phase plots, but the presented code generates the Bode plots using built-in standard functions.

5.5.4.2 Active Low-Pass Filter

Transfer function of the first-order active low-pass filter shown in Figure 5.14 is

$$\begin{aligned} G(s) &= -\frac{Z_f(s)}{Z_s(s)} = -\frac{1}{R_s}\cdot\left[\frac{1}{R_f}+\frac{1}{sC}\right]^{-1} = -\frac{1}{R_s}\frac{sR_fC}{R_f+sC} \\ G(s) &= -\frac{R_f}{R_s}\cdot\frac{1}{1+sR_fC} \end{aligned} \tag{5.46}$$

We notice that the frequency response of the active filter includes a gain component.

5.5.5 High-Pass Filters

A high-pass filter transfers signals above the cutoff frequency but presents a higher transfer impedance to signals below the cutoff frequency.

In mechatronics, high-pass filters are used in a variety of applications due to their ability to allow high-frequency signals to pass while attenuating lower frequencies. These filters are particularly useful in systems where high-frequency signal integrity is needed.

High-pass filters are often employed in sensor signal processing to eliminate low-frequency noise and drift that can obscure important high-frequency data. For instance, in accelerometer and gyroscope applications, high-pass filters help in distinguishing between the relevant rapid movements and unwanted slow drifts or vibrations.

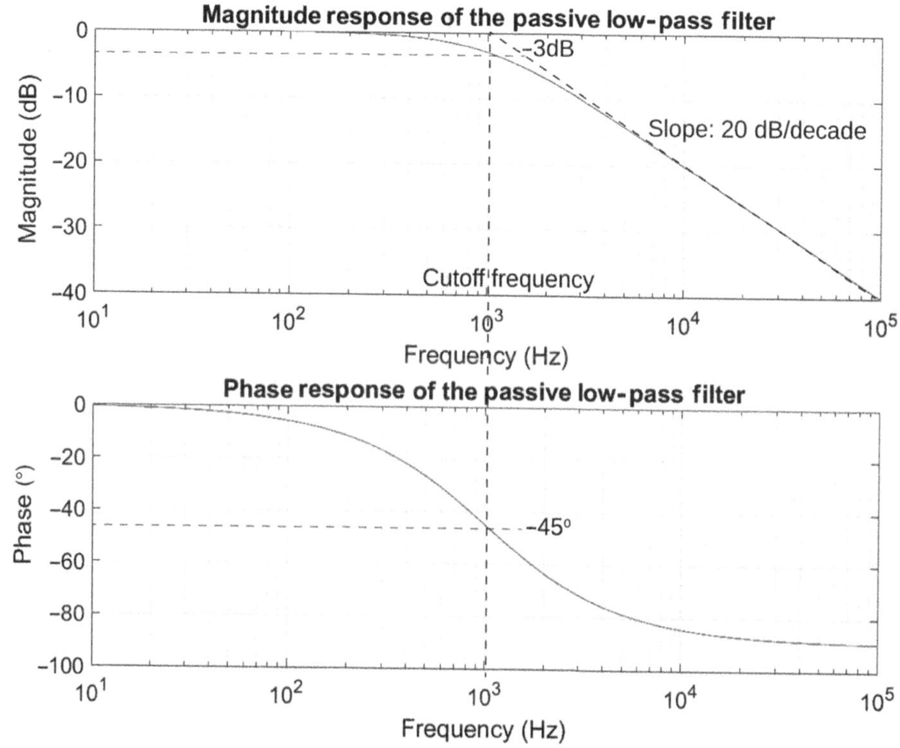

Figure 5.15 Bode plots for Example 3.9

In communication systems within mechatronics, high-pass filters are essential for ensuring the clarity and reliability of high-frequency communication signals. They help to eliminate lower-frequency interference that can distort the signal, thereby improving the overall performance of wireless communication systems.

Additionally, high-pass filters play a crucial role in audio processing within mechatronic systems. They are used to filter out low-frequency hums and rumbles, thereby enhancing the clarity and quality of the audio signal. This is particularly beneficial in robotics and automation systems where precise audio feedback is required.

High-pass filters are also utilized in the design of control systems to improve the responsiveness and stability of the system. By removing low-frequency noise from the control signals, these filters ensure that the system responds more accurately and quickly to changes in the input.

5.5.5.1 Passive High-Pass Filter

A passive high-pass filter can be made by just swapping the resistor and capacitor in the low-pass filter as shown in Figure 5.16. Using the s-domain voltage–current relationships of the circuit elements given in equation 5.42, the circuit equations are obtained as follows:

$$\begin{aligned} V_o(s) &= R \cdot \frac{V_{in}(s)}{R + \dfrac{1}{sC}} = \frac{sRC}{1 + sRC} V_{in}(s) \\ G(j\omega) &= \frac{V_o(j\omega)}{V_{in}(j\omega)} = \frac{j\omega RC}{1 + j\omega RC} \end{aligned} \tag{5.47}$$

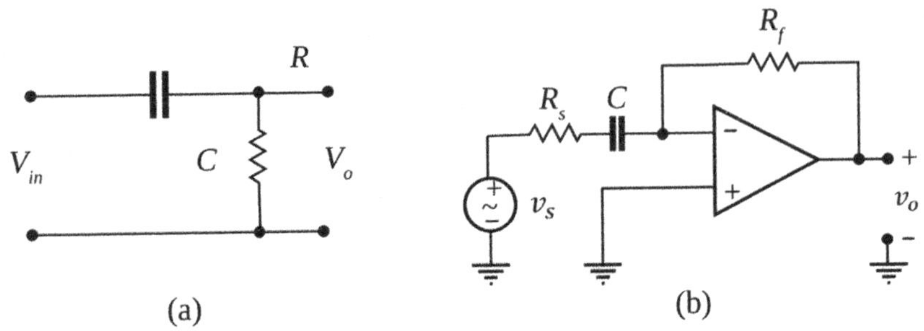

Figure 5.16 Passive and active high-pass filters (a: Passive; b: Active)

The cutoff frequency is when $\omega RC = 1$. The magnitude and phase angle of the frequency response are

$$|G(j\omega)| = \frac{\frac{\omega}{\omega_o}}{\sqrt{1+\left(\frac{\omega}{\omega_o}\right)^2}}$$
$$\varphi = 90 - \tan^{-1}\left(\frac{\omega}{\omega_o}\right) \tag{5.48}$$
$$\omega_o = \frac{1}{RC}$$

By inspection of the equation above, we see that the frequency response would be zero at $\omega = 0$ and would asymptotically approach 1 as ω approaches infinity. The phase shift changes from 90° to zero. The Bode plot of the first-order high-pass filter gain rises to the cutoff frequency with 20 dB/decade slope. At the cutoff frequency, the gain is −3 dB. At frequencies above the cutoff frequency, the gain approaches zero on the decibel scale.

5.5.5.2 Active High-Pass Filter

The first-order active high-pass filter is similar to the op-amp differentiator circuit with a resistor connected in series to the capacitor. The analysis is similar to the low-pass filter.

$$G(s) = -\frac{Z_f(s)}{Z_s(s)} = -R_f \cdot \left[R_s + \frac{1}{sC}\right]^{-1} = -\frac{sCR_f}{1+sCR_s}$$
$$G(j\omega) = -\frac{j\omega CR_f}{1+j\omega CR_s} \tag{5.49}$$

5.5.6 Band-Pass Filters

Band-pass filter transfers a desired frequency range but presents higher transfer impedance to signals below and above the cutoff frequencies, sometime called corner frequencies. It essentially "filters" out frequencies that are too low or too high, letting pass through only the desired band of frequencies.

Band-pass filters play a crucial role in mechatronics by helping to isolate and process specific frequency ranges, which is essential for accurate signal analysis and control. In practice, band-pass

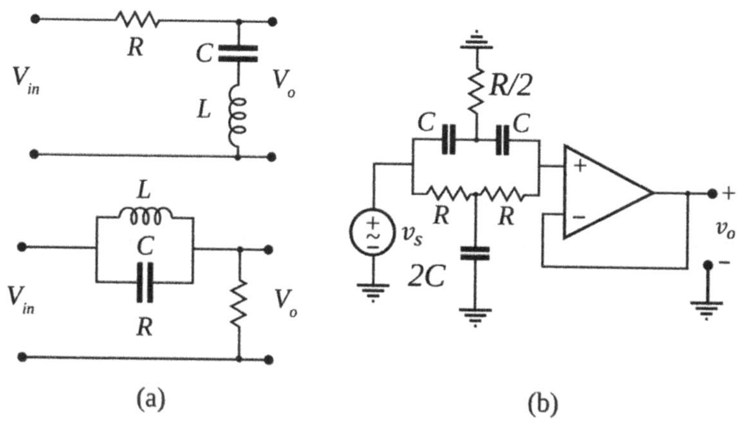

Figure 5.17 Passive and active band-pass filters (a: Passive; b: Active)

filters are utilized to adjust to a specific frequency band or to eliminate unwanted high- and low-frequency noise from sensor signals.

Band-pass filters are particularly important in applications like robotics, where precise measurements are critical. In mechatronic systems that involve wireless communication, band-pass filters help to select the desired signal frequency and eliminate interference from other signals. In feedback loops band-pass filter removes high-frequency noise and low-frequency drift, improving the stability and performance of control systems. In vibration test and monitoring, band-pass filters shown in Figure 5.17 can isolate specific vibration frequencies that show potential issues or faults.

5.5.6.1 Passive Band-Pass Filter

Passive band-pass filters are essentially resonance circuits that have two cutoff frequencies. Therefore, at least two reactive elements (inductors and/or capacitors) must be used to achieve the desired frequency response. Higher-order filters have multiple sections cascade connected.

Band-pass filters can be designed using various circuit topologies. The passive band-pass filter circuit presented in Table 3.1 can be regarded as a voltage divider. In this configuration, the input voltage is applied across the series RLC circuit, while the output voltage is measured across the resistor. The circuit equations in the phasor domain are

$$V_o(j\omega) = \frac{R}{R + jX} \cdot V_{in}(j\omega) = \frac{R}{R + j\left(\omega L - \dfrac{1}{\omega C}\right)} V_{in}(j\omega)$$

$$|G(j\omega)| = \frac{V_o(j\omega)}{V_{in}(j\omega)} = \frac{R}{\sqrt{R^2 + \left(\omega L - \dfrac{1}{\omega C}\right)^2}} \tag{5.50}$$

$$\varphi = -\tan^{-1}\frac{\omega L - \dfrac{1}{\omega C}}{R}$$

At zero frequency (DC) the gain is zero, which is expected because the series capacitor acts as an open circuit. At the resonance frequency, the imaginary part of the denominator equals zero and the transfer impedance becomes resistive.

$$\omega_0 = \frac{1}{\sqrt{LC}} \text{ rad/s or } f = \frac{1}{2\pi\sqrt{LC}} \text{ Hz} \tag{5.51}$$

At ω_0, the gain reaches its maximum value of 1. As the frequency increases, the magnitude of the gain approaches zero again because the inductor presents increases the impedance at higher frequencies.

To find the cutoff frequencies ω_1 and ω_2 at which the power is half its maximum value, set the magnitude of the total impedance equal to $\sqrt{2}R$ and solve the equation for ω (Alexander & Sadiku, 2013).

$$\sqrt{R^2 + \left(\omega L - \frac{1}{\omega C}\right)^2} = \sqrt{2}R$$
$$\omega_{1,2} = \pm\frac{R}{2L} + \sqrt{\left(\frac{R}{2L}\right)^2 + \frac{1}{LC}} \tag{5.52}$$

The cutoff frequencies and the resonance frequency are related by the following equation:

$$\omega_0 = \sqrt{\omega_1 \cdot \omega_2} \tag{5.53}$$

Bandwidth is defined as the frequency range between the upper and lower cutoff frequencies.

$$B = \omega_2 - \omega_1 = \frac{R}{L} \tag{5.54}$$

In a resonant circuit, the "sharpness" of the resonance indicates its selectivity, which is quantitatively expressed by the quality factor Q. At the resonance frequency, the inductor and capacitor exchange the reactive energy while the resistor dissipates active energy. The quality factor is defined as the ratio of peak energy stored to the energy lost per cycle of oscillation.

$$Q = 2\pi \frac{\frac{1}{2}LI^2}{\left(\frac{1}{2}RI^2\right)\cdot\frac{1}{f_0}} = \frac{(2\pi f_0)L}{R} \tag{5.55}$$

Other expression of the quality factor is

$$Q = \omega_0 \frac{L}{R} = \frac{1}{\omega_0 RC} = \frac{\omega_0}{B} \tag{5.56}$$

In filters that must be tuned to a specific frequency, the quality factor should be high. If a filter must pass a wider range, the quality factor is lower. A higher Q indicates a narrower passband. If the quality factor is high (typically, $Q \geq 10$), the cutoff frequencies are symmetrical with respect to the resonance frequency and can be approximated as

$$\omega_1 \cong \omega_0 - \frac{B}{2}; \omega_2 \cong \omega_0 + \frac{B}{2} \tag{5.57}$$

5.5.6.2 Active Band-Pass Filters

A band-pass active filter can be designed by combining a low-pass and high-pass filter. In the configuration depicted in Table 5.1, both reactive elements are capacitors. The transfer function of the active band-pass filter is calculated using the expression of the op-amp gain in the frequency domain.

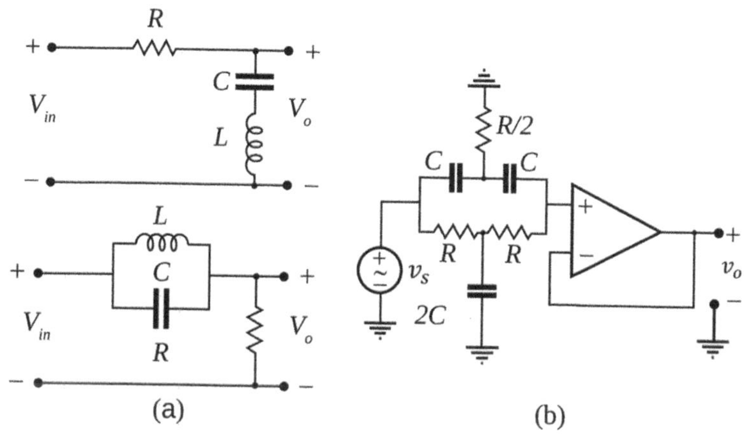

Figure 5.18 Band-reject (notch) filters (a: Passive; b: Active)

$$G(s) = -\frac{Z_f(s)}{Z_s(s)} = -\left[\frac{1}{R_f} + sC_f\right]^{-1} \cdot \left[R_s + \frac{1}{sC_s}\right]^{-1} = \frac{sR_fC_s}{(1 + sR_fC_f)(1 + sR_sC_s)}$$
$$G(j\omega) = -\frac{j\omega R_fC_s}{(1 + j\omega R_fC_f)(1 + j\omega R_sC_s)} \tag{5.58}$$

5.5.7 Band-Reject Filters

Band-reject filters, also known as band-stop and notch filters, reduce a narrow frequency band and pass all other frequencies. Both passive and active stop-band filters are RC circuits with two cutoff frequencies. They can be designed by combining a high-pass and a low-pass filter in parallel as shown in Figure 5.18.

The simplest passive band-reject filter is made by swapping the resistor and LC section of a series RLC band-pass filter. The transfer function of the series band-reject filter depicted in Table 5.1 is

$$H(j\omega) = \frac{V_o}{V_{in}} = \frac{j\left(\omega L - \frac{1}{\omega C}\right)}{R + j\left(\omega L - \frac{1}{\omega C}\right)}$$
$$|H(j\omega)| = \frac{\omega L - \frac{1}{\omega C}}{\sqrt{R^2 + \left(\omega L - \frac{1}{\omega C}\right)^2}} = \frac{\omega^2 LC - 1}{\sqrt{(\omega RC)^2 + (\omega^2 LC - 1)^2}} \tag{5.59}$$

The magnitude is 1 at $\omega = 0$ (DC) and approaches 1 at high frequencies. At the resonance frequency ω_0, the magnitude is zero since the LC portion of the filter short circuits the output.

$$\omega_0 = \sqrt{LC} \tag{5.60}$$

Bandwidth and quality factors are the same as in band-pass filter.

$$\omega^2(RC)^2 - 4\omega RC + 1 = 0$$
$$\omega_{1,2} = 2RC \pm \sqrt{(2RC)^2 - 4} \tag{5.61}$$

Example 5.10 A notch filter has the following circuit parameters.

$$R = 3.3\,\Omega, L = 20\,\mu\text{H}, C = 2\,\mu\text{F}$$

Find

a) The center frequency
b) The half-power (cutoff) frequencies
c) Bandwidth
d) The quality factor

Solution

a) Center frequency:

$$f_0 = \frac{1}{2\pi\sqrt{LC}} = \frac{1}{2\pi\sqrt{(20\times10^{-6})\times(2\times10^{-6})}} = 25{,}000\text{ Hz} = 25\text{ kHz}$$

b) Half-power frequencies:

$$\omega_{1,2} = \pm\frac{R}{2L} + \sqrt{\left(\frac{R}{2L}\right)^2 + \frac{1}{LC}}$$

$$\frac{R}{2L} = \frac{3.3}{2\times20\times10^{-6}} = 82{,}500$$

$$\sqrt{\left(\frac{R}{2L}\right)^2 + \frac{1}{LC}} = \sqrt{(12{,}500)^2 + \frac{1}{(20\times10^{-6})\times(2\times10^{-6})}} = 178{,}343$$

$$\omega_1 = 178{,}343 - 82{,}500 = 95{,}843$$

$$\omega_2 = 178{,}343 + 82{,}500 = 260{,}843$$

$$f_1 = \frac{95{,}843}{2\pi} = 15{,}253\text{ Hz}; f_2 = \frac{260{,}843}{2\pi} = 41{,}514\text{ Hz}$$

Bandwidth: 41,514 – 15,253 = 26,261 Hz

a) Quality factor:

$$Q = 2\pi f_0\frac{L}{R} = 2\pi\times25{,}000\frac{20\times10^{-6}}{3.3} = 0.952$$

5.6 High-Order Filters

The low-pass and high-pass filters discussed in the preceding section each contain a single reactive component, such as a capacitor or an inductor. They have one cutoff frequency, and their stop-band gain drops at a rate of 20 dB/decade (roll-off rate). Band-pass and band-stop filters have two cutoff frequencies and a 20 dB/decade roll-off rate in each stop-band region.

The order of a filter indicates the number of reactive elements (capacitors and inductors) or poles in its design. Higher-order filters include more reactive elements and have a steeper roll-off rate, which separates different frequency bands more effectively. Each additional order adds an extra 20 dB/decade (or 6 dB/octave) to the filter's roll-off rate. Therefore, higher-order filters have a more rapid transition from the passband to the stop-band. This means they can more effectively separate the desired signals from unwanted noise or interference. Furthermore, they provide greater attenuation of frequencies outside the passband, making them more efficient at blocking unwanted distortion on signals.

Higher-order filters can more precisely target specific frequency ranges, which is particularly useful in applications that require sharp frequency cutoffs. They generally have lower ripple in the passband and stop-band, leading to smoother signal processing and higher fidelity.

In mechatronic control systems, higher-order filters can improve stability by more effectively dampening oscillations and reducing overshoots. They offer more design parameters, allowing for customization to meet specific application requirements.

Higher-order passive filters are designed by cascading multiple lower-order filter stages and how active high-order filters are made. Active high-order filters can be created by combining multiple first-order and second-order filters in a cascaded configuration. Special filter configurations, such as the Sallen–Key, Butterworth, and Chebyshev filters, are also available for optimizing the design of higher-order filters.

5.6.1 Sallen–Key Filter

The Sallen–Key filter is a popular electronic filter topology invented by R.P. Sallen and E.L. Key in 1955. It is commonly used in analog signal processing (Zumbahlen, 2008). This filter can be configured to operate as a low-pass, high-pass, band-pass, or band-stop filter. It uses a minimal number of components (typically two resistors and two capacitors) along with an op-amp. They provide stable filter performance with predictable behavior and allow for a wide range of frequency and gain settings.

Sallen–Key filters are widely used in audio electronics, instrumentation, and various communication systems. They are especially valued for their ease of design and implementation.

5.6.2 Butterworth Filters

Butterworth filters are designed to have a flat frequency response in the passband and a smooth transition to zero in the stop-band. This type of filter is known for its maximally flat magnitude response, meaning it has no ripples in the passband. The Butterworth filter achieves this characteristic by a polynomial function in its transfer function, which ensures that all derivatives of the magnitude response are zero at the cutoff frequency. This results in a smooth and reliable performance, making Butterworth filters ideal for applications requiring a clean and consistent signal, such as audio processing and communication systems. They are commonly employed in scenarios where a smooth frequency response is crucial, preventing any alterations or distortions in the signal within the desired frequency range.

5.7 Modulation

Signal modulation is the process of altering the properties of a signal such as its amplitude, frequency, and phase in order to encode information it carries. This allows for the efficient data exchange between internal components of a system and communication through an external network over various communication channels, such as electrical wires, coaxial cables, radio waves, infrared light, and fiber optics. Common types of signal modulation are amplitude modulation (AM), frequency modulation (FM), phase modulation (PM), PWM, and pulse code modulation (PCM).

AM and FM are essential modulation techniques used in analog communication systems, and they are based on varying the amplitude and frequency of the carrier wave. PWM and PCM are both crucial techniques used in mechatronics for different purposes.

PWM is used to control the power delivered to an electronic device by rapidly switching the power on and off. Changing the duty ratio, which is the ratio of the on- and off-time of the pulse train, causes the average voltage to change. This technique is commonly used in mechatronics for motor control, light dimming, and temperature regulation.

PCM is a method developed to digitally represent analog signals. In mechatronics, PCM is often used for transmission and digital processing of analog sensor data. PCM converts analog signals obtained from sensors such as temperature, pressure, and velocity into a digital format that can be processed by microcontrollers or DSP circuits. Digital data can be stored or transmitted to DACs to control motors and various actuators. In mechatronics, key steps in PCM are sampling, quantization, and analog-to digital conversion. We will discuss these steps in the following sections.

Both PWM and PCM play crucial roles in the design and operation of mechatronic systems, enabling precise control and efficient data communication.

5.8 Summary

This section outlines signal conditioning techniques, including how analog and digital signals are mathematically represented, as well as methods of amplification and filtering. Signal conditioning forms a critical foundation in mechatronic systems, where diverse elements such as sensors, actuators, and electronic circuits interact to perform intended tasks.

- Mechatronic systems integrate various components including electronic sensors, digital logic ICs, power electronics, and actuators. Signal sources produce low-power analog or digital signals that often require conditioning to correct distortions, reduce noise, and match impedances for effective processing. Signal conditioning circuits amplify signals, filter noise, and may perform mathematical operations such as addition, subtraction, integration, or differentiation to improve system response and reduce output error.
- Signals in mechatronics represent physical quantities like velocity, force, pressure, and temperature, typically converted by sensors into analog or digital signals. Analog signals vary continuously over time, while digital signals are discrete-time sequences with constant values over

sampling intervals. Some sensors produce digital signals directly, such as rotational sensors generating pulses or optical sensors detecting positions.

- Periodic signals repeat their values at regular intervals characterized by amplitude and frequency. Sinusoidal signals are described by sine or cosine functions with angular frequency and phase shift parameters. Nonsinusoidal periodic signals such as square, rectangular, triangular waves, and pulse trains are common in mechatronics. The Fourier series decomposes these signals into sums of sinusoidal components (harmonics) at integer multiples of the fundamental frequency, facilitating analysis and processing.
- An example demonstrates calculating Fourier series components of a rectangular pulse train with 5 V amplitude and 10 Hz frequency, using MATLAB® code to compute coefficients via the sinc function and reconstruct the signal. Increasing the number of harmonics improves reconstruction accuracy.
- Aperiodic signals, such as single pulses, are common outputs in mechatronics for event triggering and system testing. Singularity functions like the unit impulse (Dirac delta), unit step, and unit ramp functions model switching signals and abrupt changes. The unit impulse function, with infinite amplitude at a single point and integral equal to one, is fundamental in DSP and sampling theory. The unit step function represents a signal switching on at a defined time, and the unit ramp function models a linearly increasing signal. These functions underpin signal analysis and control system design.
- While Fourier series applies to periodic signals, the Fourier transform generalizes this to aperiodic signals, representing them as integrals of sinusoidal components over a continuous frequency spectrum. This transformation enables frequency-domain analysis of signals, useful for identifying dominant frequencies and characterizing signal behavior. MATLAB® scripts illustrate computing and plotting the Fourier transform of a rectangular pulse, showing its frequency spectrum.
- Sensors and transducers generate electrical signals proportional to physical phenomena. Active sensors produce voltages directly, while passive sensors (e.g., variable resistors, thermistors) require additional circuitry for signal output. Complex signal sources are often modeled using Thevenin's and Norton's theorems, which simplify two-terminal linear circuits into equivalent voltage or current sources with series or parallel resistances. These models aid in analyzing and designing signal conditioning circuits.
- An example describes a rotational position sensor using a potentiometer as a voltage divider, deriving its Thevenin equivalent circuit where the output voltage is proportional to the rotation angle and equivalent resistance varies nonlinearly with angle.
- Sensor outputs are typically low-level signals requiring amplification before processing. Ideal amplifiers increase signal magnitude without distortion, characterized by voltage, current, and power gains, often expressed in decibels (dB). Amplifiers are two-port devices with input and output impedances; high input impedance and low output impedance are desirable for signal integrity.
- Cascading amplifiers require careful impedance matching to avoid gain reduction due to voltage dividers formed by input and output resistances.
- Op-amps are fundamental analog components composed of multiple transistors and passive elements, typically packaged as ICs. They have high gain, infinite input impedance, and zero output impedance in ideal models. Op-amps have two input terminals (inverting and noninverting) and one output terminal. They are versatile in signal conditioning for amplification, filtering, integration, differentiation, and comparison.

- An op-amp without feedback acts as a comparator, outputting saturated voltage levels depending on input signal relative to a reference. Positive feedback introduces hysteresis, preventing oscillations near threshold levels, useful in control systems and analog-to-digital conversion.
- Input signal applied to the inverting terminal through resistor; feedback resistor connects output to the inverting input. Gain depends on resistor ratio and introduces a 180° phase shift. Virtual ground concept applies, assuming zero voltage difference between inputs. Practical op-amps have finite open-loop gain, affecting closed-loop gain slightly.
- Input signal applied to the noninverting terminal; feedback loop connects output to inverting terminal with resistors. Gain is positive and depends on resistor ratio. Input impedance is ideally infinite, advantageous for buffering signals.
- Summing amplifier adds multiple input signals weighted by resistor ratios, producing an inverted sum at output. Useful for combining sensor outputs.
- Differential amplifier amplifies difference between two input signals, combining the inverting and noninverting configurations. Sensitive to resistor mismatch and common-mode signals, limiting input resistance and accuracy.
- A specialized differential amplifier with high input impedance, low output impedance, and high CMRR, essential for accurate sensor signal conditioning. Typically implemented with three op-amps and precision resistors, often available as ICs.
- Integrator and differentiator circuits perform the mathematical operations of integration and differentiation. In an IC, the feedback resistor of an inverting amplifier is replaced by a capacitor, resulting in an output proportional to the integral of the input. These circuits are applied in signal processing and control to determine quantities such as distance from velocity. A differentiator circuit is constructed by placing a capacitor in series with the input and a resistor in the feedback loop, which produces an output corresponding to the derivative of the signal with gain.
- Filters remove undesired frequency components to enhance signal quality and system stability. They are characterized by frequency response, transfer function, and cutoff frequencies, visualized via Bode plots showing magnitude and phase. Filters are classified as low-pass, high-pass, band-pass, and band-reject based on frequency response. Transfer functions use complex frequency variables; magnitude in dB and phase angle describe filter behavior. Roll-off rate indicates attenuation steepness beyond cutoff frequency, increasing with filter order.
- Passive filters use resistors, capacitors, and inductors; no power supply needed; simple and reliable but bulky for inductors at high frequencies; no signal gain; lower noise. Active filters employ op-amps and RC circuits; provide gain and tunability; no inductors needed; compact and flexible; better selectivity and impedance matching.
- Low-pass filters pass signals below cutoff frequency; attenuate higher frequencies. They are used for noise reduction, smoothing PWM signals, and machinery monitoring.
- High-pass filters pass signals above cutoff frequency; block lower frequencies. Applications include removing low-frequency noise in sensors, communication, and audio devices.
- Band-pass filters pass signals within a frequency band; block frequencies outside this range. They are used for isolating vibration frequencies, selecting communication channels, and improving control system stability.
- Band-reject (notch) filters reject narrow frequency bands while passing others. They are combination of high-pass and low-pass filters in parallel.
- Higher-order filters use multiple reactive elements for steeper roll-off rates and improved selectivity. Cascading lower-order filters or using configurations like Sallen–Key, Butterworth, and Chebyshev filters optimize performance. Butterworth filters provide maximally flat passband response, ideal for audio and communication applications.

- Signal modulation alters amplitude, frequency, or phase to encode information for internal communication or external transmission. Key types include AM, FM, PWM, and PCM. PWM controls power delivery by varying pulse duty cycle, commonly used in motor control, lighting, and temperature regulation. PCM digitizes analog sensor signals for processing and transmission, involving sampling, quantization, and analog-to-digital conversion.

5.9 Check Your Understanding

5.9.1 Self-Assessment Quiz

1 Which of the following is a periodic signal?
- **a.** Unit step
- **b.** A rectangular pulse
- **c.** Square wave
- **d.** Sinc function

2 Thevenin equivalent of a signal source consists of
- **a.** A current source and a series impedance
- **b.** A current source and a parallel impedance
- **c.** A voltage source and a parallel impedance
- **d.** A voltage source and a series impedance

3 The ideal op-amp responds only to the
- **a.** Difference of the input signals
- **b.** Positive input signal
- **c.** Negative input signal
- **d.** Sum of the two input signals

4 Negative feedback is applied to an op-amp by connecting
- **a.** A passive component between its output terminal and its inverting (negative) input terminal
- **b.** An active component between its output terminal and its inverting (negative) input terminal
- **c.** A passive component between its output terminal and its noninverting (positive) input terminal
- **d.** An active component between its output terminal and its noninverting (positive) input terminal

5 Between which terminals of an ideal op-amp a virtual short circuit is said to exists?
- **a.** Two supply terminals
- **b.** Input and output terminals
- **c.** Two input terminals
- **d.** Output and ground

6 If the positive input terminal is connected to ground, a virtual ground appears on the
- **a.** Output terminal
- **b.** Negative input terminal

c. Compensation terminal
d. Positive supply terminal

7 In an ideal op-amp with negative feedback,
a. The input impedance is zero
b. The output impedance is infinity
c. Output current is zero
d. Two input terminals are at the same voltage

8 Given the series resistance 5 kΩ and the feedback resistance 10 kΩ, the gain of an inverting amplifier is
a. 0.5
b. 2
c. 15
d. 5

9 Given the series resistance 5 kΩ and the feedback resistance 10 kΩ, the gain of a noninverting amplifier is
a. 2
b. 0.5
c. 3
d. 1.5

10 If the feedback resistor of an op-amp is replaced with a capacitor, the circuit becomes
a. A differentiator
b. An integrator
c. Unstable
d. A summing amplifier

11 A differentiator circuit has a
a. Series capacitor and feedback resistor
b. Series resistor and feedback capacitor
c. Series capacitor and a feedback inductor
d. Series resistor and feedback inductor

12 An ideal voltage follower has
a. Zero input resistance
b. Infinite input resistance
c. Infinite gain
d. Infinite output resistance

13 The gain of a filter at a cutoff frequency is
a. 3 dB
b. −3 dB
c. 20 dB
d. 6 fB

14 In a first-order low-pass filter, the slope of the asymptote of the gain plot above the cutoff frequency is
  a. 20 dB/decade
  b. 40 dB/decade
  c. 0 dB/decade
  d. −20 dB/decade

15 Which one of the following properties is desirable for an instrumentation amplifier?
  a. Low input resistance
  b. High common-mode rejection ratio
  c. High differential gain
  d. Both b and c

Answers:

1: c, 2: d, 3: a, 4: a, 5: c, 6: b, 7: a, 8: b, 9: c, 10: b, 11: a, 12: b, 13: b, 14: d, 15: d

5.9.2 Short Answer Questions

1 What is the primary function of signal conditioning circuits in mechatronic systems?

2 How do signal conditioning circuits enhance the quality of signals in mechatronic systems?

3 What are the key functions of signal conditioning circuits in feedback control systems?

4 Define a periodic signal and give an example.

5 What is the significance of the Fourier series in analyzing nonsinusoidal periodic signals?

6 How does an operational amplifier (op-amp) contribute to signal conditioning in mechatronic systems?

7 What are the different types of filters used in mechatronic systems?

8 Explain the role of Thevenin's and Norton's theorems in simplifying the analysis of signal conditioning circuits.

9 What is the difference between analog and digital signals in mechatronic systems?

10 How do integrator and differentiator circuits function in mechatronic systems?

5.9.3 Essay Questions

1 How do signal conditioning circuits enhance the quality of signals in mechatronic systems? Discuss the key functions and their importance in system performance.

2 Explain the Fourier series representation of periodic signals and its significance in analyzing and processing nonsinusoidal signals in mechatronic systems.

3 How do operational amplifiers contribute to signal conditioning in mechatronic systems? Discuss their various configurations and applications.

4 What are the different types of filters used in mechatronic systems, and how do they improve signal quality and system stability? Provide examples of their applications.

5 How do Thevenin's and Norton's theorems simplify the analysis and design of signal conditioning circuits in mechatronic systems? Discuss their practical applications with examples.

6 How do modulation techniques like PWM and PCM facilitate efficient data exchange and control in mechatronic systems? Discuss their principles and applications.

7 How have advancements in signal amplification techniques impacted the design and performance of mechatronic systems? Discuss the role of ideal and practical amplifiers.

8 How do integrator and differentiator circuits function in mechatronic systems? Discuss their mathematical principles and practical applications.

9 Why are higher-order filters preferred in certain applications? Discuss their advantages and the design considerations involved.

10 How does MATLAB® facilitate the analysis and design of signal conditioning circuits in mechatronic systems? Provide examples of its applications in Fourier series and filter design.

5.9.4 Problems

1 Two identical amplifiers are tandem connected to amplify a signal with 10 mV amplitude. The signal source has 50 Ω internal resistance. The amplifiers have 500 Ω input resistance, 100 Ω output resistance and 10 dB voltage gain. Find the overall gain and the amplitude of the output voltage.

2 The op-amp in voltage follower configuration has an open-loop gain of 150,000, an input resistance of 10 kΩ, and an output resistance of 100 Ω. Find the voltage gain using the nonideal model of the op-amp.

3 The open-loop gain of an op-amp chip is 100,000. It is connected in an inverting amplifier configuration with 2.2 k series and 33 k feedback resistors. Find the gain.

4 The input signal of a differentiator is a 1-kHz triangular wave of amplitude 0.1 V. Select a resistance and capacitance such that the amplitude of the output is 5 V. Sketch the output waveform.

5 Design a differentiator to amplify 0–0.5 V DC sensor signal to 0–5 V logic level.

6 Design an active filter to obtain quasi-sinusoidal waveform from 100 Hz square wave with amplitude 5 V. Plot the output waveform.

References

Alexander, C. K. & Sadiku, M. N. O., 2013. *Fundamentals of Electric Circuits*. 5th ed. New York, NY: McGraw-Hill.

Chapra, S. C. & Canale, R. P., 2006. *Numerical Methods for Engineers*. 5th ed. New York, NY: McGraw Hill.

Dally, J. W., Riley, W. F. & McConnell, K. G., 1984. *Instrumentation for Engineering Measurements*. 2nd ed. New York: John Wiley.

Horowitz, P. & Hill, W., 1989. *The Art of Electronics*. 2nd ed. Cambridge: Cambridge University Press.

Rizzoni, G. & Kearns, J., 2007. *Fundamentals of Electrical Engineering*. 6th ed. New York: McGraw-Hill.

Sedra, A. S., Smith, K. C., Carusone, T. C. & Gaudet, V., 2020. *Microelectronic Circuits*. 8th ed. New York, Oxford: Oxford University Press.

Smith, S. W., 2003. *Digital Signal Processing - A Practical Guide for Engineers and Scientists*. Amsterdam: Newnes.

Zumbahlen, H., 2008. *Basic Linear Design Handbook*. Amsterdam: Newnes/Elsevier.

6

Magnetic Circuits and Static Devices

OBJECTIVES

- Explain the concept of magnetic circuits and their analogy with electric circuits.
- Describe the properties of different magnetic materials.
- Present solution techniques for magnetic circuit problems.
- Present an introduction to static electromagnetic devices such as transformers and inductors.
- Discuss practical applications of magnetic circuits and static devices in mechatronics.

Electric motors, generators, transformers, actuators, and magnetic sensors rely on interactions between electric circuits and magnetic fields. Electric current flowing on a conductor creates a magnetic field in the surrounding space. A magnetic field changing in time produces an electric potential across the conductors placed around it.

6.1 Fundamental Concepts

6.1.1 Magnetic Field Intensity

In 1820, Hans Christian Oersted discovered in Denmark that a current in a wire affected a compass needle. This observation led several physicists to investigate the relationship between electric current and magnetic field. Experiments have shown that the magnetic field intensity represented by H is proportional to the current and inversely proportional to the radius r of the circle defining the field line. For a long straight wire, the magnetic field intensity is

$$H = \frac{I}{2\pi r} \tag{6.1}$$

The unit of magnetic field strength is A/m. The magnitude of the magnetic field intensity vector described in equation 6.1 is independent of the medium surrounding the wire. Its value is the ratio of the current and the perimeter of the circle of radius r.

6.1.2 Magnetic Induction

Magnetic induction, also known as magnetic flux density, is represented by vector B. This vector describes the density of magnetic field lines passing through a given area. To include the magnetic properties of the environment surrounding the wire, vector H is multiplied by a constant called magnetic permeability.

$$\vec{B} = \mu\vec{H} \tag{6.2}$$

The value of the permeability of vacuum in the standard international unit system (SI) is $\mu_0 = 4\pi \times 10^{-7}$. Permeability of all materials is defined by multiplying with a constant named relative permeability represented as μ_r. Relative permeability of air and nonmagnetic materials like wood, plastic, glass, and any other material that a magnet does not attract is close to unity. Any substance that a magnet attracts is called ferromagnetic. The relative permeability of ferromagnetic materials is very high, typically in the order of thousands.

6.1.3 Magnetic Flux

In the early years of the magnetic field theory, scientists were visualizing magnetism as field lines that form a magnetic flux. Although magnetic field does not actually flow through matter, the concept of magnetic flux has been helpful in the analysis and design of electromagnetic devices. Magnetic flux is expressed in weber in the international unit system (SI).

Magnetic flux is the scalar (dot) product of the B-vector and the A-vector, which is the normal vector of the area where the magnetic field takes place. If the magnetic field vectors are in different directions at every point of the surface, magnetic flux at each point is obtained from the scalar product of the flux density vector at that point and the normal vector perpendicular to the infinitesimal surface element with magnitude dA.

$$\phi = \iint \vec{B} \cdot d\vec{A} \tag{6.3}$$

If the flux density is constant at every point in the same angle α with respect to the normal direction of the surface, the expression of flux is reduced to a simple multiplication.

$$\phi = B \cdot A \cos\alpha \tag{6.4}$$

In most practical magnetic devices, the induction vector is perpendicular to the surface, and the expression is further simplified since $\cos\alpha = 1$. Substituting equations 6.1 and 6.2 in 6.4, the expression of magnetic flux becomes

$$\phi = B \cdot A = \frac{\mu A I}{2\pi r} \tag{6.5}$$

In the international unit system (SI), flux density is in the dimension of Weber per square meter (Wb/m^2). The General Conference on Weights and Measures named the standard unit of flux density Tesla (T) in 1960 in honor of Nikola Tesla, who made enormous contributions to the development of electric machines. In commercial specifications, the older unit Gauss is still being used. 1 T is equal to 10,000 Gauss (G) (Giancoli, 2020).

In the construction of practical devices, the magnitude of the magnetic flux is increased by using tightly wound coils with a large number of turns. In the case of an N-turn coil, each turn links the same flux. In practice, it is convenient to define the flux linkage as

$$\lambda = N\phi = \frac{\mu N A}{2\pi r} I \tag{6.6}$$

6.1.4 Ampere's Circuital Law

Six years after Oersted published his experiments, French mathematician and physicist André-Marie Ampere suggested the mathematical equation to calculate the magnetic field strength around a current-carrying wire, known as Ampere's circuital law.

Ampere's circuital law states that the line integral of the magnetic field intensity (H) around a closed path is equal to the total current enclosed by that path. Mathematically, this can be expressed as

$$\oint H \cdot d\ell = \sum I_{enclosed} \tag{6.7}$$

This law is fundamental in the study of electromagnetism, as it provides a relationship between electric currents and the magnetic fields they generate. It is particularly useful for calculating the magnetic field in systems with symmetrical current distributions, such as solenoids and toroidal coils.

Ampere's law is based on the principle that a current-carrying conductor creates a magnetic field in the space around it. The direction and magnitude of this field can be determined using Ampere's circuital law, which takes into account the geometry of the conductor and the distribution of the current.

Magnetic field is a vector quantity that has both magnitude and direction. In the early nineteenth century, scientists described the direction of magnetic field around a current-carrying wire by observations, mostly using compass needles, small permanent magnets, and iron powder. While the magnetic field occurs at every point in the space, it was easier to visualize it by field lines. Current flowing through a straight line creates circular magnetic field lines on a plane perpendicular to the wire. The direction of the field vectors reverses if the current flows in the opposite direction.

Note that the directions of current and magnetic field are arbitrary. The first scientists, who used a compass to sense the existence of a magnetic field, assumed the direction of the compass needle showing the North Pole as the positive direction for the magnetic field. To find the direction of the magnetic field around a wire carrying current, we can use the right-hand rule. According to this rule, wrap your fingers around the wire, the thumb being in the direction of current flow. Your fingers will show the direction of the magnetic field vectors.

6.1.5 Faraday's Law of Induction

Electromagnetic induction is one of the most important principles in mechatronics. The operation of all moving electromechanical devices is based on the induced voltage across a coil due to a time-varying magnetic field. British experimental scientist Michael Faraday discovered that moving a coil supplied from a liquid battery in another stationary coil induced a current in the stationary coil. James Clerk Maxwell described this experiment mathematically and named it Faraday's law of induction. Energy transferred by electromagnetic induction is named electromotive force (EMF) and expressed as the time derivative of the total magnetic flux enclosed by an-turn coil, which we defined above as flux linkage λ.

$$e = \frac{d\lambda}{dt} = N\frac{d\phi}{dt} \tag{6.8}$$

In 1834, Heinrich Lenz described the direction of the induced EMF and current resulting from electromagnetic induction. Based on the Lenz law, the direction of the electric current induced in a conductor by a changing magnetic field is such that the magnetic field created by the induced current opposes changes in the initial magnetic field. The Lenz law adds a negative sign to the right side of equation 6.8.

American physicist Joseph Henry's early discovered electromagnetic induction in 1831, about the same time as Michael Faraday. However, because Faraday published his results before Henry, he is credited with the discovery of magnetic induction. The unit of self-induction was assigned "henry" to honor Joseph Henry's contributions to the research and development in the field of electromagnetics.

6.2 Magnetic Circuits

Electromechanical systems such as transformers, machines, actuators, and transducers have complex magnetic structures linking multiple coils. Analysis of the magnetic flux distribution in such devices is generally simplified by modeling the structure like electric circuits and applying circuit analysis methods.

A magnetic circuit is an analogy to an electric circuit, used to analyze static or low-frequency magnetic fields. It simplifies the understanding of complex interactions in magnetic systems, enabling straightforward computation and assessment of magnetic fields within materials. If the spatial distribution of the magnetic field is needed, advanced numerical techniques such as finite differences, finite elements, and finite volumes can be employed. These methods offer precise solutions by discretizing the magnetic domain into smaller elements and solving the field equations within these elements iteratively. They are particularly useful for complex geometries and inhomogeneous materials where analytical solutions are impractical. Within the context of this textbook, magnetic circuit equations offer adequate precision for designing electromechanical transducers and electric machines.

Electromechanical devices typically contain one or more coils wound around a ferromagnetic core. The purpose of the core is to keep the magnetic flux in a specified path and enhance its magnitude. Figure 6.1 illustrates a single coil placed on a toroidal ferromagnetic core with a narrow air gap. This structure is commonly used as an inductor. Transformers, practical transducers, and electric machines have several windings and more complex core structures where the magnetic flux completes its path through several ferromagnetic sections and air gaps with varying geometry.

6.2.1 Magnetomotive Force (MMF)

Ampere's circuital law states that the line integral of magnetic field intensity H around a closed path equals the total current enclosed by the path. When the magnetic field intensity is uniform within

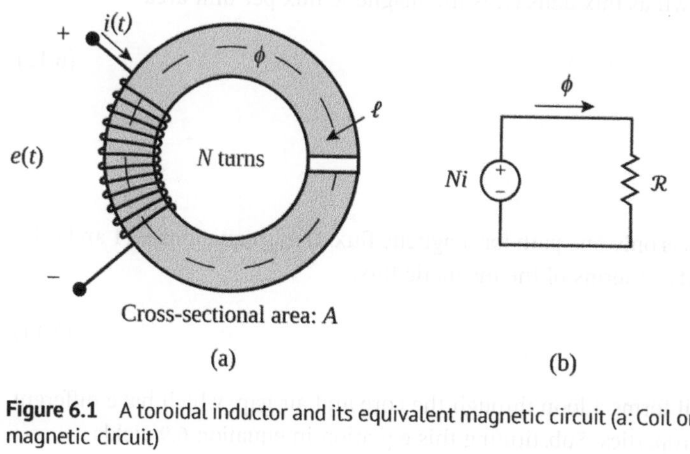

Figure 6.1 A toroidal inductor and its equivalent magnetic circuit (a: Coil on a toroidal core; b: Corresponding magnetic circuit)

the ferromagnetic core, and the fringe flux in the air gap is negligible, Ampere's law can be simplified for a magnetic device consisting of m coils and n core segments with varying magnetic properties by replacing the integral with summation operations.

$$\sum_{j=1}^{m} N_j I_j = \sum_{k=1}^{n} H_k \cdot \ell_k \tag{6.9}$$

The inductor shown in Figure 6.1 (a) has a single coil of N turns where the current I flows. As a result, the left side of the equation becomes NI. In the context of magnetic systems, the product of the number of turns and the current passing through the coil is defined as magnetomotive force (MMF).

$$\mathcal{F} = Ni \tag{6.10}$$

MMF is the driving force of the magnetic field, analogous to the EMF in electric circuits.

Equation 6.9 indicates that the magnetic field intensity is independent of the magnetic properties of the core material. Magnetic induction B is calculated by multiplying the field intensity by permeability, which reflects the characteristics of the medium where the magnetic field exists. The permeability is represented by μ.

$$B = \mu H \tag{6.11}$$

The permeability of vacuum, denoted as μ_0, is a fundamental physics constant with a value of $4\pi \times 10^{-7}$ H/m. This constant serves as the reference for specifying the permeability of other materials.

$$\mu = \mu_0 \mu_r \tag{6.12}$$

The permeability of a material determines how a material responds to an applied magnetic field. Permeability can vary widely among different substances. The permeability of air is practically the same as vacuum. Nonmagnetic materials, like wood, glass, plastics, aluminum, copper, and any metal that a magnet does not attract, have relative permeability near unity, making them less effective in channeling magnetic fields. Ferromagnetic materials have significantly higher relative permeability, which intensifies magnetic induction and allows them to direct the magnetic flux within a specified path. These characteristics make them suitable for use in electromechanical devices.

Magnetic induction, also known as flux density, is the magnetic flux per unit area.

$$B = \frac{\phi}{A} \tag{6.13}$$

6.2.2 Reluctance

The structure in Figure 6.2 (a) has only one path for magnetic flux. Using equations 6.11 and 6.13, we can express the field intensity in terms of the magnetic flux.

$$H = \frac{B}{\mu} = \frac{\phi}{\mu A} \tag{6.14}$$

The flux generated by the coil forms a loop through the core and air gap, which have different average length and magnetic properties. Substituting this equation in equation 6.9 yields

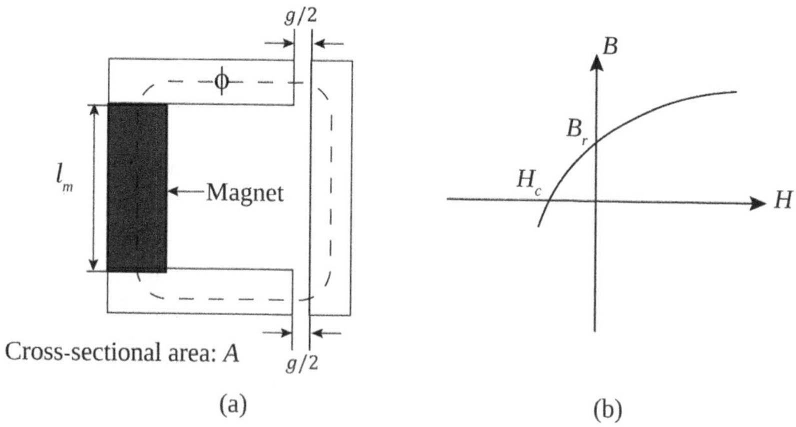

Figure 6.2 A magnetic circuit driven by permanent magnet (a: Magnetic structure; b: B-H curve)

$$\mathcal{F} = Ni = \phi\left[\frac{\ell_c}{\mu_c\mu_0 A_c} + \frac{g}{\mu_0 A_g}\right] \tag{6.15}$$

where ℓ is the average length of the magnetic flux and A is the cross-sectional area perpendicular to the flux. If the core material is in the linear region of its B–H characteristic, its permeability μ_c can be assumed to be constant (not a function of the current). Consequently, the term in the brackets is constant. Equation 6.15 indicates that in a linear magnetic circuit, the MMF is proportional to the magnetic flux.

In magnetic circuits, MMF is analogous to the EMF, and ϕ is analogous to the electric current. The term "magnetic reluctance" denoted as $\mathcal{R}$ is defined by analogy with the electric resistance of a conductor wire, as a measure of the opposition of the medium to the magnetic flux.

The reluctance of a segment of a magnetic circuit of length ℓ and cross-sectional area A perpendicular to the flux and permeability μ is

$$\mathcal{R} = \frac{\ell}{\mu A} \tag{6.16}$$

The following equations represent the reluctance of the ferromagnetic core and the air gap:

$$\mathcal{R}_c = \frac{\ell_c}{\mu_c\mu_0 A_c}$$
$$\mathcal{R}_g = \frac{g}{\mu_0 A_g} \tag{6.17}$$

By substituting the reluctances in equation 6.15, we obtain the equivalent of Ohm's law in magnetic systems. Figure 6.1 (b) is the magnetic circuit corresponding to the solenoid illustrated in Figure 6.1 (a).

Any magnetic structure with a linear core material can be analyzed using a magnetic equivalent circuit formed by connecting MMFs and reluctances. Magnetic circuits allow the use of DC circuit analysis methods (Kraus, 1991). Table 6.1 summarizes the analogy between variables used in electric and magnetic circuits.

Table 6.1 Analogy between electric and magnetic circuits

Magnetic Circuits	Electric Circuits
Magnetic flux ϕ	Current I
Magnetomotive force (MMF) $\mathcal{F} = NI$	Electromotive force (EMF) E
Reluctance $\mathcal{R} = \dfrac{\ell}{\mu A}$	Resistance $R = \dfrac{\ell}{\sigma A}$
Permeability μ	Conductivity σ
$\sum \phi_k = 0$	Kirchhoff's current law (KCL) $\sum I_k = 0$
$\sum \mathcal{F}_k = 0$	Kirchhoff's voltage law (KVL) $\sum V_k = 0$
$\mathcal{F} = \mathcal{R}\phi$	Ohm's law $V = RI$

6.2.3 Fringe Flux

Fringe magnetic flux, also known as magnetic flux leakage, occurs when the magnetic flux lines do not follow the intended path within a magnetic circuit. This phenomenon is often observed in systems with air gaps, where some portion of the magnetic flux crosses the air gap but completes its path outside the main core.

Fringe magnetic flux usually arises due to imperfections in the magnetic circuit such as discontinuity in the magnetic path that cause flux lines to stray into air gaps and variations in permeability due to flux leakage. Sharp edges and corners in the magnetic structure can influence the flux distribution.

Fringe magnetic flux can have several implications on the performance of a magnetic circuit including reduced efficiency, localized heating due to eddy currents, and measurement errors in instruments relying on magnetic flux detection.

In practical engineering applications, fringe magnetic flux is often minimized by using magnetic shields to guide the flux lines, designing smooth transitions and avoiding sharp edges, and using materials with uniform permeability.

Example 6.1 The electromagnetic device illustrated in Figure 6.1 has a toroidal ferromagnetic core of radius $R_1 = 45$ mm and $R_2 = 55$ mm with an air gap of $g = 1$ mm. The thickness of the core is $d = 20$ mm, and relative permeability of the ferromagnetic material is $\mu_r = 3{,}000$. The 100-turn coil is supplied from a DC source, and the current flowing through the coil is 2 A.

a) Calculate the reluctances of the ferromagnetic core and air gap.
b) Determine the flux in the magnetic circuit.
c) Calculate the flux density in the core.

Solution

The core is made from a toroidal ferromagnetic material with an air gap. The relative permeability of the ferromagnetic material is given as 3,000, and the permeability of air is approximately equal to $\mu_0 = 4\pi 10^{-7}$. The structure has only one path for magnetic flux. The equivalent circuit can be drawn as shown below. Note that the reference directions for the MMF and flux are selected similarly to the passive notation we used for electric circuits. That is, the flux arrow leaves the MMF from the positive sign.

$$\phi = \frac{NI}{\mathcal{R}_1 + \mathcal{R}_2}$$

$$\mathcal{R}_1 = \frac{l_c}{\mu_0 \mu_r A}; \mathcal{R}_2 = \frac{g}{\mu_0 A}$$

$$l_c = 2\pi \frac{R_1 + R_2}{2} - g \Longrightarrow l_c \cong \pi(R_1 + R_2) = 0.313 \text{ m}$$

$$A = d(R_2 - R_1) = 20 \times (55 - 45) \times 10^{-6} = 2 \times 10^{-4} \text{ m}^2$$

Substituting the corresponding numbers into the expression, we find

$$\text{Reluctance of the core: } \mathcal{R}_c = \frac{0.313}{300 \times (4 \times 10^{-7}) \times 2 \times 10^{-4}} = 0.415 \times 10^6$$

$$\text{Reluctance of the air gap: } \mathcal{R}_g = \frac{0.001}{(4 \times 10^{-7}) \times 2 \times 10^{-4}} = 3.979 \times 10^6$$

Note that although the average length of the core is much longer than the air gap, the reluctance of the core is about one-tenth of the air gap. In many engineering calculations, 10% tolerance is acceptable. Consequently, in most design problems, the reluctance of the ferromagnetic core is negligible compared to the reluctance of the air gap. In this example, without neglecting the reluctance of the iron core, the flux is obtained

$$\phi = \frac{NI}{\mathcal{R}_1 + \mathcal{R}_2} = \frac{100 \times 2}{(0.415 + 3.979) \times 10^6} = 45.517 \times 10^{-6} \text{ Wb}$$

The core and the air gap have the same cross-sectional area. Assuming the leakage flux is negligible, the flux density in the air gap is found:

$$\phi = \frac{B}{A} = \frac{45.517 \times 10^{-6}}{2 \times 10^{-4}} = 0.228 \text{ T}$$

6.2.4 Self-Inductance

Faraday's law of induction states that the EMF induced across the terminals of a coil is equal to the rate of change of total magnetic flux linking N turns of the coil. If the coil is tightly wound, we can assume that each turn of the coil links the same flux. In magnetic circuits, the flux linkage of a coil is defined as the product of the number of turns in the coil and the magnetic flux it encloses. Magnetic flux in the core is a function of the current flowing through the coil. In electromechanical devices, magnetic flux varies over time due to changes in current, or the movement of the magnetic core, or both.

$$\lambda = N\phi \tag{6.18}$$

$$e = -\frac{d\lambda}{dt} = -N\frac{d\phi}{dt} \tag{6.19}$$

The negative sign indicates that the induced EMF opposes the change in current, in accordance with Lenz's law.

In a linear magnetic field where the flux density B is proportional to the field intensity H, the flux linked by the coil is proportional to the current flowing through it. The constant that relates the flux linkage to the current flowing through the coil is defined as the self-inductance. Hence, self-inductance of the coil is defined as the ratio of the flux linkage to the current of the coil.

$$\lambda = Li \Rightarrow L = \frac{\lambda}{i} = \frac{N\phi}{i} \tag{6.20}$$

In a magnetic circuit, self-inductance can be calculated using the concept of reluctance. In practice, a small portion of the flux completes its path without linking some turns of the coil. If such leakage flux is negligible, the self-inductance of a coil can be approximated in terms of the reluctance of the magnetic circuit and the number of turns in the coil.

$$\begin{aligned} e &= -\frac{d}{dt}\left(N\frac{Ni}{\mathcal{R}}\right) = -\left(\frac{N^2}{\mathcal{R}}\right)\frac{di}{dt} \\ L &= \frac{N^2}{\mathcal{R}} \end{aligned} \tag{6.21}$$

Self-inductance determines the ability of a coil or an inductor to induce an EMF in itself due to a change in current passing through it.

6.2.5 Circuits with Permanent Magnets

Permanent magnets are widely used in mechatronic devices due to their ability to retain a significant amount of magnetization without the need for continuous electrical power. These magnets create a constant magnetic flux, which can be harnessed for various applications, such as in motors, generators, and magnetic sensors.

In circuits with permanent magnets, the magnetic field is established by the magnet itself, eliminating the need for an external power source to maintain the magnetic flux. The design of such circuits often involves careful consideration of the magnet's properties, such as its coercivity and remanence, to ensure optimal performance.

Permanent magnets can be integrated into magnetic circuits to provide a stable and predictable magnetic field. This is particularly useful in applications where consistent performance is required over long periods, without the need for regular maintenance or power consumption.

The interaction between the permanent magnet and the surrounding magnetic circuit components is defined by the principles of magnetic flux and reluctance. Permanent magnet motors and generators are more reliable and durable because they do not have slip rings, commutators, and brushes to transmit electric current to moving parts. Furthermore, permanent magnets, are less susceptible to degradation over time compared to electromagnets. This makes them ideal for use in critical applications where sustained magnetic performance is essential.

Figure 6.2 (a) displays a magnetic circuit powered by a permanent magnet instead of a coil. The magnetic flux closes its path through a ferromagnetic material and to air gaps, each of length $g/2$ the upper left and right quadrants of the hysteresis loop for a permanent magnet material.

Permanent magnet materials such as Alnico, Samarium–cobalt, and neodymium–iron–boron are known for their high coercivity and remanence, as shown in Figure 6.2 (b), making them ideal for applications requiring stable and consistent magnetic fields over extended periods. Alnico magnets, composed primarily of aluminum, nickel, and cobalt, offer excellent temperature stability and are often used in high-temperature applications. Samarium–cobalt magnets provide strong magnetic performance and are resistant to oxidation, making them suitable for applications in harsh environments. Neodymium–iron–boron magnets are among the strongest available, offering superior magnetic strength, which is desirable in compact and high-performance devices.

6.2.6 Magnetically Coupled Circuits

Magnetically coupled circuits are two or more electrical circuits that interact through a shared magnetic field. This coupling allows the transfer of energy from one circuit to another without direct electrical contact. Such systems are fundamental in various applications, including inductors, transformers, and certain types of sensors.

In magnetically coupled circuits, the mutual inductance between the circuits is a key parameter that governs their interaction. This mutual inductance depends on factors such as the geometry of the coils, the distance between them, and the properties of the core material. When a current flows through one coil, it generates a magnetic field that induces a voltage in the adjacent coil, reflecting the principle of electromagnetic induction.

The efficiency and effectiveness of magnetically coupled circuits are influenced by the design and placement of the components. Engineers must consider aspects such as coil alignment, core materials, and shielding to optimize performance and minimize the undesired interference. These considerations are crucial in applications ranging from power transfer in transformers to signal isolation in communication systems.

In electric and electronic circuits, when two circuit elements affect each other (with or without contacts) through the magnetic field generated by one of them, they are said to be magnetically coupled. In many electromechanical devices, two or more circuits interact through magnetic field or, in other words, through *magnetic induction.*

Consider two coils wound on a magnetic core as shown in Figure 6.3 (a) where ϕ_{l1} and ϕ_{l2} are called the leakage fluxes of each coil. $\phi_{12} = \phi_{21}$ is the flux that links both coils.

Magnetic flux created by the coil on the left is the sum of the leakage flux linking only coil 1 and the flux produced in the core that links the other coil too.

$$\phi_1 = \phi_{l1} + \phi_{12} \tag{6.22}$$

The EMF induced across the left coil is

$$e_1 = N_1 \frac{d\phi_1}{dt} = N_1 \frac{\partial \phi_1}{\partial i_1} \frac{di_1}{dt} \tag{6.23}$$

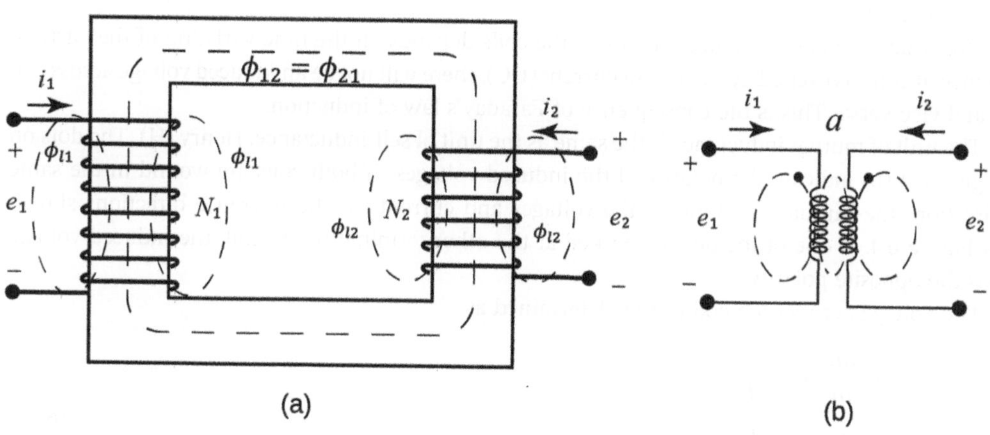

Figure 6.3 Mutually coupled coils (a: Schematic representation of a non-ideal transformer; b: Corresponding circuit diagram)

The partial derivative of the total magnetic flux linking coil 1 with respect to the current flowing in the same coil is defined as self-inductance.

$$L_{11} = \frac{\partial \phi_1}{\partial i_1} \tag{6.24}$$

Coil 2 on the right side of the core is

$$\phi_2 = \phi_{l2} + \phi_{12} \tag{6.25}$$

Self-inductance of coil 2 is

$$L_{22} = \frac{\partial \phi_2}{\partial i_2} \tag{6.26}$$

The voltage induced on coil 2 because of the flux produced by coil 1 can be written in terms of the mutual flux linking both coils.

$$M_{21} = N_2 \frac{\partial \phi_{12}}{\partial i_1} \tag{6.27}$$

Subscript 21 indicates that the inductance relates the voltage induced in coil 2 to the current in coil on the left. Thus, the open-circuit mutual voltage (or induced voltage) across coil on the right is

$$e_2 = M_{21} \frac{di_1}{dt} \tag{6.28}$$

Similarly, the mutual inductance between coil 2 and coil 1 is

$$M_{12} = N_1 \frac{\partial \phi_{21}}{\partial i_2} \tag{6.29}$$

The EMF induced across the terminals of coil 2 is

$$e_1 = M_{12} \frac{di_2}{dt} \tag{6.30}$$

The mutual inductance of coil 2 with respect to coil 1 is equal to the mutual inductance of coil 1 with respect to coil 2.

Note that a voltage is induced in any of the coils depends on the time variation of the current. Hence, if coil 1 is excited by a constant current (DC), there will not be an induced voltage across coil 2 and vice versa. This is the consequence of Faraday's law of induction.

The unit of mutual inductance is the same as the unit of self-inductance, Henry (H). The dots on Figure 6.4 (b) indicate the polarity of the induced voltages. If both coils are wound in the same direction, the mutual flux induces the voltages and currents in the reference directions shown in Figure 6.4. If one of the dots is marked at the other terminal of the coil, the induced voltage has the opposite polarity.

The voltage across each coil can be determined as

$$\begin{aligned} e_1 &= L_1 \frac{di_1}{dt} \pm M \frac{di_2}{dt} \\ e_2 &= L_2 \frac{di_2}{dt} \pm M \frac{di_1}{dt} \end{aligned} \tag{6.31}$$

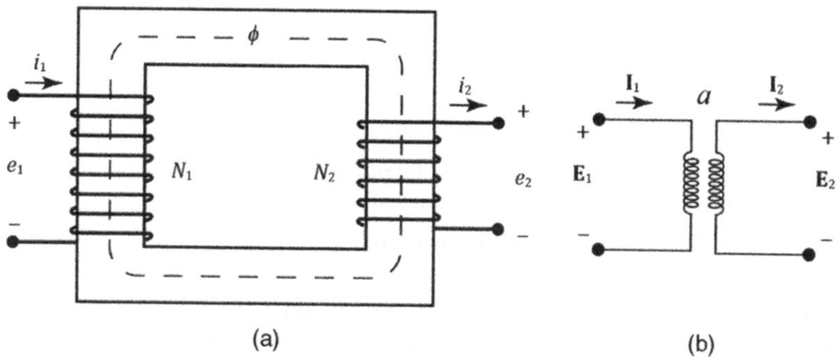

Figure 6.4 Ideal transformer and its equivalent circuit (a: Ideal transformer without leakage flux; b: Circuit diagram of an ideal transformer)

The voltages across the mutually coupled coils can be written in matrix form below:

$$\begin{bmatrix} e_1 \\ e_2 \end{bmatrix} = \begin{bmatrix} L_1 & \pm M \\ L_2 & \pm M \end{bmatrix} \frac{d}{dt} \begin{bmatrix} i_1 \\ i_2 \end{bmatrix} \tag{6.32}$$

The sign before the mutual inductance term is positive if the dots are on the same side of the coils, and negative if the dots are on the opposite terminals.

6.3 Properties of Ferromagnetic Materials

Ferromagnetic elements such as iron, cobalt, and nickel have unpaired electrons that generate a magnetic moment. Such elements and their alloys exhibit strong magnetic properties. When exposed to an external magnetic field, ferromagnetic materials tend to align their magnetic moments in the direction of the field, resulting in a highly magnetized state.

Ferromagnetic materials exhibit nonlinear characteristics and hysteresis properties as illustrated in Figure 6.5. Hysteresis occurs due to the delayed response of magnetic induction B to changes in magnetic field strength H. When a ferromagnetic material is placed in a changing external magnetic field, its magnetization changes with a delay instead of immediately aligning with the field. This results in a hysteresis loop in the graph of magnetic field intensity (H) versus magnetic flux density (B).

Figure 6.5 *B–H* characteristic of a ferromagnetic material

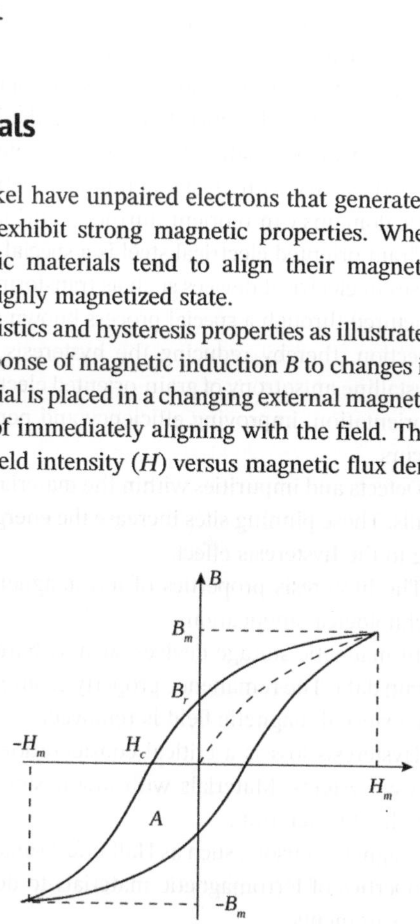

The hysteresis loop is a graphical representation of the relationship between the magnetic flux density (B) and the magnetizing force (H) in ferromagnetic materials. It illustrates several key features.

When a ferromagnetic material is magnetized for the first time or has not been magnetized for a long time, the initial magnetization curve traces the path of increasing magnetic flux density as the magnetizing force is applied. Ferromagnetic materials have significantly higher permeability than nonmagnetic materials. They are more efficient in confining magnetic fields and enhancing the induction (Umans, 2003).

After reaching the maximum, if the field intensity H is reduced to zero, the material retains a certain amount of magnetization known as remanence or residual magnetism, represented by B_r in Figure 6.5. This occurs because some of the magnetic domains remain aligned in the direction of the previously applied field (Kraus, 1991).

To demagnetize the material, a reverse magnetizing force must be applied. The magnitude of this force required to reduce the magnetization to zero is termed coercivity, denoted as H_c. This property is particularly important in the design of permanent magnets and magnetic storage devices, where high coercive force materials are preferred for their ability to retain magnetization in the presence of external fields. The hysteresis loop is critical in applications where the retention of magnetic properties is necessary, such as in permanent magnets and various electromagnetic devices.

As the material undergoes cyclic magnetization, the area within the hysteresis loop represents the energy loss per cycle due to hysteresis. This energy is converted to heat within the material and is a critical factor in applications involving alternating magnetic fields. In transformers and AC machines, the hysteresis loop causes part of the core loss known as "hysteresis loss." Electrical steel is a special alloy designed to minimize the hysteresis loop and reduce hysteresis loss.

Ferromagnetic materials exhibit magneto-crystalline anisotropy, meaning that their magnetic properties vary with crystallographic direction. This anisotropy affects the ease with which magnetic domains can reorient, further contributing to hysteresis.

Grain-oriented electrical steel is a special type of steel alloy designed specifically to reduce core losses in electrical devices such as transformers and large electric machines. This material is manufactured through a special process known as cold annealing, which aligns the grains in a specific direction, thereby reducing the hysteresis loop and minimizing hysteresis loss. The magneto-crystalline anisotropy of grain-oriented electrical steel allows for better control of magnetic domain reorientation, improving efficiency and performance in applications where cyclic magnetization occurs.

Defects and impurities within the material act as pinning sites that hinder the motion of domain walls. These pinning sites increase the energy required to reorient the domains, thereby contributing to the hysteresis effect.

The hysteresis properties of ferromagnetic materials have significant implications for various technological applications.

In magnetic storage devices such as hard drives and magnetic tapes, hysteresis is exploited to retain data. The remanence property ensures that the stored information remains intact even when the external magnetic field is removed.

Hysteresis loss is a critical consideration in the design of transformers and inductors used in power systems. Materials with low hysteresis loss are preferred to minimize energy dissipation and heat generation.

Magnetic sensors, such as Hall-effect sensors and magneto-resistive sensors, utilize the hysteresis properties of ferromagnetic materials to detect changes in magnetic fields and provide accurate measurements.

Frequency of the applied magnetic field affects the B–H characteristics of ferromagnetic materials. As the frequency increases, the hysteresis loop tends to widen, indicating greater energy loss per cycle. This is largely due to the lag in domain wall motion and the eddy currents induced within the material. Eddy currents are loops of electric current induced within conductors by a changing magnetic field, and they generate opposing magnetic fields that complicate the magnetization process. At higher frequencies, these eddy currents become more pronounced, leading to increased hysteresis loss and overall energy dissipation.

Moreover, the skin effect, which causes the alternating magnetic field to penetrate less into the material, becomes significant at higher frequencies, further altering the effective B–H characteristics. This dependency on frequency must be carefully considered in the design and application of ferromagnetic materials, particularly in devices like transformers, inductors, and magnetic storage systems that operate under varying frequencies.

6.3.1 Electric Steel

Electric steel, also known as silicon steel, is a specialized steel that has been alloyed with silicon to enhance its magnetic properties. It is widely used in the cores of transformers, inductors, and other electrical applications due to its high permeability, low coercivity, and low core losses. The addition of silicon increases the electrical resistivity of the steel, which helps to reduce eddy current losses. This makes electric steel particularly suitable for applications operating at lower frequencies where minimizing energy dissipation is crucial.

In the context of transformers and inductors, electric steel is often employed in the laminated form. The lamination process involves stacking thin sheets of silicon steel separated by a thin insulating layer. This technique effectively increases the resistance to eddy currents, thereby reducing the associated losses. Laminated electric steel cores are essential in maintaining the efficiency of transformers by minimizing heat generation and improving overall performance.

Moreover, grain-oriented silicon steel, which has grains aligned in the direction of the magnetic flux, exhibits even lower hysteresis losses and higher efficiency. This type of steel is particularly advantageous in power transformers, where reducing core losses directly translates to energy savings and reduced operating costs.

Despite its advantages, electric steel's performance can be limited at higher frequencies due to the increasing significance of eddy currents. In such scenarios, alternative materials like ferrite cores become more suitable. However, for applications operating at lower frequencies, electric steel remains a preferred choice due to its excellent magnetic properties and cost-effectiveness.

6.3.2 Ferrite Cores

Ferrite cores are an alternative to electric steel at higher-frequency devices. They are composed of iron oxide mixed with other metallic elements such as manganese, zinc, and nickel, which give them their unique magnetic properties. Unlike electrical steel, ferrite cores exhibit low electrical conductivity, which minimizes eddy current losses. This characteristic makes them ideal for high-frequency applications, such as inductors, transformers in switching power supplies, and electromagnetic interference suppression.

Ferrite cores are known for their high permeability and low coercivity, which result in a narrow hysteresis loop and thus low hysteresis loss. Their performance is highly dependent on the composition and manufacturing process, which can be tailored to meet specific frequency requirements. In high-frequency applications, the reduction in eddy current loss is particularly beneficial, as it minimizes energy loss and heat generation.

Moreover, ferrite cores are often used in various electromechanical devices where efficient magnetic energy transfer is essential. The ability to maintain magnetic properties at high frequencies makes ferrite cores indispensable in applications like signal transformers, broadband transformers, and noise filters. Their versatility and efficiency contribute significantly to the reliability and performance of modern electronic systems.

Ferrite core electromagnets are widely used due to their excellent magnetic properties, but they do come with some limitations:

1) **Saturation Limit**: Ferrite cores can become magnetically saturated at relatively lower magnetic field strengths compared to metallic cores like iron. This limits the maximum strength of the electromagnet.
2) **Brittleness**: Ferrite is a ceramic material, which makes it brittle and prone to cracking or breaking under mechanical stress.
3) **Temperature Sensitivity**: Ferrite cores may lose their magnetic properties at very high temperatures. They also have a lower Curie temperature (the temperature above which they lose ferromagnetic properties) compared to some metal cores.
4) **Lower Magnetic Flux Density**: While ferrites are effective for high-frequency applications, they have a lower maximum magnetic flux density compared to ferromagnetic metal cores, which can affect their performance in certain applications requiring strong magnetic fields.
5) **Conductivity Issues**: Ferrites have low electrical conductivity, which helps to reduce eddy current losses at high frequencies, but this can also limit their use in low-frequency, high-power applications.
6) **Size and Design Constraints**: Ferrite cores are not as versatile as metallic cores in terms of shapes and sizes, which can limit their applications in certain designs.

Let me know if you'd like to explore these points further or discuss alternatives to ferrite cores!

6.3.3 Core Losses

Core losses in magnetic circuits are primarily due to two mechanisms: hysteresis loss and eddy current loss. Hysteresis loss is associated with the energy dissipated in reversing the magnetization of the material during each cycle of the alternating magnetic field. It is directly proportional to the area of the hysteresis loop of the material, meaning that materials with a narrow hysteresis loop exhibit lower hysteresis losses. On the other hand, eddy current loss arises from currents induced in the core material itself by the changing magnetic field (Figure 6.6). These currents circulate within the core and generate heat, leading to energy dissipation.

The choice of core material significantly impacts the overall efficiency of magnetic circuits. For instance, soft magnetic materials like silicon steel and amorphous alloys are commonly used in transformers and motors due to their low hysteresis loss and high magnetic permeability. However, at higher frequencies, ferrite cores become a preferred choice due to their low electrical conductivity, which effectively minimizes eddy current losses.

In practical applications, optimizing the core geometry and material composition is crucial for reducing core losses. Laminated cores are often employed to limit eddy currents by increasing the resistance path for these currents, thereby reducing the overall losses. Additionally, the use of grain-oriented steel can further enhance efficiency by aligning the grains in the direction of the magnetic flux, minimizing hysteresis loss (Juvinall & Marshek, 2020).

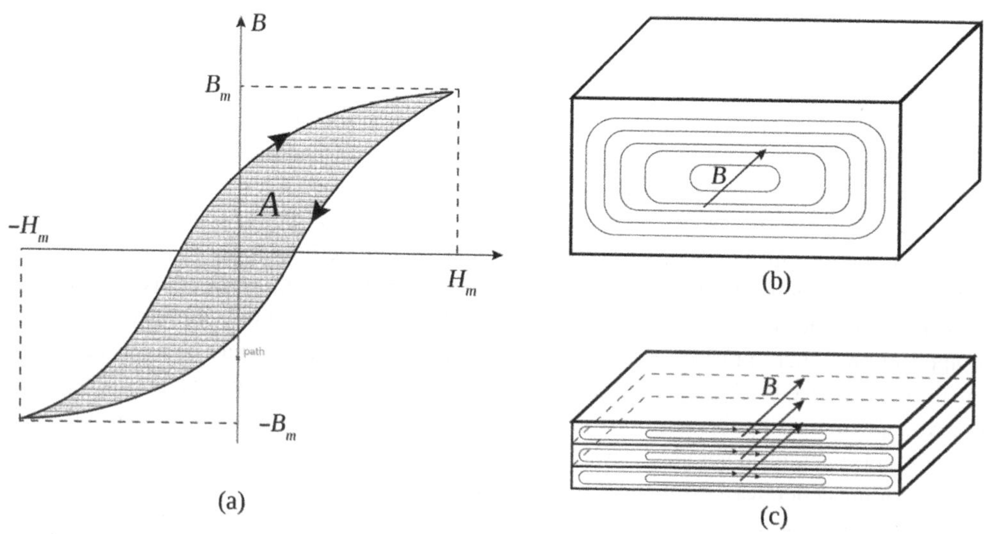

Figure 6.6 Hysteresis and eddy current effects in electrical steel (a: Hysteresis curve of a ferromagnetic material; b: A block of steel core; c: Laminated steel core)

6.4 Inductor

Inductors are essential components in power converters, where they serve to store and regulate energy in the form of a magnetic field. In switch-mode power supplies and DC–DC converters, inductors help smooth out current fluctuations, filter noise, and maintain continuous current flow, which is critical for efficient energy transfer. Their ability to temporarily store energy enables them to buffer changes in load and input voltage, contributing to stable and reliable power conversion (Pressman et al., 2009).

The performance of an inductor in a power converter depends largely on its core material, geometry, and winding configuration. Ferrite and powdered iron cores are commonly used for high-frequency converters due to their low core losses and effective suppression of eddy currents. Proper selection and optimization of the inductor minimize losses and improve the overall efficiency of the power converter, making them indispensable in modern mechatronic and electrical systems.

6.5 Transformer

Transformers play a crucial role in mechatronic systems, acting as indispensable elements in power supply circuits and power converters. Their ability to transfer electrical energy between circuits through electromagnetic induction makes them vital for ensuring efficient and adaptable power delivery in various applications.

Transformers are devices that transfer electrical energy between two or more circuits via electromagnetic induction. This transfer of energy typically involves increasing or decreasing voltage levels, which is fundamental for power transmission and distribution. The broad functionality and types of transformers ensure they meet the diverse needs of electrical engineering. There are several types of transformers, including step-up, step-down, isolation transformers, and auto-transformers, each engineered for specific purposes.

Step-up transformers are used to increase the voltage levels, while step-down transformers decrease them. These transformations are essential in mechatronic systems where power needs to be efficiently transmitted over distances or adapted to suit various components and devices.

Special transformers are used for isolation to separate different sections of a circuit, ensuring that failure or faults in one part do not affect the other. Isolation transformers are particularly important in safety-critical mechatronic applications.

Impedance-matching transformers are designed to ensure optimal power transfer between circuits with different impedance levels, enhancing efficiency and performance.

Variable transformers, also known as autotransformers or variacs, allow for adjustable voltage output. They provide versatile control over voltage, making them crucial in laboratory and industrial applications where precise voltage regulation is necessary. Variacs may be adjusted either manually or with a feedback-controlled electric motor. Controlled transformers can precisely adjust output voltage, increasing the versatility of mechatronic systems that require higher power.

Transformers operate on Faraday's law of induction. A time-varying current flowing through the primary winding generates a magnetic flux in the ferromagnetic core, inducing an EMF across the secondary coil. This principle allows transformers to convert electrical energy to magnetic energy and back to electrical energy without a direct electrical connection between primary and secondary windings.

The core of a transformer is typically made of laminated silicon steel, which enhances efficiency by minimizing energy losses due to eddy currents and hysteresis. These design considerations are crucial for ensuring the reliable and effective operation of transformers in mechatronic systems.

Transformers are used extensively in mechatronic systems to achieve various functions, including:

- **Power Supply Circuits**: Ensuring stable and reliable power delivery to component parts and devices.
- **Power Converters**: Adapting the voltage levels to meet the specific requirements of different elements within the system.
- **Signal Processing**: Facilitating the handling and manipulation of electrical signals for control and monitoring purposes.
- **Impedance Matching**: Ensuring efficient power transfer between circuits with varying impedance levels.
- **Isolation**: Protecting sensitive components and ensuring operational safety by isolating different parts of the circuit.

The operation of a transformer requires a time-varying current applied to the primary side. A transformer consists of two or more coils sharing the same magnetic flux as shown in Figure 6.4. The coil connected to a source is called the primary winding, and the coil connected to a load is called the secondary winding. The coils may have independent terminals or a common terminal. The voltages and currents associated with each winding are also referred to as primary and secondary variables.

Transformers operate on the principle of Faraday's law of induction. A current that changes in time flowing through the primary winding creates a time-varying magnetic flux in the ferromagnetic core. This flux induces an EMF across the secondary coil without a direct electrical connection. The interaction between the primary and secondary windings occurs by means of the mutual magnetic field. Thus, a transformer converts electric energy to magnetic energy and then back to electric energy.

The core of the transformer is generally made of laminated silicon steel to enhance its efficiency by reducing energy losses due to eddy currents and hysteresis.

6.5.1 Ideal Transformer

Ideal transformers operate based on the following assumptions:

- The flux is confined to the core, hence linking all turns of both coils.
- The core has linear *B–H* characteristics and infinite permeability.
- The coils have no resistance.
- The core does not cause any magnetic loss.

The EMF induced across each winding is found by applying Faraday's law of magnetic induction:

$$e_1 = N_1 \frac{d\phi}{dt}$$

$$e_2 = N_2 \frac{d\phi}{dt} \tag{6.33}$$

The ratio of the EMF across each winding is strictly equal to the ratio of the number of turns of each winding.

$$\frac{e_1}{e_2} = \frac{N_1}{N_2} = a \tag{6.34}$$

The ratio of the primary voltage to the secondary voltage a is called the *turn-ratio* or *transformation ratio*. Since the ideal transformer does not cause any energy loss, the power absorbed by the primary winding is equal to the power the secondary winding delivers to the load. Hence,

$$e_1 i_1 = e_2 i_2 \Longrightarrow \frac{i_1}{i_2} = \frac{e_2}{e_1} = \frac{1}{a} \tag{6.35}$$

In most transformer applications, the primary voltage is in sinusoidal form. In sinusoidal steady-state operation, we use phasor quantities for all variables. The impedance seen from the primary side of an ideal transformer is

$$Z_1 = \frac{E_1}{I_1}$$

$$Z_2 = \frac{E_2}{I_2} \tag{6.36}$$

The ratio of the primary and secondary impedances is obtained as

$$\frac{Z_1}{Z_2} = \frac{E_1}{I_1}\frac{I_2}{E_2} = a^2 \tag{6.37}$$

Therefore, a transformer changes the impedance seen from its primary side to its secondary side by the square of the transformation ratio. This is a useful property to match the impedance of two separate networks.

Transformers are used in many practical applications, from milliwatts to megawatts. The most common use of transformers is to change the voltage level from higher to lower (step-down) or from lower to higher (step-up). For example, an adapter transformer reduces the utility voltage of 120 V to the low-voltage level in the range of 5–20 V required for electronic devices.

Since the ratio of the primary current to the secondary current is equal to the inverse of the voltage transformation ratio, a transformer steps up the current as it steps down the voltage and vice versa. This property allows obtaining high current levels while decreasing the voltage. Such transformers are particularly used in welding machines.

Impedance transformers are used in electronic circuits to transfer maximum power from one circuit to another by matching their output and input impedances.

6.5.2 Practical Transformers

The equations derived for ideal transformers apply approximately to practical transformers. In a practical transformer, the coils have resistances, and the magnetic core has eddy current and hysteresis losses. The permeability of the core material is not infinite. A portion of the magnetic flux in each coil, called leakage flux, closes its path outside the core. Figure 6.7 illustrates a two-winding nonideal transformer and its equivalent circuit.

These practical considerations mean that transformer efficiency is never perfect and that design choices must balance the influences of resistance, core losses, and leakage flux. Engineers employ various techniques, such as using laminated cores and high-permeability materials, to minimize energy losses and optimize performance for specific applications. Additionally, the equivalent circuit of a nonideal transformer provides a valuable tool for analyzing real-world behavior, allowing for predictions of voltage drops, efficiency, and regulation under load conditions.

Beyond traditional fixed-ratio transformers, the technology has evolved to include specialized types designed to perform particular tasks in various environments.

6.5.3 Variable Transformers

Variable transformers, also known as autotransformers and variacs, are a specific type of transformer that allows for adjustable voltage output. Unlike fixed-ratio transformers, variacs can be manually regulated to provide a variable voltage to the load, which makes them highly versatile in laboratory and industrial applications. They are commonly used in testing and prototyping environments where precise control over voltage is required.

The design of a variac typically includes a movable contact that slides along a winding, thus varying the number of turns involved in the voltage transformation process. This adjustable feature enables users to fine-tune the output voltage according to their specific needs. Variable transformers are essential tools for engineers and technicians who need to simulate different operating conditions or gradually increase voltage to prevent damage to sensitive components.

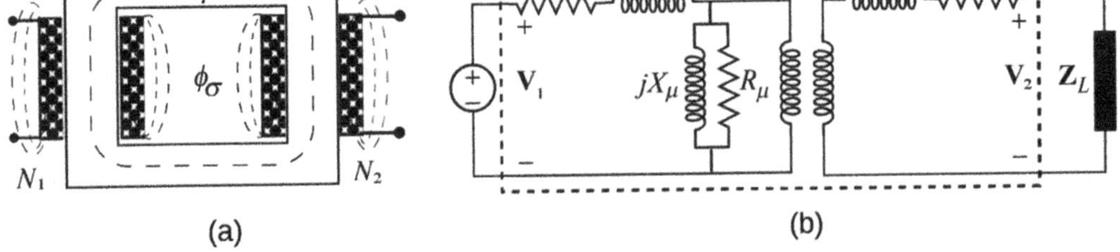

Figure 6.7 Nonideal transformer (a: Physical structure of a single-phase transformer; b: Equivalent circuit)

In addition to their primary function, variacs can serve as voltage regulators, providing stable output in situations where the input voltage may fluctuate. This capability ensures the consistent performance of equipment and enhances the reliability of critical systems.

6.5.4 High-Frequency Transformers and Inductors

High-frequency transformers and inductors are essential components in modern electronic circuits, particularly in applications such as switching power supplies, radio-frequency amplifiers, and signal processing devices. Unlike their low-frequency counterparts, these devices operate efficiently at frequencies ranging from tens of kilohertz to several megahertz. To minimize core losses and maintain signal integrity at high frequencies, designers often use ferrite cores or powdered iron materials, which offer low eddy current and hysteresis losses (Mohan, 2012).

Windings in high-frequency transformers and inductors are typically constructed with special techniques to reduce skin effect and proximity effect losses. These measures help ensure that the components maintain high efficiency and stable operation even under rapid switching conditions. Careful attention to layout, shielding, and insulation is also required to prevent unwanted electromagnetic interference and maintain performance in demanding environments.

6.6 Summary

Magnetic circuits and static devices form the foundation of many electromechanical systems such as electric motors, generators, transformers, actuators, and sensors. These systems operate based on the interaction between electric currents and magnetic fields, where electric current generates magnetic fields, and changing magnetic fields induce electric potentials in conductors.

- Magnetic field intensity (H) is proportional to the current and inversely proportional to the radius of the circle of the field line around a conductor. It is measured in amperes per meter (A/m) and is independent of the surrounding medium. Magnetic induction or magnetic flux density (B) accounts for the magnetic properties of the environment and is calculated by multiplying H by the magnetic permeability (μ) of the medium. Magnetic permeability of vacuum (μ_0) is a fundamental constant, and materials have relative permeability (μ_r) that varies widely, with ferromagnetic materials having very high values, typically in the thousands.
- Magnetic flux (Φ), measured in Webers (Wb), represents the quantity of magnetic field lines passing through a surface. It is the scalar product of the magnetic flux density vector and the area vector perpendicular to the surface. Flux linkage (λ) is the product of the number of coil turns (N) and the magnetic flux, reflecting the total flux linking the coil.
- Ampere's circuital law relates the magnetic field intensity around a closed path to the total current enclosed, enabling the calculation of magnetic fields in symmetrical current distributions. Faraday's law of induction states that a time-varying magnetic flux induces an EMF in a coil, with the induced EMF opposing changes in flux as described by Lenz's law. The unit of self-inductance, the henry (H), honors Joseph Henry's contributions to electromagnetics.
- Magnetic circuits simplify the analysis of complex magnetic structures by analogy with electric circuits. Magnetic circuits consist of coils wound around ferromagnetic cores, often with air gaps. The MMF, analogous to EMF, is the product of coil turns and current (NI). Magnetic reluctance, analogous to electrical resistance, quantifies opposition to magnetic flux and depends on core

length, cross-sectional area, and permeability. The magnetic circuit obeys an equivalent of Ohm's law: MMF equals flux times reluctance. This analogy is summarized in Table 6.1.

- Magnetic circuits can be analyzed using DC circuit methods, and the reluctance of air gaps is often dominant compared to ferromagnetic cores.
- Fringe magnetic flux occurs when flux lines stray outside the intended magnetic path, especially near air gaps or discontinuities. This leakage can reduce efficiency, cause localized heating, and introduce measurement errors. Engineering solutions include magnetic shielding and smooth design transitions to minimize fringe flux.
- A toroidal ferromagnetic core with given dimensions and relative permeability, carrying a DC current through a coil, illustrates calculation of reluctances of the core and air gap, magnetic flux, and flux density. The reluctance of the air gap is significantly higher than that of the core, often justifying neglecting the core's reluctance in design calculations.
- Self-inductance relates the flux linkage to coil current and quantifies a coil's ability to induce EMF in itself due to changing current. It can be calculated using magnetic circuit reluctance and coil turns. Magnetically coupled circuits involve two or more coils interacting through shared magnetic flux. Mutual inductance quantifies the voltage induced in one coil due to a change in current in another. The mutual inductance is symmetric between coils, and the polarity of induced voltages depends on coil winding directions. Voltages in coupled coils can be represented in matrix form, incorporating self and mutual inductances.
- Ferromagnetic materials such as iron, cobalt, and nickel exhibit strong magnetic properties due to unpaired electrons. They have nonlinear *B–H* characteristics and display hysteresis, where magnetic induction lags changes in magnetic field intensity. The hysteresis loop shows key features: initial magnetization, remanence (residual magnetism), coercivity (required reverse field to demagnetize), and energy loss per cycle (hysteresis loss). These properties are critical in permanent magnets, magnetic storage, transformers, and sensors. Grain-oriented electrical steel reduces hysteresis loss and improves efficiency in AC applications. Frequency influences hysteresis and eddy current losses, with higher frequencies increasing energy dissipation due to lag and induced currents. Eddy currents and skin effect complicate magnetization at high frequencies.
- Electric steel, or silicon steel, is widely used in transformers and inductor cores due to its high permeability, low coercivity, and reduced eddy current losses due to silicon alloying and lamination. Grain-oriented silicon steel further improves efficiency by aligning grains with magnetic flux. However, electric steel is less suitable for high-frequency applications due to eddy current losses.
- Ferrite cores, composed of iron oxide mixed with metals like manganese and zinc, have low electrical conductivity, minimizing eddy current losses and making them ideal for high-frequency devices such as switching power supplies and noise filters. They exhibit high permeability and low coercivity, resulting in narrow hysteresis loops and low losses. Limitations include lower saturation flux density, brittleness, temperature sensitivity, and size constraints compared to metal cores.
- Core losses arise from hysteresis and eddy currents. Materials with narrow hysteresis loops and high resistivity, such as silicon steel and ferrites, minimize these losses. Laminated cores increase resistance to eddy currents, reducing losses. Grain orientation further decreases hysteresis loss, enhancing transformer and motor efficiency.
- Transformers transfer electrical energy between circuits through electromagnetic induction, enabling voltage step-up, step-down, isolation, impedance matching, and variable voltage control. They consist of primary and secondary coils wound on a ferromagnetic core, typically laminated silicon steel for efficiency. Ideal transformers assume no losses, perfect flux linkage, and

infinite core permeability, with voltage ratios proportional to coil turn ratios and power conserved between primary and secondary. Practical transformers include coil resistance, core losses, and leakage flux, modeled by equivalent circuits. Variable transformers (variacs) allow adjustable voltage output, useful in testing and voltage regulation.

- Transformers are essential in power supply circuits, power converters, signal processing, impedance matching, and isolation in mechatronic systems, ensuring efficient and adaptable power delivery.

6.7 Check Your Understanding

6.7.1 Self-Assessment Quiz

1 Magnetic field intensity (*H*) is measured in Weber (Wb).
True False

2 Magnetic circuits can be analyzed using DC circuit methods.
True False

3 Ferrite cores are ideal for high-frequency devices due to their low electrical conductivity.
True False

4 The mutual inductance between two coils is not symmetric.
True False

5 Grain-oriented silicon steel is less suitable for high-frequency applications due to eddy current losses.
True False

6 Magnetic flux (Φ) is measured in Tesla (T).
True False

7 The permeability of vacuum (μ_0) is a fundamental constant.
True False

8 The reluctance of the air gap is often negligible compared to the reluctance of the ferromagnetic core.
True False

9 The hysteresis loop represents the relationship between magnetic flux density (*B*) and magnetizing force (*H*) in ferromagnetic materials.
True False

10 Ferrite cores have higher saturation flux density compared to metal cores.
True False

11 What is the unit of magnetic flux (Φ)?
- **a.** Amperes per meter (A/m)
- **b.** Weber (Wb)
- **c.** Henry (H)
- **d.** Tesla (T)

12 Which law relates the magnetic field intensity around a closed path to the total current enclosed?
- **a.** Faraday's law of induction
- **b.** Ampere's circuital law
- **c.** Lenz's law
- **d.** Ohm's law

13 What is the product of the number of coil turns (N) and the magnetic flux (Φ) called?
- **a.** Magnetic field intensity
- **b.** Magnetic induction
- **c.** Flux linkage
- **d.** Magnetomotive force

14 Which material is widely used in transformer and inductor cores for its high permeability and low coercivity?
- **a.** Ferrite
- **b.** Silicon steel
- **c.** Copper
- **d.** Aluminum

15 What is the primary function of transformers in power supply circuits?
- **a.** To generate magnetic fields
- **b.** To transfer electrical energy between circuits
- **c.** To store electrical energy
- **d.** To measure electrical resistance

16 What is the product of coil turns (N) and current (I) called in magnetic circuits?
- **a.** Magnetic flux
- **b.** Magnetomotive force (MMF)
- **c.** Magnetic induction
- **d.** Flux linkage

17 Which material is known for its high coercivity and remanence, making it ideal for stable magnetic fields over extended periods?
- **a.** Silicon steel
- **b.** Ferrite
- **c.** Neodymium–iron–boron
- **d.** Copper

18 What does the hysteresis loop in ferromagnetic materials indicate?
- **a.** The relationship between electric current and magnetic field
- **b.** The energy loss per cycle due to hysteresis
- **c.** The magnetic flux density in a vacuum
- **d.** The permeability of nonmagnetic materials

19 Which law states that a time-varying magnetic flux induces an electromotive force (EMF) in a coil?
- **a.** Ampere's circuital law
- **b.** Ohm's law
- **c.** Faraday's law of induction
- **d.** Lenz's law

20 What is the primary advantage of using laminated cores in transformers?
- **a.** To increase magnetic flux density
- **b.** To reduce eddy current losses
- **c.** To enhance coercivity
- **d.** To improve temperature stability

Answers:

1: False; 2: True; 3: True; 4: False; 5: True; 6: False; 7: True; 8: False; 9 True; 10: False; 11: b; 12: b; 13: c; 14: b; 15: b; 16: b; 17: c; 18: b; 19: c; 20: b.

6.7.2 Short Answer Questions

1 What is the unit of magnetic field intensity (H)?

2 How is magnetic induction (B) calculated?

3 What is the significance of the permeability of vacuum (μ_0)?

4 Define magnetic flux (Φ).

5 What does Ampere's circuital law relate to?

6 Explain Faraday's law of induction.

7 What is the analogy between magnetic circuits and electric circuits?

8 What is fringe magnetic flux, and how can it be minimized?

9 Describe the concept of self-inductance.

10 What are the key properties of ferromagnetic materials?

11 How does grain-oriented electrical steel improve efficiency in AC applications?

12 What are the advantages of ferrite cores in high-frequency devices?

13 What causes core losses in magnetic circuits?

14 How do transformers transfer electrical energy between circuits?

15 What are the primary functions of transformers in power supply circuits?

6.7.3 Essay Questions

1 Discuss the fundamental concepts of magnetism, including magnetic field intensity, magnetic induction, and magnetic flux. How do these concepts relate to the operation of electromechanical systems such as electric motors and transformers?

2 Explain Ampere's circuital law and Faraday's law of induction. How do these laws contribute to the understanding and design of magnetic circuits and devices?

3 Analyze the analogy between magnetic circuits and electric circuits. How does this analogy simplify the analysis of complex magnetic structures in electromechanical systems?

4 Describe the phenomenon of fringe magnetic flux. What are its implications on the performance of magnetic circuits, and what engineering solutions can be implemented to minimize its effects?

5 Examine the properties of ferromagnetic materials, including their B–H characteristics and hysteresis. How do these properties influence the design and performance of magnetic devices such as transformers and sensors?

6 Compare electric steel and ferrite cores in terms of their magnetic properties and applications. What are the advantages and limitations of each material in different frequency ranges?

7 Discuss the concept of self-inductance and mutual inductance in magnetically coupled circuits. How do these concepts apply to the operation of transformers and other inductive devices?

8 Evaluate the impact of core losses, including hysteresis and eddy current losses, on the efficiency of magnetic circuits. What materials and design techniques can be used to minimize these losses?

9 Explain the role of transformers in power supply circuits and power converters. How do transformers enable voltage step-up, step-down, isolation, impedance matching, and variable voltage control in mechatronic systems?

10 Analyze the design and operation of variable transformers (variacs). How do they provide adjustable voltage output, and what are their applications in laboratory and industrial settings?

6.7.4 Problems

1 A long straight wire carries a current of 5 A. Calculate the magnetic field intensity (H) at a distance of 0.1 m from the wire.

2 A solenoid with 500 turns per meter carries a current of 2 A. The relative permeability of the core material is 1,000. Calculate the magnetic induction (B) inside the solenoid.

3 A rectangular loop with dimensions 0.2 m by 0.3 m is placed in a uniform magnetic field of 0.5 T. The plane of the loop is perpendicular to the magnetic field. Calculate the magnetic flux (Φ) through the loop.

4 A toroidal ferromagnetic core has an average radius of 0.05 m and a cross-sectional area of 0.001 m^2. The relative permeability of the core material is 2,000. Calculate the reluctance ($\mathcal{R}$) of the core.

5 A coil with 200 turns is wound on a magnetic core with a reluctance of 5,000 A/Wb. Calculate the self-inductance (L) of the coil.

6 Two coils are wound on a common magnetic core. Coil 1 has 100 turns, and coil 2 has 150 turns. The mutual flux linking both coils is 0.002 Wb when the current in coil 1 is 2 A. Calculate the mutual inductance (M) between the coils.

7 A transformer has a primary coil with 500 turns and a secondary winding with 100 turns. If the primary voltage is 240 V, calculate the secondary voltage.

8 A magnetic core has a hysteresis loss of 10 W and an eddy current loss of 5 W at a frequency of 50 Hz. Calculate the total core loss at this frequency.

9 A magnetic circuit has a core with a cross-sectional area of 0.002 m^2 and a magnetic flux of 0.004 Wb. Calculate the magnetic flux density (B) in the core.

10 A coil with 50 turns is placed in a magnetic field that changes from 0.2 to 0.8 T in 0.1 seconds. The area of the coil is 0.01 m^2. Calculate the induced electromotive force (EMF) in the coil.

References

Giancoli, D. C., 2020. *College Physics: Principles with Applications.* New York, NY: McGraw-Hill.

Juvinall, R. C. & Marshek, K. M., 2020. *Fundamentals of Machine Component Design.* 7th ed. Hoboken, NJ: John Wiley.

Kraus, J. D., 1991. *Electromagnetics.* 4th ed. New York: McGraw Hill.

Mohan, N., 2012. *Power Electronics: A First Course.* Hoboken, NJ: Wiley.

Pressman, A. J., Billings, K. & Morey, T., 2009. *Switching Power Supply Design.* New York, NY: McGraw-Hill.

Umans, S. D., 2003. *Fitzgerald and Kingsle's Electric Machinery.* 7th ed. New York, NY: McGraw Hill.

7

Electromechanical Energy Conversion

OBJECTIVES

- Explain the fundamental principles of converting electrical energy to mechanical energy and vice versa, which is the core of mechatronic systems.
- Describe the broad variety of devices involved in electromechanical energy conversion, such as motors, actuators, generators, transducers, and electromagnetic sensors.
- Provide an understanding of how energy flows in electromechanical devices, including the bidirectional nature of energy conversion.
- Introduce the analysis of energy balance, losses (electrical and mechanical), and efficiency in electromechanical systems.
- Present practical examples and problems to reinforce the concepts.

Conversion from electrical to mechanical energy or from mechanical to electric energy is the core of mechatronic systems. A broad variety of devices fall into this category. All motors and actuators use electrical energy to produce mechanical motion. Generators, various transducers, and electromagnetic sensors convert mechanical motion to electric voltage and current.

All practical electromechanical devices convert electric energy to mechanical work in a magnetic field. In principle, electrostatic fields are also capable of generating force or torque; however, practical application requires the use of high-voltage levels to achieve forces or torques of significant magnitude (Zorpette, 2024). This chapter focuses on energy conversion in magnetic field. If the content of this chapter seems challenging, review Chapter 2 for fundamental concepts about magnetic fields and Chapter 6 for magnetic circuits.

7.1 Energy Flow in Electromechanical Devices

Electromechanical systems consist of coils that generate magnetic flux and a mechanical structure driven by electromagnetic forces. In electric machines, actuators, relays, electromagnetic transducers and sensors, and electrical and mechanical variables are interrelated through dynamic equations. Electromechanical systems can convert electrical energy into mechanical energy, or vice versa. However, each system is typically designed to operate more efficiently in one direction (Fitzgerald et al. 1992).

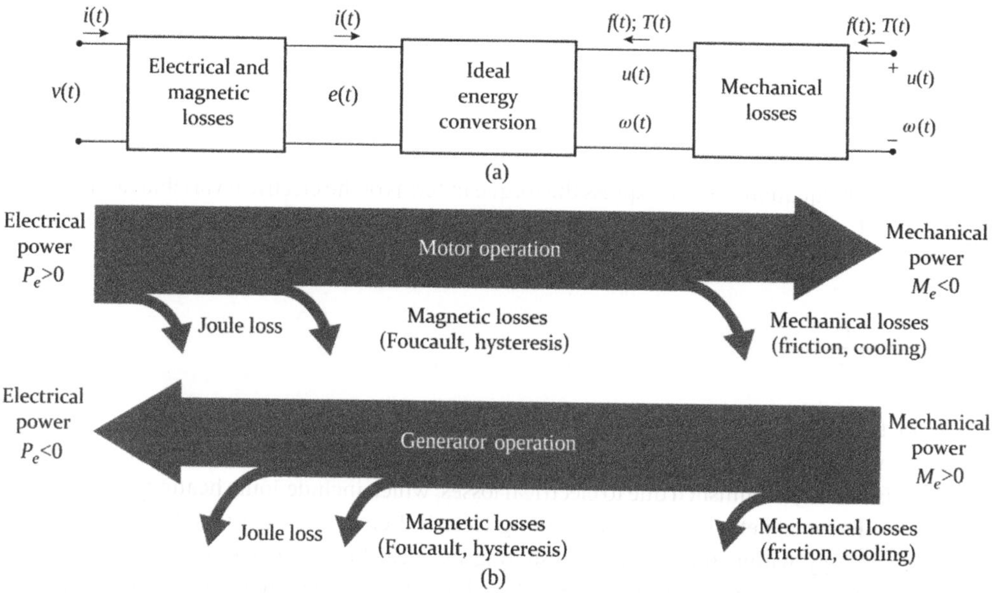

Figure 7.1 Energy flow in an electromechanical system (a: Block representation of energy components; b: Energy flow in motor and generator operations)

Figure 7.1 illustrates energy flow in an electromechanical energy conversion system. Reference directions are chosen according to the passive convention, where absorbed power is positive and delivered power is negative.

Energy flow in electromechanical systems is bidirectional. In principle, when energy flows from the mechanical side to the electrical side, the system functions as a generator. If the system converts electrical energy into mechanical energy, it functions as a motor. Any electromechanical system can function as either a generator or a motor, regardless of the type of mechanical motion or energy level involved. For example, a loudspeaker connected to an audio amplifier vibrates the membrane to produce sound waves, operating as a motor. The same device can also act as a microphone when a sound wave moves the membrane and the moving coil generates an electrical signal, functioning as a generator. Rotary machines can function as motors or generators. Motors turn into generators due to changes in stored mechanical energy as speed decreases. This feature saves energy in electric and hybrid vehicles and large industrial systems by allowing electronic converters to return energy to the power supply.

During motor operation, both voltage and current can be either positive or negative, resulting in positive electrical input power. On the mechanical side, velocity and torque must have opposite signs to ensure that the mechanical output power is negative. In generator operation, torque (or force) and the motion are in the same direction, making mechanical input power positive and electrical output power negative. The expressions below summarize motor and generator operation modes in linear and rotary motion.

$$\text{Motor operation} \quad \begin{array}{l} v>0;\, i>0 \quad \text{or} \quad v<0;\, i<0 \Longrightarrow P_e = vi > 0 \\ \omega>0; T<0 \quad \text{or} \quad \omega<0;\, T>0 \Longrightarrow P_m = \omega T < 0 \end{array}$$

$$\text{Generator operation} \quad \begin{array}{l} v>0;\, i<0 \quad \text{or} \quad v<0;\, i>0 \Longrightarrow P_e = vi < 0 \\ \omega<0; T<0 \quad \text{or} \quad \omega>0;`\, T>0 \Longrightarrow P_m = \omega T > 0 \end{array} \tag{7.1}$$

In the figure, electrical and mechanical losses are separated. The middle block performs ideal energy conversion where the sum of the mechanical and electrical input powers is zero.

$$ei + \omega T = 0 \tag{7.2}$$

By solving this equation, we can express the torque in terms of the electrical variables and angular velocity in rad/s. In the standard international unit system, torque is measured in Newton-meter (Nm).

$$T_e = -\frac{ei}{\omega} \tag{7.3}$$

In a practical electromechanical conversion system, both electrical and mechanical losses are considered when analyzing energy flow. During motor operation, the electric power supplied to the system is initially diminished due to electrical losses, which include Joule heating and magnetic losses in the ferromagnetic core. The remaining electrical energy is then wholly converted into mechanical energy within the block, representing an ideal lossless machine. A portion of the mechanical energy is subsequently dissipated as heat due to friction in the bearings. Larger machines typically employ natural or forced cooling systems, such as internal fans or water or oil circulation pumps, to expel the energy converted to heat due to losses.

In generator operation, energy flows in the opposite direction from the mechanical to the electrical side. In either case, the delivered energy is equal to the input energy minus the total losses (Umans, 2003).

Example 7.1 Regenerative Brake in a Hybrid Car

Hybrid and electric vehicles (EV) have an electronically controlled brake system that combines a regenerative brake system with conventional friction brakes. When the driver applies the brakes, the electric motor reverses its operation, acting as a generator. Regenerative braking converts the vehicle's kinetic energy into electrical energy, which is then stored in the battery for later use. This reduces wear on traditional brake components and improves energy efficiency. If additional braking force is needed beyond what regenerative braking provides, conventional friction brakes (discs and pads) are activated. These brakes work like those in traditional vehicles to ensure safety and stopping power. Instead of direct mechanical action, electronic signals control the braking process. Sensors track pedal pressure and vehicle speed, while the electronic control unit (ECU) determines the optimal levels of regenerative and friction braking.

Problem

A hybrid car has a 110 HP electric motor and a 207.2 V battery bank. When the brakes are applied, the electric motor runs at 1,400 rpm and the battery charge current becomes 187 amps. Assuming a total loss of 8% of the motor power, find the mechanical power the brake system absorbs and calculate the torque turning the generator shaft.

Solution

During braking, the motor switches to generator. Electrical output power becomes

$$P_e = vi = 207.2 \times 187 = 38{,}746\ W$$

Total loss is calculated as 8% of the rated motor power, using the conversion factor 1 HP = 746 W.

$$P_{loss} = \frac{8}{100} 110 \times 746 = 6{,}565\ W$$

Mechanical power absorbed equals electrical output total plus loss.

$$P_m = P_e + P_{loss} = 38{,}746 + 6{,}565 = 45{,}311\ W$$

At 1,400 rpm, the motor angular velocity is

$$\omega = \frac{2\pi n}{60} = 146.6\ \mathrm{rad/s}$$

The torque applied to the shaft is calculated using equation 7.1 for the generator operation mode.

$$T = \frac{P_m}{\omega} = \frac{45{,}311}{146.6} = 309\ N \cdot m.\therefore$$

7.2 Energy Balance

Energy output of an electromechanical system is always less than the energy input due to the electrical and mechanical losses. In Figure 7.1 (a), the central block represents the conceptual, idealized system that excludes all the losses. Energy conversion in an ideal electromechanical system, where input equals output, results from changes in stored magnetic field energy. The conceptual lossless component functions as an energy storage system (Umans, 2003). The equation below describes the energy balance.

$$\begin{bmatrix} \text{Electrical} \\ \text{Energy} \\ \text{Input/Output} \end{bmatrix} + \begin{bmatrix} \text{Mechanical} \\ \text{Energy} \\ \text{Input/Output} \end{bmatrix} = \begin{bmatrix} \textbf{Change} \\ \text{in Stored} \\ \text{Magnetic Energy} \end{bmatrix} + \begin{bmatrix} \text{Energy} \\ \text{Converted to heat} \\ \text{Due to losses} \end{bmatrix}$$

According to the passive notation, input energy is positive, and output energy is negative. During motor operation, the energy supplied from an external source is positive, and the mechanical energy driving a load is negative. In generator operation, mechanical energy provided by an external force or torque is positive and the electrical energy delivered to an electric load is negative.

The energy balance concept can be represented mathematically using differential energy terms as shown below:

$$dW_e + dW_m = dW_f + dW_{loss} \tag{7.4}$$

In time interval dt, the differential electrical energy terms is

$$dW_e = eidt \tag{7.5}$$

Similarly, the mechanical energy terms for linear and rotary motions are

$$\begin{aligned} &\text{Linear motion} \quad dW_m = f_e u dt \\ &\text{Rotary motion} \quad dW_m = T_e \omega dt \end{aligned} \tag{7.6}$$

In the energy storage part of the electromechanical system, according to Faraday's law, the voltage induced across the coils equals the time rate of change of the magnetic field. The differential change in the stored magnetic energy can be found by substituting the induced voltage into equation 7.5.

$$e = \frac{d\lambda}{dt} \Longrightarrow dW_f = i d\lambda \tag{7.7}$$

Losses are electrical and mechanical energy converted to heat and dissipated to the surrounding environment. In an electromechanical energy conversion system, losses include Joule heating effect on the wires, magnetic losses in the ferromagnetic core, friction between the moving parts, and energy consumed by active cooling elements.

Joule loss is proportional to the current flowing through the coils.

$$P_{Joule} = Ri^2 = \frac{v^2}{R} \tag{7.8}$$

In systems where mechanical components move against each other, bearings and sliders experience viscous friction, which results in energy losses proportional to the square of the velocity.

$$\begin{array}{ll} \text{Linear motion (slider)} & P_{friction} = \beta v^2 \\ \text{Rotary motion (bearing)} & P_{friction} = \beta \omega^2 \end{array} \tag{7.9}$$

Efficiency of an energy conversion system is defined as the ratio of the output energy (or power) to the input energy (or power). In the generator operation, input power is mechanical, and output power is electrical. In motor operation, the input and output variables are reversed. Efficiency is determined by using the expressions below:

$$\eta = \frac{W_{out}}{W_{in}} = \frac{W_{in} - W_{loss}}{W_{in}} = 1 - \frac{W_{loss}}{W_{in}} \tag{7.10}$$

In practice, efficiency is often given in percent value, which is obtained by simply multiplying the decimal figure by 100.

$$\%\eta = 100 \times \frac{W_{out}}{W_{in}} \tag{7.11}$$

Unlike thermodynamic cycles where entropy changes, in ideal electromechanical conversion, all mechanical energy is available for conversion to electrical energy, and vice versa. As a result, the electromechanical energy conversion process is significantly more efficient compared to heat engines, such as turbines and internal combustion engines, which operate based on thermodynamic principles.

Efficiency of a system can be determined by experimentally measuring the input and output powers. Since energy is the integral of power over a given time interval, the ratio of output power to the input power gives efficiency.

Generally, calculating the electrical and mechanical losses in a system can be complex and often not necessary. Instead, the total loss of the system can be determined by evaluating its efficiency.

$$W_{loss} = W_{in}(1 - \eta) \tag{7.12}$$

Example 7.2 A DC motor draws 1.5 amps from a 12-V supply, driving a mechanical load at a speed of 1,200 rpm with a torque of 0.13 Nm. Calculate the total loss in the motor.

Solution

Electrical power supplied to the motor:

$$P_e = vi = 12 \times 1.5 = 18\ W$$

To calculate the mechanical power, we need to convert rotations per minute to radians per second.

$$\omega = \frac{2\pi n}{60} = \frac{2\pi \times 1{,}200}{60} = 125.66\ \text{rad}/s$$

Mechanical power driving the load is

$$P_m = \omega T = 125.66 \times 0.13 = 16.33\ \text{W}$$

The efficiency is

$$\eta = \frac{16.33}{18} = 0.90$$

Total loss can be calculated by substituting power for energy in equation 5.12.

$$P_{loss} = P_e(1-\eta) = 18 \times (1-0.9) = 1.8\ W \therefore$$

7.3 Energy Stored in Magnetic Field

In electromechanical energy conversion systems, the magnetic field acts as a medium for the conversion of energy between electrical and mechanical forms through changes in stored energy.

The instantaneous power supplied to the coil is the product of the voltage $e(t)$ across the coil terminals and the current $i(t)$ flowing through the coil. Assuming that the coil and core are lossless, the energy stored in the magnetic field is equal to the integral of the instantaneous power supplied to the coil.

$$P = \int (e \cdot i) dt \tag{7.13}$$

The expression of the field energy in the core is derived by substituting $e = d\lambda/dt$ in the equation above:

$$W_m = \int \left[\frac{d\lambda}{dt} i\right] dt = \int i d\lambda \tag{7.14}$$

This equation shows that magnetic energy is represented by the area between the λ–I curve and the vertical axis. The properties of magnetic materials are typically provided in the form of $\lambda(i)$. Calculation of the integral in this equation is often challenging.

Coenergy, a complement to magnetic field energy, is often used to compute forces and torques in electromechanical systems.

Coenergy is defined as the energy required to establish the flux linkage in the coil while keeping the current constant. It is represented by the area between the current axis and the λ–i curve. In mathematical terms, coenergy can be expressed as the integral of the flux linkage with respect to the current.

$$W_c = \int \lambda \, di \tag{7.15}$$

From Figure 7.2, we can deduct the relationship between energy and coenergy.

$$W_c = \lambda i - W_m \tag{7.16}$$

In a magnetic circuit, assuming the λ–i characteristic of the core is linear, the flux linkage is directly proportional to the current. The proportionality constant, by definition, is the self-inductance of the coil, represented by L.

$$\lambda = Li \tag{7.17}$$

The EMF induced across the terminals of a linear inductance is

$$e = \frac{d\lambda}{dt} = L\frac{di}{dt} \tag{7.18}$$

Therefore, energy stored in the magnetic field is

$$W_m = \int \left(L\frac{di}{dt} i \right) dt = \int Lidi = \frac{1}{2}Li^2 \tag{7.19}$$

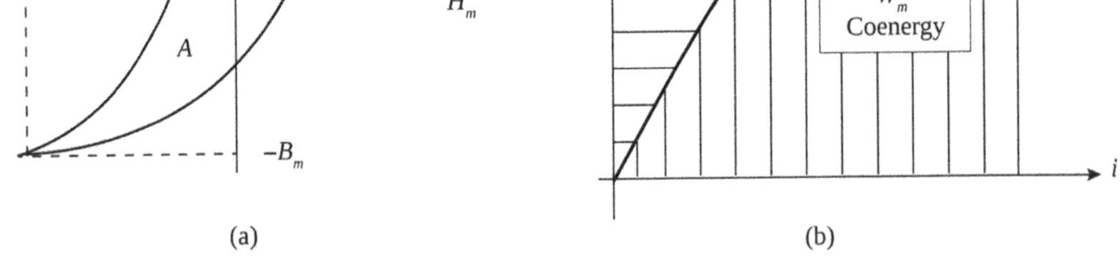

Figure 7.2 Magnetic energy and coenergy (a: Hysteresis curve; b: Simplified characteristic to show areas corresponding to energy and coenergy)

Energy stored in a linear magnetic circuit depends on the field intensity, permeability, and the core geometry.

$$W_m = \frac{1}{2}N\phi i = \frac{1}{2}NiBA = \frac{1}{2}Ni\mu HA$$

$$Ni = H\ell \tag{7.20}$$

$$W_m = \frac{1}{2}\mu H^2(A\ell)$$

Energy density is a useful concept in the design of electromechanical elements. Energy density is derived by dividing magnetic energy with the volume of the core material $A\ell$.

$$\hat{W} = \frac{1}{2}\mu H^2 = \frac{B^2}{2\mu} \tag{7.21}$$

7.4 Lorentz Force

Consider the classical example of a segment of rigid conductor of length ℓ freely sliding on two parallel conductor rails as shown in Figure 7.3. In Figure 7.3 (a), a sensitive galvanometer with ideally infinite internal resistance is connected across the fixed rails. Any friction and contact resistance between the conductor and rails is negligible. The permanent magnet produces magnetic induction B perpendicular to the conductor at all points. For a differential movement of the conductor from the original position, the magnetic flux will change by

$$d\lambda = BA = B\ell dx \tag{7.22}$$

The voltage across the moving conductor can be obtained by applying Faraday's law.

$$e = -\frac{d\lambda}{dt} = -Bl\frac{dx}{dt} \tag{7.23}$$

According to Lenz's law, as the conductor moves from left to right, it increases the area, resulting in a negative induced electromotive force (EMF). If the initial area decreases with the motion, the EMF becomes positive. This operation corresponds to the generator mode.

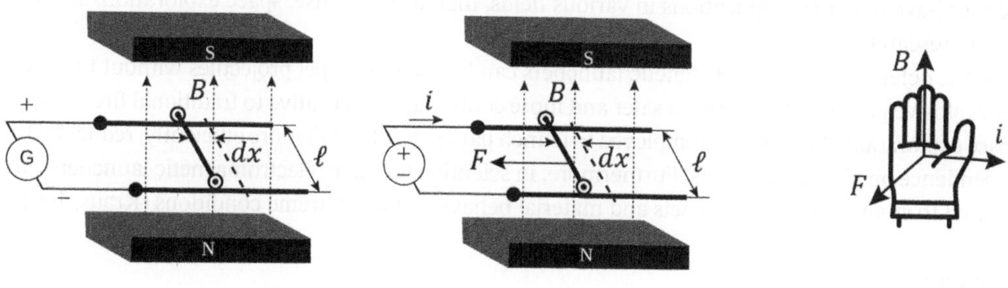

Figure 7.3 Rigid conductor moving in a magnetic field (a: Galvanometer measuring the induced EMF as the bar moves; b: Voltage source applied to move the bar; c: Right-hand rule)

In Figure 7.3 (b), a voltage source provides a constant current through the conductor. Assuming that all losses in the energy balance equation are negligible, differential change of the magnetic energy equals the work done by force to move the wire over the distance dx.

$$dW = Fdx = i \cdot d\lambda = i(B\ell dx) \tag{7.24}$$

By dividing both sides of the above expression by dx, we obtain the formula for the force exerted by the magnetic field on a current-carrying wire.

$$F = B\ell i \tag{7.25}$$

In most electromechanical devices, the magnetic flux is perpendicular to the moving wire. If not, the force is obtained by the cross-product of the vectors.

$$\vec{F} = \left(\vec{B} \cdot \vec{\ell}\right) i \tag{7.26}$$

The current is a scalar quantity since it is the time rate of change of electric charge. The magnitude of the vector $\vec{\ell}$ is equal to the length of the conductor in the magnetic field, and its direction is the same as the current flowing through the conductor.

This force, known as the Lorentz force, is crucial in the operation of various electromechanical devices. The direction of the force can be intuitively visualized using the right-hand rule illustrated in Figure 7.3. In Figure 7.3 (c) where the thumb is in the direction of the current and the fingers point the direction of the magnetic field, the palm points to the direction of the force exerted on the conductor. This intuitive method aids in visualizing how various electromechanical devices and systems function.

Example 7.3 Electromagnetic Launcher

An electromagnetic launcher, also known as a railgun, is a device that uses the Lorentz force to propel objects at high speeds. It consists of two parallel conductive rails and an armature bridging them, moving in a perpendicular magnetic field. When an electric current flows through the rails, it generates a Lorentz force that accelerates the armature along the rails, launching the attached projectile.

The principle of the electromagnetic launcher is based on the same physics principle that governs other electromechanical devices. Electric current flowing through the rails interacts with the magnetic field, producing a force that acts on the armature. By increasing the current or strength of the magnetic field, the launcher can achieve greater acceleration and higher launch velocities. These devices have potential applications in various fields, including defense, space exploration, and scientific research.

In the defense field, electromagnetic launchers can be used to propel projectiles without the need for explosive propellants, offering a safer and more controllable alternative to traditional firearms. In space exploration, they could be employed to launch payloads into orbit more efficiently, reducing the dependence on chemical rockets. Furthermore, in scientific research, electromagnetic launchers can be used to study high-speed impacts and material behavior under extreme conditions (Kraus, 1991).

Problem

Figure 7.4 shows a schematic structure of a torpedo launcher. As the cylindrical conductor bar rolls on two conducting rails, a current flows toward the picture. Because the current is perpendicular to

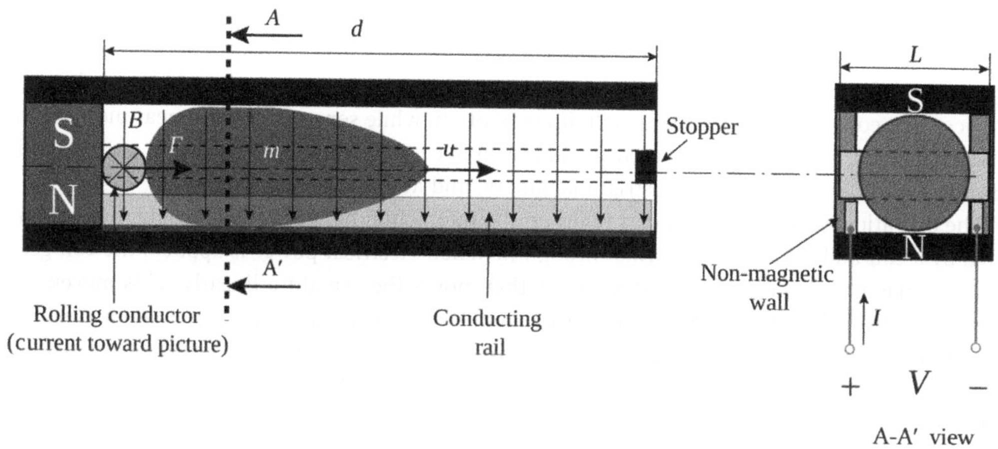

Figure 7.4 Torpedo launcher

the magnetic field, the Lorentz force moves it to the right according to the right-hand rule. The torpedo of mass m accelerates until the rolling bar hits the stoppers. Using the following parameters, calculate the current required to increase the velocity of the torpedo to 150 km/h when it leaves the launcher.

$$B = 2\,T; d = 25\,m; L = 1.5\,m; m = 50\,\text{kg}$$

Solution

By ignoring all frictions, the following equation can be utilized to calculate the acceleration of a missile propelled by a cylindrical conductor due to the Lorentz force.

$$F = BLI = ma$$

Initially, the missile is at rest and reaches the velocity of 250 km/h when it exits the launcher. This speed is equal to $(150 \times 10^3)/3{,}600 = 41.6$ m/s. Since the force is constant, the acceleration time can be determined by dividing the distance to the average speed.

$$\Delta t = \frac{d}{u} = \frac{25}{41.6} = 0.6\,s$$

To reach the desired speed, the missile must accelerate at the rate

$$a = \frac{u}{\Delta t} = \frac{41.6}{0.6} = 69.3\,m/s^2$$

The current is calculated by substituting the obtained acceleration value and given parameters into the first equation.

$$I = \frac{ma}{BL} = \frac{50 \times 69.3}{2 \times 1.5} = 1{,}155\,\text{A}.\therefore$$

Such a high current is obtained by stepping down AC voltage and rectifying by a diode bridge.

7.5 Linear Transducers

Linear transducers can function as either an actuator or a sensor. Actuators are electromechanical devices that convert electrical energy into linear motion, while sensors convert linear motion into electric signals. They are widely used in various applications where precise linear movement is involved, such as in robotics, industrial machinery, and automation systems.

The operation of linear transducers relies on the interaction between the magnetic field generated by a coil and the current flowing through it. When electrical power is applied, the coil generates a magnetic field, which creates a force that moves the armature linearly. This movement can be controlled accurately by adjusting the input current, making linear actuators ideal for on–off control and various applications requiring controlled linear movement. Linear sensors function based on Faraday's law of induction. A voltage is produced across a coil when exposed to a changing magnetic field caused by linear motion.

7.5.1 Moving Coil Devices

Moving coil devices are electromechanical systems that generate motion utilizing the Lorentz force that arises due to the interaction of a magnetic field with an electric current. Practical examples include moving coil meters, loudspeakers, and galvanometers and any type of transducer where a coil moves in a magnetic field.

In moving coil meters, a current-carrying coil is placed within a magnetic field, causing it to experience a Lorentz force. This results in an angular rotation of a pointer, attached to the coil that is directly proportional to the current passing through it. Such precise measurements are essential in various scientific and industrial applications.

Galvanometers, used for detecting and measuring small electric currents, also rely on the Lorentz force. The coil, suspended within a magnetic field, moves in response to the current, providing a visual indication of the current's magnitude.

Similarly, loudspeakers operate based on the same principle, where the electric current through the coil interacts with the magnetic field to move a cone that produces sound waves proportional to the current. A strong permanent magnet produces the magnetic field, and the output voltage of an audio amplifier provides the current. When the cone vibrates due to a sound wave and the coil terminals are connected to the input of an audio amplifier, the loudspeaker functions as a microphone.

Dynamic microphones are sensitive moving coil transducers that convert sound waves into electric signals. The core element of a dynamic microphone is a lightweight diaphragm attached to a coil. The coil is generally made of fine wire, usually copper, that winds around the diaphragm's perimeter. When the diaphragm moves, so does the coil, traversing through the magnetic field created by the permanent magnet. This movement generates an electrical signal proportional to the diaphragm's motion. The generated signal is then transmitted through the microphone's output to an amplifier or recording device.

Dynamic microphones utilize the principles of electromagnetic induction. When the coil moves within the magnetic field, it cuts across magnetic flux lines, inducing an EMF according to Faraday's law of induction. This induced EMF causes an electric current to flow through the coil, creating an electrical representation of the sound waves.

Example 7.4 The structure of a typical loudspeaker is schematically illustrated in Figure 7.5. Suppose the N–S permanent magnet creates the flux density B in the circular air gap of average

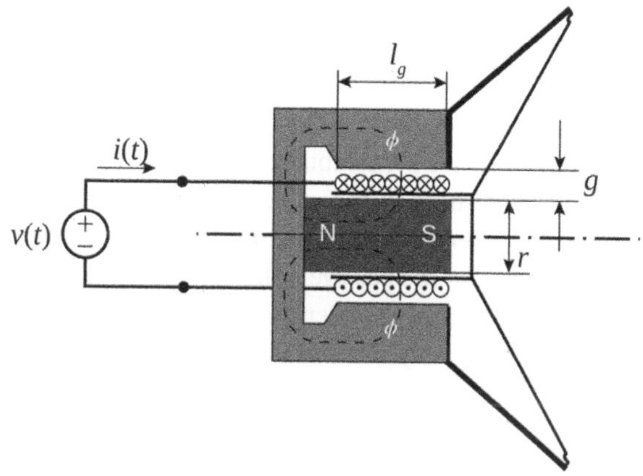

Figure 7.5 Loudspeaker example

radius r. The coil has N turns, and its electrical resistance is 4 Ω. The function below describes the voltage applied across the coil terminals

$$v(t) = 2\sin(1{,}000\, t)$$

Given $N = 50$ turns, $B = 1.2\ T$, and $r = 20$ mm, find the time variation of the force acting on the speaker cone.

Solution

Since the magnetic flux density in the air gap is perpendicular to the current flowing through the wire at all points, we can use the scalar equation below:

$$F = B\ell i(t)$$

The length of the coil wire is

$$\ell = N(2\pi r) = 50 \times 2 \times \pi \times \left(20 \times 10^{-3}\right) = 6.28\ \text{m}$$

The current flowing through the coil is determined using Ohm/s law.

$$i = \frac{v}{R} = \frac{2}{4}\sin(1{,}000\, t) = 0.5\sin(1{,}000\, t)$$

The force acting on the coil is obtained from equation 7.25.

$$F = B \times \ell \times [0.5\ \sin(1{,}000\, t)] = 1{,}2 \times 6.28 \times 0.5\sin(1{,}000t\,) = 3.768\sin(1{,}000\, t)$$

Note that the variation of the force in time is the same as the variation of the applied; therefore there is no distortion. This is because the magnetic flux is independent of the current, or linear. This is the property of a good speaker.

7.5.2 Moving-Armature Devices

In many practical electric machines, actuators, and transducers, the coil that produces the magnetic field is fixed, and an iron piece in the magnetic core moves. Devices such as electromagnets, solenoids, relays, and an important class of electric machines fall in this group.

In applications involving moving-armature actuators, magnetic force and torque play significant roles. These actuators convert electrical energy into mechanical motion depending on the change of energy stored in the magnetic field.

Figure 7.6 (a) shows an electromagnet, which has a fixed core and a moving armature, both made from a ferromagnetic material that has extremely large relative permeability. *N*-turn coil placed on the static part of the electromagnet produces a magnetic flux ϕ in the core.

The electromagnet shown in Figure 7.6 is supposed to operate as an actuator; therefore, its input is electric power, and the output is mechanical. According to the passive notation, the force f_e and displacement dx are shown in the same direction. When mechanical energy is the output, the force and displacement must have opposite signs. We can reasonably anticipate that the force will be negative.

$$f_e dx < 0 \Rightarrow f_e < 0 \tag{7.27}$$

Assuming that coil resistance, all frictions, and external forces are negligible, total electric energy is equal to the sum of the mechanical energy (or work) and the change in magnetic field energy (dW_m).

$$eidt = f_e dx + dW_m \tag{7.28}$$

The block diagram in Figure 7.6 (b) illustrates the energy balance of the electromagnet with electrical input and mechanical output. The electrical input power results in a net change of energy in the magnetic field W_m and the work done by the mechanical system. The differential change in the energy stored in the magnetic field is the difference between the electric and mechanical powers.

$$dW_m = eidt - f_e dx \tag{7.29}$$

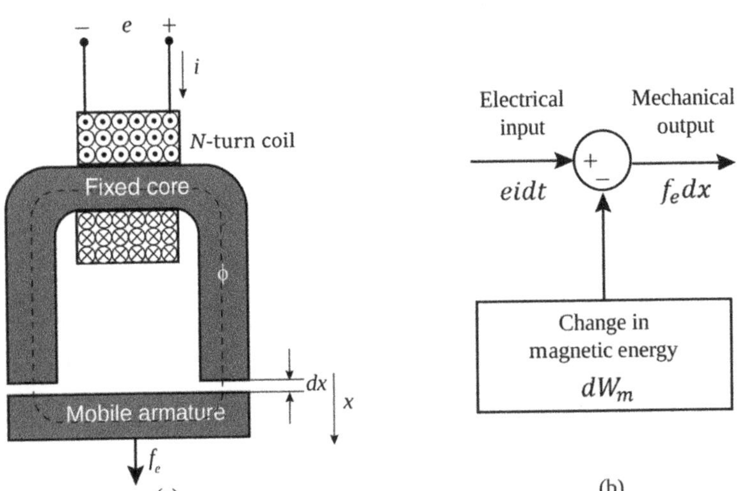

Figure 7.6 A moving armature linear actuator: Electromagnet (a: Schematic structure; b: Block diagram to show energy flow)

Substituting $e = d\lambda/dt$ according to Faraday's law, the differential change in magnetic energy becomes

$$dW_m = i\frac{d\lambda}{dt}dt - f_e dx \tag{7.30}$$

or

$$dW_m(\lambda, x) = id\lambda - f_e dx \tag{7.31}$$

The magnetic field energy is a function of two independent state variables: the current i and the displacement x.

We can express the complete differential of the magnetic field energy (W_m) as a function of the flux linkage (λ) and the displacement (x). This differential can be written as

$$dW_m(\lambda, x) = \left(\frac{\partial W_m}{\partial \lambda}\right)d\lambda + \left(\frac{\partial W_m}{\partial x}\right)dx \tag{7.32}$$

Here, the term ($\partial W_m/\partial \lambda$) represents the partial derivative of the magnetic field energy with respect to λ while the other variable is kept constant. Comparing equations 7.31 and 7.32 yields

$$f_e = -\frac{\partial W_m(\lambda, x)}{\partial x}(\lambda = constant) \tag{7.33}$$

$$i = \frac{\partial W_m(\lambda, x)}{\partial \lambda}\ (x = constant) \tag{7.34}$$

Calculating the force with flux linkage and displacement is usually more complex than using current and displacement. The force can be alternatively expressed in terms of the coenergy.

$$W'_m(i, x) = i\lambda - W_m(\lambda, x) \tag{7.35}$$

$$dW'_m(i, x) = d(i\lambda) - dW_m(\lambda, x) \tag{7.36}$$

Substituting equation 7.29, we can write

$$dW'_m = id\lambda + \lambda di - (id\lambda - f_e dx) = id\lambda + f_e dx \tag{7.37}$$

Coenergy is a state function of two independent variables i and x. Therefore, its total differential is

$$dW'_m = \left(\frac{\partial W'_m}{\partial i}\right)di + \left(\frac{\partial W'_m}{\partial x}\right)dx \tag{7.38}$$

The expression of the force and current is derived by comparing each term of the last two equations.

$$f_e = \frac{\partial W'_m(i, x)}{\partial x}(i = constant) \tag{7.39}$$

$$i = \frac{\partial W'_m(i, x)}{\partial i}\ (x = constant) \tag{7.40}$$

Magnetic force acting on the armature is simply the derivative of the coenergy with respect to the displacement.

In linear magnetic circuits where the λ–i curve is a straight line, the coenergy is equal to the magnetic energy. If the magnetic device operates in the linear region of the magnetic characteristic, the energy can be assumed equal to coenergy.

$$W'_m = W_m \tag{7.41}$$

$$f_e = \frac{\partial W'_m}{\partial x} \tag{7.42}$$

This property simplifies the calculations if the system can be considered linear.

7.5.3 Applications of Linear Actuators

Linear actuators, including electromagnets and solenoids, are integral components in modern mechatronic systems, offering versatile and reliable solutions for a wide range of applications.

7.5.3.1 Electromagnet

An electromagnet is a type of magnet in which the magnetic field is produced by the flow of electric current. Unlike permanent magnets, electromagnets can be switched on and off, and the strength of the magnetic field can be adjusted by changing the amount of current passing through the coil. This makes electromagnets extremely versatile and useful in various applications.

Electromagnets are used in industrial machinery to move heavy objects, control machinery, and facilitate manufacturing processes. Their ability to exert strong magnetic forces when turned on and to be switched off makes them perfect for tasks that require precise control and flexibility.

In consumer electronics, electromagnets are essential components in devices such as relays and magnetic locks. These devices rely on the ability to create and manipulate magnetic fields to perform their functions efficiently.

Electromagnets play a crucial role in scientific research, particularly in particle accelerators, where they help guide and control the paths of charged particles. Additionally, they are employed in medical devices like MRI machines, which utilize magnetic fields to generate detailed images of the inside of the human body.

The electromagnet illustrated in Figure 7.6 has a core made from a soft magnetic material of relative permeability μ_c. The average length of the magnetic flux in the core is ℓ_c. The cross-sectional area of the core perpendicular to the flux is A, and the coil has N turns. We will derive an expression for the force exerted on the moving armature as a function of the air gap length and coil current.

To obtain the expression of the force, we need to find the stored magnetic energy or coenergy. The magnetic flux completes its path through magnetic material and two air gaps. Therefore, the magnetic equivalent circuit consists of two reluctances and the MMF connected in series. Magnetic flux can be determined by solving the equivalent magnetic circuit. The reluctances of the core and air gap will be determined as explained in Chapter 4 using the equation $\mathcal{R} = \ell/\mu A$:

Assuming the magnetic flux remains constant at every point, the average reluctances are approximately determined as follows:

$$\mathcal{R}_c = \frac{\ell_c}{\mu_0 \mu_c A_c} \tag{7.43}$$

For relatively small air gap, fringe (leakage) flux is negligible. Therefore, the cross section of the air gap is approximately equal to the cross section of the core. Denoting the variable length of the air gap as x, the total length of the two air gaps will be $2x$ and the permeability of the air gap is μ_0.

$$\mathcal{R}_a = \frac{2x}{\mu_0 A_c} \tag{7.44}$$

The magnetic flux is found by applying the analogy of KVL and Ohm's law for magnetic circuits. It is important to note that ferromagnetic materials have relative permeability in the order of thousands. Consequently, the reluctance within the magnetic core is usually negligible compared to the air gap's reluctance.

$$\phi = \frac{Ni}{\mathcal{R}_c + \mathcal{R}_a} = \frac{Ni}{\dfrac{\ell_c}{\mu_0 \mu_c A_c} + \dfrac{2x}{\mu_0 A_c}} \cong \frac{\mu_0 A_c Ni}{2x} \tag{7.45}$$

The flux linked by the N-turn coil is $\lambda = N\phi$. Assuming that the current is not large enough to saturate the core material, the magnetic circuit is linear. Hence, energy is equal to the coenergy.

$$W = W' = \frac{1}{2}\lambda i = \frac{1}{2}\frac{\mu_0 A_c (Ni)^2}{2x} \tag{7.46}$$

The following equation can be used to calculate the magnetic force acting on the movable armature:

$$f_e = \frac{\partial W'_m}{\partial x} = \frac{\partial}{\partial x}\left[\frac{1}{2}\frac{\mu_0 A_c (Ni)^2}{2x}\right] = -\frac{\mu_0 A_c (Ni)^2}{4x^2} \tag{7.47}$$

The negative sign indicates that the force is acting in the opposite to the direction of displacement shown in the figure.

Example 7.5 Electromagnetic Gripper

The electromagnet shown in Figure 7.7 is designed to pick up and handle objects in a warehouse. Using the specified dimensions and material properties provided below, determine the maximum weight that can be lifted when the gripper is positioned approximately 5 mm from the object and the coil current is 500 mA.

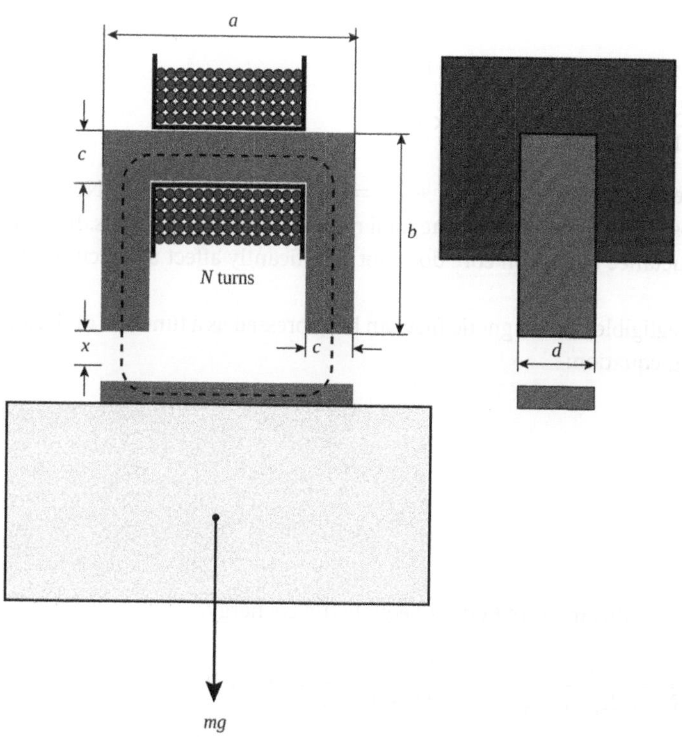

Figure 7.7 Electromagnetic gripper

$$a = 50\,\text{mm}; b = 40\,\text{mm}; c = 10\,\text{mm}; d = 15\,\text{mm}; e = 2\,\text{mm}$$
$$N = 75\,\text{turns}; \mu_o = 4\pi 10^{-7}; \mu_c = 3{,}000$$

Solution

The magnetic flux completes its path through three components with different parameters: a U-shaped magnetic core (1), a moving magnetic plate (2), and two air gaps (3).

1) U-shaped static core:

$$\ell_{f1} = (a-c) + 2\left(b - \frac{c}{2}\right) = \left[(50-10) + 2\left(40 - \frac{10}{2}\right)\right] \times 10^{-3} = 0.110\,\text{m}$$

$$A_{f1} = c \cdot d = \left(10 \times 10^{-3}\right) \times \left(15 \times 10^{-3}\right) = 150 \times 10^{-6}$$

$$\mathcal{R}_{f1} = \frac{\ell_{f1}}{\mu_0 \mu_c A_{f1}} = \frac{0.130}{4\pi 10^{-7} \times 3{,}000 \times 150 \times 10^{-6}} = 0.973 \times 10^6$$

2) Moving steel plate (armature)

$$\ell_{f2} = (a-c) = (50-10) \times 10^{-3} = 0.04\,\text{m}$$

$$A_{f2} = d \cdot e = \left(15 \times 10^{-3}\right) \times \left(2 \times 10^{-3}\right) = 30 \times 10^{-6}$$

$$\mathcal{R}_{f2} = \frac{\ell_{f2}}{\mu_0 \mu_c A_{f2}} = \frac{0.04}{4\pi 10^{-7} \times 3{,}000 \times 30 \times 10^{-6}} = 0.354 \times 10^6$$

3) Air gaps

$$\ell_g = 2x$$

$$A_g = c \cdot d = \left(10 \times 10^{-3}\right) \times \left(15 \times 10^{-3}\right) = 150 \times 10^{-6}$$

$$\mathcal{R}_g = \frac{\ell_g}{\mu_0 A_g} = \frac{2x}{4\pi 10^{-7} \times 150 \times 10^{-6}} = \left(10{,}610 \times 10^6\right)x$$

Total reluctance of the magnetic circuit is $\mathcal{R}_{f1} + \mathcal{R}_{f2} + \mathcal{R}_g = (0.584 + 2{,}650x) \times 10^6$. Note that the reluctance of the two air gaps is much greater than the total reluctance of the iron parts. In most practical cases, ignoring the reluctance of an iron core does not significantly affect the accuracy of the result.

If the fringe (leakage) flux is negligible, the magnetic flux can be expressed as a function of the air gap length x using the following equation:

$$\phi = \frac{Ni}{\mathcal{R}_{f1} + \mathcal{R}_{f2} + \mathcal{R}_g}$$

The flux linkage of the coil is

$$\lambda = N\phi$$

Assuming the magnetic circuit is linear, stored energy equals the coenergy.

$$W_m = W'_m = \frac{1}{2}\lambda i = \frac{1}{2}\frac{(Ni)^2}{\mathcal{R}_{f1} + \mathcal{R}_{f2} + \mathcal{R}_g} \cong \frac{1}{2}\frac{(750 \times 0.5)^2}{2{,}650 \times 10^6}\frac{1}{x} = 663 \times 10^{-6}\frac{1}{x}$$

The force acting on the moving part is the derivative of the coenergy with respect to the displacement x.

$$F = \frac{\partial W'_m}{\partial x} = -\frac{663 \times 10^{-6}}{x^2}$$

For $x = 5$ mm, we find $F = 13.8$ N.

The negative sign indicates that the force is acting in the opposite direction to the reference chosen for displacement x. The force required to lift 20 kg is $20 \times 9.81 = 196.2$ N. For $x = 5$ mm, the minimum current to attract this load is

$$i = \sqrt{\frac{196.2 \times (5 \times 10^{-3})^2}{0.08 \times 10^{-3}}} = 7.83\, A \therefore$$

7.5.3.2 Solenoid

A solenoid is composed of a coil, a ferromagnetic core, and a movable metal plunger or armature. Like the electromagnet, it converts electrical energy into linear mechanical movement. When an electric current passes through the wire, it generates a magnetic field, causing the armature to move inside the coil. This movement can be used to perform various tasks such as closing or opening a valve, activating a switch, or creating linear motion.

Solenoid actuators are widely used in industrial processes for controlling valves, switches, and other mechanisms. Their ability to provide precise control and quick response makes them ideal for applications such as regulating the flow of liquids and gases in pipelines. Industries such as chemical processing and water treatment rely on solenoid valves for accurate and efficient fluid control.

In heating, ventilation, and air conditioning (HVAC) systems, solenoid actuators are employed to manage the flow of air and coolant, ensuring the efficient operation of heating and cooling systems. Their reliability and rapid response times make them suitable for maintaining the desired environmental conditions.

The automotive industry makes extensive use of solenoid actuators in systems such as fuel injection, transmission control, and braking. These actuators contribute to the performance, safety, and efficiency of modern vehicles.

Solenoid actuators are crucial in medical equipment, where they are used to control various mechanisms in diagnostic and therapeutic devices. Their precise control and reliability are essential for medical applications, ensuring accurate and safe operation.

In consumer electronics, solenoid actuators are used in devices such as printers, cameras, and home appliances. Solenoid valves in dishwashers, laundry washers, refrigerators, and carpet cleaners control water intake and discharge to regulate the water level. Their ability to perform rapid and precise movements makes them indispensable for the functionality of these gadgets.

Solenoids are key components in robotics and automation, where they provide the necessary movements and controls for robotic arms, grippers, and other automated systems. Their versatility enables the creation of complex and efficient robotic solutions.

Environmental and agricultural systems use solenoid actuators for tasks such as irrigation control, environmental monitoring, and farming equipment. These actuators contribute to the sustainable and efficient management of resources in these sectors.

Example 7.6 Solenoid Actuator

The solenoid actuator shown in Figure 7.8 is constructed from a soft magnetic material with a sufficiently high relative permeability to make its reluctance negligible. It has a 75-turn coil with 8 Ω resistance. When the coil is not energized, the spring holds the armature at $x = 5$ mm distance. Determine the net force acting on the armature when the solenoid is energized from a 12-V DC power supply for the parameters given below. Plot the net force as a function of the displacement x using MATLAB®.

$$d = 15\text{ mm}, g = 0.5\text{ mm}, a = 5\text{ mm},$$

Mass of the armature: $M = 10$ g
Spring constant: $k = 50$ N/m
Viscous friction constant of the slider bushing: $\beta = 0.1$ Nm/s

Solution

To determine the force acting on the armature as a function of the displacement, we must derive the equation for the flux distribution within the actuator. Since the device is cylindrical, the magnetic flux symmetrically surrounds the core. Dashed lines on the figure represent the average path of the flux.

Since the core reluctance is negligible, the magnetic equivalent circuit consists of the MMF NI and the reluctances of two air gaps. The air gap reluctances are calculated below.

Air gap around the armature:

Area around the section of the armature inserted in the circular bushing is the side surface of a cylinder given by

$$A_1 = \pi d a$$

The length of the air gap (ℓ) is the thickness of the nonmagnetic bushing g. Substituting the area and distance in the reluctance equation, we obtain

$$\mathcal{R}_{g1} = \frac{g}{\mu_o \pi d a}$$

Air gap between the left side of the armature and the stationary part:

The air gap between the left side of the moving armature and the cylindrical central piece of the fixed body is the area of the circle and the length equals the displacement x.

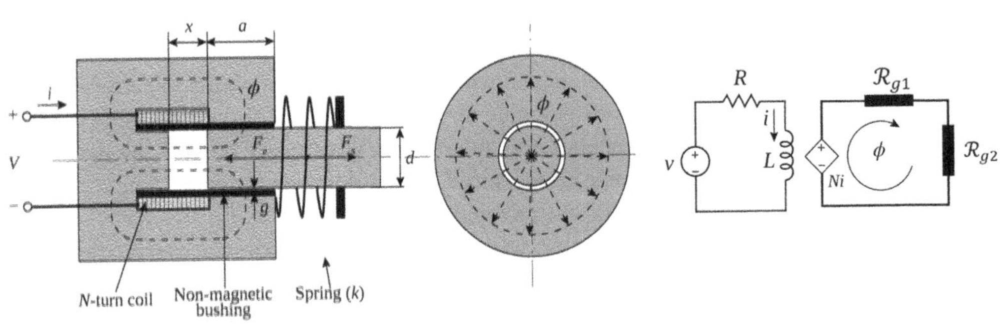

Figure 7.8 Solenoid actuator and its magnetic equivalent circuit

$$A_2 = \frac{\pi d^2}{4}$$

$$\mathcal{R}_{g2} = \frac{x}{\mu_o \frac{\pi d^2}{4}} = \frac{4x}{\mu_o \pi d^2}$$

$$\mathcal{R}_t = \mathcal{R}_{g1} + \mathcal{R}_{g2} = \frac{g}{\mu_o \pi da} + \frac{4x}{\mu_o \pi d^2} = \frac{gd + 4ax}{\mu_o \pi a d^2}$$

Because the reluctance of the ferromagnetic parts is neglected, the magnetic field is linear. Thus, energy is equal to coenergy.

$$W = W' = \frac{1}{2} L i^2$$

The inductance of the coil can be determined from the reluctance.

$$L = \frac{N^2}{\mathcal{R}_t} = \frac{N^2 \mu_o \pi a d^2}{gd + 4ax}$$

Coenergy:

$$W' = \frac{1}{2} \frac{N^2 \mu_o \pi a d^2}{gd + 4ax} i^2$$

The force acting on the moving part is

$$F = \frac{\partial W'}{dx} = \frac{\partial}{\partial x}\left[\frac{1}{2}\frac{N^2 \mu_o \pi d^2 a}{gd + 4ax} i^2\right] = -\frac{1}{2}\left(N^2 \mu_o \pi d^2 a i^2\right)\frac{4a}{(gd + 4ax)^2}$$

After simplifications, we obtain

$$F_e = -2\mu_o \pi (Nda)^2 \frac{i^2}{(gd + 4ax)^2}$$

Net force acting on the armature is determined by including the force F_s exerted by the spring (Figure 7.9).

$$F_{net} = F_s + F_e = kx - 2\mu_o \pi (Nda)^2 \frac{i^2}{(gd + 4ax)^2}$$

MATLAB® Code to Calculate and Plot the Force as a Function of the Displacement.

```
% Given parameters
N=75;
d=15e-3;
g=0.5e-3;
a=5e-3;
k=50;
mu_0=pi*4.e-7;

% Define the x values
x = linspace(0, 0.005, 1000);
```

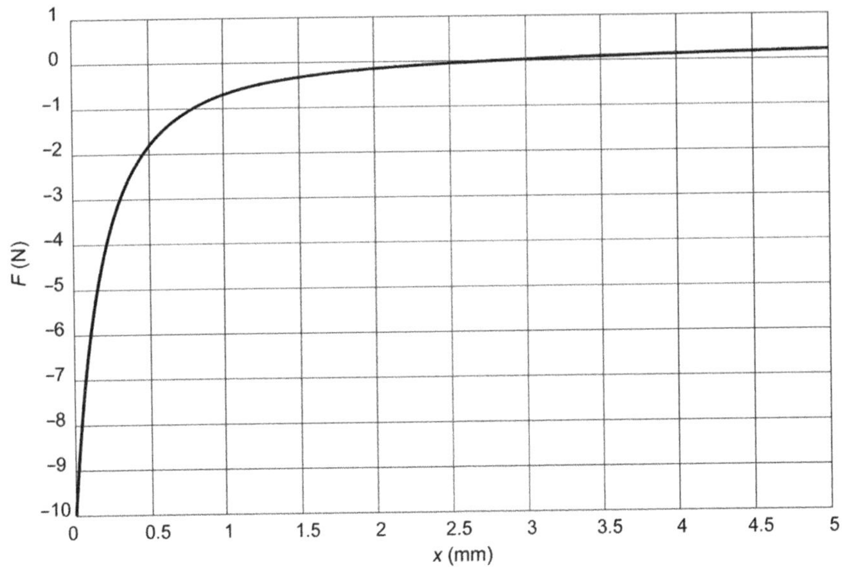

Figure 7.9 Net force acting on the armature

```
% Calculations
V=12.;
R=8.;
I=V/R;
K=-2*mu_0*pi*(N*d*a)^2;
Fe=K*I^2./(g*d+4*a*x).^2;
Fs=k*x;
Ft=Fe+Fs;
% Print result
formatSpec = 'For x=0: F= %4.2f N \nFor x=5mm: F= %4.2f N\n';
fprintf(formatSpec,Ft(1),Ft(1000))

For x=0: F= -9.99 N
For x=5mm: F= 0.20 N

% Plot the force versus displacement
figure;
plot(x*1000, Ft, 'LineWidth', 2);
grid on;
% Add labels and title
xlabel('x (mm)');
ylabel('F (N)');
title('Net Force');
xlim([0 5])
ylim([-10 1])
```

7.6 Rotary Transducers

Rotary transducers are devices designed to convert electrical energy into rotational work or the reverse. The stationary component of a rotary transducer is often called stator, and the moving component is called rotor. In motor operation mode, the rotary transducer functions as an actuator. The magnetic field generated by either a permanent magnet or coils causes rotation of the moving part. In rotary actuators, torque generation occurs either through the Lorentz force acting on current-carrying conductors or due to variations in magnetic reluctance. They can also function as an angular motion sensor generating electric signal as a function of the angular velocity.

Figure 7.10 displays two types of rotary actuator. In Figure 7.10 (a), a permanent magnet creates magnetic flux in the core and air gap. A coil with N turns is positioned diametrically in the slots on the rotor. In Figure 7.10 (b), the current flowing through a coil placed on the stator produces magnetic flux. The rotor does not have any coil or permanent magnet. This type of transducer is known as "reluctance motor." Both transducers have a limited range of motion since the torque becomes zero at certain positions. However, they highlight key aspects of rotary actuators, sensors, and electric machines.

Rotary actuators are utilized in mechanisms that need accurate control of angular movement and torque or an angular motion signal. They have numerous applications, including automation and robotics, due to their reliability and simplicity. Analysis of rotary actuators is essential for understanding the functioning of rotating machinery, which will be addressed in the next chapter.

7.6.1 Moving Coil Rotary Actuator

Figure 7.11 (a) is a simplified diagram of a moving coil rotary transducer where N-turn coil is placed in a slot on the rotor. A permanent magnet produces the magnetic field required for its operation.

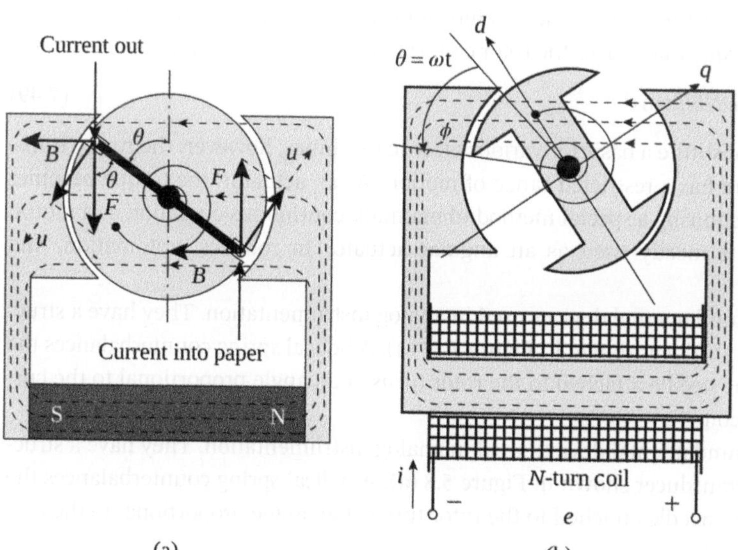

Figure 7.10 Two types of rotary transducers (a: Permanent magnet moving coil transducer; b: Electromagnet driven variable reluctance transducer)

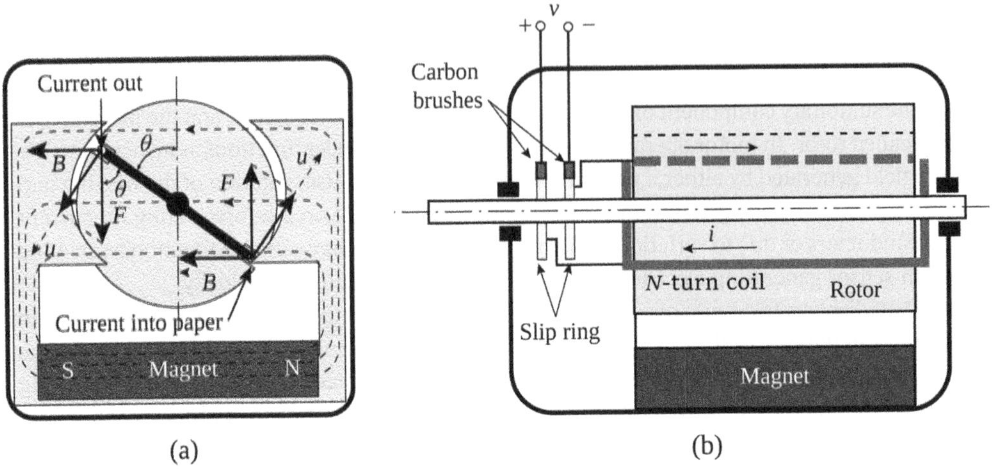

Figure 7.11 Tachometer. (a) Front view; (b) Side cross section

The rotor is cylindrical; therefore, the distance of the air gap remains approximately constant, ignoring the effect of the slots. Assuming that the magnetic field is uniform around the air gap, the Lorentz force produces a force acting on the conductor through which a current flows.

$$F = N(B\ell i) \tag{7.48}$$

The direction of the force can be determined by using the right-hand rule. Suppose that the coil is connected to an external power source, which produces a current that flows out of the figure in the upper wire and toward the figure in the lower wire. Since the magnetic induction vector is in horizontal direction perpendicular to the current, the force vector is vertical. The tangential force components on each side of the wire create a torque that moves the rotor. By denoting the coil radius as r and diameter $d = 2r$, the torque that moves the rotor can be calculated using the following formula:

$$T = 2Nr\ell Bi \sin\theta = DNBi\ell \sin\theta \tag{7.49}$$

Moving coil actuators constitute a basis for various electric machines; however, the rotary transducer depicted in the figure has a restricted range of motion. As an actuator, the torque becomes zero at certain positions, requiring a special method to maintain continuous operation as a motor. The illustrated structure is mostly used as an angular actuator in robotics, automation, and instrumentation.

Moving coil meters are fundamental devices used in analog instrumentation. They have a structure similar to the rotary transducer shown in Figure 7.10 (a). A helical spring counterbalances the magnetic torque so that the needle attached to the rotor turns at an angle proportional to the current flowing through the coil.

Moving coil meters are fundamental devices used in analog instrumentation. They have a structure similar to the rotary transducer shown in Figure 5.8 (a). A helical spring counterbalances the magnetic torque so that the needle attached to the rotor turns at an angle proportional to the current flowing through the coil.

Moving coil meters are widely used due to their reliability and precision. They can be found in various applications, including voltmeters, ammeters, and galvanometers, each designed to measure different electrical quantities by utilizing the same underlying principle. The ability of moving coil meters to provide precise measurements makes them indispensable in many fields, from electronics to industrial automation.

7.6.2 Moving Coil Rotary Sensor

When an external driver moves the rotor, a voltage is induced across the coil according to Faraday's induction law. As illustrated in the figure, the magnetic flux enclosed within the coil is

$$\phi = BA = D\ell B \cos\theta \tag{7.50}$$

Ignoring coil resistance, the expression of the voltage across the load resistor is obtained by applying Faraday's magnetic induction law.

$$v = N\frac{d\phi}{dt} = -ND\ell B \sin\theta \frac{d\theta}{dt} \tag{7.51}$$

Rotary sensors produce sinusoidal voltage during continuous rotation, but a special connection system is needed to transfer current between the moving coil and the external circuit. To facilitate this connection, a slip ring and brush system is employed. This system consists of copper or brass rings mounted on the shaft with insulation and a set of stationary carbon brushes.

Moving coil rotary sensor shown in Figure 7.11 generates a voltage proportional to the angular velocity ω. Time function of the generated voltage for continuous operation is obtained by substituting $\theta = \omega t$.

$$v = V_m \sin\omega t = (-ND\ell B\omega) \sin\omega t \tag{7.52}$$

This voltage can be calibrated to measure rotational speed accurately, creating a tachometer. Tachometers are analog sensors used for engine speed monitoring, industrial machinery control, and other applications requiring precise speed measurement.

Example 7.7 Tachometer Design

We are designing a tachometer based on the conceptual drawing illustrated in Figure 7.11 to produce a sinusoidal voltage with an amplitude of 10 V when the shaft rotates at 3,000 rpm. We will use an existing magnetic structure where the rotor diameter is 15 mm and the axial length is 10 cm. The permanent magnet creates 1.2 T flux density in the air gap.

Solution

Since the dimensions of the magnetic structure are known and the flux density is given, we need to determine the number of turns to generate 10 V amplitude at 5,000 rpm rotation speed. First, we must convert the rotations per minute to angular velocity in radians per second.

$$\omega = \frac{2\pi n}{60} = \frac{2\pi \times 3{,}000}{60} = 314.16 \text{ rad/sec}$$

From equation 7.52, we can write

$$N = \frac{V_m}{D\ell B\omega} = \frac{10}{0.015 \times 0.1 \times 1.2 \times 314.16} = 17.68$$

For practical reason, the number of turns must be integer. Therefore, we can round up to 18 turns. The error due to rounding will be canceled in calibration.

7.6.3 Moving-Armature Rotary Actuator

Figure 5.8 (b) shows a rotary actuator constructed from a soft magnetic material. A coil on the stator generates a magnetic field, which completes its path through the air gap and the rotor. The rotor lacks any winding or permanent magnet. As the rotor turns, the reluctance of the air gap varies.

The torque developed in a moving-armature rotary actuator can be examined through the application of the energy balance equation. In the idealized energy conversion part without electrical and mechanical losses, the energy balance equation can be written by replacing the force with torque and displacement with the rotation angle in equation 7.28 obtained for linear actuators.

$$eidt = T_e d\theta + dW_m \tag{7.53}$$

Torque may be defined using the magnetic energy and coenergy, following the same methodology applied to linear actuators.

$$T_e = -\frac{\partial W_m(\lambda, \theta)}{\partial \theta}(\lambda = constant) \tag{7.54}$$

$$T_e = \frac{\partial W'_m(i, \theta)}{\partial \theta}(i = constant) \tag{7.55}$$

Example 7.8 Reluctance Motor

The rotary actuator in Figure 5.8 (b) is constructed using a magnetic material with high relative permeability, which makes its reluctance negligible compared to the air gap. Determine the torque for the following parameters.

$$r = 2\text{ cm}, \ell = 5\text{ cm}, \delta = 0.5\text{ mm}, i = 2.5A, N = 200\text{ turns}$$

Solution

Magnetic energy is mainly stored within the air gap between the rotor and the stator denoted by θ in the figure. Because the rotor diameter is not constant, the air gap is greater at all other points; hence, the magnetic energy is negligible. We can find magnetic flux using the magnetic energy density $B^2/2\mu_o$. Magnetic energy is expressed by multiplying the energy density by the volume of the air gap defined by θ.

$$W_m = \frac{B^2}{2\mu_o}(2r\theta\ell\delta) = \frac{B^2 r\ell\delta}{\mu_o}\theta$$

In the air gap, coenergy equals energy. We can express magnetic coenergy by substituting $B = \mu_o Ni/2\delta$ in the equation above.

$$W'_m = W_m = \frac{\mu_o (Ni)^2 r\ell}{4\delta}\theta$$

Torque is obtained using equation 7.54 for constant current during the infinitesimal small motion.

$$T_e = \frac{\partial W'_m(i, \theta)}{\partial \theta} = \frac{\mu_o (Ni)^2 r\ell}{4\delta}$$

Given $r = 1.5\text{ cm}, \ell = 5\text{ cm}, \delta = 0.5\text{ mm}, i = 2.5\,A, N = 200$ turns

$$T_e = \frac{4\pi 10^{-7}(200 \times 2.5)^2 1.5 \times 10^{-2} \times 5 \times 10^{-2}}{4 \times 0.5 \times 10^{-3}} = 0.118\text{ Nm} \therefore$$

7.7 Multiple Input Multiple Output (MIMO) Electromechanical Systems

MIMO systems are integral to various high-performance electromechanical applications, where multiple coils contribute to the magnetic field to produce motion. In most electric machines, a combination of windings generates a magnetic field, and interconnected coils produce torque. An electromagnetic robotic gripper utilizes multiple coils and magnetic cores to generate controlled magnetic fields, enabling the precise manipulation of objects.

Consider an electromechanical system that has n electrical and m mechanical ports. Suppose that this system performs rotary motion, which is more common in practice. The magnetic flux linkages can be expressed as the functions of the currents and rotation angles of mechanical parts.

$$\begin{aligned} \lambda_1 &= f_1(i_1, i_2, i_3, \ldots i_n; \theta_1, \theta_2, \theta_3 \cdots \theta_m \\ \lambda_2 &= f_2(i_1, i_2, i_3, \ldots i_n; \theta_1, \theta_2, \theta_3 \cdots \theta_m \\ &\vdots \\ \lambda_n &= f_n(i_1, i_2, i_3, \ldots i_n; \theta_1, \theta_2, \theta_3 \cdots \theta_m \end{aligned} \tag{7.56}$$

The differential magnetic energy described in equation 5.31 can be modified for rotary MIMO systems as follows:

$$dW_m(\lambda_1, \lambda_2, \lambda_3, \ldots \lambda_n; \theta_1, \theta_2, \theta_3 \cdots \theta_m) = \sum_n i_j d\lambda_j - \sum_n T_k \theta_k \tag{7.57}$$

Linear (or linearized) magnetic systems are described by an inductance matrix that contains self and mutual inductances.

$$\begin{bmatrix} \lambda_1 \\ \lambda_2 \\ \vdots \\ \lambda_n \end{bmatrix} = \begin{bmatrix} L_{11} & L_{12} & \cdots & L_{1n} \\ L_{21} & L_{22} & & L_{2n} \\ \vdots & & \ddots & \vdots \\ L_{n1} & & \cdots & L_{nn} \end{bmatrix} \begin{bmatrix} I_1 \\ I_2 \\ \vdots \\ I_n \end{bmatrix} \tag{7.58}$$

In a linear system, energy equals coenergy and is determined by integrating $\boldsymbol{\lambda I}$.

$$W_m = W'_m = \int \boldsymbol{\lambda} d\boldsymbol{I} = \int [\boldsymbol{LI}] d\boldsymbol{I} = \frac{1}{2}[\boldsymbol{I}]^t [\boldsymbol{L}][\boldsymbol{I}] \tag{7.59}$$

To simplify, an electromechanical system with two electrical ports and one mechanical port is considered. By adapting equation 7.55, the following result is obtained:

$$T_e = \frac{\partial W'_m(\lambda_1, \lambda_2, \theta)}{\partial \theta} \ (i = constant) \tag{7.60}$$

Torque is determined in terms of the inductance matrix as below:

$$T_e = \frac{1}{2}[\boldsymbol{I}]^t \frac{\partial [\boldsymbol{L}]}{\partial \theta} [\boldsymbol{I}] \tag{7.61}$$

Example 7.9 The schematic diagram in Figure 7.12 illustrates a transducer that activates a hydraulic motion transmission system. The device can be considered as two solenoids combined in the same magnetic structure. The armature of mass m can slide in a plastic ring of thickness δ and viscous friction constant β. The core is made from a very high-permeability soft magnetic material. The coils are identical with N turns and resistance R. When both coil currents are zero, two springs with constant k keep the armature horizontally centered in the air gap. However, when the currents differ, the armature moves by x.

The windings of the solenoid are connected to two digital outputs of a microcontroller (for example, an Arduino® board). At logic 1 level, the voltage across the output and ground terminals is 5 V.

a) Derive an equation to calculate the electromagnetic force.
b) Write a dynamic equation for displacement x as a function of the coil currents.
c) Given the following parameters, plot the displacement of the armature in time when the microcontroller outputs turn from 0 to 10 at $t = t_0$.

$$d = 5\text{ mm}, \delta = 0.25\text{ mm}, \ell = 15\text{ mm}, a = 8\text{ mm}, N = 100, R = 4\Omega$$

Solution

The core is made from soft magnetic material with very high permeability. When the air gap is much smaller compared to the diameter of the armature, the reluctance of iron parts and leakage flux are negligible. The following expressions give the reluctances of three air gaps:

$$\mathcal{R}_1 = \frac{\delta}{\mu_o[\pi d(a-x)]}$$

$$\mathcal{R}_2 = \frac{\delta}{\mu_o[\pi d\ell]}$$

$$\mathcal{R}_3 = \frac{\delta}{\mu_o[\pi d(a+x)]}$$

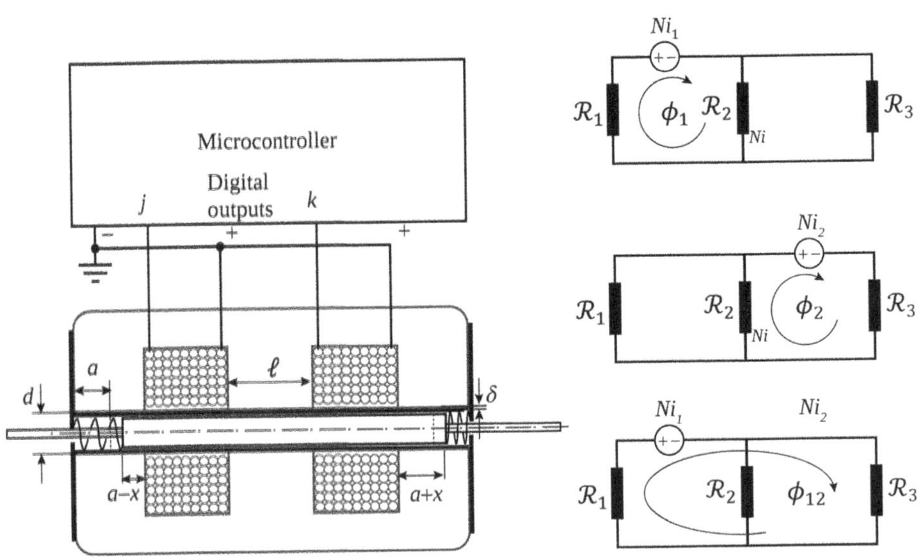

Figure 7.12 Solenoid with two coils

The magnetic equivalent circuit of the transductor is drawn for each magnetic flux components in Figure 7.12 (b) according to the superposition principle. The self-inductance of each coil can be determined by analyzing the equivalent circuit with the MMF of the other coil is turned off, which is equivalent to replacing this MMF by a short circuit.

$$\phi_1 = \frac{Ni_1}{\mathcal{R}_1 + \mathcal{R}_2 \| \mathcal{R}_3} = \frac{Ni_1}{\mathcal{R}_1 + \dfrac{\mathcal{R}_2\mathcal{R}_3}{\mathcal{R}_2 + \mathcal{R}_3}} = \frac{Ni_1(\mathcal{R}_2 + \mathcal{R}_3)}{\mathcal{R}_1\mathcal{R}_2 + \mathcal{R}_1\mathcal{R}_3 + \mathcal{R}_2\mathcal{R}_3}$$

$$\phi_2 = \frac{Ni_2}{\mathcal{R}_3 + \mathcal{R}_1 \| \mathcal{R}_2} = \frac{Ni_2}{\mathcal{R}_3 + \dfrac{\mathcal{R}_1\mathcal{R}_2}{\mathcal{R}_1 + \mathcal{R}_2}} = \frac{Ni_2(\mathcal{R}_1 + \mathcal{R}_2)}{\mathcal{R}_1\mathcal{R}_2 + \mathcal{R}_1\mathcal{R}_3 + \mathcal{R}_2\mathcal{R}_3}$$

Using the fundamental property of linear magnetic systems $\lambda = Li$,

$$L_1 = \frac{\lambda_1}{i_1} = \frac{N\phi_1}{i_1} = \frac{N^2(\mathcal{R}_2 + \mathcal{R}_3)}{\mathcal{R}_1\mathcal{R}_2 + \mathcal{R}_1\mathcal{R}_3 + \mathcal{R}_2\mathcal{R}_3}$$

$$L_2 = \frac{\lambda_2}{i_2} = \frac{N\phi_2}{i_2} = \frac{N^2(\mathcal{R}_1 + \mathcal{R}_2)}{\mathcal{R}_1\mathcal{R}_2 + \mathcal{R}_1\mathcal{R}_3 + \mathcal{R}_2\mathcal{R}_3}$$

By definition, the mutual inductance between two coils is the flux created by one of them linking the other.

$$\phi_{21} = \phi_{12} = \frac{\mathcal{R}_2}{\mathcal{R}_2 + \mathcal{R}_3} = \frac{Ni_1\mathcal{R}_2}{\mathcal{R}_1\mathcal{R}_2 + \mathcal{R}_1\mathcal{R}_3 + \mathcal{R}_2\mathcal{R}_3}$$

Mutual inductance between coil 1 and coil 2 is the flux of coil 1 lined by coil 2 divided by the current flowing through coil 1.

$$M_{12} = M_{21} = \frac{N^2\mathcal{R}_2}{\mathcal{R}_1\mathcal{R}_2 + \mathcal{R}_1\mathcal{R}_3 + \mathcal{R}_2\mathcal{R}_3}$$

To find the inductances based on displacement x, substitute the reluctances with their expressions involving displacement. After arithmetic simplifications, we obtain the following equations:

$$L_1 = \frac{\mu_o \pi d N^2 a^2}{\delta(2a + \ell)}\left(1 - \frac{x}{a}\right)\left(1 + \frac{1}{a} + \frac{x}{a}\right)$$

$$L_2 = \frac{\mu_o \pi d N^2 a^2}{\delta(2a + \ell)}\left(1 + \frac{x}{a}\right)\left(1 + \frac{1}{a} - \frac{x}{a}\right)$$

$$M_{12} = M_{21} = \frac{\mu_o \pi d N^2 a^2}{\delta(2a + \ell)}\left(1 - \frac{x}{a}\right)\left(1 + \frac{x}{a}\right)$$

Force acting on the armature is determined by applying equation 7.61 for linear displacement in two-input, one-output system.

$$F_e = \frac{1}{2}[i_1 \; i_2]\frac{\partial}{\partial x}\begin{bmatrix} L_1 & M_{12} \\ M_{21} & L_2 \end{bmatrix}\begin{bmatrix} i_1 \\ i_2 \end{bmatrix}$$

After taking the derivatives of L_1, L_2, and M_{12} with respect to x, we obtain the following expression for the electromagnetic force acting on the armature.

$$F_e = \frac{\mu_o \pi d N^2}{2\delta(2a + \ell)}\left[(\ell + 2)i_1^2 - (\ell - 2x)i_2^2\right] + 4xi_1i_2$$

The dynamic equation, combining the system's electromagnetic and mechanical aspects, is computed using MATLAB® and Simulink® to plot its response when switch 1 is activated.

$$m\frac{d^2x}{dt^2} + \beta\frac{dx}{dt} + kx = \frac{\mu_o \pi d N^2}{2\delta(2a+\ell)}\left[(\ell + 2x)i_1^2 - (\ell - 2x)i_2^2\right] + 4xi_1i_2$$

The microcontroller output voltages are zero at logic 0 and 5 V at logic 1. Therefore, the currents can be calculated using coil resistance.

$$i_1 = \frac{v_j}{R}; i_2 = \frac{v_k}{R}$$

For $i_1 = i_2 = 0$, the electromagnetic force is zero, and the springs center the armature horizontally. When the output j turns to logic 1 while output k remains 0, the expression of force becomes

$$m\frac{d^2x}{dt^2} + \beta\frac{dx}{dt} + kx = \frac{\mu_o \pi d N^2}{2\delta(2a+\ell)}(\ell + 2x)\frac{25}{R}$$

7.8 Electrostatic Energy Conversion

Electrostatic energy conversion involves the transformation of electrostatic energy into mechanical or other forms of energy. This principle is based on the use of electric fields generated by charged bodies. It has applications in various fields such as energy harvesting, sensors, actuators, and microelectromechanical systems (MEMS).

Electrostatic energy conversion relies on the interaction between electric charges and electric fields. The fundamental equation governing this interaction is Coulomb's law, which states that the force F between two charges is directly proportional to the product of the charges and inversely proportional to the square of the distance between them.

$$F = k_e\frac{q_1q_2}{r^2} \tag{7.62}$$

where k_e denotes Coulomb's constant (8.9875×10^9 N (m$^2 \cdot$ C^2)), q_1 and q_2 are the magnitudes of the charges, and r is the distance between the charges.

Although the principles of electrostatic attraction and repulsion have been understood for centuries, it is only in recent years that electromechanical devices utilizing electrostatic forces have become feasible and practical. The main reason is that the energy density achievable with an electrostatic system in air is many orders of magnitude lower than the density achievable with an electromagnetic system. This lower energy density presents a significant challenge when scaling electrostatic motors to the macro level.

Years before Michael Faraday discovered magnetic induction and motors operating on magnetic field principles were invented, Benjamin Franklin built the first electrostatic motor in 1748 (ETHW, 2015).

Benjamin Franklin's electrostatic motor demonstrated the practical application of electrostatic principles on a macroscopic scale as early as 1747. This historical accomplishment inspired modern efforts to develop macro-scale electrostatic motors that could potentially offer higher efficiency and use more abundant materials compared to conventional electromagnetic motors. Electrostatic motors require high voltage to create the intense electric fields necessary for operation. Contemporary prototypes employ power electronics drive systems to obtain high-voltage DC at the kilovolt level. In addition, a low-viscosity dielectric fluid in which the rotors spin enhances the electrostatic field energy (Zorpette, 2024).

Electrostatic energy conversion has a wide range of applications in mechatronics. Sensors, actuators, and MEMS are among major applications. Electrostatic actuators provide precise positioning in optical instruments and inkjet printers. These actuators enable fine control over mechanical movements by applying electrostatic forces to move components within the system. Additionally, they can be constructed with reduced mass and moment of inertia to enable quick movements and increased acceleration, particularly required in sensors, microphones, and earphones.

MEMS accelerometers, pressure sensors, and gyroscopes are examples of electrostatic sensors. These devices use electrostatic principles to measure and convert physical changes into electrical signals, providing accurate data for numerous applications.

In electromechanical energy conversion, energy stored in a magnetic field has been historically dominant because it provides higher power density at relatively lower voltage levels by using higher currents. To achieve comparable power density, electrostatic devices require significantly higher voltage level.

Electrostatic energy conversion eliminates the need for ferromagnetic materials and copper coils, thereby reducing the cost and weight of devices. However, significant challenges remain, including limitations in energy density, high-voltage requirements, and dielectric material constraints.

Advancements in power electronic converters and materials science are essential for improving electrostatic energy conversion systems. The development of new dielectric materials with increased permittivity can significantly enhance electrostatic energy storage capabilities, resulting in higher energy density.

The dynamic behavior of an electrostatic device can be examined through a methodology similar to that employed for analyzing electromagnetic energy conversion systems. The energy balance equation for angular motion, assuming no electrical and mechanical losses, is as follows:

$$T_e d\theta + v dq = dW_e + 0 \Longrightarrow T_e d\theta = dW_e - v dq \tag{7.63}$$

Electric charge q and energy stored in the electrostatic field W_e are functions of the voltage v and rotation angle θ.

$$\begin{aligned} dW_e(\theta, v) &= \left(\frac{\partial W_e}{\partial \theta}\right) d\theta + \left(\frac{\partial W_e}{\partial v}\right) dv \\ dq(\theta, v) &= \left(\frac{\partial q}{\partial \theta}\right) d\theta + \left(\frac{\partial q}{\partial v}\right) dv \end{aligned} \tag{7.64}$$

Substituting these two equations into equation 7.63, we can write

$$T_e d\theta = \left(\frac{\partial W_e}{\partial \theta} - v\frac{\partial q}{\partial \theta}\right) d\theta + \left(\frac{\partial W_e}{\partial v} - v\frac{\partial q}{\partial v}\right) dv \tag{7.65}$$

As shown in Figure 7.13, the electrostatic field energy is defined as the area between the q–v curve and the vertical axis. Expressing electric charge as a function of the voltage is usually more convenient. Therefore, electrostatic field energy is obtained as below:

$$W_e = \int v dq = vq - \int q dv \tag{7.66}$$

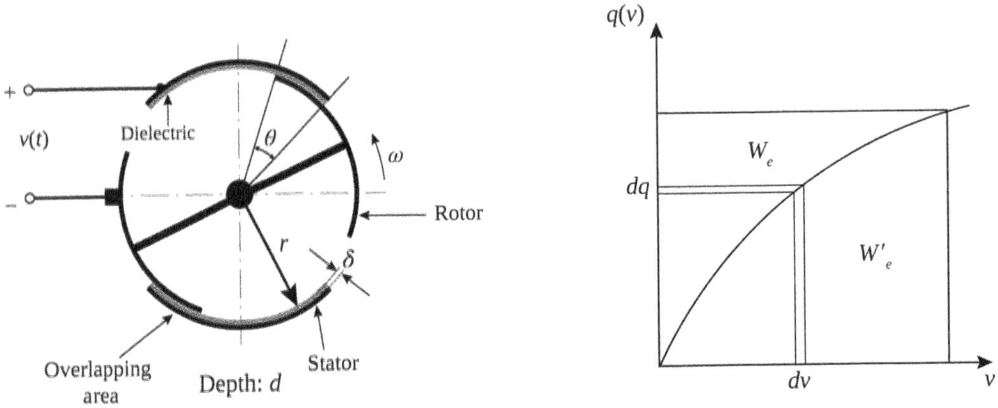

Figure 7.13 Electrostatic actuator

Partial derivative of the electrostatic field energy is

$$\frac{\partial W_e}{\partial v} = q + v\frac{\partial q}{\partial v} - q \Longrightarrow \frac{\partial W_e}{\partial v} - v\frac{\partial q}{\partial v} = 0 \tag{7.67}$$

We can express the torque by substituting this property in equation 7.65 and dividing both sides with $d\theta$.

$$T_e = \frac{\partial W_e}{\partial \theta} - v\frac{\partial q}{\partial \theta} \tag{7.68}$$

Using the concept of coenergy yields the simpler equation below:

$$T_e = -\frac{\partial W_e'}{\partial \theta} \tag{7.69}$$

If the characteristic of the dielectric is linear, energy equals coenergy and can be expressed in terms of capacitance.

$$W_e = W_e' = \frac{1}{2}C(\theta)v^2 \tag{7.70}$$

Therefore, the torque can be calculated using the derivative of the capacitance with respect to the rotation angle.

$$T_e = -\frac{1}{2}\frac{\partial C(\theta)}{\partial \theta}v^2 \tag{7.71}$$

Example 7.10 Figure 7.13 shows an electrostatic actuator that consists of cylindrical metal plates on the stator and rotor isolated from each other by silicone of relative permittivity $\epsilon_r = 12$. Given the following parameters, determine the torque on the rotor when a 2 kV DC voltage is applied across the stator and rotor.

$$r = 50\text{ mm}, d = 150\text{ mm}, \delta = 0.1\text{ mm}, \epsilon_o = 8.85 \times 10^{-12}\text{ F/m}$$

The stator and rotor separated by a dielectric form a cylindrical capacitor that varies with the rotation angle θ. The formula for the capacitance of a cylindrical capacitor is below:

Solution

The capacitance between two cylindrical plates can be calculated using the following formula.

$$C = \frac{2\pi\epsilon_r\epsilon_0 d}{\ln\left(\frac{r+\delta}{r}\right)}$$

Assuming that the electric field is homogeneous between the rotor and stator plates, the capacitance of the sections of the rotor and stator overlapping by the angle θ can be approximated proportionally to the capacitance of the full cylindrical capacitor.

$$C_\theta = C\frac{\theta}{2\pi}$$

According to equation 7.71, the magnitude of the torque generated by two electrostatic forces acting diametrically on the rotor can be calculated as follows:

$$T_e = 2\left(\frac{1}{2}\frac{\partial C(\theta)}{\partial \theta}V^2\right) = \frac{\epsilon_0\,\epsilon_r\,d}{\ln\left(\frac{r+\delta}{r}\right)}V^2$$

Substituting the given parameters, we can proceed to calculate the torque on the rotor.

$$\frac{8.85 \times 10 - 12\;12 \times 0.150}{\ln\left(\frac{50+0.1}{50}\right)}\left(2 \times 10^3\right)2 = 0.073\text{ Nm}$$

The torque remains constant with respect to the rotation angle but increases proportionally with the relative permittivity of the dielectric material and the square of the voltage. To enhance torque, it is essential to utilize a specialized dielectric material with high relative permittivity and dielectric strength. Simultaneously, the voltage must be elevated to the maximum level that the insulation can withstand.

7.9 Design Examples

Example 7.11 Electromagnetic Gripper Design
Design an electromagnet to pick up and handle steel parts for an assembly line robot. Determine the dimensions and select materials for the structure shown in Figure 7.14 such that the electromagnet can lift a maximum weight of 50 kg when it is positioned approximately 5 mm from the object.

$$a = 50\text{ mm}; b = 40\text{ mm}; c = 10\text{ mm}; d = 15\text{ mm}; e = 2\text{ mm}$$
$$N = 75\text{ turns}; \mu_o = 4\pi 10^{-7}; \mu_c = 3{,}000$$

Solution

The magnetic flux completes its path through three components with different parameters: a U-shaped magnetic core (1), a moving magnetic plate (2), and two air gaps (3).

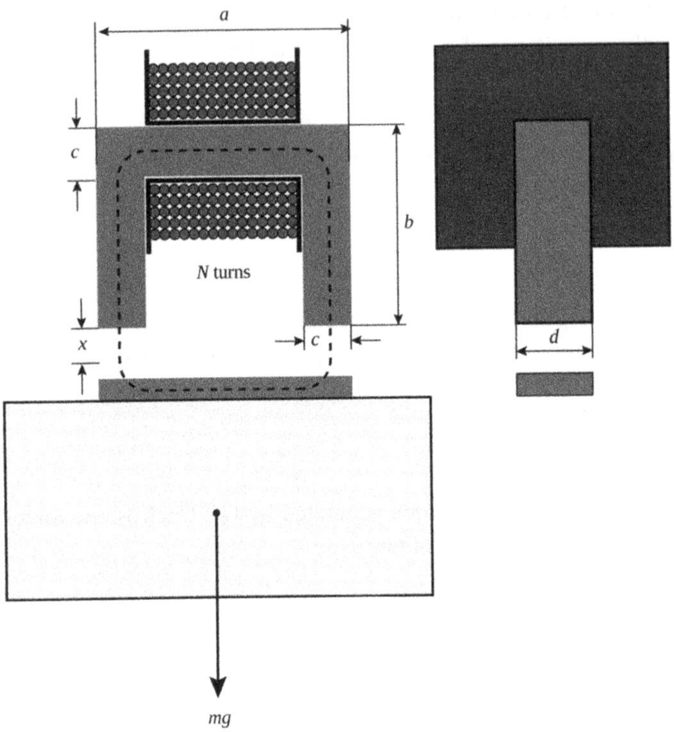

Figure 7.14 Electromagnetic gripper

1) U-shaped static core:

$$\ell_{f1} = (a-c) + 2\left(b - \frac{c}{2}\right) = \left[(50-10) + 2\left(40 - \frac{10}{2}\right)\right] \times 10^{-3} = 0.110\ \mathrm{m}$$

$$A_{f1} = c \cdot d = \left(10 \times 10^{-3}\right) \times \left(15 \times 10^{-3}\right) = 150 \times 10^{-6}$$

$$\mathcal{R}_{f1} = \frac{\ell_{f1}}{\mu_0 \mu_c A_{f1}} = \frac{0.130}{4\pi 10^{-7} \times 3{,}000 \times 150 \times 10^{-6}} = 0.973 \times 10^{6}$$

2) Moving steel plate (armature)

$$\ell_{f2} = (a-c) = (50-10) \times 10^{-3} = 0.04\ \mathrm{m}$$

$$A_{f2} = d \cdot e = \left(15 \times 10^{-3}\right) \times \left(2 \times 10^{-3}\right) = 30 \times 10^{-6}$$

$$\mathcal{R}_{f2} = \frac{\ell_{f2}}{\mu_0 \mu_c A_{f2}} = \frac{0.04}{4\pi 10^{-7} \times 3{,}000 \times 30 \times 10^{-6}} = 0.354 \times 10^{6}$$

3) Air gaps

$$\ell_g = 2x$$

$$A_g = c \cdot d = \left(10 \times 10^{-3}\right) \times \left(15 \times 10^{-3}\right) = 150 \times 10^{-6}$$

$$\mathcal{R}_g = \frac{\ell_g}{\mu_0 A_g} = \frac{2x}{4\pi 10^{-7} \times 150 \times 10^{-6}} = \left(10{,}610 \times 10^{6}\right)x$$

Total reluctance of the magnetic circuit is $\mathcal{R}_{f1} + \mathcal{R}_{f2} + \mathcal{R}_g = (0.584 + 2{,}650x) \times 10^6$. Note that the reluctance of the two air gaps is much greater than the total reluctance of the iron parts. In most practical cases, ignoring the reluctance of an iron core does not significantly affect the accuracy of the result.

If the fringe (leakage) flux is negligible, the magnetic flux can be expressed as a function of the air gap length x using the following equation:

$$\phi = \frac{Ni}{\mathcal{R}_{f1} + \mathcal{R}_{f2} + \mathcal{R}_g}$$

The flux linkage of the coil is

$$\lambda = N\phi$$

Assuming the magnetic circuit linear, stored energy equals the coenergy.

$$W_m = W'_m = \frac{1}{2}\lambda i = \frac{1}{2}\frac{(Ni)^2}{\mathcal{R}_{f1} + \mathcal{R}_{f2} + \mathcal{R}_g} \cong \frac{1}{2}\frac{(750 \times 0.5)^2}{2{,}650 \times 10^6}\frac{1}{x} = 663 \times 10^{-6}\frac{1}{x}$$

The force acting on the moving part is the derivative of the coenergy with respect to the displacement x.

$$F = \frac{\partial W'_m}{\partial x} = -\frac{663 \times 10^{-6}}{x^2}$$

For $x = 5$ mm, we find $F = -13.8$ N. The negative sign indicates that the force is acting in the opposite direction to the reference chosen for displacement x. The force required to lift 20 kg is $20 \times 9.81 = 196.2$ N. For $x = 5$ mm, the minimum current to attract this load is

$$i = \sqrt{\frac{196.2 \times (5 \times 10^{-3})^2}{0.08 \times 10^{-3}}} = 7.83\,A \therefore$$

7.10 Summary

This chapter explains electromechanical energy conversion principles, focusing on the principles and applications of devices that convert electrical energy to mechanical energy and vice versa. The core of mechatronic systems involves various electromechanical devices such as motors, actuators, and generators, which are essential for energy conversion processes in both industrial and consumer applications.

- Electromechanical energy conversion is fundamental to mechatronic systems, where devices such as motors and actuators convert electrical energy into mechanical motion, while generators and transducers convert mechanical motion into electrical energy. This process primarily occurs within a magnetic field, with electrostatic fields being less common due to the high-voltage requirements for practical applications.
- Electromechanical systems consist of coils generating magnetic flux and mechanical structures driven by electromagnetic forces. The energy flow in these systems is bidirectional; they can function as either motors or generators depending on the energy direction. For instance, in a

loudspeaker, electrical energy causes the membrane to vibrate (motor mode), while in the microphone mode, sound waves induce electrical signals. In motor operation, energy flows from electrical to mechanical, while in generator operation, energy flows from the mechanical to electrical side.

- In practical systems, energy losses occur due to electrical (Joule heating, magnetic losses) and mechanical factors (friction). The efficiency of an electromechanical system is defined as the ratio of output to input energy, highlighting the importance of minimizing losses to enhance performance.
- Energy balance in electromechanical systems is crucial, as the output energy is always less than the input due to losses. The chapter presents equations representing the relationship between electrical energy input/output, mechanical energy input/output, and changes in stored magnetic energy.
- The Lorentz force is a key concept in the operation of electromechanical devices, particularly in moving coil systems. Moving coil devices, such as loudspeakers and galvanometers, utilize the Lorentz force to produce motion.
- Moving armature devices are a type of electromechanical system where the coil that produces the magnetic field is fixed, and an iron piece in the magnetic core moves. These devices include electromagnets, solenoids, relays, and various electric machines. They convert electrical energy into mechanical motion by changing the energy stored in the magnetic field.
- Electromechanical energy converters can make either linear or rotary motion. Rotary devices convert electrical energy into rotational work, with applications in automation and robotics. The chapter explains the mechanics of these devices, including the importance of torque generation.

7.11 Check Your Understanding

7.11.1 Self-Assessment Quiz

1 What is the primary principle behind electromechanical energy conversion

- **a.** Conservation of energy
- **b.** Conversion of electrical energy to mechanical energy and vice versa
- **c.** Generation of magnetic fields
- **d.** Minimization of energy losses

2 Which device converts mechanical motion into electrical energy?

- **a.** Motor
- **b.** Actuator
- **c.** Generator
- **d.** Transducer

3 What type of field is primarily involved in electromechanical energy conversion?

- **a.** Electrostatic field
- **b.** Magnetic field
- **c.** Gravitational field
- **d.** Electric field

4 How is efficiency calculated in electromechanical systems?
   a. Ratio of input energy to output energy
   b. Ratio of output energy to input energy
   c. Difference between input and output energy
   d. Sum of input and output energy

5 What is the significance of energy balance in electromechanical devices?
   a. Ensuring output energy is greater than input energy
   b. Minimizing energy losses
   c. Examination of energy conversion
   d. Ensuring output energy is less than input energy

6 What are the key components of an electromagnet?
   a. Coil and ferromagnetic core
   b. Coil and movable metal plunger
   c. Electrical contacts and armature
   d. Coil and permanent magnet

7 How does a solenoid operate to control valves and switches?
   a. By generating a magnetic field that moves the armature
   b. By generating an electric field that moves the armature
   c. By generating a gravitational field that moves the armature
   d. By generating a mechanical force that moves the armature

8 What role does the moving armature play in a relay?
   a. Generating magnetic flux
   b. Opening or closing electrical contacts
   c. Controlling valves
   d. Producing mechanical motion

9 What are some common applications of moving-armature devices?
   a. Industrial machinery and consumer electronics
   b. Scientific research and medical devices
   c. Both A and B
   d. None of the above

10 How do magnetic force and torque contribute to the operation of moving-armature actuators?
   a. By converting electrical energy into mechanical motion
   b. By converting mechanical motion into electrical energy
   c. By generating magnetic fields
   d. By minimizing energy losses

11 How is torque calculated in a DC motor?
   a. By measuring the current and voltage
   b. By measuring the speed and power
   c. By measuring the force and radius
   d. By measuring the magnetic flux and area

12 What factors affect the efficiency of a rotary actuator?
   a. Electrical losses
   b. Mechanical losses
   c. Both A and B
   d. None of the above

13 How does regenerative braking work in hybrid vehicles?
   a. By converting electrical energy into mechanical motion
   b. By converting mechanical motion into kinetic energy
   c. By converting electrical energy into kinetic energy
   d. By converting kinetic energy into electrical energy

14 What is the purpose of rotary transducers?
   a. Ensuring output energy is greater than input energy
   b. Minimizing energy losses
   c. Converting rotary motion into an electric signal or electric energy into rotation
   d. Ensuring output energy is less than input energy

15 How is force generated in rotary actuators?
   a. By generating an electric current
   b. By changing the energy stored in the magnetic field
   c. By generating a mechanical force
   d. By generating an electrostatic field

Answers:

1: a; 2: c; 3: b; 4: b; 5: c; 6: b; 7: a; 8: d; 9: c; 10: a; 11: b; 12: c; 13: d; 14: c; 15: b.

7.11.2 Short Answer Questions

1 How is efficiency defined in electromechanical systems?

2 What is the core principle behind electromechanical energy conversion?

3 What are the key components of an electromagnet?

4 What are the main types of losses in electromechanical systems?

5 What role does the moving armature play in a relay?

6 How is torque calculated in a DC motor?

7 How does a solenoid operate to control valves and switches?

8 What is the Lorentz force, and how does it apply to electromechanical devices?

9 How do motors and generators differ in terms of energy flow?

10 How do electromechanical systems operate bidirectionally?

7.11.3 Essay Questions

1 What is the significance of energy balance in electromechanical devices?

2 What is the role of magnetic circuits in electromechanical devices?

3 How do hysteresis and eddy currents affect core losses in magnetic circuits?

4 What are the advantages of using electric steel and ferrite cores in electromechanical applications?

5 What are some common applications of moving-armature devices?

6 How do magnetic force and torque contribute to the operation of moving-armature actuators?

7 How does regenerative braking work in hybrid vehicles?

8 What is magnetomotive force (MMF), and how is it related to magnetic flux?

9 How does the permeability of a material affect its magnetic properties?

10 What is the significance of the hysteresis loop in ferromagnetic materials?

11 How do moving coil devices like loudspeakers and microphones operate?

12 What are the limitations of ferrite cores in high-frequency applications?

13 How is the energy balance maintained in electromechanical systems?

14 What factors affect the efficiency of a rotary actuator?

15 How does regenerative braking work in hybrid vehicles?

16 What is the energy balance in rotary transducers?

17 How is magnetic force generated in rotary actuators?

7.11.4 Drill Problems

1 A hybrid car has a 110 HP electric motor and a 207.2 V battery bank. When the brakes are applied, the electric motor runs at 1400 rpm and the battery charge current becomes 187 amps. Assuming a total loss of 8% of the motor power, calculate the torque turning the generator shaft.

2 A DC motor draws 1.5 amps from a 12-V supply, driving a mechanical load at a speed of 1,200 rpm with a torque of 0.13 Nm. Calculate the total motor losses.

3 An electromagnetic device has a toroidal ferromagnetic core of radius $R_1 = 45$ mm and $R_2 = 55$ mm with an air gap of $g = 1$ mm. The thickness of the core is $d = 20$ mm and relative permeability of the ferromagnetic material is $\mu_r = 3{,}000$. The 100-turn coil is supplied from a DC source, and the current flowing through the coil is 2 A.

a. Calculate the reluctances of the ferromagnetic core and air gap.
b. Calculate the flux in the magnetic circuit and the flux density in the core.
c. Determine the flux density in the air gap assuming the leakage flux is negligible.

4 A loudspeaker coil with $N = 50$ turns and radius $r = 20$ mm is powered by a 1 kHz sinusoidal signal with 2 V amplitude. If the air gap flux density is $B = 1.2$ T, determine how the force on the speaker cone varies over time.

5 An electromagnet consists of a coil with 500 turns and a ferromagnetic core. When a current of 2 A flows through the coil, it generates a magnetic flux of 0.01 Wb in the core. Calculate the force exerted by the electromagnet on a movable armature with an area of 0.02 m^2.

6 A relay uses moving armature to open and close electrical contacts. The coil has 200 turns and generates a magnetic flux of 0.005 Wb when a current of 1 A flows through it. Calculate the time it takes for the armature to move 2 mm and switch the contacts if the armature has a mass of 0.01 kg and experiences a force of 0.1 N.

7 An electromechanical system converts electrical energy into mechanical motion with an input power of 100 W and an output power of 80 W. Calculate the efficiency of the system and identify the types of losses that may occur.

8 A moving-armature device operates with an electrical energy input of 150 J and an output mechanical energy of 120 J. Calculate the energy losses and discuss the factors that contribute to these losses.

9 A rotary actuator converts electrical energy into rotational work with an input power of 200 W and an output power of 160 W. Calculate the efficiency of the actuator and discuss the types of losses that may occur.

10 A hybrid car has a 150 HP electric motor and a 300 V battery bank. When the brakes are applied, the electric motor runs at 1,800 rpm and the battery charge current becomes 250 amps. Assuming a total loss of 10% of the motor power, calculate the mechanical power the brake system absorbs and the torque turning the generator shaft.

11 A rotary transducer operates with an electrical energy input of 200 J and an output mechanical energy of 160 J. Calculate the energy losses and discuss the factors that contribute to these losses.

12 A rotary actuator consists of a coil with 400 turns and a ferromagnetic core. When a current of 3 A flows through the coil, it generates a magnetic flux of 0.02 Wb in the core. Calculate the torque exerted by the actuator on a movable rotor with a radius of 0.05 m.

7.11.5 Design Problems

1 Design a tachometer to produce a sinusoidal voltage with an amplitude of 12 V when the shaft rotates at 4,000 rpm. Given the rotor diameter is 20 mm and axial length 12 cm, and the permanent magnet creates 1.5 T flux density in the air gap, determine the number of turns required for the coil.

2 Design an electromagnet to pick up and handle steel parts for an assembly line robot. Determine the dimensions and select materials for the structure such that the electromagnet can lift a maximum weight of 50 kg when it is positioned approximately 5 mm from the object.

3 Design a solenoid actuator to control a valve in an industrial process. Specify the coil parameters, core material, and armature dimensions to achieve a precise and quick response. Calculate the force exerted by the solenoid when a current of 2 A flows through the coil.

4 Design a rotary actuator to drive a robotic arm. Determine the coil parameters, magnetic field strength, and rotor dimensions to achieve a torque of 0.5 Nm at a speed of 1,000 rpm. Calculate the efficiency of the actuator and discuss the types of losses that may occur.

5 Design an electromagnetic shaker to simulate vibrations for testing electronic components. Specify the coil parameters, core material, and armature dimensions to achieve a wide frequency range and high force output. Calculate the power required to generate a vibration amplitude of 5 mm at a frequency of 50 Hz.

6 Design an electromagnetic launcher to propel objects at high speeds. Specify the coil parameters, magnetic field strength, and armature dimensions to achieve a launch velocity of 100 m/s. Calculate the force exerted by the launcher on the armature and discuss the factors that contribute to the launch velocity.

7 Design an electrostatic actuator to provide precise positioning in an optical instrument. Specify the electrode dimensions, dielectric material, and voltage levels to achieve a displacement of 1 mm. Calculate the force exerted by the actuator and discuss the limitations of electrostatic energy conversion.

References

ETHW, 2015. *Benjamin Franklin's Electric Motor.* [Online] Available at: https://ethw.org/Benjamin_Franklin's_Electric_Motor [Accessed 9 5 2026].

Fitzgerald, A. E., Kingsley, C. & Umans, S. D., 1992. *Electric Machinery.* 5th ed. New York, NY: McGraw Hill.

Kraus, J. D., 1991. *Electromagnetics.* 4th ed. New York: McGraw Hill.

Umans, S. D., 2003. *Fitzgerald and Kingsle's Electric Machinery.* 7th ed. New York, NY: McGraw Hill.

Zorpette, G., 2024. *Electrostatic Motors Reach the Macro Scale.* [Online] Available at: https://spectrum.ieee.org/electrostatic-motor [Accessed 17 11 2024].

8

Rotating Electric Machines

OBJECTIVES

- Explain the fundamental principles of electric machines.
- Explain the bidirectional nature of energy conversion in electric machines.
- Describe the basic structure and operation of electric machines.
- Outline the main types of rotating electric machines relevant to mechatronic and robotic systems.
- Discuss the key features, advantages, and disadvantages of each motor type, focusing on aspects important for design and analysis in mechatronics and robotics.
- Provide simplified models and performance characteristics for each machine type.
- Highlight practical control methods and starting techniques for motors
- Summarize the impact of modern materials and technologies.

Electric machines are electromechanical devices that convert electric energy into continuous mechanical work or vice versa. Most electric machines used in mechatronics produce rotational motion. Since electromechanical energy conversion is bidirectional, all electric machines can function as either a motor or a generator. A particular electric machine is typically designed to function as either a motor or a generator to optimize efficiency. However, during motor or generator operation, the energy flow can reverse depending on the operating conditions. An electric motor acts as a generator when mechanical power surpasses electrical power, such as in regenerative brake systems where the energy flow reverses during slowing down. Hence, it is often more appropriate to use the term "electric machine" rather than specifying either motor or generator.

In this chapter, we focus on the motor operation and also analyze situations where a motor can generate electric energy during varying transient conditions. Bidirectional energy flow is one of the biggest advantages of electric motors in electric vehicles (EVs) and high-power industrial systems.

All electric machines operate based on Ampere's and Faraday's laws. They all have a magnetic circuit where magnetic flux is created either by coils or by permanent magnets (PMs). The stationary part of a machine is called *stator*, and the rotating part is called *rotor*. In an electric motor, the generated torque arises either from the alignment of the rotor with the magnetic flux of the stator (alignment torque) or through the interaction between the magnetic fluxes produced in both the stator and the rotor (interaction torque).

Electric motors are essential parts of mechatronic and robotic systems. A comprehensive study of electric machines is beyond the scope of this textbook. This chapter focuses on highlighting the key features relevant to the analysis and design of mechatronic and robotic systems. Thorough analysis and design techniques can be found in classical electrical machinery references such as Fitzgerald et al. (1992), Chapman (1991), and Gieras (2017).

8.1 Variable Reluctance Motors

Variable reluctance (VR) motors produce torque by aligning the rotor with the magnetic field generated by the stator windings. The air gap between the rotor and the stator is not constant, and as the rotor moves, the reluctance of the air gap increases until it reaches its maximum. As a result, the unlike traditional motors, which rely on electromagnetic coils or PMs on the rotor, the reluctance motor operates by the magnetic reluctance principle. This means the rotor is attracted to the magnetic fields in the stator, causing it to rotate.

Reluctance motors are known for their simplicity and cost-effectiveness, as they do not require expensive rare-earth magnets or complex winding arrangements (Gieras, 2017). This makes them ideal for applications where cost and robustness are critical. The motor's efficiency and performance are highly dependent on the design of the stator and rotor, which must be precisely engineered to optimize the magnetic flux paths.

These motors are utilized in various industries, including automation, robotics, and EVs, due to their reliable operation and ease of maintenance. In EVs, reluctance motors offer the advantage of reduced weight and increased efficiency, contributing to better overall performance and energy savings.

The control of reluctance motors involves sophisticated electronic systems that manage the timing and strength of the magnetic fields, ensuring smooth and precise operation. Advances in power electronics and control algorithms have significantly improved the performance and versatility of reluctance motors, making them a viable option for a wide range of modern technological applications.

8.1.1 Reluctance Torque

Moving-armature rotary actuators produce torque by decreasing the reluctance of the air gap in the overlapping area between the rotor and the stator, known as "reluctance torque." If the rotor starts moving when the stator coil is energized, energy stored in the magnetic field increases as the rotor turns and reaches its maximum when the rotor and stator poles are aligned. This type of actuator cannot complete a full rotation unless specific operating conditions are met.

Figure 8.1 shows two positions of the rotor. When the rotor and stator magnetic fields are aligned, the reluctance is at its minimum (a). When the stator and rotor magnetic field axes are perpendicular (b), the air gap length is greatest, resulting in the maximum reluctance of the magnetic circuit.

$$\mathcal{R} = \frac{2g}{\mu_o A} \tag{8.1}$$

In this relative position, the inductance of the stator winding is maximum.

$$L_{max} = \frac{N^2}{\mathcal{R}} = \frac{N^2 \mu_o A}{2g} \tag{8.2}$$

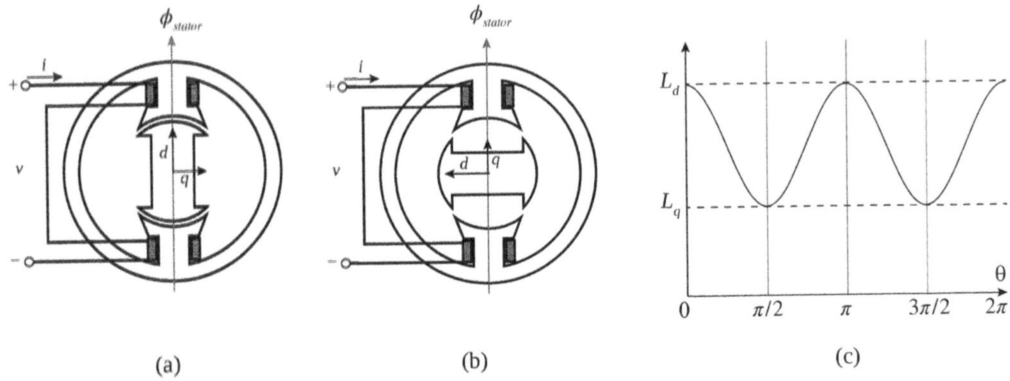

Figure 8.1 Inductance variation as a function of the rotation angle (a: Rotor axis aligned with the pole axis; b: Rotor axis perpendicular to the pole axis; c: Variation of the inductance with the rotor angle)

Suppose that the rotor rotates with an angular velocity of ω_m. The angle between the stator and the rotor magnetic axes can be expressed as

$$\theta = \omega_m t + \delta \tag{8.3}$$

At every rotation, the inductance varies between a maximum and minimum value. The function associated with this variation depends on the geometry of the rotor and stator pole surfaces. Electric machines are designed such that this variation is as close as possible to a sinusoidal waveform. If it is not sinusoidal, the dominant first harmonic is a sinusoidal function (Kocabiyikoğlu, 2021). Figure 8.1 (c) illustrates the fundamental component of the inductance variation as a function of the rotation angle. The mathematical function of the inductance is, therefore, a sinusoidal function of frequency $2\omega_m$ since in every rotation the air gap becomes minimum and maximum twice.

$$L(\theta) = L_{min} + L_{max}\cos(2\theta) \tag{8.4}$$

In the terminology of electric machinery, the angle corresponding to the maximum reluctance is referred to as the direct axis, while the angle associated with the minimum reluctance is known as the quadrature axis.

The direct axis inductance (L_d) and quadrature axis inductance (L_q) in a VR motor are defined based on the maximum and minimum values of the inductance as the rotor rotates. Specifically, the direct axis inductance corresponds to the position where the stator and rotor magnetic axes are aligned and the reluctance is at its minimum, resulting in the maximum inductance. Conversely, the quadrature axis inductance corresponds to the position where the axes are perpendicular and the reluctance is at its maximum, resulting in the minimum inductance. Mathematically, these inductances are given by

$$L_d = \frac{1}{2}(L_{max} + L_{min}); L_q = \frac{1}{2}(L_{max} - L_{min}) \tag{8.5}$$

Here, L_{max} is the maximum inductance when the rotor and stator poles are aligned (direct axis), and L_{min} is the minimum inductance when they are misaligned (quadrature axis). This distinction is essential for understanding the torque production and performance characteristics of VR motors.

In Chapter 7, we demonstrated that the torque developed in rotary actuators is the derivative of magnetic coenergy with respect to the rotation angle, with the assumption that the current remains constant during infinitesimal motion. The torque developed in the VR motor can be obtained by substituting equation 8.5 in equation 8.4 and taking the derivative of the resulting inductance function.

$$T_e = \frac{1}{2} i^2 \frac{\partial L}{\partial \theta}\bigg|_{i = constant} = -\frac{1}{2} i^2 (L_d - L_q) \sin[2(\omega_m t + \delta)] \tag{8.6}$$

Suppose that a sinusoidal source creates a current $i = I_m \cos \omega t$ through the stator coil. By including this current in the torque equation, we obtain

$$T_e = -\frac{1}{2}(L_d - L_q) I_m{}^2 \cos^2 \omega t \cdot \sin(2\omega_m t + 2\delta) \tag{8.7}$$

The torque equation is reformulated below by using the identity $2\cos^2\alpha = 1 + \cos 2\alpha$

$$T_e = -\frac{1}{2}(L_d - L_q) I_m{}^2 \frac{1}{2}(1 + \cos 2\omega t) \sin(2\omega_m t + 2\delta)$$

In AC analysis, the effective (or root mean square [rms]) values of the currents and voltages are more commonly used than their maximum values. Remember from ASIC circuit theory that the effective (rms) value of a sinusoidal current is calculated as $I_{rmsr} = I_m/\sqrt{2}$. By substituting the effective value of the current and performing arithmetic manipulations, we obtain

$$T_e = -I_{rms}^2 (L_d - L_q) \frac{1}{2} [\sin(2\omega_m t + 2\delta) + \sin(2\omega_m t + 2\delta) \cos 2\omega t] \tag{8.8}$$

An alternative form of the torque equation is obtained by utilizing the trigonometric identity $2 \sin \alpha \cdot \cos \beta = \sin(\alpha + \beta) + \sin(\alpha - \beta)$.

$$T_e = -\frac{1}{2} I_{rms}^2 (L_d - L_q) \{ \sin[2(\omega + \omega_m)t + 2\delta] - \sin[2(\omega - \omega_m)t - 2\delta] + \sin(2\omega_m t + 2\delta) \} \tag{8.9}$$

Average torque over one turn is obtained by integrating the time function of the torque and dividing by the period.

$$T_{av} = \frac{1}{2\pi} \int_0^{2\pi} T_e(\omega t) d\omega t \tag{8.10}$$

The integral of a sinusoidal function over one period is zero. Therefore, the average torque can be nonzero if the angular speed of the rotor ω_m equals the angular frequency of the stator current ω and 2δ is not equal to the integer multiples of π $(\omega_m = \omega)$ and $2\delta \neq n\pi$ $(n = 0, 1, 2, ...)$

$$T_{av} = -\frac{1}{2} I_{rms}^2 (L_d - L_q) \sin(2\delta) \tag{8.11}$$

Clearly, torque becomes maximum when $\delta = 45°$. Hence, δ is known as torque angle.

8.1.2 Synchronous Reluctance Motor

Operation of synchronous reluctance motors is based on the generation of reluctance torque that aligns the rotor pole with the nearest stator pole. We saw in the previous section that the necessary condition to achieve a nonzero average torque over one complete turn is that the mechanical rotation speed of the rotor equals the angular frequency of the stator current. Since the rotor can turn at a specific angular speed dictated by the frequency of the AC supply, this type of motor is called a synchronous motor.

Synchronous speed solely depends on the number of rotor poles *2p* and the frequency f_s of the supply source. The following equation gives the synchronous speed (N_s) in RPM (revolution per minute):

$$N_s = \frac{120 f_s}{2p} \tag{8.12}$$

where f_s is the frequency of the stator current and p is the number of pole pairs. Unlike other electric motor types, the speed of synchronous motors is independent of the mechanical load they drive. They are often used in applications requiring precise rotation speed such as analog electric clocks, timers, and mechanical switch controllers. Synchronous reluctance motors are typically designed with multiple poles to match the desired speed with the supply power frequency. If the motor is supplied from utility power, the rotation speed is an integer multiple of the utility frequency. Conversely, a variable frequency electronic converter can be utilized to adjust the speed of a synchronous reluctance motor across a wide range, beginning from extremely low speeds.

Example 8.1 Calculate the synchronous speed of a four-pole synchronous reluctance motor supplied from 60 Hz utility power.

Solution

A four-pole motor has two pairs of poles. Using equation 8.12, we obtain

$$N_s = \frac{120 \times 60}{4} = 1{,}800 \text{ RPM} \therefore$$

Example 8.2 A 24-pole synchronous motor powered by an electronic timer drives an analog clock. Calculate the period of the square wave such that the motor speed is one rotation per minute.

Solution

Equation 8.12 is rearranged to determine the frequency for the given number of poles.

$$f_s = \frac{2pN_s}{120} = \frac{24 \times 1}{120} = 0.2 \text{ Hz}$$

The period of the waveform generated by the astable timer should be

$$T = \frac{1}{f_s} = \frac{1}{0.2} = 5\,s \therefore$$

In practice, starting a synchronous motor requires technical considerations. The obvious method is driving the motor by using an external prime mover until it reaches synchronous speed. This method was utilized in early synchronous motor applications where an auxiliary motor or even a mechanical system would drive the synchronous motor up to its operating speed. Once the motor reaches synchronous speed, it is connected to its supply source, allowing it to maintain its speed and operate efficiently. This approach is not suitable for small mechatronics applications.

Another method involves the use of damper windings, also known as "amortisseur windings," which are short-circuited bars embedded in the rotor. These windings function similarly to squirrel-cage rotor bars found in induction motors, which will be described later. At startup, the motor operates like an induction motor with the damper windings providing the necessary starting torque. Once the synchronous speed is reached, the damper windings become inactive, and the motor transitions to synchronous operation.

Variable frequency drives (VFDs) can also be employed. VFDs control the frequency of the supply voltage, gradually increasing it from zero to the operating frequency. This method allows for a controlled and smooth acceleration of the motor to synchronous speed, providing precise control over the motor's startup process.

8.1.3 Switched Reluctance Motor

The stator of a switched reluctance motor (SRM) consists of multiple poles, each equipped with windings that can be energized sequentially. The rotor, on the other hand, is designed with salient poles that can align with the stator poles. Unlike traditional motors, the rotor of an SRM does not contain any coils or PMs; it relies entirely on the magnetic fields generated by the stator windings.

The number of rotor poles is typically less than the number of stator poles. SRM with six stator and four rotor poles as illustrated in Figure 8.2 is common. When the stator windings are energized, they create a magnetic field that interacts with the rotor aligning only one pair of stator and rotor poles.

Sequential excitation causes a magnetic pull on the rotor, creating rotation. The timing and sequence of the stator windings' excitation are controlled by electronic systems, which ensure that the rotor moves smoothly and efficiently. The alignment of the rotor with the energized stator poles produces torque, and the continuous switching of the stator windings leads to sustained rotation.

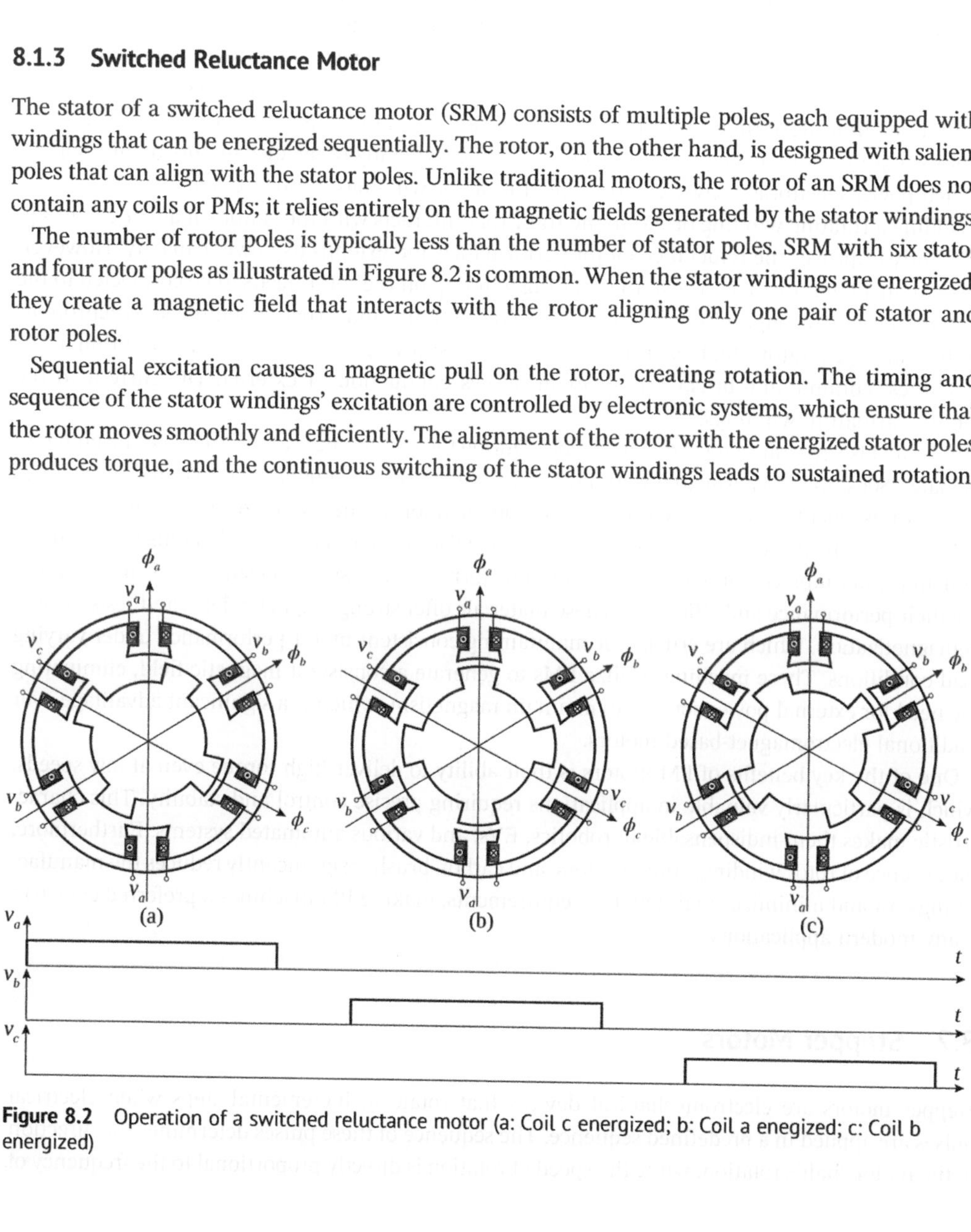

Figure 8.2 Operation of a switched reluctance motor (a: Coil c energized; b: Coil a enegized; c: Coil b energized)

SRMs are robust and capable of operating at high speeds, making them particularly suitable for industrial and automotive applications where reliable and high-performance motors are essential.

SRMs are known for their robustness and ability to operate at high speeds, making them suitable for industrial and automotive applications.

8.1.4 Advantages and Disadvantages of Reluctance Motors

Reluctance motors have a simple design without coils or PMs in the rotor, which makes it easier to manufacture and maintain stationary coils on the stator to create the magnetic field. The absence of rotor windings and magnets lowers the cost of materials and construction. The straightforward design increases the reliability of the motor, reducing potential failure points. Reluctance motors can operate at high speeds, which is useful for electronic, industrial, and automotive applications. However, reluctance motors have some disadvantages. They can exhibit torque ripples, potentially causing vibrations and noise during operation. Their performance may be impacted by magnetic saturation, limiting torque and efficiency in certain conditions. Synchronous type reluctance motors require additional elements for starting, while switching reluctance motors require sophisticated power electronic converters to supply the stator windings in precise sequences.

Adding a rotating winding or PM to the rotor that interacts with the stator's magnetic field significantly improves the reluctance motor performance and extends its power capacity. However, this enhancement involves a cost: the rotating winding on the rotor needs to be connected to the external circuit through either a slip ring or commutator and brush assemblies. Synchronous motors and generators that operate on AC utilize slip rings and brushes. In traditional DC machines, commutator–brush arrangement allows conduction of external DC current to the rotating armature windings.

PMs can also generate a rotor magnetic field. During the development of the initial DC machines in the late 1800s, the PMs available were not sufficiently strong to produce the required mechanical power. Consequently, the use of electromagnets on the rotor was necessary to address this limitation.

The use of advanced PM materials, including special alloys and rare-earth elements like neodymium and samarium–cobalt, has significantly transformed the design of electric machines, improving their performance and efficiency. These materials offer strong magnetic fields and resistance to demagnetization, which are crucial for maintaining consistent motor performance under varying load conditions. These machines utilize PMs to generate a consistent magnetic field, eliminating the need for external power sources to maintain magnetism, which is a significant advantage over traditional electromagnet-based motors.

One of the key benefits of PM motors is their ability to deliver high torque even at low speeds, which is particularly valuable in applications requiring precise control and stability. This characteristic makes them indispensable in robotics, EVs, and various automated systems. Furthermore, the absence of rotor winding, commutators, and carbon brushes significantly reduces the manufacturing cost and minimizes maintenance requirements, making PM machines a preferred choice in many modern applications.

8.2 Stepper Motors

Stepper motors are electromechanical devices that rotate in incremental steps when electrical pulses are applied in a predefined sequence. The sequence of these pulses determines the direction of the motor shaft's rotation, while the speed of rotation is directly proportional to the frequency of the pulses. Stepper motors are designed to rotate a specific angle for each pulse. This property allows precise positioning and repeatability of movement.

The frequency of the pulse (pulse per second) determines the speed of its shaft. High-quality stepper motors can achieve an accuracy between 3% and 5% per step, and this error does not accumulate from one step to the next. Stepper motors do not require feedback loop for precise speed and position control like DC and AC servomotors. This feature simplifies the system architecture and reduces the production and maintenance costs.

Stepper motors are characterized by the number of steps per revolution (resolution) or the step angle. The torque produced by a stepper motor depends on the step rate, the current flowing through the stator windings, and the drive design or type. Stepper motors are suitable in mechatronic applications where precise rotation is required. Various stepper motor designs with specific characteristics have been developed. Figure 8.3 displays three common types of stepper motor structures known as VR, PM, and hybrid stepper motors (Shakeel, 2020; Fiore, 2020).

The stator of a stepper motor is made from high-permeability soft magnetic material such as electrical steel or cast iron. High-permeability material confines the magnetic flux in paths defined by the stator structure. The stator of all stepper motor types contains pairs of poles with individual windings. Each tooth on the stator and rotor is defined as a pole. Groups of stator poles determine the number of phases. Stepper motors typically have two phases as illustrated in Figure 8.3. However, three- and five-phase stepper motors also exist.

The torque developed in a stepper motor is calculated using the same method as reluctance motors. Figure 8.4 shows the typical torque–frequency (speed) curve of a stepper motor. The torque and speed values referred to in stepper motor characteristics are defined below (Oriental Motor, n.d.).

- **Holding Torque**: The torque the motor produces at rest when the rated current flows through its windings.
- **Pull-in Torque**: The values of torque for which the motor can start, stop, or reverse in synchronism with the input pulses.

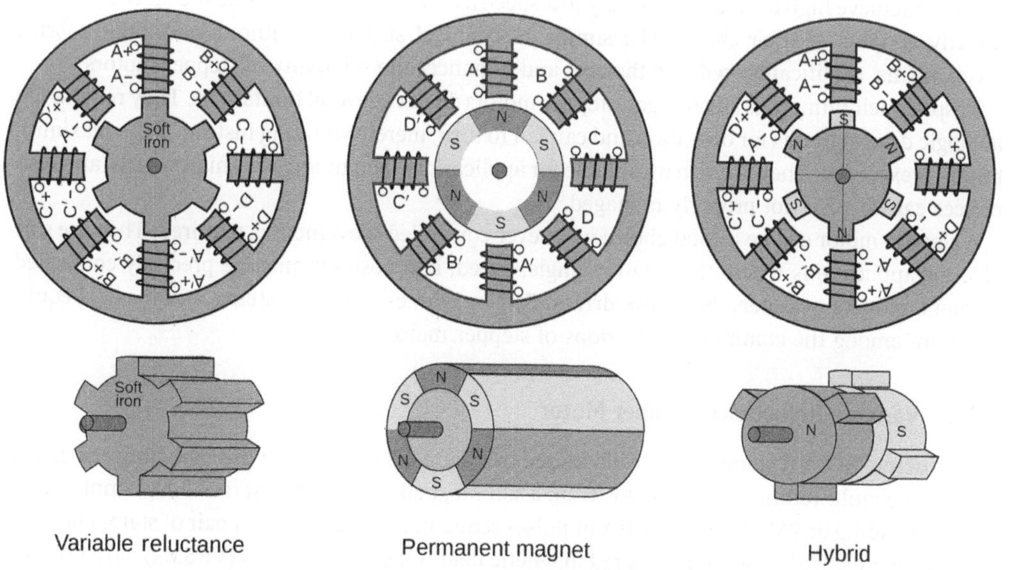

Figure 8.3 Common stepper motor types

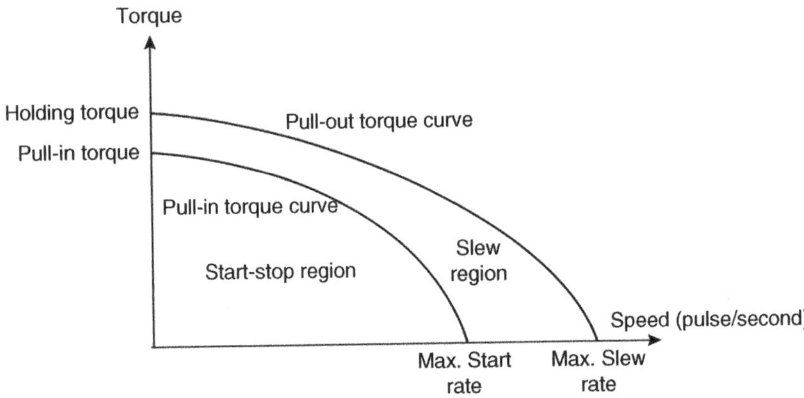

Figure 8.4 Torque–frequency characteristic of a stepper motor

- **Start/Stop Region:** The values at which the motor can start, stop, or reverse instantly.
- **Slew Region:** The region where the motor can operate without losing synchronism.
- **Maximum Start Rate:** The maximum starting frequency with no load applied.
- **Maximum Slew Rate:** The maximum operating frequency of the motor with no load applied.
- **Pull-in Curve:** Torque versus frequency curve where the motor can start or stop instantaneously, without losing synchronism, with load.
- **Pull-out Torque**: The maximum torque the motor can provide without stalling.
- **Maximum Starting Speed**: The maximum frequency at which the motor can start or stop instantaneously with a load applied without losing synchronism.
- **Maximum Slew Rate:** The maximum operating frequency of the motor with no load applied.

The rotation speed of stepper motors can be adjusted over a wide range by simply changing the input pulse frequency. Another advantage of stepper motors over DC and AC servomotors is their ability to achieve high torque at extremely low speeds; therefore, they can drive the mechanical load directly coupled to their shafts. The simple mechanical structure, without commutator–brush mechanisms, significantly reduces the cost and enhances the reliability of stepper motors.

Despite their numerous advantages, stepper motors do have several limitations. They require digital logic driver circuits for operation and can fail to turn, merely vibrating instead, if the maximum torque is exceeded. They are also unsuitable for applications demanding very high speeds, and resonances can occur if not properly managed.

A stepper motor can be a good choice whenever controlled movement is required. They are suitable for applications where the rotation angle, speed, and position must be precisely controlled. Printers, plotters, scanners, hard disk drives, CNC machines, robotic systems, and medical equipment are among the common applications of stepper motors.

8.2.1 Variable Reluctance Stepper Motor

The structure of a VR stepper motor illustrated in Figure 8.3 is similar to the switching reluctance motor. The multi-toothed rotor is made from a soft magnetic material (cast iron for example) without any winding or PM. When DC current pulses sequentially energize each pair of stator coils, the nearest rotor teeth align with the created magnetic field. VR stepper motors (VRSMs) have the simplest and most cost-effective structure among all electric machines. They provide higher speed and resolution (number of steps per revolution), but the torque it develops is often lower. In addition, when its windings are not energized, it does not produce any torque, known as detent torque.

The VRSM is known for its simple and robust design. It consists of a multi-toothed rotor made from a soft magnetic material, along with stator poles that are equipped with windings similar to those found in a reluctance motor. When sequentially pulsed DC energizes each pair of stator coils, the nearest rotor teeth align with the magnetized stator poles. This mechanism allows for precise control of the rotor position.

The VR motor's operation relies on the principle of magnetic reluctance, which is the opposition to the passage of magnetic flux. The rotor's teeth create VR paths that the magnetic flux follows, hence the name. As the stator coils are energized in sequence, the rotor moves to minimize the reluctance, resulting in discrete steps of motion.

The primary advantages of VRSMs include their simplicity and low cost. The absence of PMs makes these motors less expensive to manufacture, and their straightforward construction enhances reliability. However, VR motors typically generate lower torque compared to their PM counterparts and may exhibit higher levels of vibration and noise.

Despite these limitations, VRSMs are widely used in applications requiring precise positioning and low-speed operation. They are commonly found in printers, plotters, and other devices where accurate and repeatable motion is essential.

8.2.2 Permanent Magnet Stepper Motors

Permanent magnet stepper motors (PMSM) have a permanently magnetized cylindrical rotor. Alternating N and S poles on the rotor are parallel to the motor shaft as illustrated in Figure 8.3 (b). The stator poles are magnetized with alternating north and south poles.

Higher Torque: PMSM generates higher torque due to the increased magnetic flux intensity provided by the magnetized rotor poles, compared to VRSMs.

The use of permanent magnets eliminates the need for complex winding arrangements, simplifying the motor's construction. The magnetized rotor design enhances the motor's reliability and robustness by providing consistent magnetic properties.

Although PMSMs offer several advantages, their production is typically more expensive due to the cost of the permanent magnets. While they provide higher torque, PMSM can exhibit higher levels of vibration and noise. Compared to other types of stepper motors, PMSM may offer lower resolution and speed.

8.2.3 Hybrid Stepper Motors

Hybrid stepper motors illustrated in Figure 8.3 (c) combine the strengths of VR and PM types. The rotor has two caps with alternating polarity axially magnetized teeth. This configuration allows high resolution, speed, and torque. This higher performance requires a more complex construction and, therefore, a higher cost (Fiore, 2020).

Hybrid stepper motors provide high resolution, high speed, and high torque compared to VR and PM stepper motors. The configuration of alternating polarity axially magnetized teeth allows for precise and fine movements. Hybrid stepper motors are capable of operating at higher speeds than the other two types of stepper motors. The combination of VR and PM elements provides higher torque capabilities.

The design and manufacturing of hybrid stepper motors are more complex compared to other types of stepper motors. The advanced construction and materials used in hybrid stepper motors lead to increased production costs.

8.2.4 Stepper Motor Drives

Stepper motor drives supply a sequence of pulses to turn stepper motors in precise movements. These drives typically consist of a control stage, which generates the pulse sequences necessary for motor operation and a power stage, which supplies current to the motor winding. The control stage may be a sequential logic circuit or a microcontroller. The power circuit can consist of either bipolar junction transistor (BJT) or metal–oxide–semiconductor field-effect transistor (MOSFET) power transistors, or it may utilize a specialized stepper motor IC chip.

There are several types of stepper motor drives, each with its own unique characteristics:

- **Constant Voltage Drives**: These drives apply a fixed voltage to the motor windings, making them relatively simple and cost-effective. However, they may not provide optimal performance at higher speeds.
- **Chopper Drives**: Chopper drives use PWM to regulate the current supplied to the motor windings, ensuring stable operation even at higher speeds. These drives are more efficient and provide better torque characteristics than constant voltage drives.
- **Micro-Stepping Drives**: Micro-stepping drives subdivide each step into smaller increments, resulting in smoother motion and higher resolution. These drives are ideal for applications requiring high precision and reduced vibration and noise.

Each type of drive offers different advantages and is suitable for various applications depending on the requirements for speed, torque, and precision.

Stepper motor stator windings can be configured in a "unipolar" and "bipolar" configurations shown in Figure 8.5. Unipolar two-phase stepper motors have five leads where the central tap of each phase winding is common. In a bipolar configuration, the windings have no central tap; therefore, bipolar stepper motors have four power supply leads.

A unipolar driver circuit provides the motor with single-polarity pulses. In contrast, bipolar drivers deliver pulses with two opposite polarities to the windings. An H-bridge, consisting of four power transistors, is used to reverse the polarity of the pulses.

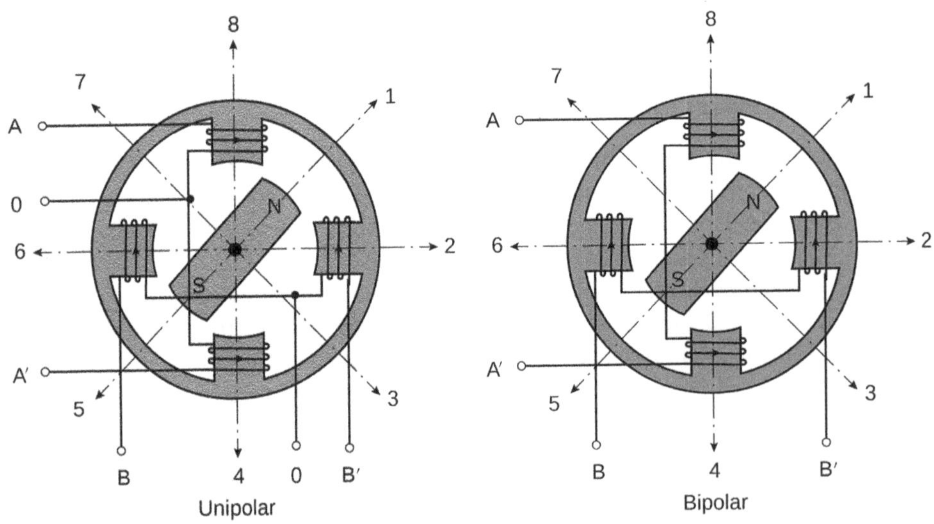

Figure 8.5 Unipolar and bipolar inding configurations

Table 8.1 Stepping mode sequences

	Wave Drive				Full-Step Drive				Half-Step Drive							
Sequence	**1**	**2**	**3**	**4**	**1**	**2**	**3**	**4**	**1**	**2**	**3**	**4**	**5**	**6**	**7**	**8**
A	1	0	0	0	1	0	0	1	1	0	0	0	0	0	1	1
B	0	1	0	0	1	1	0	0	1	1	1	0	0	0	0	0
A′	0	0	1	0	0	1	1	0	0	0	1	1	1	0	0	0
B′	0	0	0	1	0	0	1	1	0	0	0	0	1	1	1	0
Position	8	2	4	8	1	3	5	7	1	2	3	0	5	6	7	8

The most common stepper motor driving modes are known as "wave drive," "full-step drive," and "half-step drive." In a wave drive, one stator winding is energized at each step, causing the rotor to align with the nearest stator pole. Wave drive mode utilizes half of the phase windings at each step. To increase the torque, the full-step drive energizes to poles axially positioned. Half-step drive combines both the wave and full-step modes. In a half-step drive mode, energizing two adjacent stator windings positions the rotor pole precisely midway between the stator poles. Every second step, only one phase is energized, and during the other steps, one phase on each stator pole. Table 8.1 displays the excitation sequences corresponding to different drive modes.

In a micro-stepping drive, the currents in the windings vary continuously to divide one full step into several smaller discrete steps. Micro-stepping improves smoothness and resolution by dividing each step into smaller increments. It is perfect for applications needing high precision, less vibration, and reduced noise.

8.3 Synchronous Machines

A synchronous machine is an AC machine in which the rotor speed is precisely synchronized with the frequency of the supply current. These machines rely on the interaction between the electromagnetic field in the stator and the rotor, which can be created by either PMs or field windings. They are categorized into synchronous motors and generators, serving various applications where precise speed control and synchronization are critical. Synchronous motors are used for industrial systems requiring constant speed and accurate timing, while synchronous generators supply electric power grids and emergency backup systems. Their designs include cylindrical or salient pole rotors, with industrial-scale machines typically operating on three-phase power. Smaller synchronous machines used in mechatronic systems and household appliances are single phase.

Synchronous generators in electric power plants supply the interconnected grid at a fixed frequency. The electric power grid is predominantly supplied by large-scale synchronous generators driven by steam turbines, gas turbines, or hydro turbines. Smaller units, driven by combustion engines, are frequently employed as emergency generators.

Synchronous motors play a crucial role in mechatronics applications where constant speed and accuracy are critical, such as clocks and timers. Their ability to maintain constant speed makes them ideal for industrial systems that must function in synchronization.

The speed of a synchronous motor can be accurately controlled by VFCs. Some direct-drive motors are a type of PM synchronous motor driven by a microcomputer-controlled VFC to operate at various speeds needed for an automated process. Many modern appliances, where smooth

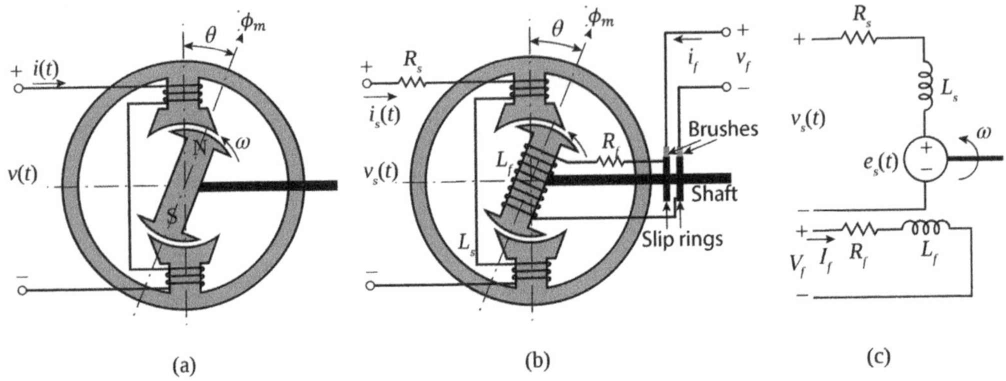

Figure 8.6 Permanent magnet and wound rotor synchronous machines (a: Permanent magnet; b: Wound rotor; c: Equivalent circuit of a wound rotor SG)

operation, endurance, and energy saving are priorities, use direct drive motors to eliminate belts, gears, and pulleys.

Figure 8.6 shows the simplified structure of a single-phase synchronous machine. The magnetic field necessary for the operation of a synchronous machine may be created by a PM (a) or a coil placed on the rotor (b). In larger synchronous machines, the rotor winding supplied by a DC current I_f produces a constant magnetic field, which rotates with the rotor. The rotor winding is connected to the external circuit through carbon brushes touching the slip rings mounted on the shaft. In smaller synchronous machines, typically up to a few kilowatts of electrical power, PM rotors are more common. The use of strong PMs simplifies the structure and reduces the manufacturing cost by eliminating the need for rotor windings, brushes, and slip rings.

In a cylindrical rotor, the radial length of the air gap is constant. Therefore, the reluctance does not change with the rotation. In salient pole machines, the reluctance depends on the angle between the rotor and stator winding axes as in previously described reluctance motors.

8.3.1 Synchronous Generators

Both PM and wound-rotor synchronous generators operate based on the same principles. The PM may be represented by a coil carrying an equivalent current to produce the same magnetic flux. To explain the operation of a synchronous generator, we will first consider the simplified structure of a single-phase PM generator shown in Figure 8.6 (a). The two-pole magnet rotates with the angular velocity ω, causing the angle between the flux and the stator pole to change over time as $\theta = \omega t + \delta$. According to Faraday's law, the voltage in the stator winding is derived by taking the derivative of the rotating flux component on its axis.

$$e_s(t) = -\frac{d}{dt}\phi_m \cos(\theta) = -\omega\phi_m \sin(\omega t + \delta) = E_m \sin(\omega t + \delta) \tag{8.13}$$

The magnitude of the generated voltage is proportional to the magnetic flux and rotation speed in radians per second. If the magnetic field is produced by a rotor coil, the flux is a nonlinear function of the current. However, in the linear region of the *B–H* characteristic of the core material, the flux can be assumed to be proportional to the field current I_f. In this case, the following expression can be used to express the amplitude of the stator voltage.

$$E_m = k\omega I_f \tag{8.14}$$

An equivalent circuit of a single-phase synchronous generator is displayed in Figure 8.6 (c). A two-pole generator produces an electric voltage, which completes one cycle at each rotation. Practical generators usually have multiple pairs of poles. Since the number of cycles per revolution equals half of the total number of poles, the frequency of the voltage generated by a synchronous generator that has in p poles and turning at N revolutions per minute (RPM) is calculated using the following formula:

$$f_s = \frac{p}{2}\frac{N}{60} = \frac{pN}{120} \tag{8.15}$$

In a multiple-pole generator, it is important to distinguish mechanical and electrical angles. When a p pole rotor completes one revolution, the generated voltage completes $p/2$ electrical cycles. Therefore, the number of electrical degrees equals $p/2$ times the number of mechanical degrees in any electric machine.

Small-sized permanent synchronous generators can be used as rotation speed sensors since the induced voltage and its frequency are exactly proportional to the rotation speed. By converting the sinusoidal voltage waveform into a square wave using a Schmitt trigger, a synchronous generator can function as a precise digital sensor.

All commercial-size synchronous generators have three-phase windings distributed around the stator as shown in Figure 8.7. Stator windings are positioned with equal angles of 120°. As the rotor turns at the angular speed of ω radians per second, voltages with the same magnitude and frequency, but shifted by the phase angle of 120° ($2\pi/3$ rad) are induced across their terminals.

$$\begin{aligned} v_a &= V_m \sin(\omega t) \\ v_b &= V_m \sin\left(\omega t - \frac{2\pi}{3}\right) \\ v_c &= V_m \sin\left(\omega t + \frac{2\pi}{3}\right) \end{aligned} \tag{8.16}$$

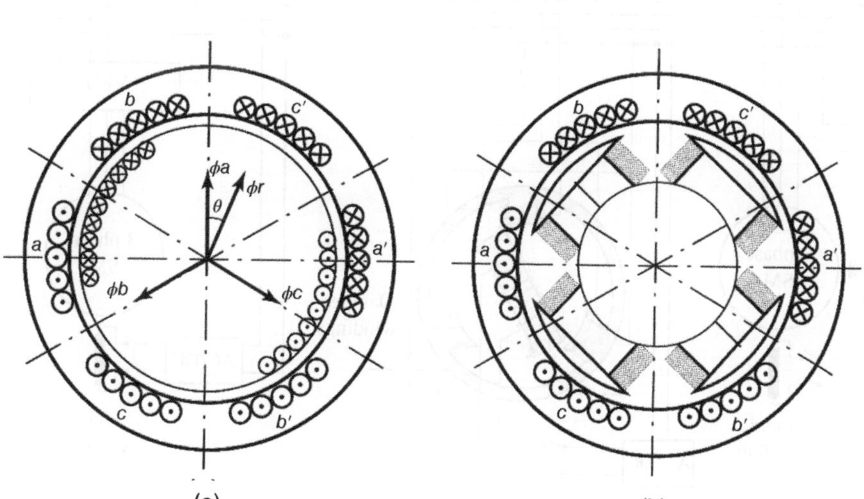

Figure 8.7 Simplified structure of three-phase synchronous generators with round-rotor (a) and salient-pole rotor (b)

In practice, there are two distinct rotor structures. In a round rotor displayed in Figure 8.7 (a), the field windings are inserted in slots around the cylindrical rotor. The air gap between the stator and the rotor is practically constant at all points. Such generators are typically used with higher-speed gas turbines, hence called turbo generators. To reduce the moment of inertia, the rotor of turbo-generators is designed with a smaller diameter and longer length. In slower systems, such as hydro-electric generation, the rotor diameter is larger, while the length is shorter. The poles are mounted around the rotor as shown in Figure 8.7 (b); hence, the air gap varies around the rotor. Such a structure is specified as a hydro generator because it is common in hydroelectric power plants.

Because the induced voltage is proportional to the time derivative of the magnetic field according to Faraday's law, in all types of generators, the voltage produced across the windings is proportional to the rotation speed if no power is delivered to an external circuit, a special operating condition described by the term "no-load operation" in practice.

The three-phase generators supplying a symmetrical load can be analyzed using a per-phase method since voltages and currents of each phase repeat the same variations with a phase angle of 120°.

8.3.2 Three-Phase Synchronous Motors

Like all electromechanical conversion systems, synchronous machines can function as either a motor or a generator. Industrial synchronous motors typically run on three-phase power. We will first examine the operation of three-phase machines because they are simpler to understand.

The stator of a three-phase motor has three identical windings positioned with 120° angles on its stator. Figure 8.8 (a) shows an empty stator with three coils to demonstrate the magnetic field variation in the air gap. When powered by a balanced three-phase source, equal amplitude currents with 120° phase-shift generate time-varying magnetic flux density orthogonal to the winding plane.

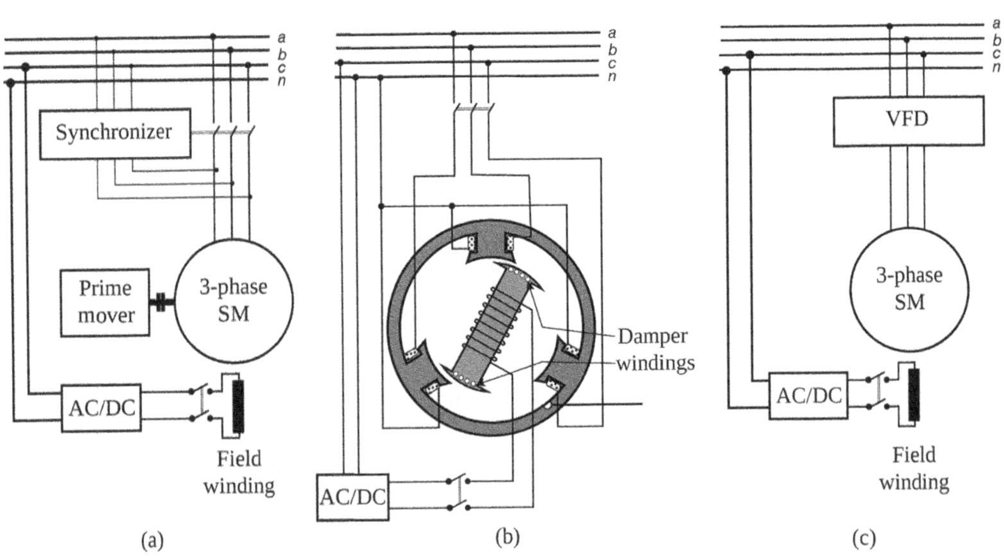

Figure 8.8 Synchronous motor starting methods (a: External prime mover; b: Damper windings; c: Variable frequency drive)

The right-hand rule indicates the direction of the flux density vector as shown in the figure. The flux density phasors are represented in polar coordinates as follows:

$$\begin{aligned} B_{aa'} &= |B_{aa'}|\angle 0^\circ = B_m \sin(\omega t)\angle 0^\circ \\ B_{bb'} &= |B_{bb'}|\angle 120^\circ = B_m \sin\left(\omega t - \frac{2\pi}{3}\right)\angle 120^\circ \\ B_{cc'} &= |B_{cc'}|\angle 240^\circ = B_m \sin\left(\omega t - \frac{4\pi}{3}\right)\angle 240^\circ \end{aligned} \tag{8.17}$$

Figure 8.8 (b) shows the time variation of the flux density vectors for each winding. To find the resultant flux density, let's express these vectors with their x and y components on the rectangular coordinate system shown in the figure.

$$\begin{aligned} B_{aa'} &= |B_m \sin(\omega t)|\vec{x} \\ B_{bb'} &= \left|B_m \sin\left(\omega t - \frac{2\pi}{3}\right)\right|\left[-\frac{1}{2}\vec{x} + \frac{\sqrt{3}}{2}\vec{y}\right] \\ B_{cc'} &= \left|B_m \sin\left(\omega t - \frac{4\pi}{3}\right)\right|\left[-\frac{1}{2}\vec{x} - \frac{\sqrt{3}}{2}\vec{y}\right] \end{aligned} \tag{8.18}$$

In this equation, $\vec{x}$ and $\vec{y}$ are the unit vectors on the x and y axes, respectively. The sum of the flux density vectors is calculated by combining the components of each vector on the x and y axes.

$$\begin{aligned} B_x &= |B_m \sin(\omega t)| - \frac{1}{2}\left|B_m \sin\left(\omega t - \frac{2\pi}{3}\right)\right| - \frac{1}{2}\left|B_m \sin\left(\omega t - \frac{4\pi}{3}\right)\right| \\ B_y &= \frac{\sqrt{3}}{2}\left|B_m \sin\left(\omega t - \frac{2\pi}{3}\right)\right| - \frac{\sqrt{3}}{2}\left|B_m \sin\left(\omega t - \frac{4\pi}{3}\right)\right| \end{aligned} \tag{8.19}$$

These equations can be simplified using the following trigonometric identities:

$$\begin{aligned} \sin A + \sin B &= 2\sin\left(\frac{A+B}{2}\right)\cos\left(\frac{A-B}{2}\right) \\ \sin A - \sin B &= 2\cos\left(\frac{A+B}{2}\right)\sin\left(\frac{A-B}{2}\right) \end{aligned} \tag{8.20}$$

The vectorial sum of the magnetic flux density vectors can be expressed as follows by performing algebraic manipulations:

$$\vec{B_r}(t) = \frac{3}{2}B_m\left[\sin(\omega t)\vec{x} - \cos(\omega t)\vec{y}\right] = \frac{3}{2}B_m\angle\left(\omega t - \frac{\pi}{2}\right) \tag{8.21}$$

The resultant vector has a constant magnitude of $1.5\,B_m$ and rotates at the angular speed of ω within the air gap.

Example 8.3 Determine the resultant flux density in the air gap of a 33-phase machine at $t = 0$ by geometric vector addition and write a MATLAB® code for visualization of the three-phase rotating magnetic field (RMF).

Solution

The values of flux density generated by each phase winding at $t = 0$ is determined by using equation 8.16.

$$B_{aa'} = B_m \sin(\omega t)\angle 0^\circ = 0$$

$$B_{bb'} = B_m \sin\left(\omega t - -\frac{2\pi}{3}\right)\angle 120^\circ = \frac{\sqrt{3}}{2}B_m\angle 120^\circ$$

$$B_{cc'} = B_m \sin\left(\omega t - -\frac{4\pi}{3}\right)\angle 240^\circ = \frac{\sqrt{3}}{2}B_m\angle 240^\circ$$

Figure 8.8 (a) shows the geometric addition of these phasors. For $\omega t = 0$, $B_r = 1.5\angle -90$.

The following MATLAB® code computes the resultant flux density in the air gap. The computation results for specific ωt values are shown in Figure 8.8 (b). This script visualizes the rotating field using the MATLAB® statement "quiver."

MATLAB® Code for Computation and Visualization of the Three-Phase Rotating Field.

```
% 3-Phase Rotating Magnetic Field Computation and Visualization
clc; clear; close all;
% Parameters
wt = linspace(0, 2*pi, 100); % Angular frequency
Bm= 1; % Amplitude
%wt = ; % Angular frequency
% Three-phase waveforms
Ba = Bm * sin(wt);
Bb = Bm * sin(wt - 2*pi/3);
Bc = Bm * sin(wt + 2*pi/3);
% Plot three-phase waveforms
x=wt;
figure;
plot(x, Ba, 'k', x, Bb, 'k--', x, Bc, 'k.', 'LineWidth', 1.5);
xlabel('wt (rad)');
ylabel('Flux density magnitudes');
%title('Three-Phase AC Waveforms');
legend('Ba', 'Bb', 'Bc');
grid on;
xlim([0 2*pi]);
ylim([-1.5 1.5])
xticks(0,pi/2:6.3); % X-axis grid spacing every 2 units
% Rotating field visualization
figure
hold on;
axis equal;
grid on;
for i = 1:length(wt)
  clf;
  hold on;
  axis equal;
  xlim([-1.5 1.5]);
  ylim([-1.5 1.5]);
  % Compute rotating vectors
```

```
    Bx = Ba(i) + Bb(i)*cos(2*pi/3) + Bc(i)*cos(-2*pi/3);
    By = Bb(i)*sin(2*pi/3) + Bc(i)*sin(-2*pi/3);
    quiver(0, 0, Bx, By, 'k', 'LineWidth', 2, 'MaxHeadSize', 0.5)
    pause(0.05);
end
```

At all times, the resultant flux density will maintain the constant magnitude of 1.5 B_m and rotate by the angular velocity of ω. Therefore, stator windings placed at an angle of 120° around the stator and supplied by balanced three-phase currents produce a magnetic field equivalent to a rotating magnet.

8.3.2.1 Starting Three-Phase Synchronous Motors

In the motor operation mode, the shaft of the synchronous machine is required to turn exactly at the synchronous speed determined by the supply frequency. The following expression of the synchronous speed in terms of the supply voltage frequency (f) and the total number of poles (p) is obtained by rearranging equation 8.15.

$$n_s = \frac{120f}{p} \tag{8.22}$$

For continuous operation, the magnetic field of the rotor windings must consistently align with the rotating field generated by the stator windings. Synchronous motors maintain a constant rotational speed regardless of varying load conditions, if they operate within their rated power capacity. Nevertheless, a synchronous motor becomes unstable if subjected to an overload.

Specific methods are required to start a three-phase synchronous motor from a standstill position as the rotor speed at the beginning is zero. These methods align the rotor's magnetic field with the rotating field generated by the stator and bring the rotor up to synchronous speed. Commonly used starting methods are explained below.

8.3.2.1.1 External Prime Mover

At the start, the rotor of the synchronous motor is stationary, making it incapable of generating the necessary torque to synchronize with the stator's RMF. The rotor must first be brought up to the synchronous speed, which is determined using equation 8.22.

One of the methods used in larger industrial systems is to accelerate the standstill rotor to synchronous speed, external motor, or prime mover as illustrated in Figure 8.8 (a). During this process, the synchronous machine first operates as a generator, inducing voltages in the stator windings in the same phase sequence as the supply network. When the frequency of the induced voltage matches the power supply frequency, the field current is adjusted to match the amplitude of the induced voltage with the power supply voltage. The power switch is turned on when both the supply and induced voltages are at the same phase. After the synchronization is achieved, the external motor is uncoupled using a mechanical or magnetic clutch.

The process explained above is known as synchronization referring to the alignment of the rotor's magnetic field with the stator's RMF. Synchronization allows the motor to operate at a constant speed determined by the supply frequency. It is also necessary to connect multiple generators in parallel or synchronize a generator with the grid, as seen in small wind turbines or larger units within power plants. The synchronization process requires additional instruments and indicators to achieve proper connection. It can be performed manually with a synchronoscope to observe the difference between the magnitude and frequency of the induced and power supply voltages or automatically using an electronic synchronizer.

Once the rotor reaches synchronous speed, the magnetic field of the rotor is aligned with the rotating field created by the stator windings. This alignment ensures steady torque production and stable operation.

8.3.2.1.2 Damper Windings

Damper windings are short-circuited bars embedded in slots located on the faces of the rotor poles as shown in Figure 8.8 (b) and short-circuited at both ends. The stator windings generate an RMF that induces voltages across these bars, leading to currents that create torque accelerating the rotor. When the rotor reaches synchronous speed, the function of the damper winding ends because the rotor speed relative to the RMF will become zero and no voltage will be induced.

Damper windings, also referred to as "amortisseur windings," are an essential component of modern synchronous machines. In addition to generating torque during starting, they also minimize transient electromechanical oscillations when the load rapidly changes.

8.3.2.1.3 Variable Frequency Drives (VFDs)

VFDs are electronic converters used to change the amplitude and frequency of AC power supply. A VFD connected between the power supply and the stator windings of a synchronous motor as shown in Figure 8.8 (c) is used to gradually increase frequency and adjust the amplitude of the motor voltage. By controlling the frequency from zero, the rotor is brought up to synchronous speed smoothly. VFDs provide precise control and are particularly advantageous in applications requiring soft starting and speed regulation.

These methods ensure that three-phase synchronous motors can transition from a standstill to synchronous operation efficiently and reliably. Each method is chosen based on the specific application and operational requirements.

8.3.3 Single-Phase Synchronous Motors

Single-phase synchronous motors have a similar structure to the synchronous reluctance motors described in Section 6.1.2. Their difference is the rotor equipped with either a field winding or a PM that produces a constant magnetic field. They can provide an average torque if the rotor turns exactly at a synchronous speed defined in equation 8.22 in terms of the stator supply frequency and the number of poles.

If the rotor field is produced by a winding, an external DC source must be connected to the rotating coil through carbon brushes that contact slip rings. This magnetic field may be produced by a PM instead of a field winding. Permanent magnet synchronous motors are more cost-effective because slip rings and brushes are eliminated, and they do not require an external power source for the rotor field.

The operation of a single-phase synchronous motor is based on the interaction between the rotor's magnetic field and the stator's RMF. For the motor to provide average torque, the rotor must turn at synchronous speed, which is determined by the stator supply frequency and the number of poles in the motor.

These motors are typically constructed as fractional or subfractional HP motors, meaning they are designed for applications that require less than one HP. Their compact structure and operation characteristics make them suitable for specific low-power tasks where precise rotation speed is required such as analog clocks and timing elements.

Single-phase synchronous motors can be started using any of the methods explained for three-phase motors. Damper windings and VFDs are common in single-phase motors. External motor or prime mover can also be used, but it is impractical for the small power applications typical of single-phase synchronous motors.

8.4 Asynchronous (Induction) Motors

Asynchronous motors, also known as induction motors, are the most widely used AC machines due to their robust structure, low cost, and ability to operate under a wide variety of load conditions. The stator of synchronous and asynchronous motors has a similar structure. Stator windings placed in slots evenly distributed around the air gap are supplied from an AC source. Asynchronous motors widely used in practice feature copper or aluminum bars embedded around the rotor and short-circuited at both ends. Such motors are referred to as "squirrel-cage induction motor" due to the visual resemblance of the rotor winding configuration. Squirrel-cage rotor structure does not require an external power supply; instead, the stator's RMF induces a voltage necessary to create torque. Figure 8.9 illustrates the simplified structure of a squirrel-cage asynchronous motor. Certain induction machines utilized in industrial applications have rotor windings that can be connected to an external circuit via a slip ring–brush assembly like synchronous machines. These rotor windings may be directly short-circuited, connected to a set of adjustable resistances, or powered by an electronic converter for advanced speed regulation. The stator and rotor of asynchronous motors are made from laminated electric steel to minimize the core losses.

Small motors that power household appliances, such as washing machines, dryers, and refrigerators, are supplied from single-phase utility power. Smaller asynchronous motors powering ceiling fans, corded kitchen tools, and various mechatronic devices are categorized as fractional and sub-fractional horsepower (hp) machines.

Larger induction motors used in commercial and industrial applications operate on three-phase power. Asynchronous motors are an essential part of industrial automation. Equipment such as conveyor systems, compressors, fans, pumps, power tools, winches, cranes, and elevators often utilize asynchronous motors. Major benefits of asynchronous motors include their simplicity, durability, robust operation in a wide range of loads, low cost, and high efficiency.

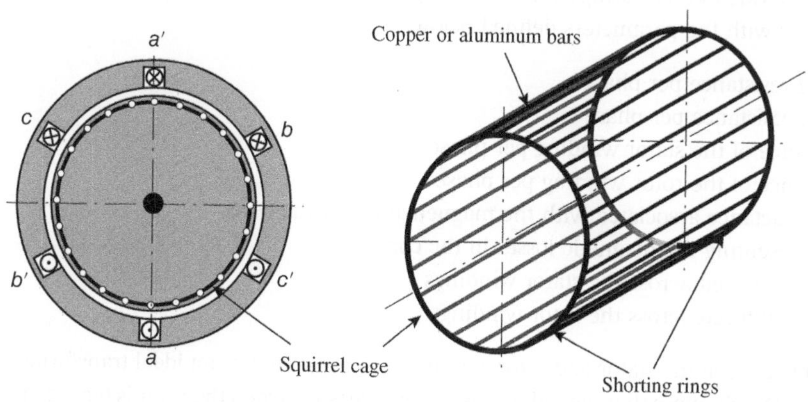

Figure 8.9 Schematic diagram of a three-phase squirrel-cage induction motor

8.4.1 Three-Phase Induction Motors

A three-phase induction motor operates on the principle of a RMF and electromagnetic induction. When three-phase alternating current is applied to the stator windings, it creates a RMF that revolves at the synchronous speed defined by the expression equation 8.22. This RMF sweeps across the standstill rotor conductors, which are not connected to any external power source. According to Faraday's law, this change in magnetic flux induces an EMF in the rotor. This EMF causes current to flow through the closed loop of the rotor coils, producing a torque that accelerates the rotor. These currents interact with the stator's magnetic field, generating a torque on the rotor via Lorentz force, causing it to turn in the direction of the stator's rotating field. As the rotor speeds up, its relative speed compared to the stator's RMF decreases. The rotor does not catch up with the stator's rotating field. If the rotor were to turn at the same speed as the stator field, there would be no induced voltage to supply the rotor magnetic field, causing it to decelerate. The rotor accelerates to just below synchronous speed and provides output power driving a mechanical load in steady-state operation. An induction motor can never reach the synchronous speed, which explains why it is also called an asynchronous motor.

The difference between the rotor's actual speed and the synchronous speed is termed as the variable "slip" (s),

$$s = \frac{n_s - n_r}{n_s} \Longrightarrow n_r = (1 - s)n_s \tag{8.23}$$

When the rotor is at rest ($n_r = 0$) slip is 1, it accelerates it approaches to zero. Slip is often expressed as a percentage of the synchronous speed. Slip typically ranges from just above zero to 5%–6%.

8.4.2 Asynchronous Motor Equivalent Circuit

In normal operation, three-phase asynchronous motors are supplied from a symmetrical power supply. Since the phase coils on the stator and rotor are made identical, the phase voltages and currents repeat with the same magnitude shifted in time by 120°. Therefore, the continuous operation of a three-phase motor can be analyzed using an equivalent circuit developed for one phase.

When the motor shaft is stationary or blocked, the asynchronous motor acts like a transformer with short-circuited secondary windings. The number of phases and frequency of the induced EMF match that of the power supplied to the stator.

Asynchronous motors are like a rotating transformer. Figure 8.10 (a) shows the equivalent circuit of an induction motor with the parameters defined below:

- R_s: Stator winding resistance per phase
- R_r: Rotor winding resistance per phase
- X_s: Leakage reactance of the stator winding per phase
- X_r: Leakage reactance of the rotor winding per phase
- X_m: Magnetizing reactance associated with the magnetization of the core
- R_c: Resistance representing the magnetic losses in the core
- E_s: Per-phase EMF induced across the stator windings
- E_r: Per-phase EMF induced across the rotor windings

Magnetic coupling between the stator and rotor windings is represented by an ideal transformer with turn ratio $a = N_2/N_1$. Suppose that the induced rotor voltage is E_{ro} when the rotor is blocked to remain stationary ($n_r = 0$; $s = 1$).

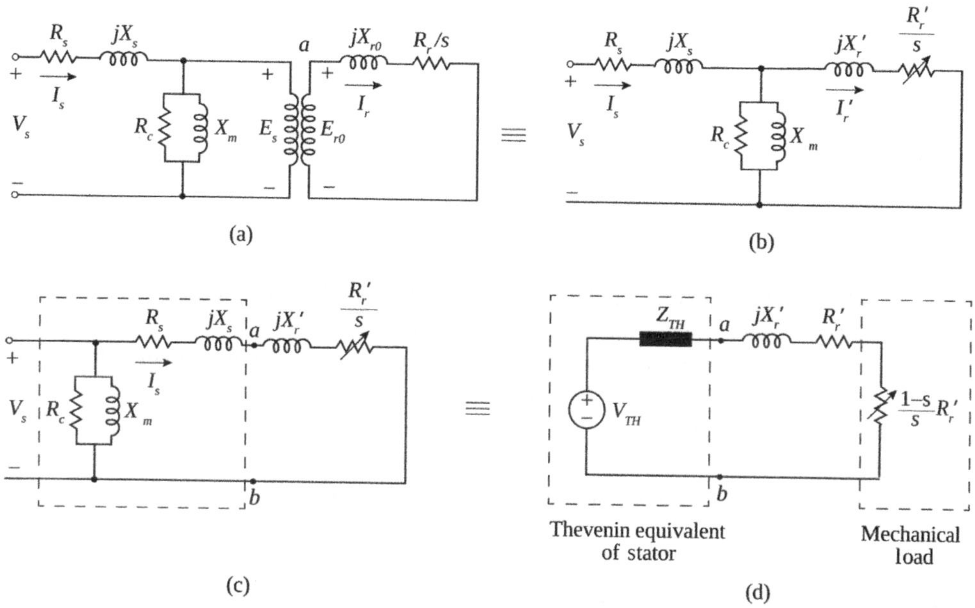

Figure 8.10 Per-phase circuit models for steady-state analysis of asynchronous motors (a: Circuit model featuring an ideal transformer; b: Circuit model with parameters referred to the stator and a parallel branch representing the core; c: Configuration with the parallel branch relocated to the stator input; d: Simplified equivalent circuit neglecting core loss and magnetization current effects)

As the rotor turns, the induced EMFs magnitude and frequency decrease in proportion to the slip, representing the rotor coils' relative speed to the rotating stator field.

$$E_r = sE_{r0}$$
$$f_r = sf_{r0} \tag{8.24}$$

As the rotor turns ($0 < s < 1$), the rotor reactance will also change proportionally to the slip. The reactance of the rotor winding is obtained by multiplying the rotor reactance at rest X_r with the slip.

$$X_r = \omega_r L_r = 2\pi(sf_{r0})L_r = sX_{r0} \tag{8.25}$$

The circuit parameters for the rotor circuit depicted in Figure 8.10 (a) are derived by considering the following expression of the rotor current:

$$I_r = \frac{sE_{r0}}{R_r + jsX_{r0}} = \frac{E_{r0}}{\dfrac{R_r}{s} + jX_{r0}} \tag{8.26}$$

E_{r0} is the voltage induced across the rotor winding when the rotor is stationary ($s = 1$). It can be evaluated using the turn-ratio a of the ideal transformer corresponding to the magnetic coupling between stator and rotor. In wound-rotor induction machines, the effective turn-ratio is determined based on the number of wires in the stator and rotor coils. However, estimating the turn-ratio in squirrel-cage motors is more complex (Yamayee & Bala, 1994).

In Figure 8.10 (b), the ideal transformer is removed, and the secondary side parameters are referred to the primary using the turn-ratio $a = N_s/N_r$. Parameters referred to the stator are

$$\begin{aligned} E_{r0}' &= aE_{r0} \\ I_r' &= \frac{I_r}{a} \\ X_r' &= a^2 X_r \\ R_r' &= a^2 R_r \end{aligned} \tag{8.27}$$

The currents flowing through the parallel combination of R_c and X_m are much smaller compared to the rated stator current. Relocating these components to the input terminals as shown in Figure 8.10 (c) simplifies the solution of complex circuit equations without introducing substantial errors in the analysis of the motor near rated load conditions.

The circuit model of an asynchronous motor can be further simplified by representing the stator side with its Thevenin equivalent as shown in Figure 8.10 (d). Additionally, R_r/s is substituted with the sum of the actual ohmic resistance and a fictional resistance that varies with the slip as indicated below:

$$\frac{R_r}{s} = R_r + \left(\frac{1}{s} - 1\right) R_r = R_r + \frac{1-s}{s} R_r \tag{8.28}$$

Thevenin voltage source is the open-circuit voltage across the rotor circuit calculated for $s = 0$. Based on the configuration in (c), open-circuit voltage is equal to the stator supply voltage. Thevenin impedance is the equivalent impedance seen from the points marked as a and b when the voltage source is replaced with a short circuit. Short-circuiting the magnetizing impedance leaves only the series-connected stator resistance and leakage reactance. The Thevenin model disregards magnetizing current and core losses.

$$\begin{aligned} V_{TH} &= V_o \\ Z_{TH} &\cong R_s + jX_s \end{aligned} \tag{8.29}$$

8.4.3 Identification and Estimation of Equivalent Circuit Parameters

The equivalent circuit of an induction motor can be constructed using physical dimensions, winding design, and material properties based on detailed design data. Industrial automation or mechatronic system designers are seldom concerned with the design details. Specifications listed in datasheets and specific tests outlined below can provide sufficient information to estimate parameters of the basic equivalent circuits outlined in Figure 8.10 (c) (Yamayee & Bala, 1994). Rotor circuit parameters identified from tests are directly referred to the stator, making the turn-ratio of stator and rotor windings irrelevant.

8.4.3.1 Stator Resistance Measurement

The stator resistance can be measured using any resistance measurement method when the motor is not powered. To obtain an accurate value for the stator resistance at the rated current level, methods such as the voltmeter-ammeter technique, a DC resistance meter, or a Wheatstone bridge are more suitable than digital instruments used for electronic circuits.

8.4.3.2 No-Load Test

This test is performed with the rotor running freely without any mechanical load other than friction ($s \approx 0$). With the rated voltage applied to the stator windings, the current I_o and active power P_o are recorded. With near-zero slip, the rotor circuit acts as an open circuit, and almost no current flows through the rotor windings. The stator resistance and leakage impedance are negligible compared to the parallel equivalent of the magnetizing reactance X_m and the resistance R_c corresponding to the core loss. The per-phase values of this resistance and reactance are calculated using the following equations:

$$
\begin{aligned}
R_c &= \frac{P_o}{3I_o^2} \\
Z_c &= \frac{U_s}{\sqrt{3}I_o} \\
X_m &= \sqrt{Z_c^2 - R_c^2}
\end{aligned} \tag{8.30}
$$

U_s is the line voltage (phase-to-phase) of the three-phase power applied to the stator.

8.4.3.3 Blocked-Rotor Test

This test is performed with the rotor locked ($s = 1$). The purpose of this test is to identify the rotor and stator resistances and leakage reactances. Applying the rated voltage to a blocked rotor causes the motor to draw excessive current, potentially damaging the motor windings. For safety, the stator voltage is increased slowly starting from zero while checking the stator current until it reaches the rated current level. Then, the line voltage U_b, line current I_b, and the input power P_b are recorded. The equations below give the total resistance and leakage reactance of the stator and rotor windings, as referred to the stator.

$$
\begin{aligned}
R_b &= \frac{P_b}{3I_b^2} \\
Z_b &= \frac{U_b}{\sqrt{3}I_b} \\
X_b &= \sqrt{Z_b^2 - R_b^2}
\end{aligned} \tag{8.31}
$$

Under normal operating conditions, the rotor frequency is proportional to the slip. Standards such as IEEE 112-2017 and IEC 60034-2-1 recommend performing the blocked-rotor test at a lower frequency to find the parameters around full load, as the slip has a lower value under those conditions (IEEE, 2017). The blocked-rotor reactance obtained at a lower test frequency f_{test} should be corrected to the rated frequency f_r.

$$
X_{b,r} = \frac{f_{test}}{f_r} X_{b,test} \tag{8.32}
$$

The value of the stator resistance is known from the DC resistance measurement test. Therefore, the rotor resistance referred to the stator is calculated by subtracting the stator resistance from the resistance estimated in the blocked-rotor test. However, separating the calculated blocked-rotor reactance into stator and rotor leakage reactances is not straightforward. However, the performance of an induction motor is slightly affected by the relative distribution of X_b between the stator and the rotor. Therefore, we can assume that the rotor leakage reactance referred to the stator is equal to the stator leakage reactance.

$$
\begin{aligned}
R_2 &= R_b - R_1 \\
X_1 &= X_2 = \frac{1}{2} X_b
\end{aligned} \tag{8.33}
$$

These tests yield enough data to construct the basic equivalent circuit of an asynchronous motor and analyze its performance and efficiency.

Example 8.4 The data below are obtained from tests on a 1 HP, four-pole, 60 Hz three-phase induction motor. Stator windings are Y-connected. Estimate the parameters of the equivalent circuit shown in Figure 8.10 (c).

Measured Variables	DC Resistance Test	No-Load Test	Blocked-rotor Test
Line Voltage (V)	2	208	35
Line Current (A)	0.8	1.5	4
Power (W)	—	50	125 (180)
Frequency (Hz)	—	60	15

Solution

DC Test

Since the stator coils are *Y*-connected, the DC test gives the total resistance of two stator windings connected in series. Stator winding resistance is found as

$$R_1 = \frac{1}{2}\frac{V_{DC}}{I_{DC}} = \frac{1}{2}\frac{2}{0.8} = 1.25\,\Omega$$

Open-Circuit (No-Load) Test

Core loss resistance R_c and magnetizing reactance X_m are estimated using the no-load test data. Phase-to-neutral no-load voltage is $V_o = 208/\sqrt{3} = 120\,\text{V}$.

The current flowing through the resistance is the real (active) component of the no-load current. This active component is responsible for the power measured in the no-load test, which constitutes the total core loss.

$$I_{Re} = \frac{P_o}{3V_o} = \frac{50}{3 \times 120} = 0.139\,A$$

$$R_c = \frac{V_o}{I_{Re}} = \frac{120}{0.139} = 863\,\Omega$$

The current flowing through the magnetizing reactance is the imaginary (reactive) component of the no-load current calculated below:

$$I_{Im} = \sqrt{I_o^2 - I_{Re}^2} = \sqrt{1.5^2 - 0.139^2} = 1.494\,A$$

Magnetizing reactance is computed by applying Ohm's law.

$$X_m = \frac{V_o}{I_{Im}} = \frac{120}{1.494} = 80.4\,\Omega$$

Blocked-Rotor Test

Total stator and rotor winding resistance is calculated using the blocked-rotor test data. The phase-to-neutral voltage applied in this test is $V_b = 35/\sqrt{3} = 20.2\,\text{V}$.

$$Z_b = \frac{V_b}{I_b} = \frac{20.2}{4} = 5.05\,\Omega$$

$$R_b = \frac{P_b}{3I_b^2} = \frac{125}{3 \times 4^2} = 2.6\,\Omega$$

Rotor resistance referred to stator is estimated by subtracting stator resistance from the total resistance.

$$R_2' = R_b - R_1 = 2.6 - 1.25 = 1.35\,\Omega$$

Total winding impedance at lower frequency is calculated from the blocked-rotor test data.

$$Z_b = \frac{U_b}{\sqrt{3}I_b} = \frac{35}{\sqrt{3} \times 4} = 5.05\,\Omega$$

$$X_{b,test} = \sqrt{Z_b^2 - R_b^2} = \sqrt{5.05^2 - 2.6^2} = 4.33\,\Omega$$

Total leakage reactance at the rated frequency is calculated

$$X_b = \frac{f_r}{f_{test}} X_{b,test} = \frac{60}{15} 4.33 = 17.32\,\Omega$$

We will assume that the total leakage reactance is split equally between the stator and the rotor.

$$X_1 = X_2 = \frac{1}{2} X_b = \frac{17.32}{2} = 8.66\,\Omega$$

This concludes estimation of the equivalent circuit parameters.

8.4.4 Performance Analysis

Performance analysis of induction motors typically evaluates stator current, torque, power, power factor, and efficiency during steady-state operation at rated voltage and frequency. Any circuit model from Figure 8.10 may be used in conjunction with the power flow illustrated in Figure 8.11.

When the motor operates without a mechanical load, the slip is close to zero, resulting in an extremely high equivalent rotor resistance R_r/s. Thus, a small current flows through the rotor wires.

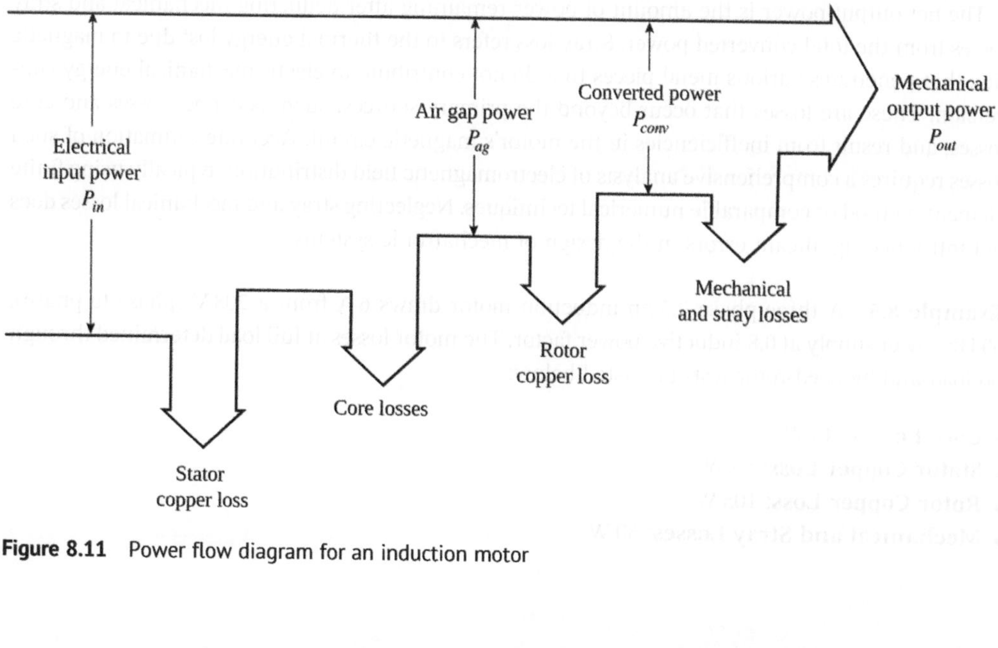

Figure 8.11 Power flow diagram for an induction motor

The electromagnetic torque developed on the rotor is just enough to overcome the total mechanical losses consisting of the viscous friction in the bearings and ventilation.

When the motor shaft rotates a mechanical load, the equivalent rotor resistance R_r/s decreases with an increasing slip. This results in an increase in rotor current. When the torque equals the sum of the load torque and rotational losses, the motor continues to run at steady-state speed, which is less than the no-load speed.

Input power can be calculated directly if the line voltage (U_s), input current (I_s), and the power factor $\cos\varphi$ are known.

$$P_{in} = 3V_s I_s \cos\varphi = \sqrt{3} U_s I_s \cos\varphi \tag{8.34}$$

Manufacturers usually list electrical parameters at rated mechanical power among motor specification data. At other load conditions, input current, power, and power factor are determined by solving the equivalent circuit and energy balance equations.

Copper losses are joule power converted to heat on the stator and rotor winding resistances.

$$\begin{aligned} P_{Cu,s} &= 3I_s^2 R_s \\ P_{Cu,r} &= 3I_r^2 R_r \end{aligned} \tag{8.35}$$

The power converted in the air gap is equal to the input power minus the stator copper loss and core loss.

$$P_{ag} = P_{in} - P_{Cu,s} - P_{core} = 3I_r^2 \left(\frac{R_r}{s}\right) \tag{8.36}$$

A small part of the electric power transferred to the rotor is converted to heat due to the actual rotor resistance, and a bigger part is converted to mechanical power. Earlier in the development of the circuit model, we decomposed the equivalent rotor resistance into two components in equation 8.28. Power converted to mechanical form is obtained by subtracting the rotor copper loss from the air gap power.

$$P_{conv} = P_{ag} - P_{Cu,r} = 3I_r^2 \left(\frac{R_r}{s}\right) - 3I_r^2 R_r = 3I_r^2 R_r \left(\frac{1-s}{s}\right) \tag{8.37}$$

The net output power is the amount of power remaining after deducting mechanical and stray losses from the total converted power. Stray loss refers to the thermal energy lost due to magnetic flux that penetrates various metal pieces that do not contribute to electromechanical energy conversion. These are losses that occur beyond the primary sources, such as copper losses and core losses, and result from inefficiencies in the motor's magnetic circuit. Accurate estimation of such losses requires a comprehensive analysis of electromagnetic field distribution, typically using finite element method or comparable numerical techniques. Neglecting stray and mechanical losses does not introduce significant errors in the design of mechatronic systems.

Example 8.5 A three-phase 2.5 hp induction motor draws 6 A from a 208 V (phase-to-phase), 60 Hz power supply at 0.8 inductive power factor. The motor losses at full load determined through no-load and blocked-rotor tests are listed below:

- **Core Losses**: 45 W
- **Stator Copper Loss**: 85 W
- **Rotor Copper Loss**: 105 W
- **Mechanical and Stray Losses**: 50 W

Find the following:

a) Air-gap power
b) Slip and speed in RPM
c) Power converted to mechanical form
d) Output power
e) Efficiency

Solution

a) The input power is

$$P_{in} = \sqrt{3}UI\cos\varphi = 1.732 \times 208 \times 6 \times 0.8 = 1{,}729\ \text{W}$$

The power transferred to the air gap is determined by subtracting the stator copper loss and core loss from the input power.

$$P_{ag} = P_{in} - P_{cu,s} - P_{core} = 1{,}729 - 85 - 45 = 1{,}599\ \text{W}$$

b) According to equation 8.36, the power transferred to the air gap equals the rotor copper loss divided by slip. Therefore, slip is the ratio of rotor copper loss to the air gap power.

$$P_{ag} = 3I_r^2\left(\frac{R_r}{s}\right) = \frac{P_{cu,r}}{s} \Rightarrow s = \frac{P_{cu,r}}{P_{ag}} = \frac{105}{1599} = 0.065$$

$$n = (1-s)n_s = (1-s)\frac{120f}{4} = (1-0.065) \times \frac{120 \times 60}{4} = 1{,}683\ \text{RPM}$$

c) Electric power converted to mechanical form is

$$P_{conv} = P_{ag} - P_{cu,r} = 1{,}599 - 105 = 1{,}494\ \text{W}$$

d) Mechanical power available on the motor shaft is determined by subtracting the rotational losses from the converted power.

$$P_{out} = P_{conv} - P_{rot} = 1{,}494 - 50 = 1{,}444\ \text{W}$$

e) Efficiency is the ratio of the output power to the input power in percentage.

$$\eta = \frac{P_{out}}{P_{in}}100 = \frac{1{,}444}{1{,}729} \times 100 = 83.5\%$$

8.4.5 Torque–Speed Characteristics

The relationship between torque and speed is an important characteristic of the induction motor for analysis and design considerations. The torque developed in the air gap is found by dividing the converted power, as given in equation 8.37, by the rotor angular velocity.

$$T = \frac{P_{out}}{\omega} = \frac{3I_r^2R_r(1-s)}{\omega s} \tag{8.38}$$

Angular velocity of the rotor is expressed in terms of the synchronous speed and slip values.

$$\omega = (1 - s)\omega_s \tag{8.39}$$

It is more convenient to express the torque in terms of the stator voltage and circuit parameters using the circuit diagram shown in Figure 8.10 (c). By substituting the rotor current in equation 8.38 with its expression derived from circuit equations and angular velocity in equation 8.39, we obtain the torque equation below:

$$T = \frac{3 \cdot V^2}{\omega_s \cdot \left[(R_s + R_r/s)^2 + (X_s + X_r)^2\right]} \frac{R_r}{s} \tag{8.40}$$

Figure 8.12 shows the torque–speed curve of a three-phase induction motor analyzed in Example 8.6. Significant operating conditions are indicated on the plot.

The National Electrical Manufacturers Association (NEMA) has published standards on induction motor parameters and characteristics. Starting torque corresponds to breakaway torque, which occurs when the rotor is blocked while the stator is energized. NEMA defines the locked-rotor torque as "the minimum torque developed by the motor at rest for all angular positions of the rotor, with rated voltage applied at rated frequency" (NEMA, 2014).

As the rotor accelerates, torque reaches a maximum value referred to as pullout or breakdown torque. The torque developed by the motor during acceleration from rest to the speed at which the breakdown torque occurs is called pull-up torque. Breakdown torque corresponds to the maximum mechanical load that an induction machine can handle (Chapman, 1991).

The rated torque of an induction machine refers to the torque it is designed to produce under rated operating conditions, such as rated voltage and frequency. The concepts of "starting torque" (also known as locked-rotor torque), breakdown torque, and pull-up torque are significant when considering the transition between standstill and full-speed operation. However, the rated torque is typically associated with the steady-state operating condition at the motor's rated load, where it delivers maximum efficiency and reliability. Standards such as those by NEMA provide guidelines to define and measure these parameters. On the torque–speed characteristic, the point C corresponds to a typical rated torque. At no-load condition, the speed approaches the synchronous speed n_s. Between points C and D, the torque–speed characteristic is approximately linear.

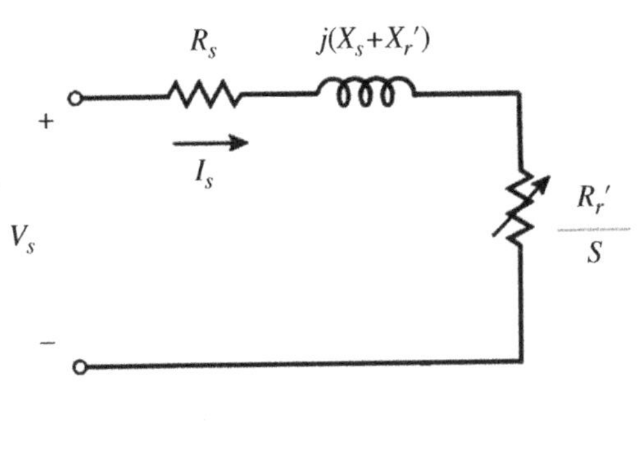

Figure 8.12 Equivalent circuit for torque calculation

Example 8.6 At steady-state operation, a 1-HP, four-pole, three-phase induction motor turns at 1,730 RPM rotor speed when supplied from 208 V, 60 Hz line voltage. Given the following test results, plot the torque–speed curve using a MATLAB© script and the simplified equivalent circuit in Figure 8.12.

Measured Variables	DC Resistance Test	Blocked-rotor Test
Line Voltage (V)	4	35
Current (A)	1.6	4
Total Power (W)	—	170
Frequency (Hz)	—	15

MATLAB® Code to Plot the Torque-Speed Curve of a Three-Phase Induction Motor.

```
% This live script computes the per-phase circuit parameters and plot
torque-speed curve of a 3-phase induction motor.
% The computations are based on the circuit model above.

% Rated specifications:
P=1; % Rated output power [hp]
U=208; % Rated stator line voltage
p=4; % Number of poles
f=60; % Rated supply frequency [Hz]

% DC test
Vdc=4;% DC voltage [V]
Idc=1.6;% DC current [A]

% Blocked-rotor test
U_b=35; % Line-to-line voltage [V]
I_b=4; % Line current [A]
P_b=170; % Total input power with blocked rotor [W]
f_b=15; % Supply frequency at blocked-rotor test [Hz]

%Estimation of Winding resistances and leakage reactances based on the
test data:
V_b = U_b/sqrt(3);              % Blocked-rotor ine voltage converted to
                                  phase voltage
Rs=(Vdc/Idc)/2;                 % Per phase stator resistance measured by
                                  DC test
Z_b=V_b/I_b;                    % Total impedance seen from stator
                                  terminals
R_b=P_b/(3*I_b^2);              % Total resistance
X_total=sqrt(Z_b^2-R_b^2);      % Total leakage reactance of stator and
                                  rotor
% at power rated frequency
Rr=R_b-R1;                  % Rotor resistance referred to stator
Xs=(f/f_b)*X_total/2;       % Stator leakage reactance assuming
                              symmetrical design
Xr = Xs;                    % Rotor leakage reactance assuming symmetrical
                              design
```

```
% Display results
fprintf('Computed Circuit Parameters: \n');
fprintf('Stator resistance (Rs): %.4f Ohms\n', Rs);
fprintf('Rotor resistance (Rr): %.4f Ohms\n', Rr);
fprintf('Stator leakage reactance (Xs): %.4f Ohms\n', Xs);
fprintf('Rotor leakage reactance (Xr): %.4f Ohms\n', Xr);

% Derived values
ws=(120*f/p)*2*pi/60;          % Synchronous speed in rad/s
s = linspace(1, 0.01,100);   % Slip range from 1% to 100%
V=U/sqrt(3);
w=(1-s).*ws;
% Torque calculation
torque = (3 * V^2 .* (Rr ./ s)) ./ ...
  (ws*((Rs + (Rr ./ s)).^2 + (Xs + Xr)^2));
%n_sync=120*f/p;  % Assume synchronous speed in RPM for plotting
n_sync=120*f/p;
n = (1 - s) * n_sync;
% Plotting the torque-speed curve:
plot(n, torque, 'k', 'LineWidth', 2)
xlabel('n (rpm)')
ylabel('Torque (Nm)')
title('Torque vs Rotor Speed')
grid on
```

Based on the test data provided, this script estimates the per-phase circuit parameters for the simplified circuit model in Figure 8.12:

Stator Resistance: 1.25 Ω
Stator Leakage Reactance: 7.2 Ω
Rotor Resistance: 2.29 Ω
Rotor Leakage Reactance: 7.2 Ω

The torque–speed curve displayed in Figure 8.13 is calculated by this script.

8.4.6 Single-Phase Induction Motors

Single-phase induction motors are widely found in everyday household and light industrial applications. Some of the most common devices that utilize single-phase induction motors include domestic appliances, ceiling and exhaust fans, HVAC units, pumps, small compressors, food processors, vacuum cleaners, garage door openers, and corded tools like drills, sanders, and saws.

Single-phase fractional HP induction motors dominate such applications because they are simple, reliable, and can run directly from single-phase utility power. They are durable, efficient, low-cost, and low-maintenance machines.

Single-phase induction motors generally have a squirrel-cage rotor, resembling those in three-phase. The stator windings are distributed in slots to produce an approximately sinusoidal space distribution of the MMF (Fitzgerald et al., 1992).

If the single-phase stator current is a sinusoidal function of time, the MMF is a function of space and time described by the following expression:

$$\mathcal{F} = \mathcal{F}_m \cos(\omega t) \cdot \cos(\theta) \tag{8.41}$$

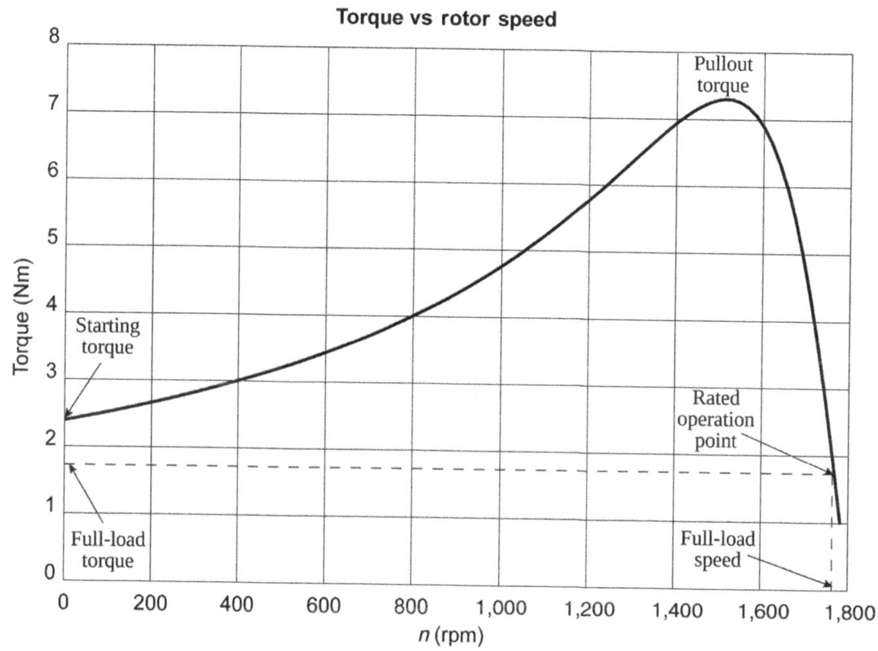

Figure 8.13 Torque–speed curve of the motor described in Example 8.6

The product of two cosine functions is identical to the sum of two cosine functions representing two waves travelling in opposite directions.

$$\mathcal{F} = \mathcal{F}_m \cos(\omega t) \cdot \cos(\theta) = \frac{1}{2}[\mathcal{F}_m \cos(\theta - \omega t) + \mathcal{F}_m \cos(\theta + \omega t)] \tag{8.42}$$

Because the torque corresponding to each of these components is in opposite direction, no starting torque can be produced. When the motor starts using an auxiliary method, its torque–speed characteristics in either direction resemble those of a polyphase motor.

8.4.7 Split-Phase Motors

Split-phase induction motors use an auxiliary stator winding along with the main winding to generate an RMF required for starting torque. The main and auxiliary windings are positioned on the stator at a 90° geometric angle as illustrated in Figure 8.14. Main and auxiliary windings generate magnetic fluxes that are perpendicular to each other. The main winding has low resistance for reducing the copper loss and high inductance to generate the magnetic flux needed for adequate torque during steady-state operation. The auxiliary winding is designed with higher resistance and lower inductance compared to the main winding resistance and high inductance to create a phase difference between the current flowing through the main winding. This phase difference generates sufficient starting torque to initiate motor rotation. A split-phase motor acts like a two-phase induction motor with unbalanced phase currents.

When the motor is powered, both windings are connected to the source. The phase difference (about 25°–30°) between the currents in the two windings creates an RMF, initiating motor

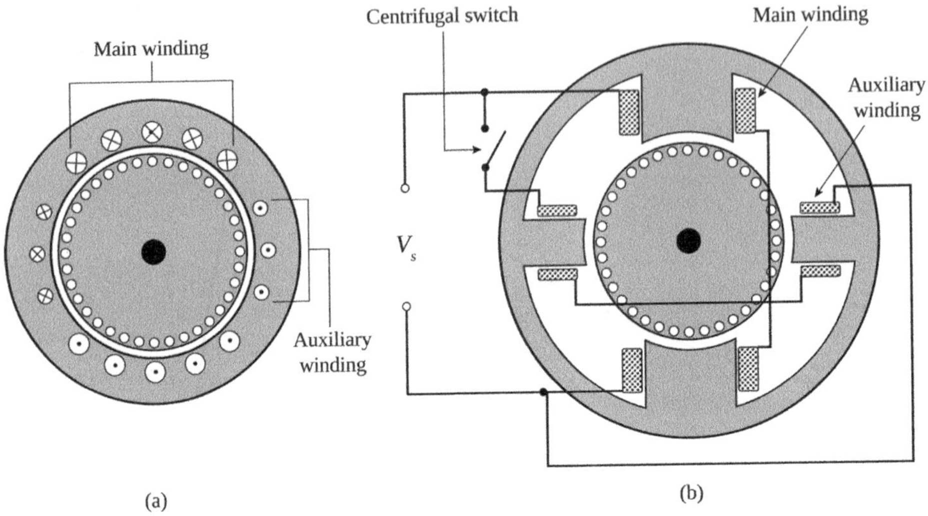

Figure 8.14 Induction motor with auxiliary winding (a: Round stator; b: Salient pole stator)

rotation. Once the rotor reaches sufficient speed, a centrifugal switch in Figure 8.14 (b) disconnects the auxiliary winding from the supply circuit, allowing the motor to operate as a standard single-phase induction motor. These motors are suitable for applications requiring moderate starting torque and are commonly found in household appliances such as fans, washing machines, and compressors.

Split-phase induction motors provide moderate starting torque in the order of 1.5–2 times full-load torque. They draw high starting current and are less efficient than capacitor-start motors, which will be explained in the next paragraph. To reverse the rotation direction, the connections of the main or auxiliary winding should be swapped while the power is off.

Split-phase motors are low cost, and their simple design requires less maintenance. Because at the start the torque is lower despite high initial current, they are suitable for light-duty tasks.

8.4.8 Capacitor-Start Motors

Capacitor-start induction motors are a type of single-phase induction motor designed to provide high starting torque. Structure of these motors resembles split-phase motors. Figure 8.15 shows the connection and phasor diagrams. A capacitor is connected in series with the auxiliary winding during the start. The capacitor introduces a phase shift, creating a stronger and more effective RMF compared to split-phase motors. Typically, a 250-μF electrolytic capacitor is placed on the stator.

When the motor starts, the capacitor allows the auxiliary winding to produce a current that is out of phase with the current in the main winding. This phase shift generates powerful starting torque, making capacitor-start motors suitable for applications requiring high initial power, such as air compressors, pumps, and other heavy-duty equipment.

Once the motor reaches a predetermined speed, a centrifugal switch or an electronic relay disconnects the auxiliary winding and the capacitor from the circuit. The motor then continues to operate as a standard single-phase induction motor, relying solely on the main winding for its operation. This design ensures efficient performance during continuous operation while still addressing the need for robust starting capacity.

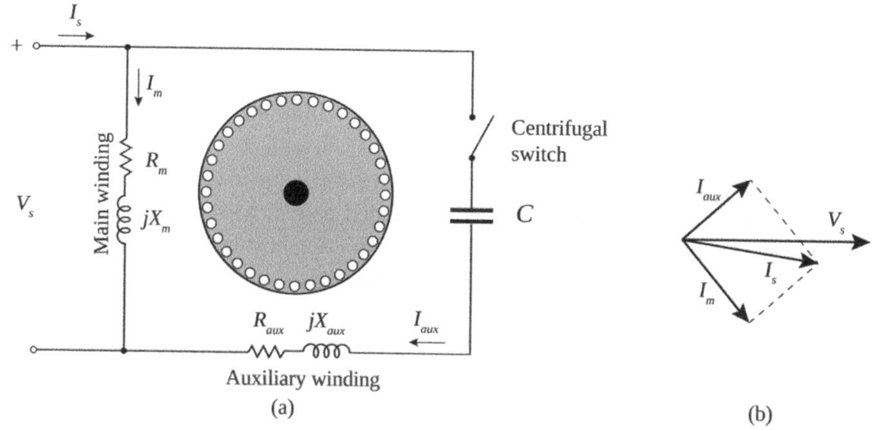

Figure 8.15 Capacitor-start induction motor (a: Equivalent circuit; b: Phasor diagram)

8.4.9 Shaded-Pole Motors

Shaded-pole induction motor is a simple, rugged, and inexpensive type of single-phase AC motor commonly used in low-power applications like fans, blowers, and small appliances, where cost-effectiveness and straightforward design are prioritized.

A typical shaded-pole motor has salient poles on the stator, each with a portion wrapped by a copper ring called a *shading coil*. The rotor is typically a squirrel-cage type. A shading coil, which is a conductive ring placed around a portion of each pole in the stator, delays the magnetic flux in the shaded portion of the pole, creating a weak RMF. This causes the rotor to start turning—from the unshaded side toward the shaded side. Figure 8.16 illustrates two common designs of shaded-pole motors.

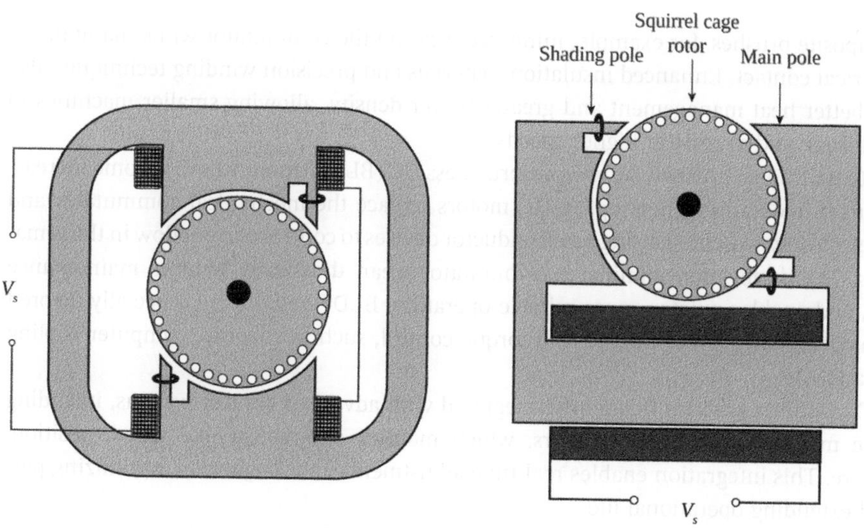

Figure 8.16 Shaded-pole induction motor

Due to their uncomplicated construction, shaded-pole motors lack auxiliary windings or capacitors, which simplifies their design and reduces manufacturing complexity. However, this simplicity comes at the cost of lower starting torque and efficiency compared to split-phase or capacitor-start motors. These motors are typically used in applications requiring minimal power, such as fans, small blowers, and office equipment like desk fans.

Shaded-pole motors typically have a power rating of less than 1/4 HP due to their lower torque at start and significantly lower efficiency (about 25%). While shaded-pole motors are not suitable for heavy-duty operations, their small size, durability, and low production cost make them ideal for a wide range of lightweight tasks in domestic and commercial settings.

8.5 DC Machines

DC machines are the oldest types of electric machines, originating in the mid-nineteenth century. Early power generation and transmission systems utilized DC technology. During this period, electricity was generated by DC generators, and DC motors were extensively used in industry. In the late 1880s, as AC transmission and distribution networks emerged, induction motors began to replace DC machines due to their simpler design, lower cost, and reduced maintenance requirements. Nevertheless, DC machines continued to be employed in industrial servomechanisms and transportation systems (including electric trains, subways, and streetcars) due to their easier control compared to AC machines and their superior torque characteristics for these specific applications.

Modern DC machines have evolved considerably from their early counterparts, incorporating new materials, advanced manufacturing techniques, and improved electronic controls to enhance efficiency and reliability. While the fundamental principles of operation remain rooted in the classic stator–rotor and commutator–brush arrangement, contemporary designs often prioritize lower maintenance, higher efficiency, and adaptability for automated environments.

One significant development in modern DC machines is the use of strong PMs, improved brush materials, and sealed bearings, which extend service life and reduce the need for routine maintenance.

Carbon-composite brushes, for example, minimize wear on the commutator while maintaining excellent electrical contact. Enhanced insulation materials and precision winding techniques also contribute to better heat management and greater power density, allowing smaller machines to deliver more torque and operate at higher speeds.

In addition to traditional brushed DC motors, brushless DC (BLDC) motors have become increasingly important in modern engineering. BLDC motors replace the mechanical commutator and brushes with electronic switching, using semiconductor devices to control current flow in the armature windings. This innovation eliminates commutator wear, drastically reduces maintenance requirements, and enables quieter, more reliable operation. BLDC motors are especially favored in applications demanding precise speed and torque control, such as robotics, computer cooling fans, drones, and EVs.

Modern DC machines are also frequently integrated with advanced control systems, including programmable microcontrollers and sensors, which monitor parameters like speed, position, and temperature. This integration enables real-time adjustments and diagnostics, optimizing performance and extending operational life.

Contemporary uses for DC machines span a broad spectrum, from compact motors powering household appliances, cordless tools, and robots to large-scale industrial drives and various types of EVs.

8.5.1 Structure of Conventional DC Machines

Conventional DC machines, like other electrical machinery previously described, consist of stator and a rotor. The key differences are the stator's function and the presence of a commutator–brush assembly. Stator is responsible for generating the magnetic field necessary to induce voltage and torque in the rotor, commonly known as the armature. In traditional designs, either windings placed on salient stator poles or strong PMs produce the magnetic field. The rotor windings are uniformly distributed in slots around its circumference. When the current flows through these windings, electromagnetic forces arise, inducing torque and causing the rotor shaft to spin.

A critical element of DC machines is the commutator–brush assembly, which serves as a rotating electrical switch. The commutator consists of copper segments, arranged side by side around the rotor shaft, forming a cylindrical surface. To insulate these segments, thin layers of mica or another durable insulating material are inserted between each copper piece. The entire assembly is tightly fitted onto the rotor shaft and machined to a perfectly smooth, round surface. Paired with the commutator, graphite brushes are pressed against its surface by springs. This contact allows electric current to flow into the rotor winding while simultaneously reversing the direction of the current depending on the angular position of the rotor windings. Such synchronized switching is essential for maintaining unidirectional torque in the rotating armature. The robust yet precise construction of the commutator is critical, as it must withstand high rotational friction and repetitive sparking.

The magnetic core is typically constructed from laminated electric steel to minimize energy losses due to hysteresis and eddy currents. Larger machines may feature durable cast iron outer shells to provide additional structural integrity.

8.5.2 Energy Conversion in a Basic DC Machine

The structure of a basic electric machine is illustrated in Figure 8.17. To simplify, we consider only a one-turn coil placed on the rotor. In practical DC machines, a number of coils are placed in slots distributed around the rotor. Windings placed on the poles create the magnetic flux; thus, we will call them *field windings*.

The rotor windings of the DC motor are called *armature windings* because the current flowing through its wires produces the electromagnetic torque to turn the shaft. In this simplified structure, armature winding forms a single loop placed diametrically on the rotor.

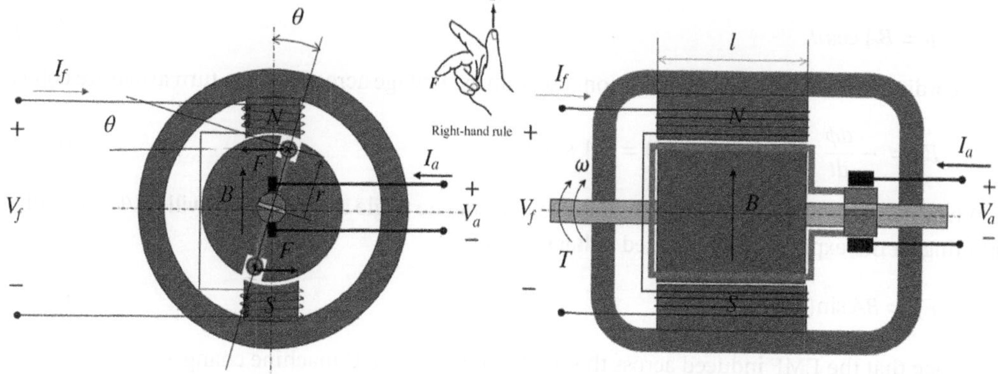

Figure 8.17 Structure and operation principle of a basic DC motor

The moving-armature windings are connected to the stationary external power supply by a brush–commutator combination. Commutator, also known as a collector, is made from conductor (usually brass) plates equally distributed around a cylindrical surface. A pair of graphite pieces called *brushes* press on the collector by a spring force sufficient to ensure good conduction without damaging the brass plates. They are called brushes because in the early days of DC machines, they were made from a bunch of wires forming a copper brush.

The core of the stator and rotor is made from high-permeability ferromagnetic material, usually electric steel. The purpose of the magnetic core is to reduce the total reluctance of the magnetic circuit such that the armature coils move in the maximum flux density in the air gap. In structural calculations, the reluctance of the core may be neglected and the whole magnetic energy can be assumed to be stored in the air gap. In practical electric machines, the time-varying magnetic flux in the core causes hysteresis and eddy current losses referred to as *core losses*. To reduce the core losses, in general, the rotor and the poles are made from laminated electrical steel, while the cylindrical external part of the core may be cast iron.

As the current flows through the armature winding, the magnetic field induces a force in the opposite direction on each side of the loop. Since the wires are always perpendicular to the magnetic flux, the force developed is

$$F = BlI_a \tag{8.43}$$

The direction of the force can be found by applying the right-hand rule shown on the figure above. The direction of the current in each side of the loop is indicated by a cross (×) and a dot (·) in the circle representing the cross section of the wire. Cross sign means that the current flows toward the figure plane, and a dot means that the current flow is from the plane to the observer. The pair of electromagnetic forces creates the torque that turns the shaft. If the radius of the rotor is r and the armature current is I_a, the torque can be calculated with the equation

$$T = 2r(B\ell I_a)\cos\theta \tag{8.44}$$

where ℓ is the length of the coil wires immersed in the magnetic field, and B is the flux density in the air gap. Since the magnetic field is constant in one direction, the force remains constant. The torque is proportional to cosθ where θ is the angle between the plane of armature coil and the magnetic flux.

As the rotor turns, a voltage defined by Faraday's law is induced across the terminals of the armature coil. The flux encircled by the armature coil is found by multiplying the air gap flux density B, the area of the coil A, and the cosine of the angle θ.

$$\phi = BA\cos\theta \tag{8.45}$$

According to Faraday's law of induction, the induced voltage across the one-turn armature coil is

$$E_a = -\frac{d\phi}{dt} = -BA(-\sin\theta) = BA\sin\theta \tag{8.46}$$

When the rotor turns at constant angular velocity ω, the angle at any instant will be $\theta = \omega t$. Substituting in the expression of induced voltage, we find

$$E_a = BA\sin(\omega t) \tag{8.47}$$

We see that the EMF induced across the rotor winding of a DC machine changes as a sinusoidal function of time.

The important role of the commutator in a DC machine is to reverse the coil terminals when the induced voltage becomes zero. This way, the current flowing on the coil wires is unidirectional, always creating forces in one direction so that the rotor can continue to turn. The commutator works as a mechanical counterpart of the electronic converter (rectifier/inverter).

The structure of practical DC machines is more complicated than the basic DC machine we have examined. The main reason is the time variation of the induced voltage and torque. There are periodic fluctuations of the induced voltage, a certain number of armature coils are distributed around the rotor, and the terminals are connected to the same number of collector plates turning with the shaft. This results in less torque variations, thus smoother operation of the machine with less mechanical vibration and noise. In addition, to increase the induced voltage and torque, each coil contains many turns.

DC motors are more complex than induction motors because they require a commutator to conduct high current to the rotating windings.

8.5.3 DC Machine Equivalent Circuit

The electromechanical behavior of a DC machine can be analyzed using the circuit model in Figure 8.18. In this equivalent circuit, the field winding is represented by its resistance R and inductance L and supplied by the DC source named v_f. Armature winding is similarly represented by the resistance R_a and inductance L_a. In addition, the voltage source e_a represents the internal velocity voltage induced on the armature windings as the rotor turns. The circuit elements corresponding to the armature winding are placed in a circle to emphasize that they are on the rotor. The external supply voltage v_a is applied to the armature winding via brushes touching the collector, represented by two small black rectangles diametrically placed on the circle. The field winding inductance is depicted perpendicular to the armature inductance, illustrating that at any given moment, the armature coil orthogonal to the magnetic field is energized due to the placement of the brushes.

The mechanical load may be directly coupled to the motor shaft or driven through a motion transmission mechanism. The mechanical side of the model is described by the total mechanical parameters referred to the motor shaft, J representing the moment of inertia, β representing the viscous friction constant, and T_L representing the torque of the mechanical load. The motor produces the torque T_e due to the interaction of the magnetic field and the current flowing through the armature windings. Note that the equivalent circuit of the motor is an electric circuit with moving elements.

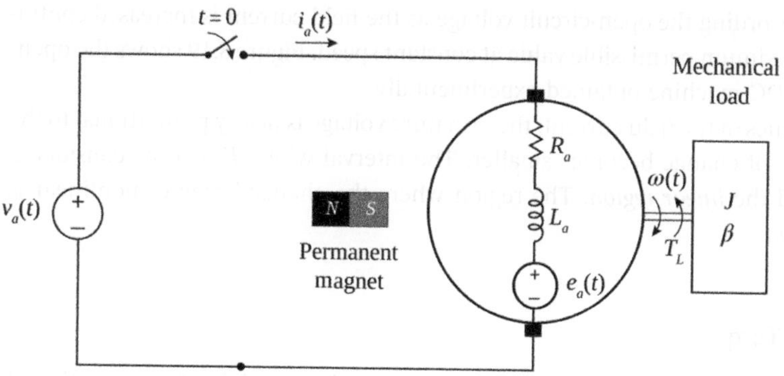

Figure 8.18 Equivalent circuit of a DC machine

8.5.4 Induced EMF

The velocity voltage, also known as the induced EMF, is the voltage generated across the armature winding of a DC machine due to its rotation in a magnetic field. This induced voltage arises because the conductors in the armature cut through magnetic flux as the rotor turns, thus generating an EMF according to Faraday's law of electromagnetic induction.

As the armature of a DC machine rotates, the conductors placed on the rotor move through the magnetic field created by the field windings. This motion results in a time-varying magnetic flux linkage with the armature coils, which induces the EMF known as velocity voltage. Due to the commutator–brush assembly, the velocity voltage is unipolar but varies periodically in time. The following expression gives the magnitude of the velocity voltage:

$$e_a = \frac{pZ}{2\pi \cdot m}\phi\omega \tag{8.48}$$

In this formula, the parameters are defined as below:

p = Number of poles
Φ = Flux per pole (in Weber)
Z = Total number of armature conductors
ω = Armature speed in radians per second
m = Number of parallel branches in the armature winding

The magnitude of the velocity voltage is directly proportional to the rotational speed of the armature and a nonlinear function of the magnetic flux generated by the field winding, due to the nonlinear characteristics of the electric steel commonly used in the cores of DC machines. Most DC machines are typically designed to operate in the linear region where the velocity voltage can be assumed to be proportional to the field winding current and rotational speed.

$$e_a = k i_f \omega \tag{8.49}$$

It is often easier to estimate velocity voltage using external variables or by performing an open-circuit test where the DC machine operates as a generator at a constant speed without supplying any load. A voltmeter with high internal resistance is connected across the armature terminals to measure the open-circuit (no-load) voltage. Since the current drawn is negligible, the measured voltage approximates the actual induced EMF, which is the velocity voltage. The open circuit characteristic is obtained by recording the open-circuit voltage as the field current is increased continuously from zero to the maximum permissible value at constant speed. Figure 8.19 shows the open-circuit characteristic of a DC machine obtained experimentally.

Note that for smaller values of the field current, the armature voltage is nearly proportional to the field current, and the rate of change becomes smaller. The interval where E_a can be considered proportional to i_f is called the *linear region*. The region where the change becomes nonlinear is called the *saturation region*.

8.5.5 Electromagnetic Torque

The torque developed on the armature can be found by applying the principle of conservation of energy to the ideal part of the motor. Suppose that the elements that cause electrical, magnetic, and mechanical losses are fictitiously separated as shown in the figure below.

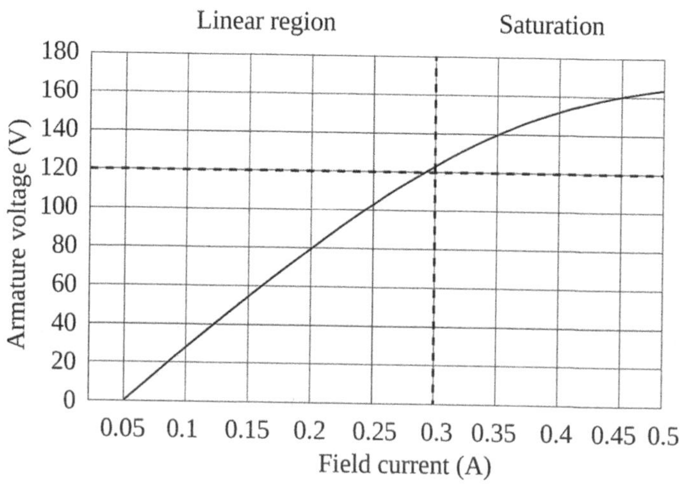

Figure 8.19 Armature voltage versus field current

The power at the electrical and mechanical sides of the ideal energy conversion system are equal.

$$e_a i_a = \omega T_e \tag{8.50}$$

If the machine operates in the linear region, $e_a = k\omega i_f$. Substituting this expression in the equation above, we obtain

$$k\omega i_f i_a = \omega T_e \Longrightarrow T_e = k i_f i_a \tag{8.51}$$

8.5.6 Dynamic Equations of DC Machines

The equivalent circuit of the DC machine depicted in Figure 8.18 consists of two electrical ports and one mechanical port. The dynamic behavior of this electromechanical system is described by differential equations obtained using KVL for the electrical circuits and Newton's law of inertia for the mechanical components. Mechanical load consists of a moment of inertia J, viscous friction β, and a load torque T_L. Mechanical system may include elements that turn at different angular speeds. In the mechanical equation, all parameters are referred to the motor shaft speed.

Field winding loop: $v_f = R_f i_f + L_f \dfrac{di_f}{dt}$

Armature winding loop: $v_a = R_a i_a + L_a \dfrac{di_a}{dt} + e_a$

Mechanical part: $T_e = J \dfrac{d\omega}{dt} + \beta\omega + T_L$

Due to the nonlinear magnetic properties of the iron core, the induced EMF depends nonlinearly on the field current, resulting in differential equations with nonlinear parameters. However, electric machines are typically designed to operate within the linear magnetic region to minimize core losses. The dynamic equations below result from substituting equation 8.49 for e_a and equation 8.51 for T_e, under the assumption that the motor operates in this linear region.

$$\begin{aligned} v_f &= R_f i_f + L_f \frac{di_f}{dt} \\ v_a &= R_a i_a + L_a \frac{di_a}{dt} + k_f i_f \omega \\ k_m i_f i_a &= J \frac{d\omega}{dt} + \beta\omega + T_L \end{aligned} \tag{8.52}$$

8.5.7 Steady-State Performance of DC Motors

Given their importance in mechatronics, we will examine the steady-state performance of DC motors. Steady-state operation refers to the condition where all transient changes have ceased, and the machine functions at a constant speed with stable DC currents and voltages. The primary variables characterizing the steady-state operation of a DC motor under continuous conditions include current, voltage, speed, and torque. Power and energy may be derived from these fundamental parameters. In static equations, we will use upper-case letters to indicate that the current and voltage remain constant.

In steady-state operation, terms with time derivatives become zero since system variables remain constant over time. The following set of algebraic equations derived by excluding the time derivative terms from the state equations below describe the steady-state performance of DC motors.

$$\begin{aligned} &V_f = R_f I_f && \text{(a)} \\ &\omega = \frac{V_a - R_a I_a}{k I_f} && \text{(b)} \\ &T_e = k I_f I_a = T_L + \beta\omega && \text{(c)} \end{aligned} \tag{8.53}$$

The performance characteristics of DC motors are greatly influenced by the interaction of the field (excitation) and armature currents. DC machines are categorized based on the type of excitation (field) circuit supply into separately excited, shunt, series, and compound machines.

Figure 8.20 illustrates the circuit configurations to supply the field winding. In a separately excited motor, the field current comes from an independent source, separate from the armature supply. In shunt motors, the field winding is parallel connected to the armature winding; hence, the same voltage supplies both the field and the armature winding. A series motor has its field winding connected in series with the armature winding. Series and shunt motors have different field winding designs to meet the specific voltage and current requirements. Shunt winding has a smaller conductor size and more turns, while series field winding has a larger conductor cross-sectional area and fewer turns. Compound motors have both shunt and series field windings, one connected in parallel, the other in series with the armature winding.

8.5.7.1 Separately Excited DC (SEDC) Motor

In equation 8.53, substituting (c) in (b) yields the expression of the speed in terms of the armature voltage, field current, and load torque.

$$\omega = \frac{1}{k I_f} V_a - \frac{R_a}{(k I_f)^2} T_e \tag{8.54}$$

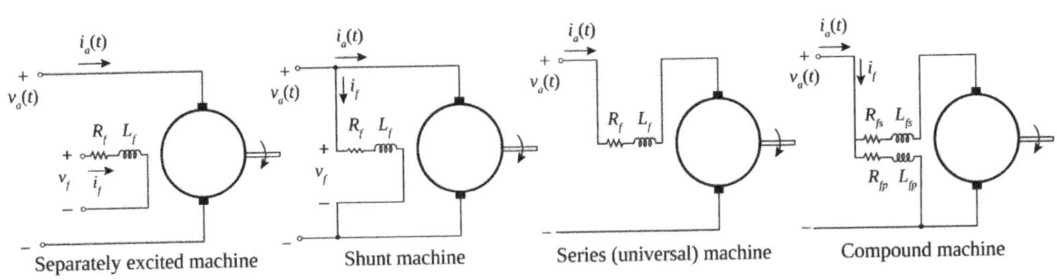

Figure 8.20 DC motor excitation circuit supply configurations

8.5.7.1.1 Torque–Speed Relationship

When the armature voltage and field current are held constant, the speed–torque characteristic of a separately excited DC (SEDC) motor forms a straight line that intersects the vertical axis at the no-load speed, with a decrease in speed proportional to the net torque. The proportionality factor decreases with the square of the field current. Consequently, greater field current leads to improved speed regulation by making speed less sensitive to torque changes.

If the torque increases excessively, the speed decreases more rapidly due to the effect of the armature current on the flux density known as armature reaction. As a result, there is a maximum torque the motor can support. Above that maximum torque, the rotor may slow down too much and even be blocked and permanently damaged.

Example 8.7 Plot the speed–torque curve for the SEDC motor described by the rated parameters below (Figure 8.21).

$$P_m = 175\text{ W}; n = 1{,}800\text{ RPM}; V_a = 120\text{ V}; I_a = 2.8\text{ A}; V_f = 120\text{ V}; I_f = 0.25\text{ A}; R_a = 1.5\,\Omega$$

Solution

We must first convert the rated speed in RPM to radians per second.

$$\omega = \frac{2\pi n}{60} = \frac{2\pi \times 1{,}800}{60} = 188.5\text{ rad}/s$$

The net torque this motor delivers is calculated below by dividing the mechanical power by the rated angular speed in radians per second.

$$T_{net} = \frac{P_m}{\omega} = \frac{175}{188.5} = 0.928\text{ Nm}$$

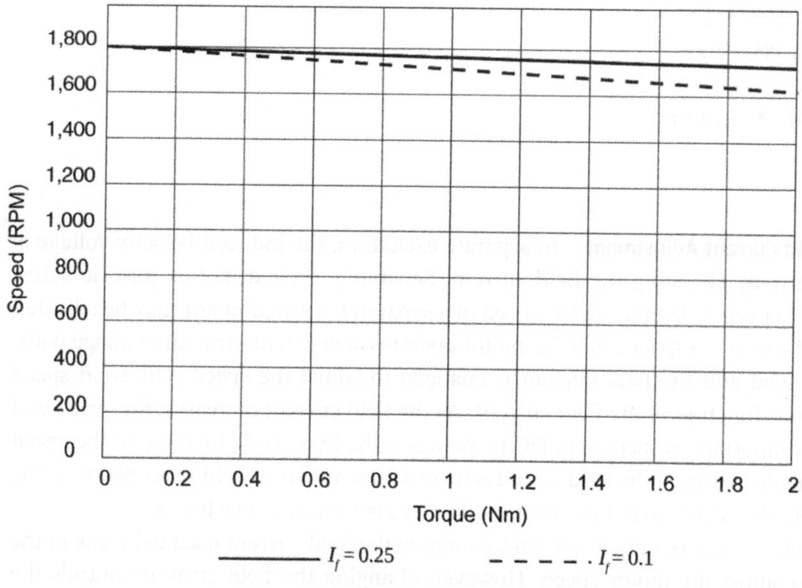

Figure 8.21 Speed–torque characteristic of the separately excited DC motor for two field currents

When the motor operates without any mechanical load, the induced EMF is $E_a \cong V_a$. Assuming the motor operates in the linear region, the velocity voltage constant is found

$$k_f I_f = \frac{E_a}{\omega} \cong \frac{120}{188.5} = 0.637$$

MATLAB® Script to Calculate and Plot the Motor Speed Based on Equation 8.54.

```
% Script to plot speed-torque curve of a separately excited DC motor
clc;
clear;

% Motor parameters (adjust as needed)
Va = 120;          % Armature voltage (V)
Ra = 1.5;          % Armature resistance (Ohm)
Kf = 0.637;          % Back EMF constant (V·s/rad)
% Torque range (Nm)
T = linspace(0, 2, 200); % From 0 to max torque
% Speed calculation (rad/s)
omega = (Va / Kf) - T*Ra / Kf^2;
% Convert to RPM
RPM = omega * 60 / (2 * pi);
% Plotting
figure;
plot(T, RPM, 'b', 'LineWidth', 2);
grid on;
title('Speed-Torque Characteristic of Separately Excited DC Motor');
xlabel('Torque (Nm)');
ylabel('Speed (RPM)');
xlim([0 max(T)]);
ylim([0 max(RPM) * 1.1];
```

Speed Control via Field Current Adjustment In separate excitation, the induced velocity voltage is adjusted independently by changing the field current. Separately excited motors provide better speed control than other types. Therefore, the speed of a separately excited motor may be adjusted by changing the field current. Suppose that the motor operates at constant armature voltage without any mechanical load and the field current is changed to adjust the speed. The shaft speed changes as a hyperbolic function of the field current. As the field current increases, the speed will decrease. One of the important properties of DC machines is the hyperbolic increase in the speed when the field current decreases. If the field circuit is turned off while the machine is operating, the speed tends to be infinite, which may be extremely dangerous for larger machines.

In the early years of electromechanical systems, changing the field current used to be one of the practical methods to adjust the motor speed. However, changing the field current controls the motor speed in a rather limited range since the speed cannot be adjusted to very low values due to the maximum current allowed for the field winding.

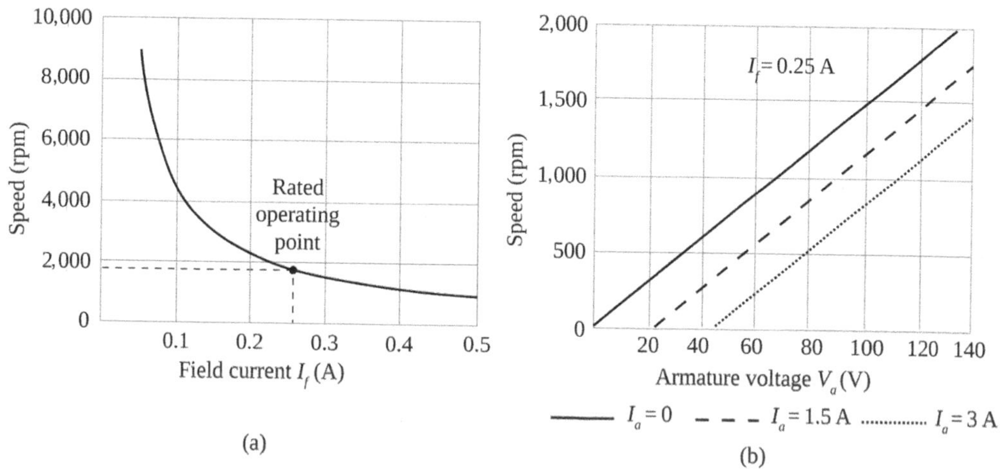

Figure 8.22 Speed control of separately excited motor via field current and armature voltage (a: Speed versus field current; b: Speed versus armature voltage)

Figure 8.22 (a) shows an experimentally obtained relationship between the speed and field current for the DC motor described in Example 8.7. The rated field current at the rated speed is 0.25 A. The data indicates that doubling the field current reduces the speed to 1,000 RPM, while halving the field current nearly doubles the speed. Thus, by adjusting the field current, the motor speed can be modified within a relatively limited range, without exceeding the electrical and mechanical limitations.

Speed Control via Armature Voltage Adjustment Equation 8.54 indicates that the angular speed of an SEDC motor is a linear function of the armature voltage at constant torque and field current. Figure 8.22 (b) displays the speed versus armature voltage curves for the motor specified in Example 8.7, measured at rated field current under various load conditions.

In modern drive systems, controlling the motor speed by changing the armature voltage is the most effective. Currently, high-power DC–DC converters can adjust voltage across a broad range, beginning from zero. Additionally, by using switch-mode converters, the motor speed can be regulated using either analog or digital feedback control systems.

8.5.7.2 Shunt Motor

8.5.7.2.1 Shunt-Excited DC Motor

In a shunt-excited DC motor, the field winding is connected in parallel (or "shunt") with the armature winding. This configuration allows both the armature and the field windings to receive the same supply voltage, but each branch draws its own current. The shunt field winding typically consists of many turns of thin wire, producing a relatively constant magnetic field due to its high resistance and low current.

One of the defining features of shunt motors is their stable speed characteristics. Because the field current is largely independent of the armature current, variations in load cause only small changes in field strength and, therefore, in motor speed. As a result, shunt-excited motors are favored in applications where a nearly constant speed is required over a range of loads.

Speed control in shunt motors can be achieved by adjusting either the field current (using a variable resistor in the field circuit) or the armature voltage, as described earlier for separately excited motors. The shunt configuration provides improved precision and predictability in speed regulation compared to series motors, making them suitable for lathes, fans, conveyors, and other systems demanding steady operation.

The overall efficiency and adaptability of the shunt-excited DC motor have ensured its continued use in many industrial drive systems, where both responsiveness and reliability are essential.

8.5.7.3 Series (Universal) Motor

Series motors are characterized by the series connection of the field winding and armature winding to the power supply. The current flowing through the field winding is the same as that through the armature. Figure 8.23 displays the equivalent circuit with the field winding connected in series to the armature coils.

Since the same current flows through both the field and armature windings, the direction of the magnetic force acting on the armature wires remains the same if the supply source is reversed. Hence, a series motor can theoretically operate on either DC or AC voltage, and it is also known as *universal motor*. In DC operation, the inductors do not affect the steady-state current. However, in AC operation, the winding impedances obtained by a combination of winding resistances and inductances affect the electrical performance. In practice, the machine is designed particularly for either AC or DC operation.

In the equivalent circuit above, there is only one electrical loop. Applying the KVL yields

$$v_a = (R_f + R_a)i_a + e_a \tag{8.55}$$

In the operation region where the magnetic field is linear, the induced velocity voltage is proportional to the armature current and the angular velocity of the rotor. Substituting $e_a = k\omega i_a$, the loop equation above becomes

$$V_a = (R_f + R_a)I_a + k\omega I_a \tag{8.56}$$

The following expression is obtained for the motor current from the electromagnetic torque equation:

$$\omega T_e = e_a I_a = k\omega I_a^2 \Longrightarrow i_a = \sqrt{\frac{T_e}{k}} \tag{8.57}$$

Speed–torque relationship can be derived by substituting the current into equation 8.56.

$$\omega = \frac{V_a - (R_f + R_a)I_a}{kI_a} = \frac{V_a}{\sqrt{kT_e}} - \frac{(R_f + R_a)}{k} \tag{8.58}$$

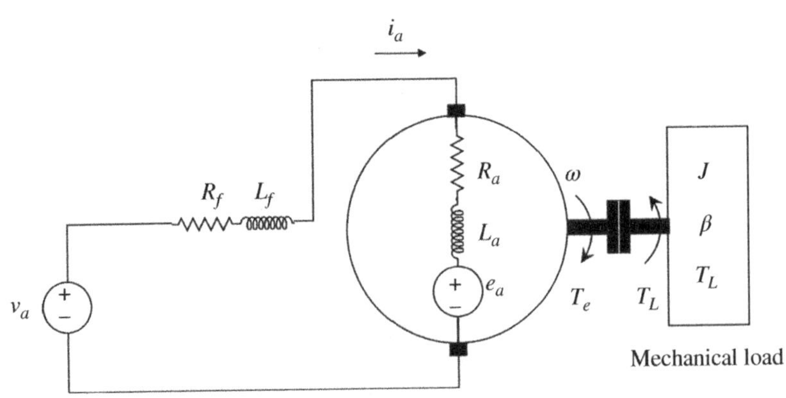

Figure 8.23 Series motor equivalent circuit

Example 8.8 Using MATLAB®, plot the speed–torque curve for the series excited DC motor described by the rated parameters below.

$$P_m = 175\,\text{W}; n = 1{,}800\,\text{RPM}; V_a = 120\,\text{V}; I_a = 2.8\,\text{A}; R_a = 1.5\,\Omega; R_f = 0.5\,\Omega$$

Solution

The rated voltage and speed are the same as in Example 8.7. Because the motor is series-excited, the rated field and armature currents are equal. The induced EMF constant is

$$k_f = \frac{E_a}{\omega I_a} \cong \frac{120}{188.5 \times 2.8} = 0.23\,\text{V} \cdot \frac{\text{s}}{\text{A} \cdot \text{rad}}$$

MATLAB® Script to Plot Speed–Torque Curve of a Series-Excited DC Motor.

```
clc;
clear;

% Motor parameters (adjust as needed)
Va = 120;          % Applied voltage (V)
Ra = 1.5;          % Total resistance (Ohm) = Ra + Rse
Rf = 0.5;
Kf = 0.637;          % Back EMF constant (V·s/rad)

% Torque vector (Nm)
T = linspace(0.1, 10, 500);    % Avoid T=0 to prevent division by zero

% Armature current: T = Kt * Ia^2 => Ia = sqrt(T / Kt)
Ia = sqrt(T / Kf );

% Speed (rad/s): ω = (Va - Ra·Ia) / (Kf·Ia)
omega = (Va - (Ra+Rf ) .* Ia) ./ (Kf .* Ia);

% Convert to RPM
RPM = omega * 60 / (2 * pi);

% Plotting
figure;
plot(T, RPM, 'k', 'LineWidth', 2);
grid on;
title('Speed-Torque Characteristic of DC Series Motor');
xlabel('Torque (Nm)');
ylabel('Speed (RPM)');
%xlim([min(T) max(T)]);
%ylim([0 max(RPM) * 1.1]);
xlim([min(T) max(T)]);
ylim([0 max(RPM) * 1.1]);
```

Advantages and Applications of Series and Universal Motors The series configuration of field and armature windings leads to several essential properties:

- **High Starting Torque**: At startup, both the armature and field winding carry the full-line current, resulting in a powerful magnetic field and, consequently, series motors deliver very high output torque. This makes series DC motors exceptionally capable of starting heavy mechanical loads.

- **Variable Speed**: The speed of a series motor decreases significantly under heavy load and increases at lighter loads. This characteristic is due to the flux produced by the series-connected field winding, which varies with armature current.
- **Simple Construction**: Series motors are structurally simpler and more rugged compared to other DC motor types, allowing for reliable operation even in demanding environments.

Given these characteristics, series motors are naturally suited to applications that require variable speed and, most importantly, substantial starting torque.

Series DC motors have historically been the workhorse of electric traction systems worldwide (Figure 8.24). Their ability to provide immense torque at low speeds is ideal for initiating the movement of massive train cars from a standstill. Even in contemporary systems where AC motors are gaining ground, many older subway systems and trams still use DC series motors because of their proven reliability and performance.

City buses and trolley systems powered by overhead lines or battery banks often use series motors for propulsion, especially in models designed before the widespread adoption of AC motor drives.

Although smaller and lighter EVs increasingly use BLDC motors for efficiency, some models of electric scooters and bicycles still employ series motors when high starting torque and cost-effectiveness are priorities.

Automobile starter motors are perhaps the quintessential example of a series DC motor application. When the ignition key is turned, the starter motor must overcome the inertia of the engine and crank it to a sufficient speed for starting. The burst of high current delivered to the series motor produces the required torque instantaneously.

Modern mechatronic systems sometimes employ hybrid solutions, leveraging the strengths of series motors for initial motion and switching to other motors for sustained or high-speed operation. For example, in EVs, a DC series motor may be used for vehicle launch and hill-climbing, with

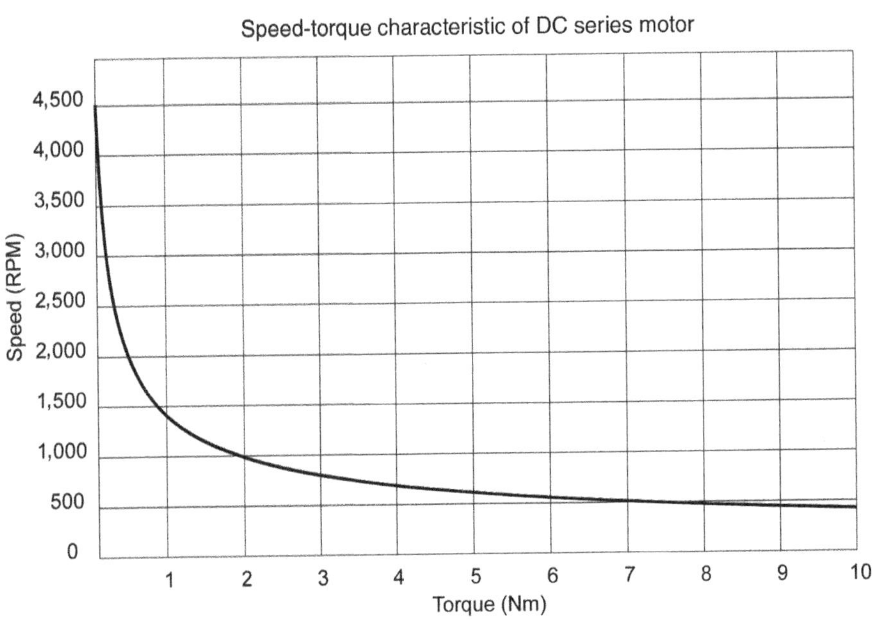

Figure 8.24 Speed–torque characteristic of DC series motor

control transferred to an efficient brushless motor during cruising. Such hybridization maximizes performance, efficiency, and longevity.

Electric winches and hoists, often found on off-road vehicles, tow trucks, and in industrial settings, employ series motors for similar reasons. Lifting or pulling heavy loads from a standstill demands a motor that can deliver high torque on demand, a role perfectly fit for the DC series motor.

Cranes, elevators, and conveyor belts in factories and warehouses utilize series motors for the rapid and reliable movement of materials. In these scenarios, the motor must accelerate the load quickly and operate at variable speeds depending on task requirements.

While modern robotics frequently use BLDC motors and stepper motors for precision, heavy-duty industrial robots, such as those used for lifting or welding, may integrate series DC motors for their high force output and resistance to mechanical stress.

Many handheld appliances, including vacuum cleaners, blenders, electric drills, and circular saws, use series motors. In cordless tools, the motor operates on a battery. Corded tools employ the version of series motors referred to as universal motor designed to operate on AC. The portability and high torque of these motors suit the variable loads and demands of power tools.

Sewing machines require motors capable of rapid acceleration and deceleration, especially when transitioning between different fabric thicknesses. Series and universal motors provide the necessary torque and speed control for these applications.

8.5.7.4 Compound Excitation

Compound DC motors combine the features of both series and shunt DC motors, integrating two sets of field windings: one connected in series with the armature, and the other in parallel (shunt) with it. This configuration allows compound motors to achieve a balance between the high starting torque of series motors and the speed regulation of shunt motors, making them uniquely versatile for a wide range of applications.

There are two main types of compound motors: cumulative-compound and differential-compound. In cumulative-compound motors, the series and shunt field windings assist each other, resulting in greater starting torque and improved load-handling capabilities. This makes them especially suitable for applications like rolling mills, presses, and elevators, where both consistent speed under varying loads and powerful torque surges are required. By contrast, differential-compound motors have field windings that oppose each other, providing exceptional speed stability but at the cost of reduced starting torque. Such motors are less common, but they find use in specific cases where precise speed control is important.

One of the main advantages of compound DC motors is their adaptability. By adjusting the relative strengths of the series and shunt fields, designers can fine-tune performance attributes, tailoring the motor to the specific needs of the task at hand. Compound motors are often found in heavy-duty cranes, conveyors, and traction systems, where variable load conditions and rapid changes in torque demand a blend of responsive power and steady operation. In many ways, compound motors bridge the gap between the brute force of series designs and the finesse of shunt configurations.

8.5.7.5 Permanent Magnet DC Motors (PMDC)

In PMDC motors, the magnetic flux is created by a set of strong PMs instead of field windings. Historically, the first DC machines were developed with PMs. But back then, materials to make strong magnets were not available. Magnetic alloys developed after the 1970s allowed the production of stronger magnets at lower cost. The iron alloy known as Alnico is one of those alloys composed of aluminum, nickel, cobalt, and iron. Later, even stronger magnets were made using rare-earth elements such as neodymium and samarium alloys (Fitzgerald et al., 1992).

In a PM motor, there is no field current; therefore, the induced EMF can be expressed only as a linear function of the angular speed $E_a = k_f\omega$. Torque produced in the air gap is simplified to $T_e = k_f I_a$. Substituting E_a in the armature loop and torque equations yields the following steady-state equations:

$$V_a = R_a I_a + k_f\omega \Longrightarrow \omega = \frac{V_a - R_a I_a}{k_f}$$
$$k_f I_a = T_L + \beta\omega \Longrightarrow I_a = \frac{T_L + \beta\omega}{k_f} \tag{8.59}$$

Most practical electromechanical applications seek the variation of motor speed with torque and electrical variables. By expressing armature current in terms of electromagnetic torque, we obtain the speed–torque characteristic shown below:

$$\omega = \frac{1}{k_f}V_a - \frac{R_a}{(k_f)^2}T_e$$

Today, most mechatronic systems use PMDC motors, which are theoretically equivalent to a separately excited motor with constant field current. External characteristics of a PMDC are comparable to the ones obtained for an SEDC motor where the field current is a constant.

PMDC motors respond quickly to changes in input voltage or load, providing precise speed and torque control. This makes them especially suited for applications that require rapid acceleration, fast stopping, or fine speed adjustments, such as in robotics or instrumentation.

The use of strong PMs allows PMDC motors to deliver a greater amount of torque for a given size and weight compared to conventional motors. This high-power density enables designers to achieve more compact and lightweight solutions without compromising on performance. PMDC motors offer a variety of advantages that make them highly valuable in both industrial and consumer applications. PMDC motors eliminate the need for field windings and their associated circuitry, resulting in a more straightforward design. This not only simplifies manufacturing but also reduces the overall size and weight of the motor, making PMDC motors particularly suitable for compact applications. PMDC motors have fewer parts that can wear out or fail. This design choice enhances long-term reliability and diminishes the frequency and complexity of maintenance tasks. In sealed designs, PM motors can be virtually maintenance free.

The absence of field windings and the reduced need for ventilation contribute to quieter operation. This benefit is particularly valuable in consumer electronics, office equipment, and applications where minimal noise is a priority.

Given their compact size, efficiency, and precise controllability, PMDC motors are widely used in a variety of domains. They are found in portable power tools, automotive systems (such as electric windows and windshield wipers), household appliances, and even in industrial automation where reliable, responsive motion is required. PMDC motors are essential in parts of small electronic devices like record players, tape recorders, and movie cameras. DC motors are convenient for such applications because they can be controlled easily, and they have better torque–speed characteristics than induction (AC) motors. In addition, obviously, they can operate on battery power.

In summary, PMDC motors combine the classic strengths of DC machines—such as reliable torque and controllability—with the advantages of modern materials and design: increased efficiency, reduced size, and lower maintenance.

8.6 Brushless DC Motors

BLDC motors eliminate the mechanical commutator and brushes. Instead, they use electronic commutation—typically achieved by ICs or microcontrollers—that sequentially energize the stator windings based on rotor position. This design greatly reduces mechanical wear, resulting in longer operational lifespans, minimal maintenance requirements, and improved reliability.

A BLDC motor operates as a synchronous motor that utilizes PMs. The rotor contains one or more PMs, while the stator has three-phase windings. Rotor position is typically detected using Hall-effect sensors or through back-EMF measurement. The electronic controller uses these signals to switch the current through the stator coils, aligning the RMF with the rotor's position to enable continuous motion.

BLDC motors offer several features that are relevant for modern mechatronic and robotic systems. Their efficiency is generally greater than brushed motors because the friction losses from brushes are eliminated. BLDC motors are also capable of reaching higher speeds and power densities, which is useful in compact designs where space and weight constraints exist. Additionally, their linear torque–speed characteristics and rapid dynamic response allow for accurate speed and position control, making them applicable to robotics, computer hard drives, drones, and EVs.

An additional advantage is noise reduction; BLDC motors produce significantly less noise due to the absence of brushes, which is particularly beneficial for household appliances, office equipment, and medical instruments. Additionally, their robust design allows them to be used in industrial environments that require reliability and long service life.

While BLDC motors have certain benefits over conventional DC motors, they require advanced electronic control systems, which result in greater system complexity and increased costs. Effective thermal management is critical, as increased power densities may lead to higher levels of heat in stator windings and electronic components.

8.7 Summary

This chapter provides an overview of electric machines, focusing primarily on their operation as motors but also covering generator functions. It highlights fundamental principles, types, structures, control methods, equivalent circuit models, and performance characteristics relevant to mechatronic and robotic systems.

- Electric machines are electromechanical devices that convert electrical energy to mechanical work or vice versa, typically producing rotational motion. They operate bidirectionally, functioning as motors or generators depending on operational conditions, such as in regenerative braking systems. The magnetic circuit, comprising stator and rotor, is central to their operation, generating torque through magnetic flux interactions.
- Common motor types include reluctance motors, stepper motors, synchronous motors, and induction (asynchronous) motors, and DC motors.
- VR motors produce torque by aligning the rotor with the stator's magnetic field, relying on variations in magnetic reluctance as the rotor moves. Their rotor lacks coils or PMs, making them simple and cost-effective, suitable for applications requiring robustness and low cost, such as automation and EVs. Control involves electronic timing of stator windings to ensure smooth rotation.
- Torque arises from the derivative of magnetic coenergy concerning rotor angle, with inductance varying sinusoidally during rotation. The torque expression depends on the current and the angle between rotor and stator magnetic axes, with maximum torque occurring at a specific torque angle δ.

- These motors operate synchronously with the supply frequency, with speed determined by pole pairs and supply frequency. Starting methods include external prime movers, damper windings acting like squirrel-cage rotors, and VFDs for smooth acceleration.
- SRMs have multiple stator poles with windings energized sequentially and a salient-pole rotor without windings or magnets. Sequential excitation creates torque through magnetic alignment, controlled by electronic systems. SRMs are robust and suitable for high-speed industrial and automotive applications.
- Reluctance motors benefit from simple rotor design, low cost, and high-speed capability but can suffer from torque ripple and require complex starting or control methods. Adding rotor windings or PMs improves performance but increases complexity and maintenance needs.
- Stepper motors rotate in discrete steps controlled by electrical pulses, allowing precise position control without feedback loops. They vary by structure into VRSM, PMSM, and hybrid stepper motors.
- Torque characteristics depend on step rate, current, and drive type. Stepper motors offer high torque at low speeds and simple mechanical design but require digital drivers and may exhibit resonance or vibration issues.
- Synchronous machines operate with rotor speed synchronized to supply frequency, using rotor fields created by PMs or field windings. They include synchronous motors and generators, widely used in industrial systems and power generation.
- Single-phase or three-phase stators with field windings or PMs on the rotor generate RMFs. Rotor types include cylindrical (constant air gap) and salient pole (variable air gap) designs.
- In generator mode, the voltage induced in stator windings depends on rotor speed and magnetic flux, with frequency related to pole pairs and rotor speed. Three-phase generators produce balanced voltages shifted by 120°.
- Starting methods include using an external prime mover to bring rotor to synchronous speed before connection, damper Windings that provide starting torque like induction motors, and gradually increasing supply frequency for a smooth start with VFDs.
- Induction (asynchronous) motors operate by inducing currents in the rotor via a rotating stator magnetic field. The rotor speed lags synchronous speed by slip, which varies with load. They are robust, cost-effective, and widely used in industrial and household applications.
- The motor can be modeled per phase with stator and rotor resistances and reactances, magnetizing reactance, and core loss resistance. Parameters can be identified through no-load and blocked-rotor tests.
- Power flow and losses (copper, core, mechanical, stray) are analyzed to determine efficiency, torque, and speed. Torque–speed characteristics show starting torque, breakdown torque, and rated torque regions.
- Single-phase motors require auxiliary windings or capacitors to generate starting torque due to the lack of a naturally RMF. Types include split-phase, capacitor-start, and shaded-pole motors with varying starting torque and efficiency.
- DC machines use commutator–brush assemblies to maintain unidirectional torque and voltage. Modern DC machines include brushless designs with improved materials and controls.
- DC machines are modeled with armature and field resistances and inductances, induced EMF, and mechanical load parameters. Dynamic equations describe electrical and mechanical behavior, often linearized for control purposes.
- Steady-state equations relate armature voltage, field current, speed, and torque. Speed–torque characteristics are linear for separately excited motors and can be controlled by adjusting field current or armature voltage.

- DC motor types are categorized based on the type of excitation (field) circuit supply into separately excited, shunt, series, and compound machines.
- PMDC motors use PMs for the field, eliminating field current and brushes, providing high torque density, efficiency, and reliability. Widely used in modern mechatronic applications.

8.8 Check Your Understanding

8.8.1 Self-Assessment Quiz

1 Stepper motors require feedback loops for precise position control.
True False

2 Synchronous machines can operate as both motors and generators.
True False

3 Induction motors are also known as asynchronous motors.
True False

4 DC machines use commutator–brush assemblies to maintain unidirectional torque and voltage.
True False

5 Permanent magnet DC motors eliminate the need for field windings and brushes.
True False

6 What is the primary principle behind electromechanical energy conversion?
a. Chemical reactions
b. Thermal dynamics
c. Magnetic flux interactions
d. Optical phenomena

7 Which type of motor relies on variations in magnetic reluctance to produce torque?
a. DC motor
b. Universal motor
c. Variable reluctance motor
d. Induction motor

8 What is the main advantage of using permanent magnets in DC motors?
a. Reduced cost
b. Increased complexity
c. High torque density
d. Lower efficiency

9 How is the speed of a synchronous motor determined?
a. By the rotor's diameter
b. By the supply frequency and number of pole pairs
c. By the load torque
d. By the stator's resistance

10 What is the primary disadvantage of reluctance motors?
   a. High cost
   b. Complex design
   c. Torque ripple
   d. Low-speed capability

Answers:

1: a; 2: c; 3: b; 4: b; 5: c; 6: b; 7: a; 8: d; 9: c; 10: a.

8.8.2 Short Answer Questions

1 What is the primary function of electric machines in mechatronics?

2 How do electric motors produce torque?

3 What is the difference between a stator and a rotor in an electric machine?

4 What are the advantages of using reluctance motors in electric vehicles?

5 How do stepper motors achieve precise positioning?

6 What is the role of damper windings in synchronous motors?

7 How do induction motors generate torque?

8 What is the main advantage of permanent magnet DC motors over traditional DC motors?

9 How do brushless DC motors differ from traditional brushed DC motors?

10 What is the significance of the torque–speed characteristic in induction motors?

8.8.3 Essay Questions

1 Discuss the advantages and disadvantages of using reluctance motors in electric vehicles. How do these motors compare to other types of electric motors in terms of efficiency, cost, and performance?

2 Explain the principle of operation of a stepper motor. How do different types of stepper motors (variable reluctance, permanent magnet, and hybrid) achieve precise positioning, and what are their respective advantages and disadvantages?

3 Describe the role of damper windings in synchronous motors. How do these windings contribute to the starting and operational stability of the motor?

4 Analyze the torque–speed characteristics of induction motors. How do these characteristics influence the selection of induction motors for various industrial applications?

5 Compare and contrast the operation of brushed and brushless DC motors. What are the key differences in their construction, performance, and typical applications?

6 Evaluate the impact of modern materials and manufacturing techniques on the performance and efficiency of DC machines. How have these advancements influenced the design and application of DC motors in contemporary engineering?

7 Discuss the significance of the torque–speed characteristic in induction motors. How does this characteristic affect the motor's performance under different load conditions?

8 Explain the concept of electromechanical energy conversion. How do electric machines convert electrical energy into mechanical work, and what are the key principles involved in this process?

9 Describe the various methods used to control the speed of DC motors. How do these methods differ in terms of complexity, cost, and effectiveness?

10 Analyze the role of power electronics in the control and operation of electric machines. How have advancements in power electronics contributed to the development of more efficient and versatile electric motors?

8.8.4 Drill Problems

1 A DC motor draws 1.5 amps from a 12-V supply, driving a mechanical load at a speed of 1,200 RPM with a torque of 0.13 N m. Calculate the total motor losses.

2 A DC motor has an armature current of 10 A and a flux per pole of 0.02 Wb. The number of conductors in the armature is 500, and the number of poles is 4. The armature diameter is 0.2 m. Calculate the torque produced by the motor.

3 An induction motor has a power factor of 0.85 and operates at 400 V with a current of 50 A. Calculate the real power consumed by the motor.

4 An electromechanical device converts 500 J of electrical energy into 450 J of mechanical energy. Calculate the efficiency of the energy conversion process.

5 A six-pole synchronous motor operates at a frequency of 60 Hz. Calculate the synchronous speed of the motor in RPM.

6 A stepper motor has 200 steps per revolution. Calculate the step angle of the motor.

7 A variable reluctance motor has the following parameters:

- Maximum inductance when the rotor and stator poles are aligned: $L_{max} = 0.08\,H$.
- Minimum inductance when the rotor and stator poles are perpendicular: $L_{min} = 0.02\,H$. The stator is supplied with an AC current of effective (rms) value $I_{rms} = 3\,A$.

The angular frequency of the stator current and the rotor speed are synchronized, and the torque angle $\delta = 45^\circ$.

a. Calculate the direct axis inductance L_d and quadrature axis inductance L_q.
b. Using the provided values, determine the average torque T_{av} produced by the motor.

8 An induction motor operates at 400 V with a current of 50 A and a power factor of 0.85. Calculate the efficiency of the motor if the mechanical power output is 15 kW.

9 A synchronous generator has 10 poles and rotates at 600 RPM. What is the frequency of the voltage generated?

10 A stepper motor has 200 steps per revolution and operates at a frequency of 1,000 pulses per second. Calculate the angular velocity of the motor in RPM and the time required to complete one full revolution.

11 A DC motor has an armature resistance of 1.5 Ω. The motor is supplied with a voltage of 24 V and drives a mechanical load with a torque of 0.5 N m. Given the motor torque constant 0.6, calculate the steady-state speed of the motor in RPM.

12 A four-pole synchronous motor operates at a frequency of 50 Hz. Calculate the synchronous speed of the motor in RPM and the slip if the rotor speed is 1,450 RPM.

13 An induction motor has a stator resistance of 0.4 Ω, rotor resistance of 0.3 Ω, and operates at a slip of 0.05. The input power is 20 kW. Calculate the power converted to mechanical form and the rotor copper loss.

8.8.5 Design Problems

1 Design a variable reluctance motor for a robotic arm joint that requires high precision and low cost. Specify the stator and rotor geometry, materials, and control strategy. Discuss how you would minimize torque ripple and optimize the inductance profile for smooth operation.
What trade-offs must be considered between cost, performance, and manufacturability?

2 A CNC machine requires precise positioning and repeatability. Compare the suitability of variable reluctance, permanent magnet, and hybrid stepper motors for this application.
a. Specify the step angle, torque requirements, and drive type.
b. Justify your selection based on performance, cost, and reliability.

3 Design a synchronous reluctance motor to drive an analog clock with a period of one rotation per minute.
a. Calculate the required number of poles and supply frequency.
b. Propose a starting method suitable for low-power applications.
c. Discuss how damper windings or VFDs could be used for smooth startup.

4 Given test data from a three-phase induction motor (DC resistance, no-load, and blocked-rotor tests), estimate the parameters for the equivalent circuit.
a. Use the provided formulas to calculate stator resistance, rotor resistance, magnetizing reactance, and core loss resistance.
b. Discuss how these parameters influence motor performance and efficiency.

5 Design a speed control system for a separately excited DC motor used in a conveyor belt.
 a. Compare field current adjustment and armature voltage control methods.
 b. Propose a feedback control strategy using modern power electronics.
 c. Analyze the impact of load changes on speed regulation.

6 Specify the design requirements for a brushless DC (BLDC) motor for a robotic manipulator.
 a. Detail the rotor and stator configuration, sensor integration, and electronic commutation method.
 b. Discuss the advantages of BLDC motors over brushed DC motors in terms of reliability, noise, and control precision.

7 An electromechanical device converts electrical energy to mechanical work.
 a. Design an experiment to measure the efficiency of the conversion process.
 b. Identify sources of losses and propose methods to minimize them in the design.

8 Compare reluctance motors, induction motors, and permanent magnet motors for use in an electric vehicle drivetrain.
 a. Analyze efficiency, torque characteristics, cost, and control complexity.
 b. Recommend a motor type and justify your choice for a specific vehicle class.

References

Chapman, S. J., 1991. *Electric Machinery Fundamentals*. 2nd ed. New York: McGraw-Hill.

Fiore, C., 2020. *Stepper Motors Basics: Types, Uses, and Working Principles*. [Online] Available at: https://www.monolithicpower.com/en/learning/resources/stepper-motors-basics-types-uses [Accessed 31 5 2025].

Fitzgerald, A. E., Kingsley, C. & Umans, S. D., 1992. *Electric Machinery*. 5th ed. New York, NY: McGraw Hill.

Gieras, J., 2017. *Electrical Machines: Fundamentals of Electromechanical Energy Conversion*. Boca Raton, FL: CRC Press.

IEEE, 2017. *IEEE Standard Test Procedure for Polyphase Induction Motors and Generators*. Schenectady, NY: Institute of Electrical and Electronics Engineers (IEEE).

Kocabiyikoğlu, Z. U., 2021. *Electromechanical Energy Conversion*. Abington, OX: CRC Press.

NEMA, 2014. *NEMA Condensed MG 1-2011: Information Guide for General Purpose Industrial AC Small and Medium Squirrel-Cage Motor Standards*. Rosslyn, VA: National Electrical Manufacturers Association.

Oriental Motor, n.d. *Basics of Stepper Motors*. [Online] Available at: https://www.orientalmotor.com/stepper-motors/technology/stepper-motor-basics.html [Accessed 1 6 2025].

Shakeel, M., 2020. *Stepper Motor Basics and Its Controlling Circuits*. [Online] Available at: https://www.academia.edu/6365820/Stepper_Motor_Basics_and_Its_controlling_Ciruits_pdf [Accessed 29 5 2025].

Yamayee, Z. A. & Bala, J. L., 1994. *Electromechanical Energy Devices and Power Systems*. New York: John Wiley.

9

Electronic Converters

OBJECTIVES

- Explain the role of power electronic converters in mechatronic systems.
- Outline the four fundamental types of power converters: AC/DC (rectifiers), DC/DC (DC choppers), DC/AC (inverters), and AC/AC (variable frequency drives).
- Explain the operation, characteristics, and applications of diodes, Zener diodes, thyristors, BJTs, MOSFETs, GTOs, and IGBTs in power electronic circuits.
- Describe converter circuit topologies and operation.
- Discuss design considerations.
- Explain the structure, operation, and control strategies of VFDs for precise AC motor control in industrial and mechatronic applications.

In mechatronic systems, power circuits are necessary to drive electric motors and actuators that convert electrical energy to motion according to the commands of analog and digital control elements.

The output voltage of signal conditioning circuits and logic controllers is in the range of a few volts, and their maximum currents are in the order of milliamps. These circuits cannot provide adequate power to drive electromechanical actuators and motors. The purpose of power circuits is to produce voltage and current to supply sufficient power for electromechanical elements.

In a mechatronic system, motors and actuators operate either in on–off or in linear (continuous) mode. The on–off operation requires an electromechanical relay or a solid-state switching circuit controlled by the output signals from electronic circuits.

Electromechanical relays are electrically operated switches that open or close when a voltage is applied across their electromagnetic coil. Typical control voltages of relays used in mechatronic systems range from 5 to 24 V. Miniature 5 V relays are common in digital circuits and microcontroller applications. Control voltage of relays used in automotive and industrial applications is 12 V. In industrial systems, 24 V control voltage is usually preferred. The maximum voltage between the relay contacts ranges from 24 V to several hundred volts. For proper operation of the system, the control voltage and current must match the specifications of the control electronics. The contact voltage must be selected based on the characteristics of the motor or actuator. The interruption of DC is always more difficult than that of AC due to the arc that occurs between the contacts. Relay manufacturers usually specify the DC and AC limits separately. Electromechanical relays electrically isolate the low-power electronic circuits from higher voltage and current power circuits.

They offer a simple solution for controlled on–off operations. However, moving parts and the arc formation between contacts make electromechanical relays less reliable than semiconductor elements. Solid-state relays use semiconductor devices such as transistors or thyristors to switch on and off the power circuit without moving parts. They offer faster switching and longer life.

Reed relays consist of a pair of magnetic strips in a glass capsule. The magnetic strips close the relay contacts when they are placed in a magnetic field. Reed relays are typically used to detect the proximity of a permanent magnet. Some applications of reed relays are door or window switches in security systems or contactless position sensors.

Velocity or position control of DC motors in continuous mode requires an adjustable voltage in a wide range. In the early years of electric power systems, a variable resistor called a rheostat was inserted in series between the motor and the power supply to adjust the voltage across the motor terminals. The voltage drop on the resistor, proportional to the load current, decreases the voltage across the motor terminals, thus reducing the rotation speed. This method is inefficient due to the heating loss in the resistor. In addition, the voltage across the motor significantly depends on the load current. At no-load, the voltage decrease is small and has a minor impact on the motor speed, but at higher loads, the impact of the resistor is greater. If the motor runs under a variable load, using a fixed resistor will result in changes to the motor's speed.

A voltage divider is an alternative to a series resistor. However, in addition to the Joule loss, the output voltage of the divider is highly dependent on the load current. Voltage dividers are appropriate only for very small load currents such as in measurement and instrumentation circuits.

Since power semiconductors were available at affordable prices, electronic converters have transformed the voltage control. A simple electronic circuit consisting of a power transistor and appropriately selected biasing resistors can operate in active mode to achieve linear voltage control. The more complex linear IC chips are available either for constant voltage regulation or to adjust the voltage across the load terminals. Three-terminal linear voltage regulators available in commercial IC chips can also be used to adjust voltage at low-power circuits as illustrated in (b).

Linear voltage control has overcome the problems of voltage variation due to the load current. However, they still cause high-power loss at larger load current if the difference between the input and output voltages is significant. Linear voltage control is a low-cost and simple method, but it has low efficiency due to the excess heat dissipated when the voltage drops across the input and output is large.

Pulse-width modulation (PWM) used in switch-mode converters is the most efficient method to control the voltage supplied to power devices. In PWM, the converter circuit produces a pulse train where the pulse width (on-time) can change. The average voltage across the load is proportional to the duty ratio, which is the ratio of the on-time to the period. DC–DC converters, also known as DC choppers, employ the PWM concept to provide an adjustable DC voltage from a fixed supply voltage.

Switch-mode DC voltage control has overcome the power loss and lower efficiency drawbacks of the linear voltage control. As displayed in the lower part of Figure 9.1, the output voltage of the converter swaps between zero and the input voltage. Over the time interval t_1, the switch is at position 1, and the output voltage is equal to the input voltage. When the switch is at position 2, the output voltage is zero. In power electronics, the fraction t_1/T is called "duty ratio." The average output voltage of a PWM converter is equal to the duty ratio times the supply voltage.

Switch-mode DC converters operate at high frequency, typically in the order of hundreds of kHz. Such a high-frequency voltage waveform usually does not affect the operation of mechatronic devices. The high-frequency harmonics may cause electromagnetic interference (EMI) on other electronic circuits in the vicinity. A large capacitor, placed at the output of the switch-mode controller, filters out the AC components of the pulse train and produces a constant DC voltage equal to the average value of the pulse train V_{av}.

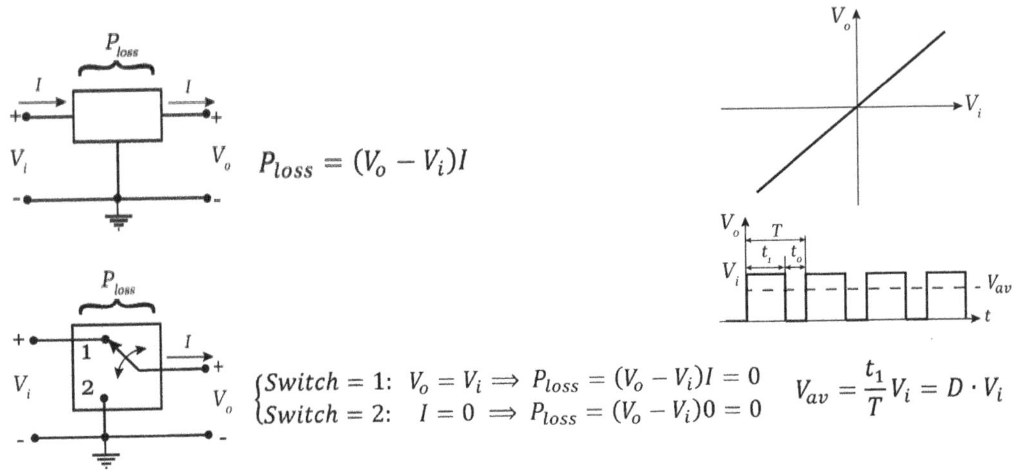

Figure 9.1 Linear and switch mode DC voltage control

In general, power converters are electronic circuits that can modify the magnitude and frequency of the supply voltage of electronic equipment, electric machines, and actuators. The four basic types of power converters are:

- AC-to-DC converters, also known as rectifiers, supply either fixed or adjustable DC from a fixed AC source such as the power outlet or an AC generator.
- DC-to-DC converters supply an adjustable DC voltage from a fixed DC source like a battery.
- DC-to-AC converters, also known as inverters, supply either fixed or adjustable AC from a fixed DC source such as a battery.
- AC-to-AC converters change the amplitude and/or frequency of an AC source voltage. Applications include AC motor drives and stabilization of variable voltage and frequency AC voltage produced by wind turbines.

Before the power transistors became commercially available and affordable, heavy-duty variable resistors called rheostats, variable autotransformers, and motor–generator sets were principal methods for power conversion. Such equipment was inefficient, large, heavy, and expensive. Furthermore, they were noisy and required more maintenance, increasing the operational cost. Starting from 1970s, progress in semiconductor device manufacturing technology and development of advanced power electronics circuits made rheostats and electromechanical systems obsolete and replaced them with electronic power converters.

9.1 Power Semiconductors

9.1.1 Diode

Diodes are two-terminal semiconductor devices that conduct current in only one direction. Ideal diode is an imaginary circuit element modeled to simplify the analysis of electronic circuits that contain real diodes. Figure 9.2 (a) shows the circuit symbol of a diode. The diode terminals are named anode and cathode. Anode is the positive voltage reference, and cathode is the negative voltage reference. The ideal diode has a piecewise linear but overall nonlinear voltage–current (V–I) characteristic. To conduct the current, the diode must be forward-biased, which means that the anode must be at a higher potential than the cathode. If the anode is at a lower potential than

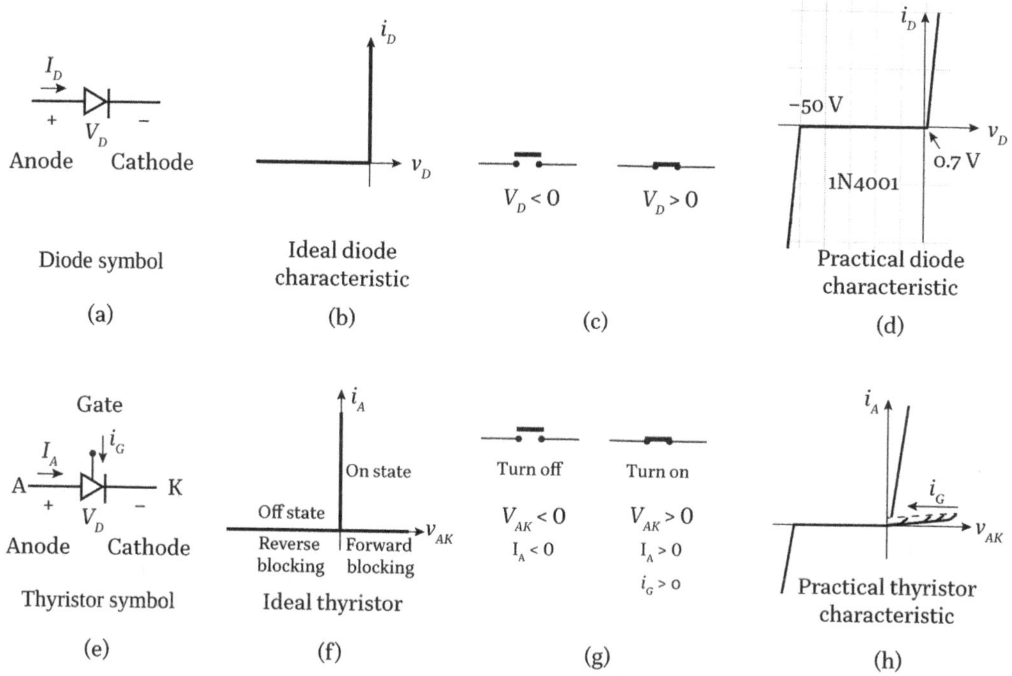

Figure 9.2 Characteristics and on–off conditions of diodes and thyristors (a: Circuit symbol of diode; b: Ideal diode characteristic; c: On-off switch representation; d: Practical diode characteristic; e: Circuit symbol of thyristor; f: Ideal thyristor characteristic; g: On-off switch representation; h: Practical thyristor characteristic

the cathode, the ideal diode does not conduct any current as seen in the voltage–current characteristic shown in Figure 9.2 (b). In this region, the diode is said to be reverse-biased, cutoff, or simply off. The ideal diode behaves like a controlled switch that turns on when forward-biased and turns off when reverse-biased. In the evaluation of a power converter circuit, a conditional switch shown in Figure 9.2 (c) is sufficient to represent a diode.

A practical diode starts conducting the current when the anode–cathode voltage is higher than a specified voltage, typically around 0.7, depending on the temperature and semiconductor material. When reverse-biased, a practical diode conducts an extremely small reverse current, in the order of microamps, until the anode–cathode voltage reaches the reverse breakdown voltage. If a power diode is subject to the reverse breakdown voltage, it is destroyed. The *V–I* characteristic of a general-purpose commercial diode is shown in Figure 9.2 (d).

Special electronic components called Zener diodes are designed to operate with a specified reverse voltage, called Zener voltage. In electronic circuits, the voltage across the terminals of a Zener diode changes within a narrow range, which makes these elements excellent voltage regulators for small currents. Zener diodes are generally used to produce a reference voltage in electronic circuits.

9.1.2 Thyristor

Thyristors are essentially controlled diodes. They have a third terminal, called a gate, for triggering the current conduction. A reverse-biased thyristor blocks the electric current similar to the diode cathode terminals to create a small DC. Figure 9.2 (e)–(h) show the circuit symbol, characteristics, and turn on–off conditions.

When forward-biased and the anode–cathode voltage is above the specified voltage, the thyristor can start conducting as soon as it is triggered by applying a voltage across the gate–cathode terminals. Once triggered, the gate current is no longer necessary; thus, a short current pulse is sufficient to turn on the thyristor. The on-state continues until the current through it remains in the forward direction. To turn the thyristor off, the current flowing through it must be below the specified holding current at least over the reverse recovery time (t_{rr}).

In AC converters, the thyristor turns off naturally when the anode current changes direction. This property makes thyristor a good choice for controlled rectifiers. In DC converters, special circuits are needed to turn the thyristor off. Thyristors are usually preferred in high-voltage, high-current power converters.

9.1.3 Switching Semiconductor Elements

Power bipolar junction transistors (BJTs), metal-oxide-semiconductor field-effect transistors (MOSFETs), gate turn-off thyristors (GTOs), and insulated gate bipolar transistors (IGBTs) are three-terminal semiconductor devices typically used in switching power converters. Since one of the terminals is a common terminal for the input and output, a three-terminal device is considered a two-port element, with one being the input and the other being the output. The voltage or current applied to the input port controls the output current. Depending on the current through or the voltage across the input terminals, a three-terminal element may be on, off, or in amplification mode.

In switching power converters, semiconductor components are either on or off during their operation. Figure 9.3 illustrates the circuit symbols of three-terminal elements and their switching characteristics. Based on the *V–I* characteristics, BJT is a current-controlled device, and MOSFET and IGBT are voltage-controlled devices. GTO turns on and off by applying a current pulse; thus, it is a pulse-controlled device.

Replacing semiconductor elements with ideal switches simplifies the analysis of power converter circuits. An ideal switch has the following properties.

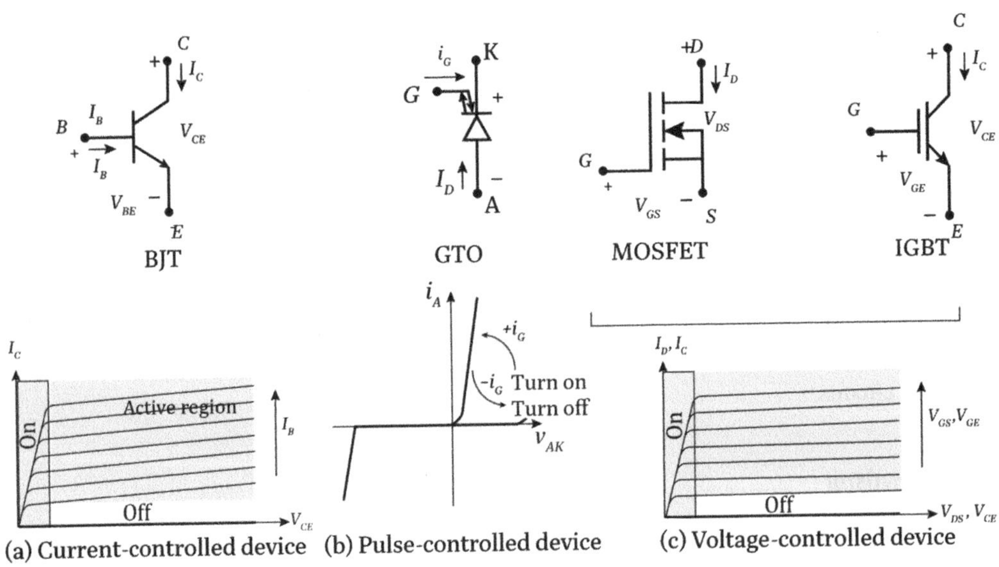

Figure 9.3 Symbols and operating characteristics of three terminal devices (a: Current controlled device; b: Pulse controlled device; c: Voltage controlled device)

- Turn on and off when triggered, without any delay.
- Conduct any current when it is on.
- Block any voltage when it is off.
- Have zero resistance during conduction.
- Have infinite resistance when the current is blocked.

The desired properties of a semiconductor switch are to approach as much as possible the ideal conditions mentioned earlier. A real semiconductor switch can be turned on and off in a short period of time, which has two practical consequences. The first one is the limitation of the switching frequency. The second is power loss during every switching process. Real switching elements have a small resistance when they are on, which causes conduction loss. Switching and conduction loss combined result in heat dissipation due to Joule's law. In addition to its impact on overall efficiency, excessive heat must be expelled to the atmosphere by a cooling system. A converter designer must consider switching and power capability in choosing the type of switching semiconductor.

9.1.3.1 Bipolar Junction Transistors (BJT)

Power BJTs are current-controlled devices and have good on-state characteristics, but they have a small current gain and long switching times, especially at turn-off (Lee & Chow, 2011). Figure 9.3 (a) shows the circuit symbol of an NPN transistor and its steady-state *V–I* characteristic. The current flowing from the base to the emitter controls the collector current. The DC gain h_{FE}, which is the ratio of the collector and base currents, is in the range of 5–10 in power BJTs. Consequently, in a higher power converter, the continuous base current becomes significant. Consequently, BJTs require complex control circuits because the controller must provide sufficient base current to turn the BJT on.

Example 9.1 A BJT with DC gain 5 is considered a switching element in a 1-kW, 50-V DC–DC converter. Calculate the base current that the control circuit must supply.

Solution

When the BJT is on, the collector current is

$$I_C = \frac{1{,}000}{50} = 20\text{ A}$$

Given the DC gain of 5, the base current is found as

$$I_B = \frac{I_C}{h_{FE}} = \frac{20}{5} = 4\text{ A} \therefore$$

A Darlington or triple Darlington configuration may be an option to achieve higher DC gain; however, the increased collector–emitter voltage in on-state and slower switching speeds are the major drawbacks.

9.1.3.2 Gate Turn-Off Thyristors (GTO)

GTO was developed to overcome the challenges thyristors faced in the turn-off process. Figure 9.3 (b) shows the circuit symbol of a GTO and its steady-state *V–I* characteristic. Similar to the thyristor, GTO turns on with a pulse current while it is forward-biased. Once turned on, the gate current is not required to maintain the on-state. Unlike the thyristor, a few microsecond long negative voltage across the gate and cathode terminals produces a gate current pulse that turns off the GTO. This device is labeled as pulse controlled in the figure.

The on-state voltage of a GTO across the anode and cathode is slightly higher than that of the thyristors. The switching speed of GTO ranges from a few microseconds to 25 μs. Because of their high-voltage and current ratings, GTOs are preferred in higher power converters. But the switching frequency is lower than that of the other semiconductor devices (up to 10 kHz).

9.1.3.3 Metal-Oxide-Semiconductor Field-Effect Transistors (MOSFETs)

MOSFETs are voltage-controlled devices and require very small gate current during switching-on period. This property simplifies the gate circuit and reduces the power loss in the control unit. Figure 9.3 (c) shows the circuit symbol of an N-channel MOSFET, which is more common in power converter configurations, and its steady-state *V–I* characteristic. To turn on a MOSFET, the voltage across the gate and source terminal V_{GS} must be above a specified threshold voltage. In addition, V_{GS} must be at an appropriate magnitude to ensure a drain current at the desired level. The gate–source voltage must be present during the on period. Because the gate is insulated from the drain and source junctions, a MOSFET practically does not draw any current from the control unit, except for the small capacitive current that flows during the on–off transitions.

The specified on-state resistance of a MOSFET between the drain and source terminals rises as the blocking voltage increases. To limit the conduction loss, devices with relatively low-voltage ratings are available (Mohan et al., 2003).

9.1.3.4 Insulated Gate Bipolar Transistors (IGBT)

The IGBT, which was introduced in the early 1980s, has become popular because of its superior characteristics (Rashid, 2011). IGBTs present a combination of the properties of BJT, GTO, and MOSFET. Figure 9.3 (c) shows the circuit symbol of an IGBT and its steady-state *V–I* characteristic. Like MOSFETs, IGBTs are voltage-controlled devices and practically do not draw current from the control circuit, except on–off transitions. Their on-state voltage across the collector–emitter terminals is small like BJT. Similar to GTO, an IGBT can block negative voltages. Commercial IGBTs have turn-on and turn-off times in the order of 1 μs; hence, they can operate at higher frequency than MOSFETs. They are available in ratings as large as 1,700 V and 1,200 A (Rashid, 2011).

9.1.3.5 Comparison of Controlled Switches

Voltage and current capabilities and switching speed of the semiconductor switching elements described earlier are compared in Table 9.1. Typically, BJTs are suitable for lower power mechatronic systems where the control current is small enough. MOSFETs are preferred in

Table 9.1 Comparison of semiconductor switching devices

Device	Voltage and Current Handling Capabilities	Switching Speed
BJT	Medium—up to 1.4 kV and a few hundred amps	Medium
MOSFET	Low—up to 1 kV and 100 A	Fast
GTO	High—up to 4.5 kV and a few kA	Slow
IGBT	Medium—up to 1.7 kV and 1.2 kA	Medium

medium-power range especially controlled by digital logic circuits and microcontrollers. GTOs and IGBTs are more cost-effective at higher power industrial-scale converters.

9.2 AC-to-DC Converters (Rectifiers)

AC-to-DC converters, also known as rectifiers, supply electronic circuits, electromechanical motion devices, battery chargers, and other DC equipment from AC utility power. Single-phase rectifiers provide DC voltage for electronic circuits, DC motors, and electromagnetic actuators in smaller mechatronic devices. Battery chargers for cordless tools and mechatronic devices also include single-phase rectifiers. Commercial and industrial mechatronic systems equipped with DC motor drives are powered by a three-phase rectifier. This section first describes single-phase rectifiers, which are more common for low-power mechatronic systems up to a few hundred watts. We then examine the structure and advantages of three-phase rectifiers in providing higher DC power in the kilowatt range.

A rectifier may provide a constant or adjustable voltage. Diodes are the essential element of rectifiers with a fixed output voltage. In practice, the output voltage of a diode rectifier may change due to the fluctuations of the AC voltage or changing load current. At low power, a regulator made with discrete electronic components such as transistor and Zener diode can regulate the output voltage. A linear IC voltage regulator chip (e.g., LM345) can provide more advanced regulation with current-limiting features. Switch-mode rectifiers that consist of transistors and diodes can adjust voltage directly at the output of the converter.

At high-power levels, a controlled rectifier with a feedback loop is necessary for voltage regulation. Thyristors are more suitable for higher voltage and higher power converters with an adjustable output voltage. Thyristor, also known as a silicon-controlled rectifier (SRC), is basically a diode with a control terminal named a gate. Another role of controlled rectifier is to provide a variable DC voltage from AC utility power to adjust the speed of DC motors. The common component of a controlled rectifier is a SRC.

There are two types of AC–DC converters classified as half-wave and full-wave rectifiers. A single diode connected in series between an AC voltage source and a load resistor is the simplest rectifier circuit, called a half-wave rectifier, because it blocks the negative half-cycle of the AC waveform. Full-wave rectifiers pass the entire period of the AC voltage, reversing the polarity of the negative half-cycle to supply a unipolar output voltage. While the purpose of the full-wave rectifier is basically the same as that of the half-wave rectifier, full-wave rectifiers have several fundamental advantages. In a full-wave rectifier, the average current in the AC source is zero; this avoids problems associated with nonzero average source currents, particularly in transformers. The output of the full-wave rectifier inherently has less voltage ripple than the half-wave rectifier. Therefore, a smaller output capacitor can be used to meet the voltage ripple requirement. In this section, we focus on full-wave rectifiers, which are common in mechatronic devices.

9.2.1 Single-Phase Full-Wave Diode Rectifiers

The simplest rectifier is a single diode connected in series between the AC source and the DC load as illustrated in Figure 9.4 (a). Since the diode conducts only in one direction, this type of rectifier, known as a half-wave rectifier, forwards the positive half-cycle to the load and blocks the negative half-cycle. The output voltage of a half-wave rectifier is unidirectional but far from DC. It requires a large capacitor to filter out the AC component of the output voltage. In addition, the DC component

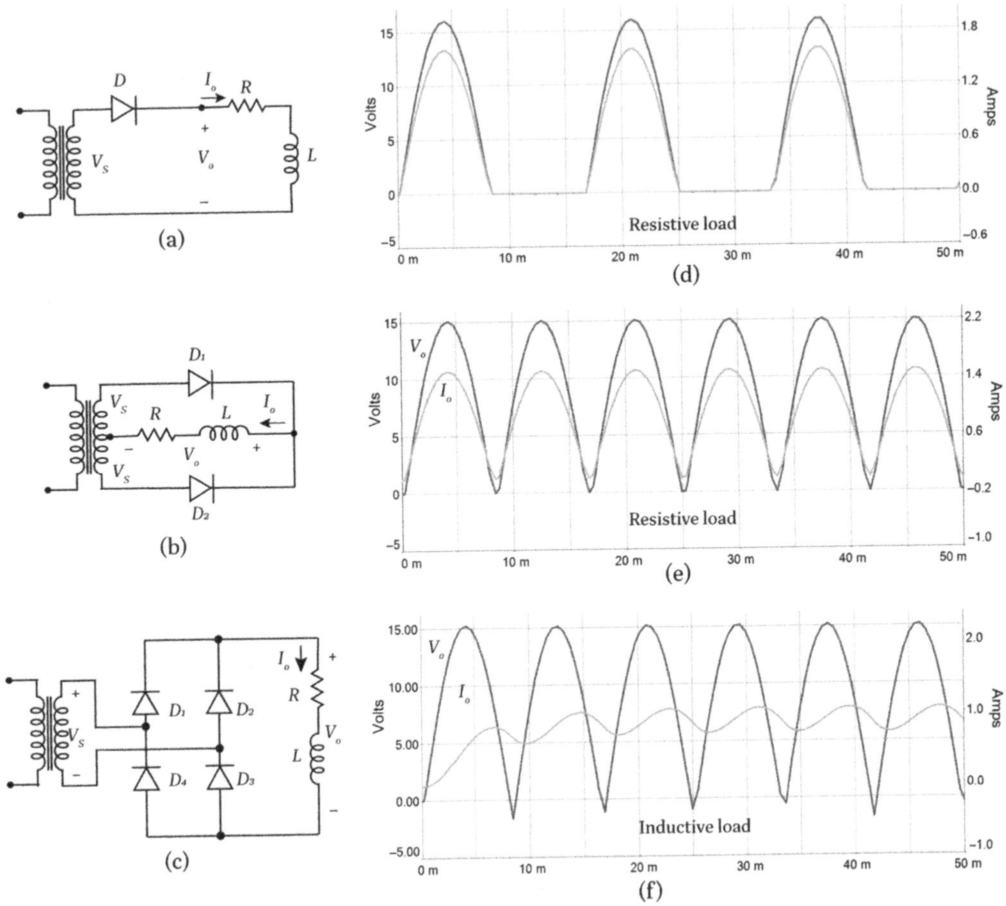

Figure 9.4 Single phase diode rectifiers and output waveforms (a: Half-wave rectifier; b: Full-wave rectifier; c: Bridge rectifier; d: Half-wave rectified voltage; e: Full-wave rectifier output voltage for resistive load; f: Full-wave rectifier output voltage for inductive load)

of the input voltage causes saturation of the supply transformer. The average value of the output voltage of a half-wave rectifier is

$$V_o = \frac{1}{2\pi}\int_0^{\pi} V_m \sin(\omega t)d(\omega t) = \frac{V_m}{\pi} = 0.318V_m \tag{9.1}$$

The effective (rms) value of the rectified output voltage is half of the AC supply voltage.

Two diodes and a transformer with center tap can achieve full-wave rectification from a one-phase AC input. The more common rectifier circuit contains four diodes connected in a bridge configuration. The bridge rectifier does not require a transformer with a center tap. However, the input transformer is still desirable because it changes the AC voltage magnitude to obtain the desired DC voltage and isolates the electronic circuit from the utility network for safety.

In mechatronic systems, a rectifier may supply a wide variety of DC loads from purely resistive to highly inductive. Electronic circuits are mostly resistive. Electromechanical elements such as motors and actuators contain large inductive components like coils with a ferromagnetic core.

Figure 9.4 shows single-phase diode rectifier circuits. The load consists of a resistance and inductance to represent the typical load conditions encountered in mechatronics.

The full-wave rectifier circuit, as Figure 9.4 (b) shows, requires a transformer with center tap on the secondary winding. In practice, most small transformers made for electronic circuits have two identical secondary windings, which may be connected in series to obtain the center tap. In this type of rectifier, only one of the two diodes can be in the forward direction. The output voltage is $+V_s$ when D_1 conducts and is $-V_s$ when D_2 conducts. Note that the fundamental frequency of the output voltage is 2ω, since two periods of the output waveform occur for every one period of the input voltage. Average output voltage of a full-wave single-phase bridge is

$$V_o = \frac{1}{\pi}\int_0^{\pi} V_m \sin(\omega t) d(\omega t) = \frac{2V_m}{\pi} = 0.636V_m \tag{9.2}$$

Kirchhoff's voltage law (KVL) shows that when one of the diodes is conducting, the maximum reverse voltage across the other diode is twice the peak voltage across the load. Hence, the selected diodes must withstand a reverse voltage about three times ($2\sqrt{2} \cdot V_s$) the rms value of the supply voltage.

Figure 9.4 (c) illustrates a single-phase bridge rectifier, which does not require a center tap on the secondary winding. Applying KVL, we can see that two diodes are forward-biased at any time. Diodes D_1 and D_3 conduct together, and D_2 and D_4 conduct together. D_1 and D_3 are forward-biased during the positive half-cycle of the AC supply voltage. The voltage across the load is $+V_S$. During the negative half-cycle, D_2 and D_4 are on and forward the supply voltage to the load by switching its polarity. In every half-cycle, two diodes in series turn on or off together. The maximum reverse voltage across each diode is half of the peak value of the source voltage $\left(\sqrt{2}/2\right)V_S$. Thus, diodes selected for a bridge rectifier can have smaller reverse breakdown voltage compared to the rectifier configuration in Figure 9.4 (b). This becomes an advantage for higher voltage rectifiers.

The load current is always positive. Since each half of the AC circulates through the load, the rectifier draws zero average current from the utility. Therefore, the current flowing through the primary winding of the transformer does not have a DC component. This is one of the benefits of full-wave rectifiers over half-wave.

Figure 9.4 (d)–(f) show the output voltage and current waveforms obtained from simulation of the bridge rectifier circuit using the NI Multisim™ software. Note that the rectified voltage is unidirectional, not a true DC. If the load is purely resistive, the variation of the load current is the same as the rectified voltage across the load (v_s/R). For an RL load, the fluctuations of the load current are smoother. As the inductance increases, the load current approaches perfect DC. A large inductance may be part of the load or added externally. It is a common practice to connect a large capacitor across the load to filter the AC components of the output voltage.

The fundamental frequency of the output voltage is 2ω, where ω is the frequency of the AC input, since two periods of the output occur for every period of the input. Fourier series of the output waveform consists of a DC term and an infinite number of harmonics. The DC component is the average value of the output voltage. Because of the symmetry, the full-wave rectified voltage contains only even harmonics (Hart, 2011).

$$\begin{aligned} v_o(\omega t) &= V_o + \sum V_n \cos(n\omega t + \pi) \qquad n = 2, 4, \ldots, \infty \\ V_o &= \frac{1}{\pi}\int_0^{\pi} V_m \sin(\omega t) d(\omega t) = \frac{2V_m}{\pi} \\ V_n &= \frac{2V_m}{\pi} \cdot \frac{2}{n^2 - 1} \end{aligned} \tag{9.3}$$

Table 9.2 Design considerations for single-phase diode rectifiers

	Half-Wave Rectifier	Full-Wave Rectifier with Center-Tapped Transformer	Full-Wave Bridge Rectifier
Transformer secondary voltage V_s	$\frac{V_m}{\pi}$	$\frac{2V_m}{\pi}$	$\frac{2V_m}{\pi}$
Effective (rms) load voltage	$\frac{V_m}{2}$	$\frac{V_m}{\sqrt{2}}$	$\frac{V_m}{\sqrt{2}}$
Form factor $\frac{V_{rms}}{V_o}$ or $\frac{I_{rms}}{I_o}$	$\frac{\pi}{2}$	$\frac{\pi}{2\sqrt{2}}$	$\frac{\pi}{2\sqrt{2}}$
Output ripple frequency	f_s	$2f_s$	$2f_s$
Peak repetitive reverse diode voltage	πV_o	πV_o	$\frac{\pi}{2}V_o$
rms voltage per transformer secondary winding	$\frac{\pi}{\sqrt{2}}V_o$	$\frac{\pi}{2\sqrt{2}}V_o$	$\frac{\pi}{2\sqrt{2}}V_o$
Diode average current	I_o	$0.50\ I_o$	$0.50\ I_o$
Peak repetitive forward diode current	πI_o	$\frac{\pi}{2}I_o$	$\frac{\pi}{2}I_o$
Diode rms current	$\frac{\pi}{2}I_o$	$\frac{\pi}{4}I_o$	$\frac{\pi}{4}I_o$

Source: Adapted from Lee and Chow (2011).

The load current is computed using the superposition principle. That is, by dividing each harmonic separately by the load impedance for the harmonic's frequency and combining the results.

As the order of the harmonic n increases, its amplitude decreases. On the other hand, the impedance of an RL load Z_n increases since the harmonic frequency is n times the fundamental frequency. The combination of decreasing V_n and increasing Z_n declines I_n rapidly for higher harmonic components. Therefore, the DC term and only a few of the AC terms are usually necessary to calculate the current in an RL load.

During the positive half period, the source current is the same as the load current. For the negative half period, the source current is equal to the negative of the load current. Since the square of the positive and negative parts are the same, the rms source current is equal to the rms value of the load current. For a highly inductive load, which is more common in mechatronic systems, the source current approaches a square wave. With low-power electromechanical systems, this is not a major problem. However, as the power of the system rises, distortion of the supply current at the AC side presents several issues. First, the power losses of the transformer increase. Second, the current harmonics reduces the power quality of the supply network and cause disturbances to other devices supplied from the same network.

The goal of a rectifier design is to achieve the required DC output voltage within a given voltage ripple constraint. Therefore, it is convenient to size the diodes and transformer ratings in terms of the output voltage V_o (Table 9.2).

9.2.2 Single-Phase Controlled Rectifier

Controlled rectifiers provide a variable DC voltage from a fixed AC source. Variable DC voltage is mainly needed for changing the speed of DC motors, adjusting light intensity of LED lightbulbs, battery charging, and various feedback control systems. In addition, when the AC input voltage or DC load current fluctuates, the DC output voltage changes too. A controllable rectifier can regulate the DC output by adding a feedback loop. Furthermore, controlled rectifiers can enable bidirectional energy flow, which allows significant energy saving in variable speed electromechanical drives, especially in industrial systems and EVs.

Controlled rectification is achieved by replacing the diodes in a rectifier circuit with any of the three-terminal switching elements described in Section 9.1. Controlled rectifiers are formed by simply replacing the diodes in the circuits displayed in Figure 9.4 with thyristors. But the same can be done using transistors too.

For the center-tapped transformer rectifier (a), T_1 is forward-biased when the supply voltage is positive or greater than the internal DC voltage of the load. T_2 is forward-biased when similar conditions occur during the negative half-cycle. In the bridge rectifier shown in (b), T_1 and T_3 will become forward-biased when the source becomes positive. Similarly, T_2 and T_4 will become forward-biased when the source becomes negative.

9.2.2.1 Resistive Load

With a resistive load, the current and voltage are at the same phase. Therefore, when the voltage across the load drops to zero, the current flowing through the thyristors becomes zero too. For a resistive load, angle $\beta = \pi$. Average voltage across the load resistor is

$$V_s = \frac{1}{\pi}\int_{\alpha}^{\pi} V_m \sin(\omega t) d(\omega t) = \frac{V_m}{\pi}(1 + \cos\alpha) \tag{9.4}$$

Average output current is simply the average voltage divided by the load resistance. The effective (rms) value of the output current is calculated by integrating the square of the average current and taking the square root of the integral. Since the waveform repeats every half-cycle, the integration limits are α and π.

$$I_{rms} = \sqrt{\frac{1}{\pi}\int_{\alpha}^{\pi}\left[\frac{V_m}{R}\sin(\omega t)\right]^2 d(\omega t)} = \frac{V_m}{R}\sqrt{\frac{1}{2} - \frac{\alpha}{2\alpha} + \frac{\sin(2\alpha)}{4\pi}} \tag{9.5}$$

9.2.2.2 Inductive (RL) Load

Consider the bridge rectifier shown in Figure 9.5 (b) with an RL load. Since the phase angle of an inductive load is lagging, the current does not become zero at the same time as the voltage. Consider the positive half-cycle where T_1 and T_3 are on. The thyristors continue conducting either when T_2 and T_4 turn on (forced turn off) or the current flowing through them falls to zero. As the waveforms in Figure 9.5 (c) and (d) show, the load current may be continuous or discontinuous.

9.2.2.3 Discontinuous Load Current

Time variation of the load current must be calculated by solving the differential equation

$$Ri(t) + L\frac{di(t)}{dt} = V_m \sin(\omega t) \tag{9.6}$$

The solution below is based on the initial condition that at the instant T_1 and T_3 are triggered the output current is zero ($i_o(\alpha) = 0$). The thyristors turn off when the current flowing through them falls to zero at $\omega t = \beta$ and remains zero until T_2 and T_4 turn on at $\omega t = \pi + \alpha$. If $\beta < \alpha + \pi$ of operation, the function representing the load current is discontinuous (Hart, 2011).

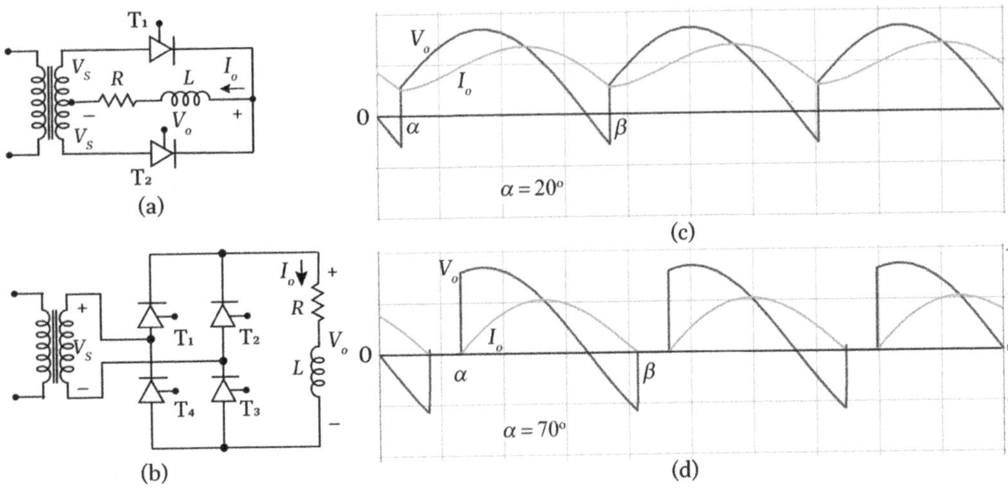

Figure 9.5 Single-phase controlled rectifiers (a: Full-wave controlled rectifier with center tap transformer; b: Controlled bridge rectifier; c: Output voltage for continuous current; d: Output voltage for discontinuous current)

$$i_o(t) = \begin{cases} \dfrac{V_m}{Z}\sin(\omega t - \theta) - \sin(\alpha - \theta)e^{-(\omega t - \alpha)/\omega\tau} & \text{for } \alpha \leq \omega t \leq \beta \\ 0 & \text{otherwise} \end{cases} \tag{9.7}$$

$$Z = R + j\omega L \qquad \tau = \frac{L}{R} \qquad \theta = \tan^{-1}\left(\frac{\omega L}{R}\right)$$

The instant when the thyristor current returns to zero corresponds to the extinction angle β. When $\omega t = \beta$, thyristor current is

$$i(\beta) = 0 = \frac{V_m}{Z}\left[\sin(\beta - \theta) - \sin(\alpha - \theta)e^{(\alpha - \beta)/\omega\tau}\right] \tag{9.8}$$

To obtain the extinction angle, this nonlinear equation must be solved numerically.

For $\alpha \leq \omega t \leq \beta$, the load voltage is equal to the source voltage. The average value of the output voltage is

$$V_o = \frac{1}{2\pi}\int_{\alpha}^{\beta} V_m \sin(\omega t)d(\omega t) = \frac{V_m}{2\pi}(\cos\alpha - \cos\beta) \tag{9.9}$$

The equations below give the time variation, average, and rms values of the load current.

$$I_o = \frac{1}{2\pi}\int_{\alpha}^{\beta} i(\omega t)d(\omega t); \qquad I_{rms} = \sqrt{\frac{1}{2\pi}\int_{\alpha}^{\beta} i^2(\omega t)d(\omega t)} \tag{9.10}$$

9.2.2.4 Continuous Load Current

The continuous current mode occurs if the current flowing through T_1 and T_3 is greater than zero at $\omega t = \pi + \alpha$. T_2 and T_4 are activated when they receive the gate pulses, and they force T_1 and T_3 to turn off by applying negative voltage across them. The solution of the KVL given in discontinuous operation mode can be modified for the continuous operation mode.

$$i_o(\pi+\alpha)\geq 0$$

$$\frac{V_m}{Z}\sin(\pi+\alpha-\theta)-\sin(\pi+\alpha-\theta)e^{-(\pi+\alpha-\alpha)/\omega\tau}\geq 0 \tag{9.11}$$

An alternative condition to check continuous current operation mode can be obtained by solving the equation above for α (Hart, 2011).

$$\alpha\leq\tan^{-1}\left(\frac{\omega L}{R}\right) \tag{9.12}$$

In a design problem, we must calculate the rms value of the load current, which also flows through the thyristors. Computer simulation and numerical integration using MATLAB® are powerful computational tools to obtain the rms value of the load current. Fourier analysis is another effective method that allows further analytical reasoning in the design process. Fourier series components of the load voltage in continuous current mode are (Hart, 2011).

$$\begin{aligned}
&v_o(\omega t)=V_o+\sum_{n=1}^{\infty}v_n(n\omega_0 t)\\
&V_0\frac{2V_m}{\pi}\cos\alpha\\
&V_n=\frac{2V_m}{\pi}\sqrt{a_n^2+b_n^2}\\
&a_n=\frac{\cos(n+1)\alpha}{n+1}-\frac{\cos(n-1)\alpha}{n-1}\\
&b_n=\frac{\sin(n+1)\alpha}{n+1}-\frac{\sin(n-1)\alpha}{n-1}\\
&n=2,4,\ldots,\infty
\end{aligned} \tag{9.13}$$

The equation to calculate rms load current is below:

$$I_{rms}=\sqrt{\left(\frac{V_0}{R}\right)^2+\frac{1}{2}\sum_n\frac{V_n^2}{R^2+(n\omega_0 L)^2}} \tag{9.14}$$

As the order of the Fourier series terms increases, the magnitudes of the harmonic components decrease and the impedances increase rapidly. Thus, the first few terms of the Fourier series are sufficient to obtain the rms current at an acceptable level for practical design calculations.

9.2.3 Three-Phase Rectifiers

Three-phase rectifiers provide a constant or adjustable DC from fixed three-phase utility power. Commercial and industrial systems require three-phase rectifiers to supply loads such as DC motors, battery chargers, and electrolysis equipment where the required load current is too high for a single-phase rectifier. Figure 9.6 (a) shows a half-wave three-phase rectifier and its output voltage waveform. In addition to allowing the use of diodes with a smaller current rating for a larger DC load, three-phase rectifiers have significantly smaller output voltage ripple than single-phase rectifiers.

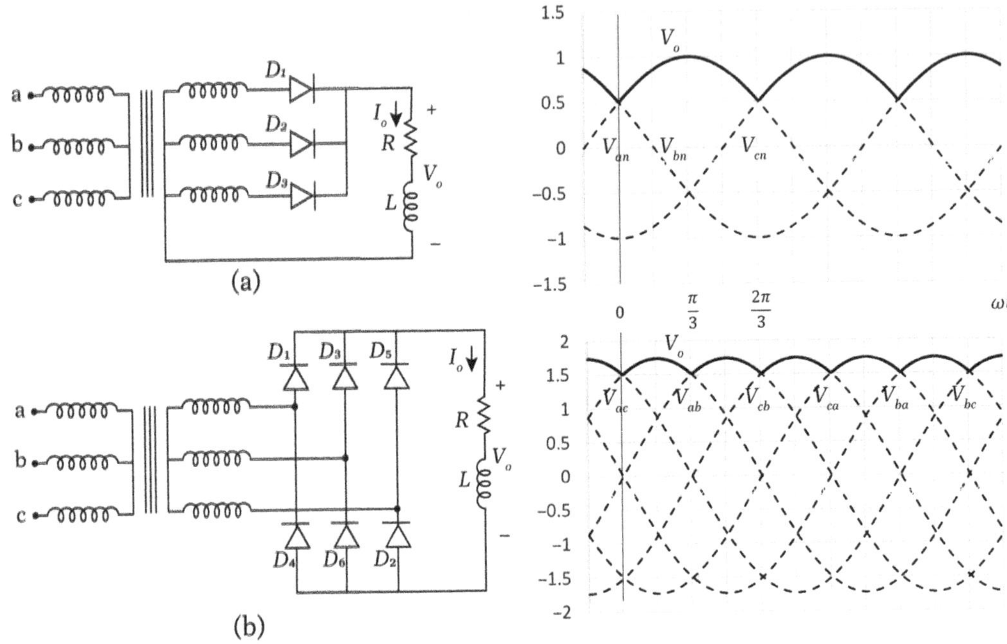

Figure 9.6 Three phase diode rectifiers (a) 3-phase half-wave rectifier and its output voltage; (b) 3-phase bridge rectifier and its output voltage)

9.2.3.1 Three-Phase Half-Wave Rectifier

The simplest three-phase rectifier type is half-wave, which combines three single-phase half-wave rectifiers shown in Figure 9.4 (a). Half-wave rectifier requires a transformer with a star connected secondary side that has a neutral terminal to allow return current. Each diode conducts when the voltage of the line it is connected is higher than that on the other two phases with respect to the neutral; hence, each diode is forward-biased over $2\pi/3$ radians (Lee & Chow, 2011).

In a half-wave three-phase diode rectifier, the average and rms values of the output voltage of this type of rectifier are

$$
\begin{aligned}
V_o &= \frac{3}{2\pi}\int_{\pi/6}^{5\pi/6} V_m \sin(\omega t) d(\omega t) = V_m \frac{3\sqrt{3}}{2\pi} = 0.827 V_m \\
V_{rms} &= \sqrt{\frac{3}{2\pi}\int_{\pi/6}^{5\pi/6} [V_m \sin(\omega t)]^2 d(\omega t)} = \sqrt{\frac{3}{2\pi}\left(\frac{\pi}{3} + \frac{\sqrt{3}}{4}\right)} = 0.84 V_m
\end{aligned}
\tag{9.15}
$$

9.2.3.2 Three-Phase Full-Wave Rectifier

The circuit in Figure 9.4 (b) is a full-wave three-phase diode rectifier. This can be regarded as the combination of two half-wave rectifiers. Therefore, it has the ability to transfer the negative half-cycle to the load by reversing the voltage sign. Note that each pair of diodes are connected between two phases. For example, diodes D_1 and D_2 are forward-biased together when the voltage across them is higher than that across the other two diode pairs. Each diode pair conduct over $\pi/3$ radians.

For a resistive load, the rms current is simply the ratio of the DC voltage to the resistance. If the load is inductive, the rms we must calculate the Fourier series components of the output current by dividing each voltage harmonic by the corresponding impedance at the harmonic's frequency.

The average output voltage and the magnitudes of the harmonics are calculated by integrating the line-to-line voltage from $\pi/3$ to $2\pi/3$ radians. To avoid confusion, in three-phase systems, we will denote the line voltages (phase-to-phase) with U. Note that if the peak value of the phase-to-neutral voltage is V_m, then $U_m = \sqrt{3}V_m$.

$$V_o = \frac{3}{\pi}\int_{\pi/3}^{2\pi/3} U_m \sin(\omega t)d(\omega t) = \frac{3}{2}U_m = 0.955U_m = 1.65\,V_m$$
$$V_n = \frac{6U_m}{\pi(n^2-1)} \quad n = 6,12,18,\ldots,\infty \tag{9.16}$$

Since each diode conducts one-third of the time, the diodes must be sized based on the following currents:

$$I_{D,avg} = \frac{1}{3}I_{o,avg};\, I_{D,rms} = \frac{1}{\sqrt{3}}I_{o,rms} \tag{9.17}$$

9.2.3.3 Controlled Three-Phase Rectifiers

The three-phase bridge rectifiers become adjustable by replacing the diodes with any type of three-terminal switching elements. Thyristors are the preferred switching elements for adjustable three-phase rectifiers because they are triggered by a set of pulses, and they turn off naturally as the voltage becomes negative in every cycle.

9.3 DC-to-DC Converters

Any type of transistor (MOSFET, IGBT, or BJT) biased to operate in on–off region can achieve the switching function required for a DC–DC converter. A timer circuit or a microcontroller can provide the switching pulse at a desired frequency and adjust the duty ratio. The major drawback of switch-mode DC voltage control is EMI due to high-frequency harmonics (Mohan, 2012).

All modern power converters utilize semiconductor devices that perform on–off switching operations. BJTs, MOSFETs, and IGBTs are among the semiconductor devices mostly used in practical power electronics circuits.

9.3.1 Switching Transistors

Power BJTs, MOSFETs, GTOs, and IGBTs are three-terminal semiconductor devices typically used in switching power converters. Since one of the terminals is a common terminal for the input and output, a three-terminal device is considered a two-port element, with one being the input and the other being the output. The voltage or current applied to the input port controls the output current. Depending on the current through or the voltage across the input terminals, a three-terminal element may be on, off, or in amplification mode.

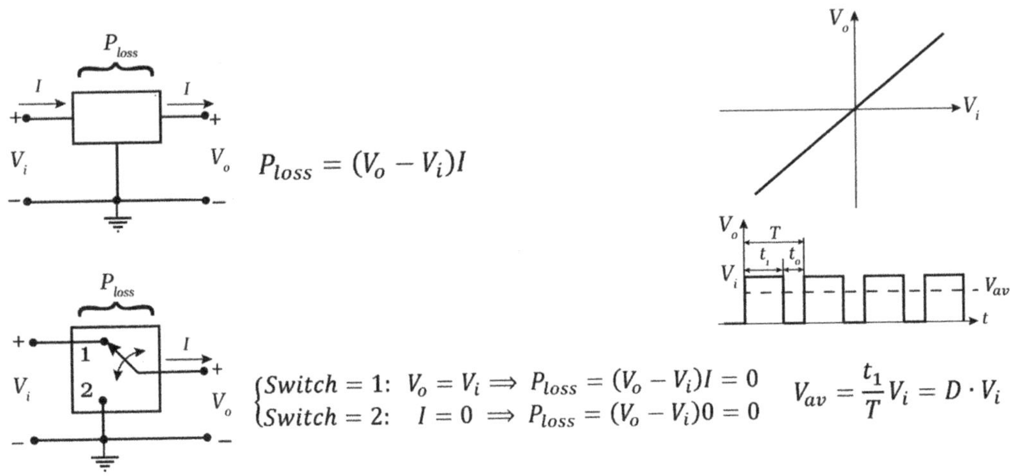

Figure 9.7 Linear and switch-mode DC voltage control

In switching power converters, semiconductor components are either on or off during their operation. Figure 9.3 illustrates the circuit symbols of three-terminal elements and their switching characteristics. Based on the *V–I* characteristics, BJT is a current-controlled device, and MOSFET and IGBT are voltage-controlled devices. GTO turns on and off by applying a current pulse; thus, it is a pulse-controlled device.

Replacing semiconductor elements with ideal switches simplifies the analysis of power converter circuits as illustrated in Figure 9.7. An ideal switch has the following properties:

- Turn on and off when triggered, without any delay.
- Conduct any current when it is on.
- Block any voltage when it is off.
- Have zero resistance during conduction.
- Have infinite resistance when the current is blocked.

The desired properties of a semiconductor switch are to approach as much as possible the ideal conditions mentioned earlier. A real semiconductor switch can be turned on and off in a short period of time, which has two practical consequences. The first one is the limitation of the switching frequency. The second is the power loss during every switching process. Real switching elements have a small resistance when they are on, which causes conduction loss. Switching and conduction loss combined result in heat dissipation due to Joule's law. In addition to its impact on overall efficiency, excessive heat must be expelled to the atmosphere by a cooling system. A converter designer must consider switching and power capability in choosing the type of switching semiconductor.

9.3.1.1 Bipolar Junction Transistors (BJT)

Power BJTs are current-controlled devices and have good on-state characteristics, but they have a small current gain and long switching times, especially at turn-off (Lee & Chow, 2011). Figure 9.3 (a) shows the circuit symbol of an NPN transistor and its steady-state *V–I* characteristic. The current flowing from the base to the emitter controls the collector current. The DC gain h_{FE}, which is the ratio of the collector and base currents, is in the range of 5–10 in power BJTs. Consequently, in a higher power converter, the continuous base current becomes significant. Consequently, BJTs require complex control circuits because the controller must provide sufficient base current to turn the BJT on.

Example 9.2 A BJT with DC gain 5 is considered as a switching element in a 1-kW, 50-V DC–DC converter. Calculate the base current that the control circuit must supply.

Solution

When the BJT is on, the collector current is

$$I_C = \frac{1{,}000}{50} = 20\ \text{A}$$

Given the DC gain of 5, the base current is found as

$$I_B = \frac{I_C}{h_{FE}} = \frac{20}{5} = 4\ \text{A} \therefore$$

A Darlington or triple Darlington configuration may be an option to achieve higher DC gain; however, the increased collector–emitter current to turn on and slower switching speeds are the major drawbacks.

9.3.1.2 Gate Turn-Off Thyristors (GTO)

GTO was developed to overcome the challenges thyristors faced in the turn-off process. Figure 9.3 (b) shows the circuit symbol of a GTO and its steady-state *V–I* characteristic. Like the thyristor, GTO turns on with a pulse current while it is forward-biased. Once turned on, the gate current is not required to maintain the on state. Unlike the thyristor, a few microseconds of long negative voltage across the gate and cathode terminals produces a gate current pulse that turns off the GTO. This device is labeled as pulse controlled in the figure.

The on-state voltage of a GTO across the anode and cathode is slightly higher than the thyristors. The switching speed of GTO ranges from a few microseconds to 25 μs. Because of their high voltage and current ratings, GTOs are preferred in higher power converters. But the switching frequency is lower than that of the other semiconductor devices (up to 10 kHz).

9.3.1.3 Metal-Oxide-Semiconductor Field-Effect Transistors (MOSFETs)

MOSFETs are voltage-controlled devices and require very small gate current during switching-on period. This property simplifies the gate circuit and reduces the power loss in the control unit. Figure 9.3 (c) shows the circuit symbol of an N-channel MOSFET, which is more common in power converter configurations, and its steady-state *V–I* characteristic. To turn on a MOSFET, the voltage across the gate and source terminal V_{GS} must be above a specified threshold voltage. In addition, V_{GS} must be at an appropriate magnitude to ensure a drain current at the desired level. The gate–source voltage must be present during the on period. Because the gate is insulated from the drain and source junctions, a MOSFET practically does not draw any current from the control unit, except for the small capacitive current that flows during the on–off transitions.

The specified on-state resistance of a MOSFET between the drain and source terminals rises as the blocking voltage increases. To limit the conduction loss, devices with relatively low-voltage ratings are available (Mohan et al., 2003).

Table 9.3 Comparison of semiconductor switching devices

Device	Voltage and Current Handling Capabilities	Switching Speed
BJT	Medium—up to 1.4 kV and a few hundred amps	Medium
MOSFET	Low—up to 1 kV and 100 A	Fast
GTO	High—up to 4.5 kV and a few kA	Slow
IGBT	Medium—up to 1.7 kV and 1.2 kA	Medium

9.3.1.4 Insulated Gate Bipolar Transistors (IGBTs)

The IGBT, which was introduced in the early 1980s, has become popular because of its superior characteristics (Rashid, 2011). IGBTs present a combination of the properties of BJT, GTO, and MOSFET. Figure 9.3 (c) shows the circuit symbol of an IGBT and its steady-state *V–I* characteristic. Like MOSFETs, IGBTs are voltage-controlled devices and practically do not draw current from the control circuit, except on–off transitions. Their on-state voltage across the collector–emitter terminals is small like BJT. Similar to GTO, an IGBT can block negative voltages. Commercial IGBTs have turn-on and turn-off times in the order of 1 μs, hence can operate at higher frequency than MOSFETs. They are available in ratings as large as 1,700 V and 1,200 A (Rashid, 2011).

9.3.1.5 Comparison of Controlled Switches

Voltage and current capabilities and switching speed of the semiconductor switching elements described earlier are compared in Table 9.1. Typically, BJTs are suitable at lower power mechatronic systems where the control current is small enough. MOSFETs are preferred in medium-power range especially controlled by digital logic circuits and microcontrollers. GTOs and IGBTs are more cost-effective at higher power industrial-scale converters (Table 9.3).

In this section, we explain the operation of three basic switching DC–DC converter circuits.

9.3.2 The Step-Down (Buck) Converter

Buck converters function as an electronic version of the simple switch illustrated in Figure 9.9. The output voltage of the buck converter is always smaller than the input voltage. Therefore, its purpose is to adjust the output voltage from zero up to the value of the fixed input voltage.

The centerpiece of any power converter is the switching element. While any kind of three-terminal electronic device can achieve switching, the circuit diagram in Figure 9.9 (a) illustrates switching by a MOSFET as an example. A digital controller module provides the voltage pulse train that turns the MOSFET on and off at the appropriate duty ratio to obtain the desired output voltage. The controller may be as simple as a transistor timer or a comprehensive microcontroller board. The capacitor filters out the AC components of the output voltage.

The second semiconductor device in the circuit is the diode. Note that the diode is connected to the source terminal of the MOSFET, and it is reverse-biased whenever the MOSFET is on. When the transistor turns off, the diode becomes forward-biased and provides a path to allow the flow of the reverse current due to the energy stored. In Figure 9.9 (a), a simple resistive load is considered to simplify the analysis. In practice, the buck converter usually supplies a rechargeable battery or electric machines that contain energy-storing elements and sources. In that case, the load resistor should be replaced by the Thevenin equivalent circuit of the load.

In addition to continuous operation ratings, other considerations are the switching speed and losses that occur during switching. The switching elements must be selected to support the on-state continuous current, off-state voltage across the output terminals, surge current, and transient voltage during the transition between on and off states.

To analyze the operation of the buck converter, in Figure 9.9 (b) and (c), the switching transistor and reverse diodes are replaced with switches. We will assume that the output capacitor is large enough to maintain the output voltage constant. Circuit in (a) corresponds to the time interval when the transistor is on and the diode is off. Based on the KVL, we see that the voltage across the inductor is $V_s - V_o$ = constant.

Applying the voltage–current relationship of an inductor, we can write

$$v_L = L\frac{di_L}{dt} = V_s - V_o = \text{constant} \Longrightarrow \frac{di_L}{dt} = \frac{\Delta i_L}{\Delta t} \tag{9.18}$$

Duty ratio defines the on and off times of the switching transistor in terms of the switching period.

$$D = \frac{t_1}{T} \Longrightarrow (\Delta t)_{on} = DT;\ \ (\Delta t)_{off} = (1 - D)T \tag{9.19}$$

Circuit in Figure 9.9 (b) corresponds to the time interval when the transistor is off and the diode is on. According to KVL, the voltage across the inductor is $-V_o$ = constant.

$$v_L = L\frac{di_L}{dt} = -V_o = \text{constant} \Longrightarrow \frac{di_L}{dt} = \frac{\Delta i_L}{\Delta t} \tag{9.20}$$

$$(\Delta i_L)_{off} = \frac{-V_o \Delta t}{L} = -\frac{V_o}{L}(1 - D)T \tag{9.21}$$

Stable operation of the switching converter requires that the inductor current at the end of the switching cycle be the same as that at the beginning. This means that the net change in inductor current over one period is zero.

$$(\Delta i_L)_{on} + (\Delta i_L)_{off} = 0 \Rightarrow \frac{(V_s - V_o)}{L}DT - \frac{V_o}{L}(1 - D)T = 0 \tag{9.22}$$

Solving this equation for the output voltage, we obtain

$$V_o = DV_i \tag{9.23}$$

Average capacitor current is zero; therefore, the average inductor current is equal to the average load current, which is found using Ohm's law for the load resistance or the output power (Figure 9.8).

$$I_o = \frac{V_o}{R} \text{ or } I_o = \frac{P_o}{V_o} \tag{9.24}$$

Inductor current increases at a constant rate when the transistor is on and decreases at another constant rate when it is off. Therefore, the time variation of the inductor current is a triangular waveform. Because of symmetry, the maximum and minimum values of the inductor current can be expressed in terms of its average. We obtain the peak inductor current by adding half of its linear deviation Δi_L to the average. Similarly, the minimum inductor current is obtained by subtracting $\Delta i_L/2$ from the average. Either the on- or off-state deviation can be used, but equation 9.21 leads to simpler expressions.

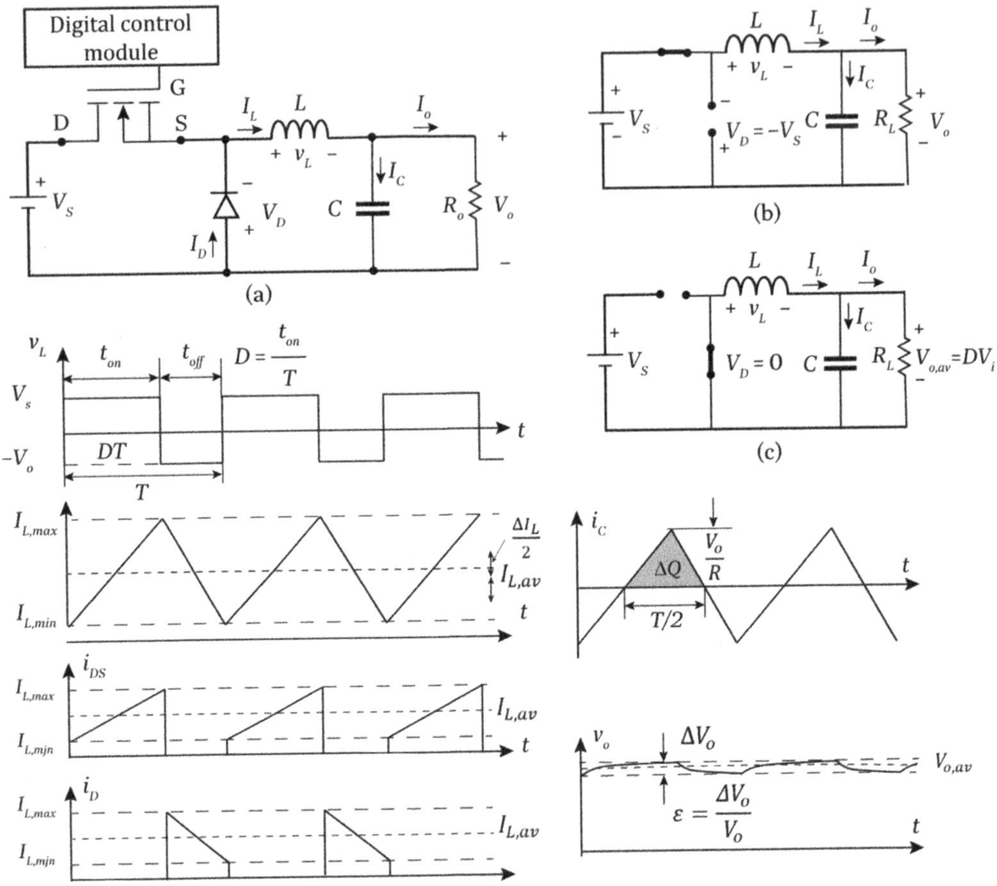

Figure 9.8 Buck converter circuit diagram and operation. (a) Circuit diagram; (b) Simplified diagram when switching device off; (c) Simplified diagram when switching diagram is on

$$I_{max} = I_L + \frac{\Delta i_L}{2} = I_o + \frac{1}{2}\frac{(1-D)}{Lf} \tag{9.25}$$

$$I_{min} = I_L - \frac{\Delta i_L}{2} = I_o - \frac{1}{2}\frac{(1-D)}{Lf} \tag{9.26}$$

To ensure the continuity of the inductor current, its minimum value must be equal to or greater than zero. The buck converter can operate in discontinuous inductor current, but the analysis method must be modified (Mohan, 2012).

Variation of the inductor current is an important criterion in converter design. In a typical design problem, input and output voltages and output power are the given requirements. The periodic fluctuation of the output voltage within a tolerance range, called voltage ripple, is one of the constraints. Power losses or efficiency and EMI limits are usually additional constraints.

In general, the input and output voltage requirements determine the duty ratio D. The designer must choose the switching frequency by considering the speed of switching elements and EMI with audio and wireless communication devices.

Equation 9.25 is useful in sizing the inductor in practical converter design. The inductance value must be sufficiently large to ensure the continuity of the inductor current. The maximum inductor current gives the current rating requirement of the selected inductor.

$$L_{min} = \frac{(1-D)R}{2f} \tag{9.27}$$

The load resistor R may not be known in a design problem, but it can be calculated from the output power and either output voltage or current using Ohm's law.

Initially, we assumed that the output capacitor is very large to maintain the output voltage constant. In practical converters, output capacitors are bulky and expensive elements; therefore, in practical design, the capacitance value must be optimized. Voltage ripple constraint is the main criterion to select the filter capacitor across the output terminals. To determine the minimum capacitance value, we must consider the static electric charge during the on and off state of the switching element. In steady state, the average electric charge stored in the capacitor is zero (Hart, 2011).

We can find the capacitor current by applying the Kirchhoff's current law (KCL) to the positive output node of the circuit in Figure 9.9.

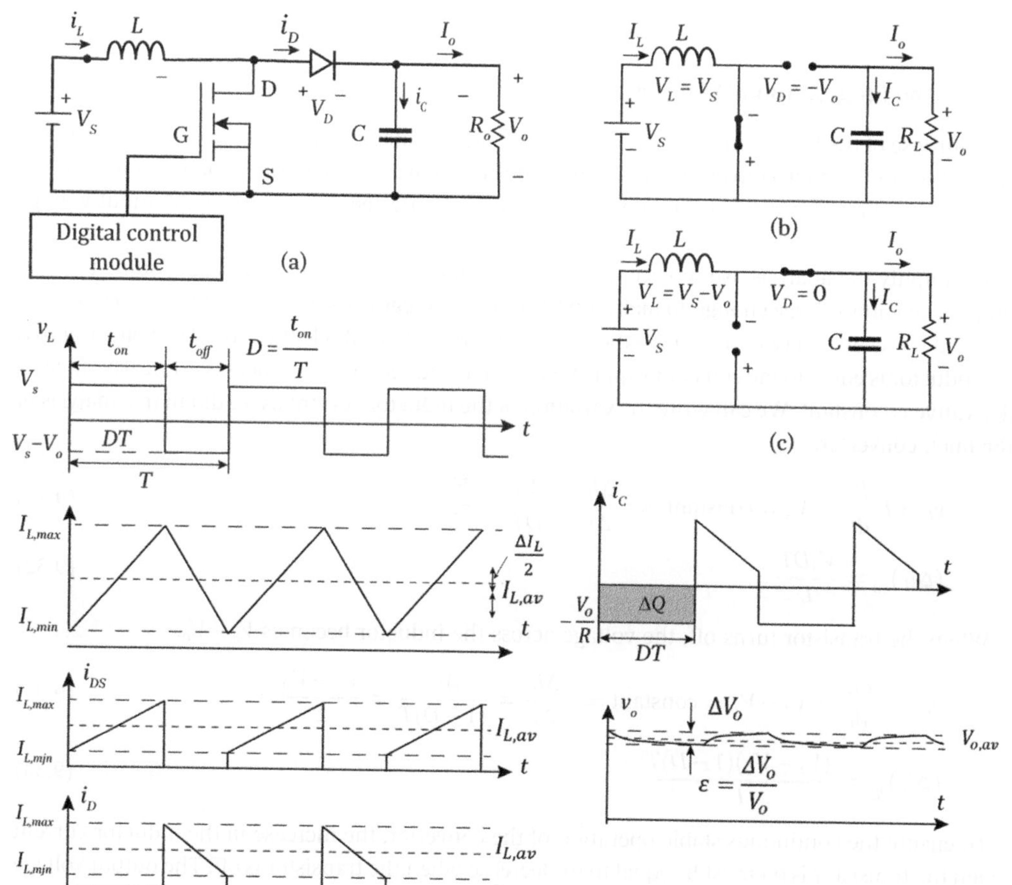

Figure 9.9 Circuit diagram and operation waveforms of a boost converter. (a) Circuit diagram; (b) Simplified diagram when switching device on; (c) Simplified diagram when switching diagram is off

$$i_c = i_L - i_R \tag{9.28}$$

The current flowing through the load resistor is around the average inductor current with a small voltage ripple, which is negligible compared to the capacitor current variation. Therefore, the capacitor current is a triangular waveform symmetrical to the time axis. The time variation of the capacitor current is obtained by shifting the inductor current plot horizontally to the horizontal axis. The capacitor charges during the on state of the switching element and supplies the load when the switch is off.

$$Q = CV_o \Rightarrow \Delta Q = C\Delta V_o$$
$$\Delta V_o = \frac{\Delta Q}{C} = \frac{1}{2C}\left(\frac{T}{2}\right)\left(\frac{\Delta i_L}{2}\right) \tag{9.29}$$

We can express the minimum capacitance to meet the voltage ripple requirement by substituting the variation of the off-state inductor current from equation 9.21 and arranging the equation.

$$C = \frac{T}{8\varepsilon Lf^2}(1-D) \tag{9.30}$$

9.3.3 The Step-Up (Boost) Converter

The purpose of the boost converter is to step up the input voltage. This is achieved by reversing the power flow of the buck converter. Note that in the circuit diagram of a simple boost converter, the transistor and diode are swapped, and the inductor is connected in series with the input voltage source.

To simplify the analysis, we replace the transistor and diode with switches and assume that the output capacitor is large enough to maintain the output voltage constant. When the transistor is on, the diode is reverse-biased. As soon as the transistor turns off, the diode turns on. The voltage across the inductor is equal to the constant supply voltage. The inductor current increases linearly since its derivative is constant. We can write the variation of the inductor current as we did in the analysis of the buck converter.

$$v_L = L\frac{di_L}{dt} = V_s = \text{constant} \Longrightarrow \frac{\Delta i_L}{\Delta t} = \frac{\Delta i_L}{DT} = \frac{V_s}{L} \tag{9.31}$$

$$(\Delta i_L)_{on} = \frac{V_s DT}{L} \tag{9.32}$$

When the transistor turns off, the voltage across the inductor becomes $V_s - V_o$.

$$v_L = L\frac{di_L}{dt} = V_s - V_o = \text{constant} \Longrightarrow \frac{\Delta i_L}{\Delta t} = \frac{\Delta i_L}{(1-D)T} = \frac{V_s - V_o}{L} \tag{9.33}$$

$$(\Delta i_L)_{off} = \frac{(V_s - V_o)(1-D)T}{L} \tag{9.34}$$

To ensure the continuous stable operation of the converter, the increase in the inductor current when the transistor is on must be equal to its decrease when the transistor is off. The output voltage is found by equating the rise and fall of the inductor current.

$$V_o = \frac{1}{1-D}V_s \tag{9.35}$$

This equation clearly shows that the output voltage of the boost converter is always equal to or greater than the input voltage, and as the duty ratio approaches unity, the output voltage rises to infinity.

As explained for a buck converter, the inductor current changes between a minimum and maximum around its average value. The average inductor current is determined by recognizing that the power supplied by the source must be the same as the power absorbed by the load resistor in the case of an ideal converter (Hart, 2011).

$$V_S I_L = \frac{V_o^2}{R} = \frac{1}{R}\left[\frac{1}{1-D}V_S\right]^2 \Rightarrow I_L = \frac{V_S}{(1-D)^2 R} \tag{9.36}$$

The minimum and maximum inductor current are important for designing the boost converter.

$$I_{L,max} = I_L + \frac{\Delta i_L}{2} = \frac{V_S}{(1-D)^2 R} + \frac{V_S DT}{2L}$$

$$I_{L,max} = I_L - \frac{\Delta i_L}{2} = \frac{V_S}{(1-D)^2 R} - \frac{V_S DT}{2L} \tag{9.37}$$

Boost converter analysis explained earlier is valid for continuous inductor current. To design a boost converter for continuous inductor current, the minimum value of the inductance must ensure that the inductor current is not below zero.

$$L_{min} = \frac{D(1-D)^2 R}{2f} \qquad f = \frac{1}{T} \tag{9.38}$$

As in the buck converter, the output capacitance is optimized to meet the voltage ripple requirement. Figure 9.10 (b) shows that the magnitude of the capacitor current is equal to the current flowing through the load resistor when the transistor is on. During this fraction of the period, the

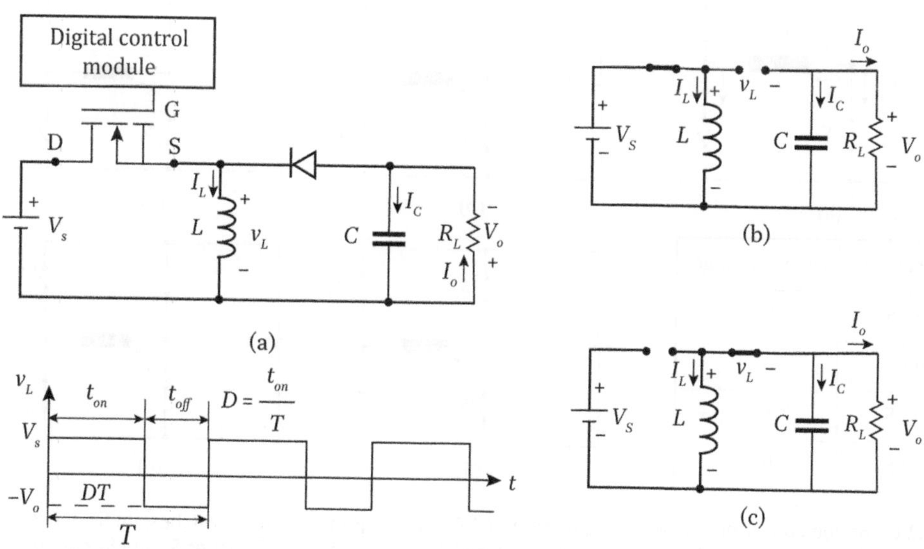

Figure 9.10 Circuit diagram and operation of the buck–boost converter (a: Circuit diagram; b: Equivalent circuit for switch on; c: Equivalent circuit for switch off)

capacitor supplies the load with the energy it has stored. The charge stored in the capacitor is equal to the integral of the current. Since the load current is approximately constant, dispensed electric charge is equal to the area of the rectangle $i_c \times DT$. At the same time, the variation of charge is equal to the capacitance time variation of the voltage. Hence, the required maximum voltage ripple determines the minimum capacitance value.

$$|\Delta Q| = \left(\frac{V_o}{R}\right) DT = C \cdot \Delta V_o \Rightarrow \Delta V_o = \frac{V_o DT}{RC} \tag{9.39}$$

Substituting the inverse of switching frequency to switching period, we can express the required minimum capacitance in terms of the voltage from this equation.

$$C_{min} = \frac{D}{\epsilon RCf} \qquad \epsilon = \frac{\Delta V_o}{V_o} \tag{9.40}$$

9.3.4 The Buck–Boost Converter

The buck–boost converter is the combination of buck and boost converters and can adjust the output voltage between zero and infinity for any fixed input voltage. Figure 9.11 (a) illustrates the circuit diagram of the buck–boost converter using a MOSFET and its simplified versions for switch on and off states.

Figure 9.11 (b) and (c) shows that when the transistor is on, the inductor voltage is equal to the supply voltage. When it is off, it becomes equal to the output voltage. Since the voltage across the inductor is constant in either switching mode, the current through it rises and falls as a straight line. The variation of the inductance current around its average value is

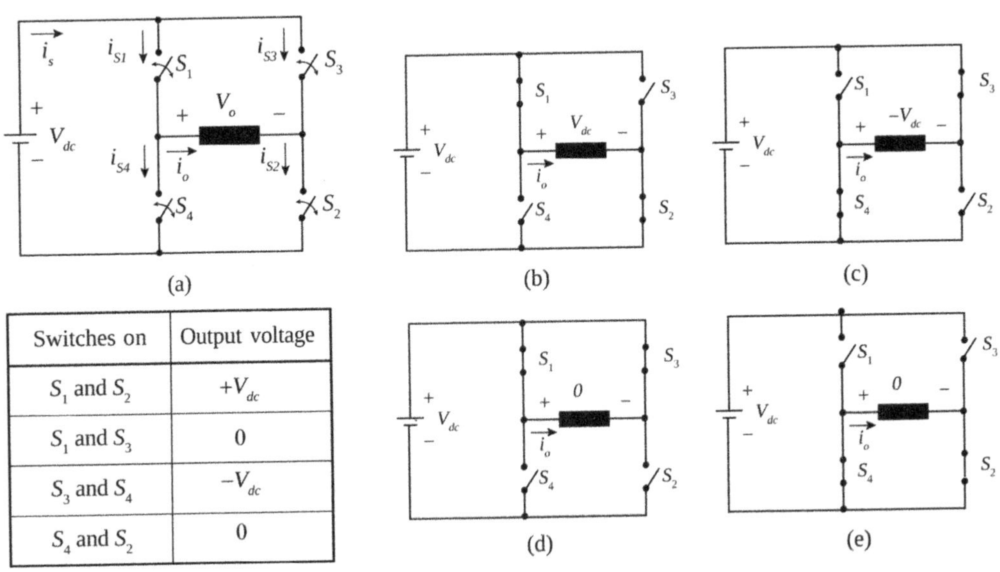

Switches on	Output voltage
S_1 and S_2	$+V_{dc}$
S_1 and S_3	0
S_3 and S_4	$-V_{dc}$
S_4 and S_2	0

Figure 9.11 Bridge converter and switching patterns. (a) Circuit diagram with four switches; (b) Switches S_1 and S_2 closed, S_3 and S_4 open; (c) Switches S_3 and S_4 closed, S_1 and S_2 open; (d) Switches S_1 and S_3 closed, S_2 and S_4 open; (e) Switches S_2 and S_4 closed, S_1 and S_3 open)

$$\Delta i_L = \frac{V_{in}}{L} DT_s = \frac{V_o}{L}(1-D)T_s \tag{9.41}$$

The output voltage of the buck–boost converter, defined below, is the combination of buck and boost converters.

$$V_o = \frac{V_{in}D}{1-D} \tag{9.42}$$

Note that for $D = 0$, $V_o = 0$ and when the duty ratio approaches unity, the output voltage increases to infinity. Obviously, the upper limit of the output voltage depends on the selected switching elements.

Average value of the inductor current can be found assuming that the converter is ideal; therefore, the input power is equal to the output power. Hence,

$$\begin{aligned} \frac{V_o^2}{R} &= (V_{in} I_{L,av})D \Rightarrow I_{L,av} = \frac{V_{in}D}{R(1-D)^2} \\ i_L &= I_{L,av} \pm \frac{\Delta i_L}{2} = V_{in}D\left[\frac{1}{(1-D)^2R} + \frac{1}{2Lf}\right] \end{aligned} \tag{9.43}$$

Minimum inductance must ensure the continuity of the inductance current.

$$L_{min} = \frac{(1-D)^2R}{2f} \tag{9.44}$$

The buck–boost converter's capacitance current waveform is identical to that of a boost converter. The expression of the minimum output capacitance in terms of the peak-to-peak output voltage ripple is the same as the boost converter.

$$C_{min} = \frac{D}{RCf} \qquad \varepsilon = \frac{\Delta V_o}{V_o} \tag{9.45}$$

9.3.5 DC–DC Converter Design Considerations

Each one of the DC–DC converters has strengths and drawbacks. Table 9.4 shows a comparison of the converter configurations described in this section (Mohan, 2012).

The buck converter can only reduce the supply voltage, but the switching frequency has significant effect on the inductance and capacitance values. Thus, it requires a smaller output capacitor to meet the same voltage ripple requirements. The inductor is smaller size and lower cost compared to the other converters. Furthermore, the transistor voltage rating is lower.

Boost converter has the ability to increase the input voltage. Buck–boost converter can produce an output voltage in a wider range, practically from the minimum to the maximum voltage the switching transistor can allow. However, the transistor voltage rating is higher compared to a buck converter, and the switching frequency has less effect on the inductance and capacitance values. Therefore, a bigger inductor and capacitor should be used, which increases the initial cost of the boost converter.

The buck–boost converter requires the highest transistor voltage and current rating. The effect of switching frequency on inductance and capacitance values is similar to the boost converter. Therefore, the buck–boost converter has the highest component cost among the three converter topologies.

Table 9.4 Design considerations for DC–DC converter configurations

		Buck	Boost	Buck–Boost
Output voltage	V_o	DV_{in}	$\frac{V_{in}}{1-D}$	$-\frac{DV_{in}}{1-D}$
Min. inductance	L_{min}	$\frac{(1-D)R_l}{2f}$	$\frac{D(1-D)^2R_l}{2f}$	$\frac{(1-D)^2R_l}{2f}$
Output voltage ripple	ε	$\frac{1-D}{8LCf^2}$	$\frac{D}{RCf}$	$\frac{D}{RCf}$
Transistor peak voltage	$\hat{V}$	V_{in}	V_o	$V_{in}+V_o$
Transistor peak current	$\hat{I}$	I_o	I_{in}	$I_{in}+I_o$
Transistor rms current	I_{rms}	$I_o\sqrt{D}$	$I_{in}\sqrt{D}$	$(I_{in}+I_o)\sqrt{D}$
Transistor average current	I_{rms}	DI_o	DI_{in}	$D(I_{in}+I_o)$
Inductance average current	$I_{L,\ av}$	I_o	I_{in}	$I_{in}+I_o$

All DC–DC converters described in this section can operate at any frequency. However, the designer must choose the switching frequency considering the reliable switching speed of the semiconductors and EMI with other electronic circuits. In systems sensitive to EMI, the converter must be appropriately shielded.

At higher frequencies, the size and cost of the inductor and capacitor become smaller. This is important because inductors and capacitors are usually the largest, heaviest, and most expensive parts in the converter circuit. Choosing a higher switching frequency would reduce the overall size and weight of the converter. However, as the switching frequency increases, the switching losses of the transistor and diode increase too.

Two types of losses occur in switching transistors. Switching loss is due to the rise and fall time of the drain–source voltage (V_{DS}) and drain current of the transistor. The equation below can be used to calculate the switching loss with sufficient accuracy for design purposes (Mohan, 2012).

$$P_{sw} = \frac{1}{2}V_{in}I_o\left(t_{rise} + t_{fall}\right)f_s \tag{9.46}$$

Conduction loss on the switching elements is based on Joules law similar to the heating loss on any resistive element. Given the on-state resistance of the semiconductor device, the conduction loss is

$$P_{cond} = D\left(R_{DS,on}{I_o}^2\right) \qquad D = \frac{t_{on}}{T}\text{: Duty ratio} \tag{9.47}$$

Excessive heat must be dispersed to the environment using a cooling system. The simplest cooling method is the use of a heatsink. If the heatsink is too large, a fan may be a more efficient cooling method.

Any type of power transistor can serve as a switching element. As a rule of thumb, at low power, range, up to several hundred watts, MOSFET may be a cost-effective option. For higher power converters in the kW range, IGBT is usually a better option.

A switching element is chosen primarily based on the maximum voltage across and the maximum continuous current that would pass through the element in the worst case. Maximum voltage across the element should not exceed the rated voltage when it is off. Effective (rms) current through the element should not exceed the rated current when it is on. Repetitive peak current should not exceed the value given in the specifications. Available elements that meet the voltage and current constraints above are compared based on the total loss (switching loss and conduction loss).

The designer should compare the performance of different switching element types, such as MOSFET, IGBT, or BJT, that are available on the market. The total material cost is one of the factors in any design problem. Cost of the cooling elements is higher if the total loss is greater. In addition, higher loss results in lower efficiency, which is undesirable. Consequently, achieving a higher efficiency is usually one of the design goals.

9.4 DC-to-AC Converters

DC-to-AC converters, also known as inverters, are essential electronic circuits for applications where AC power is needed from a DC source. In mechatronics and motor drive systems, DC-to-AC converters (inverters) are used to power AC motors and actuators from DC sources, such as batteries or solar panels. These converters enable precise control of motor speed and torque by varying the frequency and amplitude of the AC output, which is essential for applications like robotics, conveyor systems, and automated machinery. By integrating advanced control strategies, inverters help achieve efficient and flexible motion control in modern mechatronic assemblies.

Additionally, DC/AC converters allow seamless integration of renewable energy sources and mobile power supplies into mechatronic systems, expanding their versatility and energy efficiency. The ability to convert and regulate power on demand supports dynamic operation and adaptation to varying load conditions, which is critical for high-performance motor drives and intelligent automation.

A full-wave controlled rectifier can operate as an inverter when it is supplied from a DC source and connected to an AC load. In this configuration, by appropriately controlling the firing angles of the semiconductor switches (such as thyristors), the rectifier can force current to flow from the DC side back into the AC side, effectively reversing the direction of power flow. This process allows the circuit to deliver power from a DC supply into an AC grid or load, rather than just converting AC to DC. The ability to control the firing angle is what enables the device to function as both a rectifier and an inverter, depending on the direction of the desired power transfer.

DC/AC converters utilize switching devices to produce a quasi-sinusoidal or modified square-wave output, enabling the operation of AC motors and other equipment from DC power supplies. The efficiency and waveform quality of inverters depend on the control strategy and type of semiconductor switches used, making proper selection critical for the desired application.

Modern inverter designs employ sophisticated modulation techniques such as PWM and space vector modulation (SVM) to optimize output waveform quality and efficiency. These methods minimize harmonic distortion and EMI, which are crucial factors in sensitive mechatronic applications. Furthermore, advancements in digital control and real-time monitoring have enabled adaptive operation, allowing inverters to respond dynamically to changes in load and system requirements. This section presents a summary of fundamental inverter circuits. Readers seeking an in-depth analysis can consult well-known textbooks on power electronics including Hart (2011), Mohan et al. (2003), and Rashid (2011).

9.4.1 Bridge Converter

Figure 9.12 illustrates the structure and operation of a bridge converter, commonly known as an H-bridge, the most straightforward circuit design for producing AC output from a DC source. The four switches in the circuit can be any type of semiconductor switching element. In practice, BJTs are generally used for low-power applications, MOSFETs for medium-power, and IGBTs for high-power needs.

An AC output is synthesized from a DC input by turning the switches on and off in an appropriate sequence. The output voltage V_o can be $+V_{dc}$, $-V_{dc}$ or zero, depending on which switches are closed.

9.4.1.1 Square-Wave Inverter

When switches S_1 and S_2 are closed while the other two switches remain open, as illustrated in Figure 7.13 (b), the voltage across the load is equal to the supply voltage with a positive polarity. If S_1 and S_2 are opened while S_3 and S_4 are closed, the output voltage and current reverse direction. In this mode of operation, the converter generates a square-wave output of magnitude V_{dc}.

While this alternating output is nonsinusoidal, it can still serve as a suitable AC waveform for basic applications. The current waveform across the load is determined by the characteristics of the load itself. For a resistive load, the current takes on the same shape as the output voltage. However, with an inductive load, the current variation resembles a sinusoidal waveform due to the smoothing effect of inductance. In many mechatronic systems, electromechanical devices and motors typically have low resistance and relatively high inductance, which causes the current to flow in an almost sinusoidal waveform. If this is not the case, adding an external inductor in series with the load can effectively act as a filter. When using an inductive load, it's important to account for additional factors in switch design for the full-bridge circuit, since the switches need to handle currents flowing in both directions.

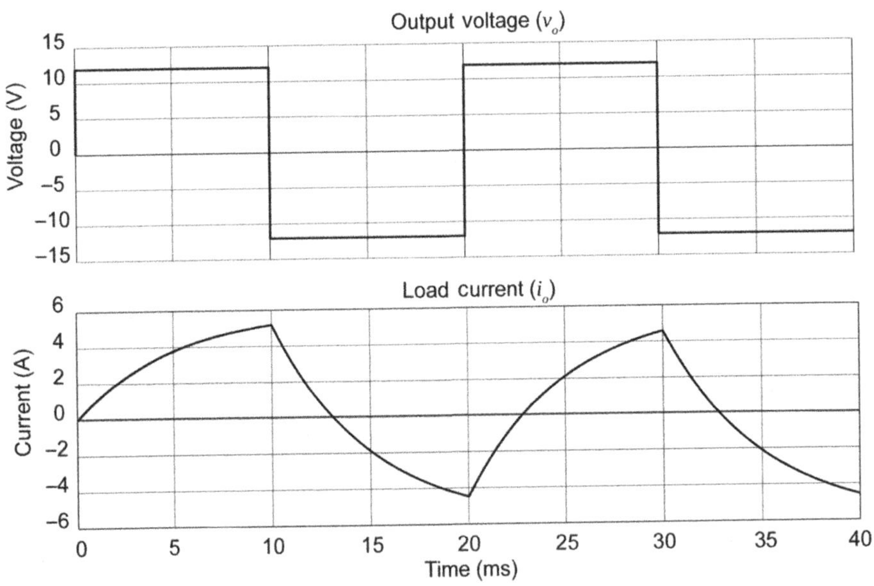

Figure 9.12 Output voltage and current waveforms of the square-wave inverter supplying an inductive load

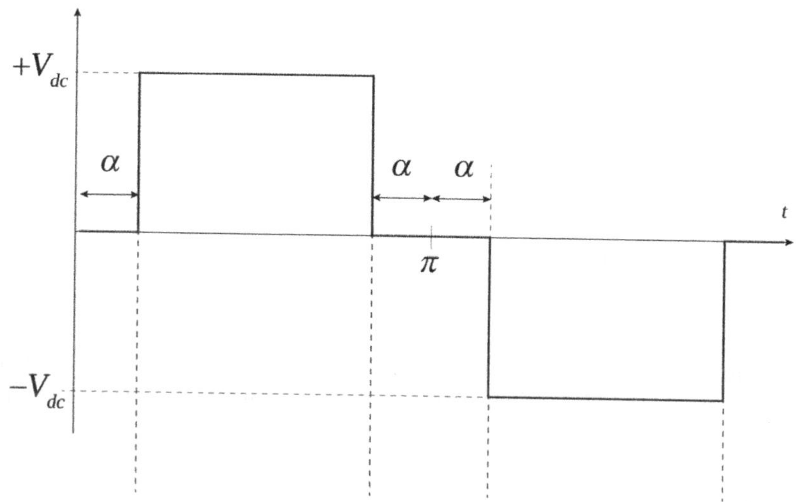

On	S_2, S_4	S_1, S_2	S_1, S_3	S_3, S_4	S_2, S_4
Off	S_1, S_4	S_3, S_4	S_2, S_4	S_1, S_2	S_1, S_2

Figure 9.13 Modified square wave and switching pattern

Suppose that switches S_1 and S_2 close at $t = 0$. The voltage across the load is $+V_{dc}$, and the load current begins to increase. The current is expressed as the sum of the forced and natural responses. For $0 \leq t \leq T/2$,

$$i_o(t) = i_f(t) i_n(t) = \frac{V_{dc}}{R} + Ae^{-t/\tau} \tag{9.48}$$

In this equation, $\tau = L/R$ represents the time constant of the load circuit. A is evaluated from the initial condition. When the circuit is first energized and the initial inductor current is zero, a transient occurs before the load current reaches a steady-state condition. At steady state, $i_o(t)$ is periodic and symmetric about zero changing between I_{min} and I_{max}. Using these two initial conditions, we can obtain the load current as

$$i_o(t) = \begin{cases} \frac{V_{dc}}{R} + \left(I_{min} - \frac{V_{dc}}{R}\right)e^{-t/\tau} & 0 < t < \frac{T}{2} \\ -\frac{V_{dc}}{R} + \left(I_{min} + \frac{V_{dc}}{R}\right)e^{-\left(t-\frac{T}{2}\right)/\tau} & \frac{T}{2} < t < T \end{cases} \tag{9.49}$$

The maximum load current expression can be obtained by substituting t with $T/2$. By symmetry, the minimum value of the output current equals the maximum current with opposite sign.

$$I_{max} = -I_{min} = \frac{V_{dc}}{R} \cdot \frac{1 - e^{-T/2\tau}}{1 + e^{-T/2\tau}} \tag{9.50}$$

Effective load current is calculated by using the definition of rms.

$$I_{rms} = \sqrt{\frac{1}{T}\int_0^T i_o^2(t)dt} = \sqrt{\frac{2}{T}\int_0^{T/2}\left[\frac{V_{dc}}{R} + \left(I_{min} - \frac{V_{dc}}{R}\right)e^{-t/\tau}\right]^2 dt} \tag{9.51}$$

An alternative method for determining the output current variations involves utilizing the Euler integration algorithm in MATLAB®. The rms value can then be computed through numerical integration techniques.

Example 9.3 A bridge inverter supplied from 12-V battery powers an inductive load consisting of 2 Ω resistor and 10-mH inductor at 50-Hz. Plot the output voltage and current waveforms and calculate their rms values using MATLAB®.

Solution

The goal is to plot the output voltage and current waveforms using a MATLAB® code. We will model the voltage waveform as an ideal square-wave switching between +12 and −12 V at 50 Hz. The load current will be computed using the Euler algorithm.

The inductive load affects current shape, but not the voltage waveform in an ideal inverter, since the voltage is imposed by the DC supply. The rms value of a symmetric square wave is simply its magnitude: $V_{rms} = V_{dc}$. Because the current waveform is obtained over two periods, the rms value can be calculated by summing the square of the discrete current values, dividing by the total time, and taking the square root.

MATLAB® Script to Plot the Output Voltage and Current Waveforms of a Bridge Inverter.

```
% Parameters
Vdc = 12; % DC supply voltage (Volts)
R = 2; % Resistance (Ohms)
L = 10e-3; % Inductance (Henries)
f = 50; % Output frequency (Hz)
T = 1/f; % Period (seconds)
t_end = 2*T; % Simulate for two cycles
dt = 1e-5; % Time step (seconds)
t = 0:dt:t_end; % Time vector
% Output voltage: square wave ±Vdc at 50 Hz
v_out = Vdc * sign(sin(2*pi*f*t));
% Initialize current vector
i_out = zeros(size(t));
i = 0; % Initial current
% Numerical integration using forward Euler method
for k = 1:length(t)-1
 di_dt = (v_out(k) - R*i) / L;
 i = i + di_dt * dt;
 i_out(k+1) = i;
end
% RMS calculations using integration
v_rms = sqrt(sum(v_out.^2) * dt / t_end);
i_rms = sqrt(sum(i_out.^2) * dt / t_end);
% Plotting
```

```
figure;
subplot(2,1,1);
plot(t, v_out, 'k', 'LineWidth', 1.5);
xlabel('Time (s)');
ylabel('Voltage (V)');
title('Output Voltage Waveform');
grid on;
subplot(2,1,2);
plot(t, i_out, 'k', 'LineWidth', 1.5);
xlabel('Time (s)');
ylabel('Current (A)');
title('Load Current Waveform');
grid on;
% Display results
fprintf('Voltage RMS: %.2f V\n', v_rms);
fprintf('Current RMS: %.2f A\n', i_rms);
```

The output voltage and current waveforms are displayed in Figure 9.12. The calculated rms values are:

Voltage rms: 12.00 V
Current rms: 3.15 A

9.4.1.2 Harmonics and Total Harmonic Distortion (THD) in Square-Wave Inverter

Square-wave inverters, while simple and cost-effective, introduce significant harmonic distortion into both voltage and current waveforms when supplying inductive loads. This harmonic content arises because the square waveform inherently contains a series of odd harmonics, which can cause additional heating in inductive components, reduce overall efficiency, and potentially interfere with sensitive electronic equipment. The presence of these harmonics is quantified using metrics such as total harmonic distortion (THD), and mitigation often requires the use of filters or alternative inverter topologies to achieve smoother output waveforms suitable for demanding applications.

The output voltage and load current of a square-wave inverter may be represented using the Fourier series. When the load is resistive, the output current waveform is also a square wave with magnitude V_{dc}/R. Due to symmetry, the output contains no DC component. A square-wave output voltage has only odd harmonics. The expression below provides the Fourier series coefficients for a square wave.

$$\begin{aligned} v_o(t) &= V_n \sum_{n=1}^{\infty} [\sin(n\omega_0 t) + \cos(n\omega_0 t)] \\ V_n &= \frac{4V_{dc}}{n\pi}; \quad I_n = \frac{V_n}{Z_n} \end{aligned} \tag{9.52}$$

Magnitudes of the current harmonics can be calculated by dividing each voltage harmonic by the impedance at the corresponding harmonic frequency.

In design calculations, the rms value of the current is often required to size the switching elements. Since each current harmonic is a sinusoidal waveform, the rms values of the harmonic components of the current are calculated by simply dividing their amplitudes by $\sqrt{2}$. The following expressions may be used to calculate the rms value using the Fourier transform components of the voltage waveform.

$$I_{rms} = \sum_{n=1,3,5,\ldots} \left(\frac{I_n}{\sqrt{2}}\right)^2 = \frac{1}{2}\sum_{n=1,3,5,\ldots} \frac{V_n^2}{Z_n^2} \tag{9.53}$$

THD is defined as the ratio of the rms value of all harmonic components of a waveform to the rms value of its fundamental frequency component. Specifically, for a voltage waveform, THD quantifies how much of the waveform's total energy is contained in harmonics as opposed to the fundamental frequency, providing a measure of distortion in the signal. This metric is commonly used to assess the quality of inverter outputs and is critical in applications where minimizing waveform distortion is important.

Mathematically, THD can be expressed as

$$THD = \frac{\sqrt{\sum_{n=2}^{\infty} (V_{n,rms})^2}}{V_{1,rms}} = \frac{\sqrt{V_{rms}^2 - V_{1,rms}}}{V_{1,rms}} \tag{9.54}$$

Here, $V_{n,rms}$ represents the rms value of the nth harmonic, and $V_{1,rms}$ is the rms value of the fundamental component. A lower THD value indicates a waveform that more closely resembles a pure sine wave, while higher THD values signify greater distortion. The THD of current is determined by substituting current for voltage in the above equation. The THD of load current is often of greater interest than that of output voltage.

9.4.1.3 Modified Square-Wave Inverter

The output voltage of bridge inverters may be adjusted by slightly changing the switching sequence, specifically by closing the bridged switches with an angular delay of α radians as shown in Figure 9.14. Effective (rms) value of the delayed square-wave voltage waveform is as follows:

$$V_{rms} = \sqrt{\frac{1}{\pi}\int_0^{\pi-\alpha} V_{dc}^2 d(\omega t)} = V_{dc}\sqrt{1 - \frac{2\alpha}{\pi}} \tag{9.55}$$

With the delay angle α as defined in the figure, the Fourier series components of the square wave with delay gaps are calculated by the equation below:

$$V_n = \frac{2}{\pi}\int_{\alpha}^{\pi-\alpha} V_{dc}\sin(n\omega_0 t)d(\omega_0 t) = \frac{4V_{dc}}{n\pi}\cos(n\alpha) \tag{9.56}$$

Since the amplitude of each harmonic component is a function of α, the harmonic content can be adjusted by selecting α properly. For example, by selecting $\alpha = 30°$, the third harmonic can be eliminated.

Delaying the switching sequences allows for control over either the amplitude or harmonic composition of the output voltage. To regulate both amplitude and harmonic content simultaneously, a more sophisticated switching strategy such as PWM can be implemented.

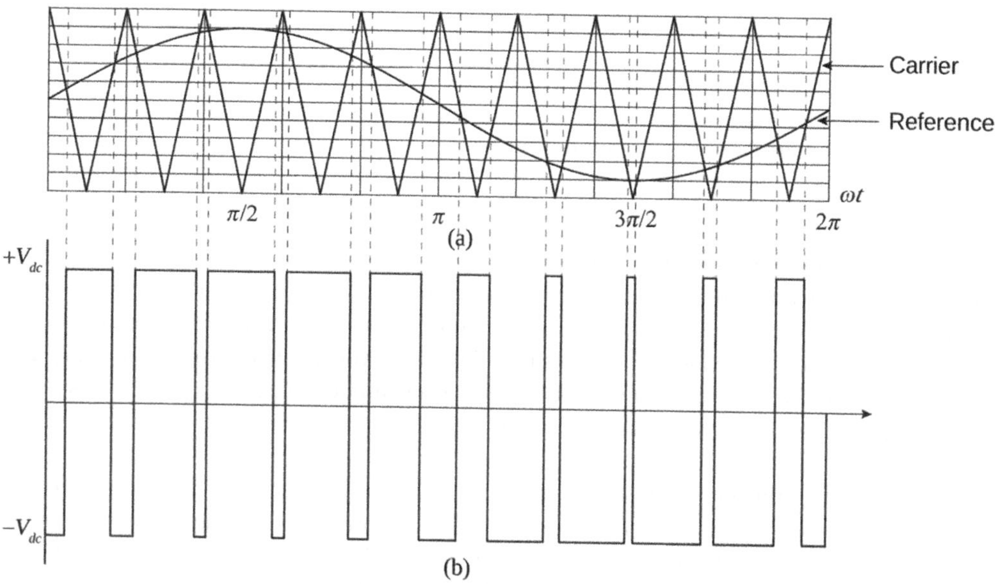

Figure 9.14 PWM waveform generation

Example 9.4 A bridge inverter produces a modified square-wave output by introducing a delay angle α in its switching sequence. The output voltage is $\pm V_{dc}$, with a delay gap of α radians in each half-cycle.

Given DC supply voltage, $V_{dc} = 24$ V and fundamental frequency, $f = 50$ Hz.

a) Calculate the rms value of the output voltage.
b) Calculate the amplitudes of the third and fifth harmonics for $\alpha = 30^\circ$.

Solution

The rms value of a modified square wave is determined by substituting the given values in equation 9.55, where α is in radians. 30° shift corresponds to $\pi/6$ radians.

$$V_{rms} = V_{dc}\sqrt{1 - \frac{2\alpha}{\pi}} = 24 \times \sqrt{1 - 2 \times \frac{\pi/6}{\pi}} \cong 20 \text{ V}$$

a) The amplitude of the nth harmonic is given in equation 9.56. For the third harmonic, $n = 3$, which results in

$$\cos(n\alpha) = \cos\left(3 \times \frac{\pi}{6}\right) = \cos\left(\frac{\pi}{2}\right) = 0$$

Therefore, for $\alpha = 30°$, the third harmonic becomes zero, regardless of the DC supply voltage. Amplitude of the fifth harmonic is

$$V_5 = \frac{4V_{dc}}{5\pi}\cos(5\alpha) = \frac{4 \times 24}{5\pi}\, cos\left(5 \times \frac{\pi}{6}\right) \cong 6.1 \text{ V}$$

Key Takeaways.

- The rms value of the output voltage decreases as the delay angle increases.
- Introducing a delay angle α in the switching sequence allows selective harmonic elimination.
- For $\alpha = 30°$, the third harmonic is eliminated.

9.4.1.4 Pulse-Width Modulation (PWM) Inverter

PWM is a widely used strategy for controlling the output waveform of inverters. In PWM, the switching devices in the inverter are turned on and off at a high frequency, with the duration of the "on" period (the pulse width) varied to shape the output voltage waveform. By adjusting the pulse widths within each cycle, the inverter can produce an output that closely approximates a sine wave, significantly reducing harmonic distortion compared to simple square-wave or modified wave strategies.

The fundamental principle of PWM is to modulate the width of each pulse so that the average voltage over a given period matches the desired output waveform. This technique not only allows precise control of the output amplitude but also enables selective attenuation of specific harmonic components. As a result, PWM improves the quality of the AC output, enhances efficiency, and minimizes heat loss. The flexibility of PWM makes it highly effective for applications requiring both amplitude and harmonic regulation in inverter circuits.

Modern inverter designs often employ various PWM schemes, such as sinusoidal PWM, where the reference signal is a sine wave compared to a high-frequency triangular carrier, or space vector PWM, which optimizes switching sequences for three-phase outputs. These methods offer improved performance in terms of waveform fidelity and electromagnetic compatibility, making PWM the preferred choice for advanced inverter applications.

The core idea of PWM is to use high-frequency switching to produce pulses whose widths correspond to the instantaneous amplitude of the desired output waveform. When the average value of these pulses is filtered (e.g., by the load or additional filters), a near-sinusoidal voltage and current can be achieved. The width of each pulse is determined by comparing a sinusoidal reference with a triangular carrier signal.

For a full-bridge inverter, there are two main switching strategies known as bipolar and unipolar switching. In bipolar switching, when the instantaneous value of the reference waveform (sinusoidal) is greater than the carrier (triangular), the output is $+V_{dc}$. Otherwise, the output voltage is $-V_{dc}$. In a unipolar switching, each terminal of the load becomes at higher or lower potential depending on the relative values of the reference and carrier.

9.4.2 Three-Phase Inverters

Three-phase inverters extend PWM concepts to supply three-phase loads, such as induction motors or industrial equipment, by generating three output voltages that are 120° apart. These systems use coordinated switching of multiple transistors or IGBTs to synthesize balanced three-phase sinusoidal waveforms. Effective control of the switching sequence and timing is crucial to minimize harmonic distortion and ensure smooth motor operation. Advanced modulation techniques, like space vector PWM, are particularly beneficial in three-phase systems for optimizing output voltage utilization and reducing switching losses.

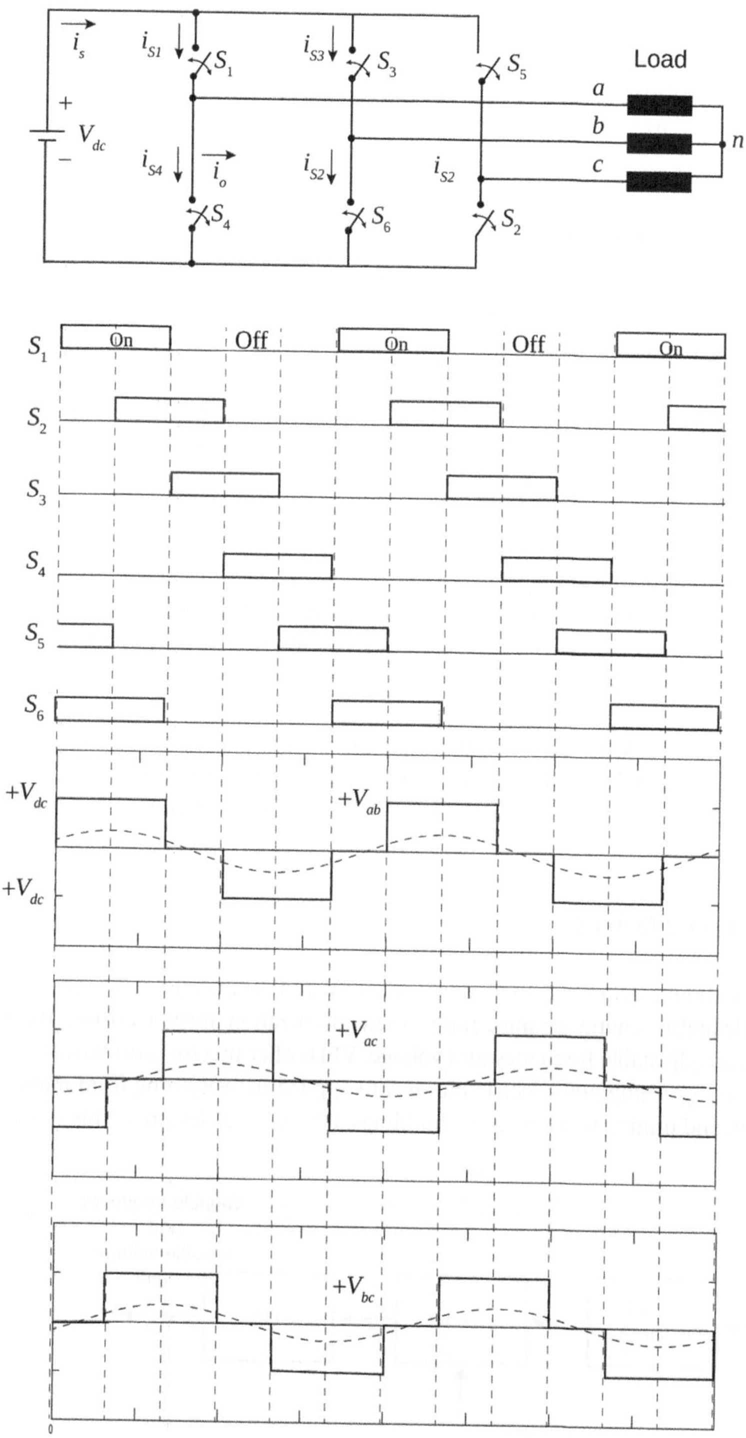

Figure 9.15 Three-phase inverter and six-pulse switching scheme

The three-phase full-wave inverter is constructed by connecting an additional pair of switches in parallel with the single-phase full bridge, as illustrated in Figure 9.15. Various switching patterns may be used to generate square or PWM waveforms. The diagram illustrates a switching sequence that generates a modified square wave for three-phase line-to-line voltages, each phase shifted by 120° relative to the others.

9.4.3 Design Considerations in Bridge Inverters

When designing bridge inverters, several factors must be considered to ensure reliable operation and optimal performance. Key aspects include the selection of appropriate switching devices, thermal management strategies, and protection circuits to prevent damage from overcurrent or voltage spikes.

Any kind of switching semiconductor may be employed in bridge configurations. The voltage and current ratings of the selected device must exceed the maximum reverse voltage and forward current that the element must withstand. In addition, the switching frequency (or delay time) must suit the selected switching frequency.

The switching frequency and control methods directly affect efficiency, waveform quality, and EMI. Switching losses are proportional to the switching frequency. Higher switching frequency results in higher losses, but smaller size components, such as inductors, capacitors, and transformers. Furthermore, higher frequency increases the possibility of EMI with other circuits in the vicinity. Proper shielding and PCB layout help minimize losses and noise in high-frequency applications.

For inductive loads, the voltage across each switching component becomes negative during parts of the waveform. To protect semiconductor devices from potential damage, a reverse-biased diode should be connected between the source (or collector) and drain (or emitter) terminals.

9.5 Variable Frequency Drives

VFDs have become a foundational technology in modern industrial and commercial motor control, enabling unprecedented flexibility, energy savings, and process optimization. By converting a fixed frequency AC supply into an adjustable frequency and voltage, VFDs offer precise control over AC motor speed and torque, transforming how electric motors are applied in everything from manufacturing, water treatment, and mining to commercial buildings, HVAC, and elevators. This report

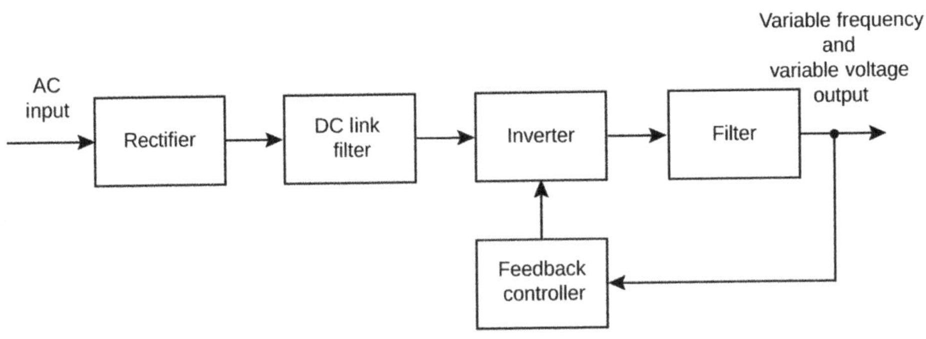

Figure 9.16 Variable frequency drive (VFD) configuration

provides an in-depth exploration of VFD structure, the major components—rectifier, DC link, inverter—their collective operation, advanced control strategies (including PWM), and the practical benefits and applications of VFDs. The discussion also addresses relevant safety and performance standards, market dynamics, and best practices and includes technical diagrams and tables for clarity.

A VFD system essentially consists of several electronic sections that work together, each handling key roles in converting and managing electrical power. The primary components of a VFD are illustrated as a block diagram in Figure 9.16.

9.5.1 Rectifier

The rectifier is the entry point of a VFD, converting the incoming AC power, either fixed or variable frequency, to a DC voltage. Utility grid provides fixed frequency (50 or 60 Hz) and voltage, while wind turbines and some grid-independent generators may deliver variable frequency and amplitude AC power. Residential low-voltage grids use single-phase, whereas typical industrial networks are three-phase. Single-phase voltage is usually rectified using a full-wave diode bridge. In most industrial VFDs, this is achieved using a six-pulse SCR (thyristor) bridge. For some higher-end or regenerative drives, MOSFETs, IGBTs, or SCRs are used to allow controlled rectification and even bidirectional power flow for regenerative brake features. Controlled rectifiers (using SCRs, MOSFETs, or IGBTs) can both rectify and invert, supporting regenerative applications and harmonic mitigation.

9.5.2 DC Link/Filter

The DC link, located between the rectifier and the inverter, consists of large electrolytic capacitors and sometimes series reactors or chokes. The main tasks here are to smooth out the rectified DC (removing AC "ripples" and harmonics), provide a stable voltage to the inverter, and buffer energy for rapid load changes or during regeneration. The capacitors ensure that the inverter receives a relatively ripple-free DC supply, essential for high-quality AC synthesis at the output. Some designs include braking resistors or regenerative modules to safely dissipate or return energy during rapid deceleration or overhauling loads.

9.5.3 Inverter

The inverter stage converts DC into an AC output with the required frequency and voltage. Modern inverters almost universally use IGBTs as switching elements. These are arranged in a three-phase bridge configuration and are switched at high speed (ranging from a few kHz up to 16 kHz or higher) to generate a series of PWM voltage pulses that closely approximate a sinusoidal AC waveform.

By adjusting the width and timing of these pulses, the VFD can finely control both the frequency and amplitude of the output voltage. To keep an AC motor running efficiently, the voltage-to-frequency ratio must stay constant to preserve air gap flux density. The use of IGBTs enables fast, efficient, and robust switching, even at high-power densities, making them crucial for modern VFD topologies.

9.5.4 Control and Interface Circuits

VFD control relies on circuitry controlled by a microprocessor, microcontroller, or DSP for precise motor and process management. The control circuit takes user commands, combining this with sensor feedback (speed, torque, current, temperature), and generates the gate signals for the inverter's power switches. It also supervises soft start/stop profiles, implements advanced control algorithms (V/Hz, vector control, etc.), and manages overload protection, diagnostics, and fault logging.

Speed of synchronous and asynchronous motors is determined by the frequency f of the power supply and the number of poles p. As explained in the previous chapter, the synchronous speed in revolution per minute (RPM) is

$$N_s = \frac{120f}{p} \tag{9.57}$$

By adjusting f, the VFD controls the speed from 0 (standstill) up to or even above the nameplate (base) speed of the motor (using field-weakening techniques for overspeed).

The torque output relates primarily to the V/f ratio. When frequency and voltage are changed proportionally, the air gap flux is kept constant, maximizing torque throughout the speed range. More sophisticated VFDs employ the following advanced vector control algorithms.

- **Scalar (V/Hz) Control:** Keeps voltage and frequency proportional. This method is robust and simple, but with limited torque accuracy at low speeds.
- **Open-Loop Vector:** Uses current/voltage feedback and sophisticated algorithms to estimate rotor position and current, enhancing dynamic torque and speed regulation without a sensor or encoder.
- **Closed-Loop Vector (Field-Oriented Control):** Employs an encoder or resolver for real rotor feedback, delivering precise torque and speed regulation over the full operating range, including full torque at zero speed.
- **Direct Torque Control (DTC):** Calculates magnetic flux and torque directly, adjusting inverter switching in real time for ultra-fast response (used in high-end ABB drives).

These control modes allow VFDs to meet the needs of constant torque loads like conveyors and hoists and variable-torque loads like pumps and fans. Most modern VFDs allow the choice of open-loop (no encoder required), sensorless vector, or closed-loop operation. For the highest performance or demanding applications (cranes, tension winders), closed-loop systems are preferred. Closed-loop control ensures the drive can maintain a precise output, regardless of variations in mechanical load or supply quality.

9.5.5 Protection and Monitoring Circuits

Protection is integrated throughout the VFD system, providing safety for the device, connected motor, and personnel. Common protection elements include overcurrent, over/undervoltage, phase loss, ground fault, and overtemperature detection. Modern drives also offer predictive maintenance functions, collecting data for trends and early warning of failures (such as bearing wear, capacitor aging, insulation degradation, and encoder faults).

9.6 Summary

Power electronic converters play a crucial role in mechatronic systems by transforming electrical energy into controlled motion through motors and actuators. These systems require power circuits capable of delivering voltages and currents significantly higher than those provided by control electronics, which typically output signals in the millivolt to volt range at milliamp current levels.

Power circuits enable on–off or continuous linear control of electromechanical components, using devices such as electromechanical relays or solid-state switches. Electromechanical relays provide electrical isolation and simple on–off control but suffer from mechanical wear and arcing issues, while solid-state relays offer faster switching and longer life due to the absence of moving parts. Reed relays, which operate via magnetic fields, are often used for proximity sensing applications. Voltage control for DC motors initially relied on resistors or voltage dividers, but these methods were inefficient and load dependent. The advent of power semiconductors allowed the development of linear voltage regulators and more efficient switch-mode converters using PWM, which adjust output voltage by varying the duty cycle of a switching element, significantly improving efficiency and reducing heat loss.

1) Power converters modify voltage magnitude and frequency to supply electronic devices, electric machines, and actuators. The four fundamental types are:
   - AC-to-DC converters (rectifiers) that provide fixed or adjustable DC voltage from AC sources.
   - DC-to-DC converters that adjust DC voltage levels.
   - DC-to-AC converters (inverters) that generate AC voltage from DC sources.
   - AC-to-AC converters that alter AC voltage amplitude and frequency.
2) Historically, power conversion relied on inefficient and bulky devices like rheostats and motor–generator sets, but advances in semiconductor technology since the 1970s have led to compact, efficient electronic converters.
3) Diodes are two-terminal devices allowing current flow in one direction, with practical diodes conducting above a threshold voltage (~0.7 V). Zener diodes are specialized for voltage regulation in reverse bias. Thyristors introduce a third terminal (gate) for controlled conduction; once triggered, they conduct until current falls below a holding value. Thyristors are suitable for high-voltage and current applications, especially in AC converters where they turn off naturally as current reverses.
4) Three-terminal devices such as BJTs, MOSFETs, GTOs, and IGBTs are widely used in switching power converters. These devices operate primarily in on/off states to function as switches controlled by voltage or current at their input ports. BJTs are current controlled with moderate switching speeds and low-current gain, requiring significant base current for operation. MOSFETs are voltage controlled, draw negligible gate current, and switch rapidly, making them preferred for medium-power applications. GTOs handle high voltages and currents but switch relatively slowly. IGBTs combine the advantages of BJTs and MOSFETs, offering voltage control, fast switching, and high-current and voltage handling capabilities, suitable for industrial-scale converters.
5) Rectifiers convert AC utility power to DC voltage for electronic circuits, DC motors, and battery chargers. Single-phase rectifiers are common for low-power applications, while three-phase rectifiers serve higher power needs.
6) Half-wave rectifiers use a single diode, passing only positive half-cycle, resulting in significant voltage ripple and transformer saturation issues. Full-wave rectifiers use either two diodes with a center-tapped transformer or a four-diode bridge configuration, providing improved DC output with reduced ripple and zero average current drawn from the AC supply, enhancing transformer efficiency.
7) Replacing diodes with thyristors enables output voltage control by adjusting the firing angle (delay angle α). This control is vital for applications requiring variable DC voltage, such as motor speed control and regulated power supplies. Load type affects conduction and extinction

angles, with resistive loads exhibiting current zero-crossing at voltage zero, and inductive loads requiring more complex analysis to determine conduction intervals and current waveforms. Both discontinuous and continuous conduction modes are analyzed using differential equations and Fourier series for design purposes.

8) Three-phase rectifiers provide higher power DC output with lower voltage ripple compared to single-phase rectifiers. Half-wave three-phase rectifiers use three diodes connected to a star transformer with neutral, while full-wave three-phase rectifiers use diode pairs connected between phases. Controlled three-phase rectifiers employ thyristors for adjustable output voltage, benefiting industrial applications with high-power demands.
9) The buck converter reduces the input voltage to a lower output voltage. It uses a switching transistor (e.g., MOSFET), diode, inductor, and capacitor. The duty ratio defines the ratio of on-time to switching period, determining the average output voltage. Inductor current is triangular, with design considerations ensuring continuous conduction and limiting current ripple. Output capacitors are sized to meet voltage ripple specifications. The switching frequency affects inductor and capacitor sizes, balancing EMI and efficiency.
10) The boost converter increases the input voltage to a higher output voltage by reversing the buck converter topology. When the transistor is on, energy is stored in the inductor; when off, the inductor releases energy to the load through the diode. Output voltage depends on the duty ratio and can theoretically approach infinity as the duty ratio approaches unity. Inductor and capacitor sizing follow similar principles as the buck converter, with continuous conduction mode ensuring stable operation.
11) Combining buck and boost converters, the buck–boost converter can produce output voltages either lower or higher than the input voltage, including polarity inversion. It operates with similar switching principles and component considerations, with output voltage depending on the duty ratio. Inductor current continuity and capacitor sizing are critical for performance.
12) Switching losses and conduction losses in semiconductor devices contribute to heat generation, requiring effective cooling solutions such as heat sinks or fans. Switching frequency selection balances component size, EMI, and efficiency. MOSFETs are cost-effective for low to medium power, while IGBTs are preferred for higher power applications. Designers must consider voltage, current ratings, switching speed, and losses to select suitable devices and optimize converter performance.
13) DC/AC converters transform DC power (from batteries, solar panels, etc.) into AC power for motors and actuators in mechatronic systems.
14) DC/AC converters enable precise control of motor speed and torque by adjusting AC output frequency and amplitude—critical for robotics, conveyors, and automated machinery.
15) Bridge converter (H-Bridge) uses four switches (BJTs, MOSFETs, or IGBTs) to synthesize AC output from DC input.
16) Square-wave inverter produces a square-wave AC output; suitable for basic applications but introduces harmonic distortion.
17) Modified square-wave inverter adjusts switching sequence to control harmonic content and output amplitude.
18) PWM inverter uses high-frequency switching to closely approximate a sine wave, reducing harmonic distortion and improving efficiency.
19) Three-phase inverters supply three-phase loads, using coordinated switching for balanced sinusoidal outputs.
20) Selection of switching devices, thermal management, protection circuits, and proper sizing for voltage/current ratings and switching frequency are crucial for reliable operation.

21) VFDs convert fixed frequency AC supply into adjustable frequency and voltage for precise AC motor control.
22) The main components of VFD are rectifier DC link/filter, inverter, control circuits, and protection.
23) VFD enables energy savings, process optimization, and flexible motor control in industrial and commercial applications.

9.7 Check Your Understanding

9.7.1 Self-Assessment Quiz

1 Electromechanical relays are more reliable than solid-state relays due to their lack of moving parts.
True False

2 Pulse-width modulation (PWM) is used in switch-mode converters to control voltage efficiently.
True False

3 The buck converter can only increase the input voltage.
True False

4 Zener diodes are used for voltage regulation in reverse bias.
True False

5 Thyristors are suitable for high-voltage and current applications, especially in AC converters.
True False

6 DC-to-AC converters (inverters) are used to power AC motors and actuators from DC sources in mechatronic systems.
True False

7 Variable frequency drives (VFDs) allow precise control of AC motor speed and torque by adjusting the frequency and amplitude of the output voltage.
True False

8 Which of the following devices is used for proximity sensing applications?
a. Electromechanical relays
b. Solid-state relays
c. Reed relays
d. Zener diodes

9 What is the primary advantage of using pulse-width modulation (PWM) in power converters?
a. Reduced mechanical wear

b. Improved efficiency and reduced heat loss
c. Increased voltage regulation
d. Simplified circuit design

10 Which type of converter is used to adjust DC voltage levels?
a. AC-to-DC converter
b. DC-to-DC converter
c. DC-to-AC converter
d. AC-to-AC converter

11 What is the main function of a buck converter?
a. Increase input voltage
b. Decrease input voltage
c. Convert AC to DC
d. Convert DC to AC

12 Which semiconductor device is known for its fast switching speed and is preferred for medium-power applications?
a. BJT
b. MOSFET
c. GTO
d. IGBT

13 Which of the following is NOT a typical application of DC-to-AC converters (inverters)?
a. Powering AC motors from batteries
b. Integrating renewable energy sources
c. Converting AC to DC for battery charging
d. Enabling precise control of motor speed

14 What is the main advantage of using PWM in inverter circuits?
a. Increased mechanical wear
b. Improved efficiency and reduced harmonic distortion
c. Higher voltage output
d. Simpler circuit design

15 Which switching device is commonly used for high-power inverter applications?
a. BJT
b. MOSFET
c. IGBT
d. Diode

16 In a bridge inverter, what determines the output voltage polarity?
a. The type of load
b. The switching sequence of the four switches
c. The value of the inductor
d. The frequency of the AC supply

17 What is the primary function of the DC link/filter in a VFD?
  a. To convert AC to DC
  b. To buffer energy and smooth DC voltage
  c. To control the motor speed
  d. To generate PWM signals

18 Which control strategy in VFDs uses an encoder for real rotor feedback?
  a. Scalar (V/Hz) control
  b. Open-loop vector control
  c. Closed-loop vector (field-oriented) control
  d. Direct torque control

19 Which of the following statements about harmonic distortion in square-wave inverters is correct?
  a. It only affects resistive loads.
  b. It can cause additional heating and reduce efficiency.
  c. It is eliminated by increasing the switching frequency.
  d. It is not a concern in mechatronic systems.

20 What is the synchronous speed formula for an AC motor controlled by a VFD?
  a. $N_s = 60f/p$
  b. $N_s = 120f/p$
  c. $N_s = f/120p$
  d. $N_s = p/120f$

Answers:

1: False; 2: True; 3: False; 4: True; 5: True; 6: True; 7: True; 8: c; 9: b; 10: b; 11: b; 12: b; 13: c; 14: b; 15: c; 16:b; 17: b; 18: c; 19: b; 20: b.

9.7.2 Short Answer Questions

1 What is the primary role of power electronic converters in mechatronic systems?

2 How do electromechanical relays differ from solid-state relays in terms of reliability and switching speed?

3 What is the main advantage of using pulse-width modulation (PWM) in switch-mode converters?

4 Describe the function of a buck converter and how it adjusts the output voltage.

5 What are the four fundamental types of power converters?

6 Explain the role of Zener diodes in voltage regulation.

7 How do thyristors differ from diodes in terms of their operation and applications?

8 What are the key characteristics of MOSFETs that make them preferred for medium-power applications?

9 How do three-phase rectifiers provide higher power DC output with lower voltage ripple compared to single-phase rectifiers?

10 What are the design considerations for selecting the switching frequency in DC–DC converters?

11 What is the primary function of a DC-to-AC converter (inverter) in mechatronic systems?

12 How do inverters enable precise control of motor speed and torque in automated machinery?

13 What are the main types of switching devices used in bridge converters, and how are they selected for different power levels?

14 Describe the difference between a square-wave inverter and a modified square-wave inverter.

15 How does pulse-width modulation (PWM) improve the output waveform quality of an inverter?

16 Explain how a full-wave controlled rectifier can operate as an inverter. What role does the firing angle play in this process?

17 How do variable frequency drives (VFDs) contribute to energy savings and process optimization in industrial motor control?

18 List and describe the main components of a VFD system.

19 What is the function of the DC link/filter in a VFD?

20 How does the inverter stage in a VFD generate adjustable frequency and voltage for AC motors?

9.7.3 Essay Questions

1 Discuss the role of power electronic converters in mechatronic systems. Explain how these converters transform electrical energy into controlled motion and the importance of power circuits in delivering adequate voltage and current to electromechanical components.

2 Compare and contrast electromechanical relays and solid-state relays. Highlight their differences in terms of reliability, switching speed, and applications in mechatronic systems.

3 Explain the concept of pulse-width modulation (PWM) and its significance in switch-mode converters. Discuss how PWM improves efficiency and reduces heat loss in power electronic converters.

4 Describe the operation and design considerations of a buck converter. Include details on how it adjusts output voltage, the role of the duty ratio, and the importance of inductor and capacitor sizing.

5 Analyze the advantages and disadvantages of different types of power converters. Focus on AC-to-DC converters, DC-to-DC converters, DC-to-AC converters, and AC-to-AC converters, and discuss their applications in various mechatronic systems.

6 Examine the characteristics and applications of different power semiconductor devices. Discuss the roles of diodes, Zener diodes, thyristors, BJTs, MOSFETs, GTOs, and IGBTs in power electronic converters.

7 Discuss the design considerations for selecting the switching frequency in DC–DC converters. Explain how switching frequency affects component size, electromagnetic interference, and efficiency, and the trade-offs involved in choosing an optimal frequency.

8 Explain the operation of three-phase rectifiers and their advantages over single-phase rectifiers. Discuss the differences between half-wave and full-wave three-phase rectifiers and their applications in high-power mechatronic systems.

9 Describe the operation and design of a boost converter. Include details on how it increases the input voltage, the role of the duty ratio, and the importance of inductor and capacitor sizing for continuous conduction mode.

10 Analyze the impact of switching and conduction losses in semiconductor devices on the performance of power electronic converters. Discuss the importance of effective cooling solutions and the factors to consider when selecting suitable devices for optimal converter performance.

11 Why is harmonic distortion a concern in square-wave inverters, and how can it be mitigated?

12 What is the role of space vector modulation (SVM) in modern inverter designs?

13 How do three-phase inverters differ from single-phase inverters in terms of output and applications?

14 What are the key design considerations when selecting switching devices for bridge inverters?

15 Why is thermal management important in inverter design? Discuss the advantages and drawbacks of cooling methods.

16 Discuss the importance of maintaining a constant voltage-to-frequency ratio in AC motor control.

17 Compare scalar (V/Hz) control, open-loop vector control, closed-loop vector control, and direct torque control in VFDs.

18 Research the types of protection and monitoring circuits are integrated into modern VFDs.

19 Discuss the effects of switching frequency on the efficiency and electromagnetic interference (EMI) in inverter circuits.

20 Why might a reverse-biased diode be necessary in bridge inverter circuits with inductive loads?

9.7.4 Problems

1 A buck converter is designed to step down a 24 V input to a 12 V output. The load requires 5 A of current. Calculate the duty ratio, the average inductor current, and the peak inductor current if the switching frequency is 50 kHz and the inductor value is 100 μH.

2 A boost converter is used to step up a 12 V input to 24 V output. The load requires 3 A of current. Calculate the duty ratio, the average inductor current, and the minimum inductance required to ensure continuous conduction mode if the switching frequency is 100 kHz.

3 A single-phase full-wave rectifier is connected to a 120 V rms AC supply. The load resistance is 50 Ω. Calculate the average output voltage, the rms output voltage, and the peak inverse voltage (PIV) of the diodes.

4 For a buck converter with an input voltage of 48 V, output voltage of 24 V, load current of 10 A, and switching frequency of 25 kHz, calculate the inductor current ripple if the inductor value is 200 μH.

5 In a boost converter, the input voltage is 15 V, output voltage is 30 V, load current is 2 A, and switching frequency is 75 kHz. Calculate the minimum capacitance required to limit the output ripple to 1%.

6 A MOSFET in a buck converter operates with an input voltage of 36 V, output voltage of 18 V, load current of 8 A, and frequency of 40 kHz. The MOSFET has an on-state resistance of 0.05 Ω and a switching time of 100 ns. Calculate the conduction loss and the switching loss.

7 A three-phase full-wave rectifier is connected to a 208 V rms line-to-line AC supply. The load resistance is 20 Ω. Calculate the average output voltage, the rms output voltage, and the current rating of the diodes.

8 A square-wave inverter supplies a load with a DC voltage of $V_{dc} = 12$ V. The load consists of a resistor $R = 2\,\Omega$ and an inductor = 10 mH. The inverter operates at a frequency of = 50 Hz.
a. Calculate the rms value of the output voltage.
b. Calculate the rms value of the load current.

9 A bridge inverter produces a modified square-wave output with a DC supply voltage $V_{dc} = 24$ V and a delay angle $\alpha = 30°$ (convert to radians for calculation).
a. Calculate the rms value of the output voltage.
b. Calculate the amplitude of the third harmonic.
c. Calculate the amplitude of the fifth harmonic.

10 Given a square-wave inverter with a fundamental rms voltage $V_{1,\,rms} = 8$ V and total rms voltage $V_{rms} = 12$ V, calculate the total harmonic distortion (THD) of the output voltage.

11 A bridge inverter supplies a load with $V_{dc} = 15$ V, $R = 3\,\Omega$, and $L = 20$ mH at $f = 60$ Hz.

a. Calculate the time constant τ of the load.

b. Calculate the maximum load.

12 A VFD is set to supply a three-phase induction motor with a frequency =40 Hz. The motor has $p = 4$ poles.

13 Calculate the synchronous speed N_s in revolutions per minute (RPM)

14 A PWM inverter uses a triangular carrier signal at $f_{carrier} = 10$ kHz and a sinusoidal reference at $f_{ref} = 50$ Hz.

a. How many pulses are generated per cycle of the reference signal?

b. What is the output rms voltage for a bipolar PWM scheme?

References

Hart, D. W., 2011. *Introduction to Power Electronics*. New York, NY: McGraw-Hill.

Lee, Y.-S. & Chow, M., 2011. *Power Electronics Handbook*. 3rd ed. Amsterdam: Elsevier.

Mohan, N., 2012. *Power Electronics: A First Course*. Hoboken, NJ: John Wiley and Sons.

Mohan, N., Underland, T. M. & Robbins, W., 2003. *Power Electronics*. 3rd ed. Hoboken, NY: John Wiley and Sons, Inc.

Rashid, M., ed., 2011. *Power Electronics Handbook - Devices, Circuits, and Applications*. 3rd ed. Amsterdam: Elsevier.

10

Motion Transmission and Conversion

OBJECTIVES

- Provide an overview of the most common motion transmission mechanisms used in modern mechatronic systems.
- Explain how transmission mechanisms transfer mechanical energy with desired characteristics of speed, direction, force, and accuracy.
- Explore conversion systems that modify the direction of movement and transform one type of motion into another (e.g., rotary to linear).
- Explain the operating principles of motion transmission and discuss practical applications.
- Provide an understanding of how motion and force are transmitted from actuators to the functional components of devices.
- Explain how motion transmission systems enable precise control of force, speed, or torque for specific applications.

Motion transmission and conversion mechanisms constitute the backbone of mechatronic systems, enabling the delivery of motion and force from actuators to the working components of devices. Whether the goal is to move a robotic arm, open a garage door, or precisely position a tool in automated manufacturing, a deep understanding of motion transmission systems is essential to the successful design and operation of mechatronic systems. This chapter provides a comprehensive overview of the most common motion transmission mechanisms used in modern mechatronic systems, their operating principles, and their practical applications.

At its core, motion transmission is about transferring mechanical energy from one component (the source) to another (the load) with the desired characteristics of speed, direction, force, and accuracy. The three fundamental types of motion in mechatronics are:

- **Translational Motion**: Straight-line movement, such as a piston moving up and down.
- **Rotary Motion**: Rotational movement, such as motor shaft turning.
- **Cyclic/Reciprocating Motion**: Repetitive back-and-forth or oscillatory movement, such as a crank–slider mechanism.

Transmission mechanisms facilitate the transfer of motion over distances, while conversion systems modify the direction of movement and transform one type of motion into another, such as converting rotary motion to linear. These systems also allow for precise adjustments of force, speed, or torque to meet specific application requirements.

10.1 Rotary Transmission

Rotary transmission mechanisms deliver or modify rotational motion and torque from motors or actuators. Gear trains are a series of interlocking gears that transmit rotation, change speed or torque, and can even reverse direction. Types include spur, helical, bevel, and planetary gears (Juvinall & Marshek, 2020). Figure 10.1 provides an overview of several widely utilized rotary gear systems.

10.1.1 Spur Gear

A spur gear illustrated in Figure 10.1 is the simplest and most common type of gear used in rotary motion transmission. It consists of a cylindrical wheel with straight, evenly spaced teeth projecting radially from its circumference. When two spur gears mesh, their teeth engage directly, transferring rotational motion and torque efficiently along parallel shafts. Spur gears are widely used due to their straightforward design, reliable performance, and ease of manufacture, appearing in everything from clocks to industrial machinery. However, they are best suited for applications where noise and vibration are not critical, as the sudden engagement of teeth can lead to increased sound at higher speeds.

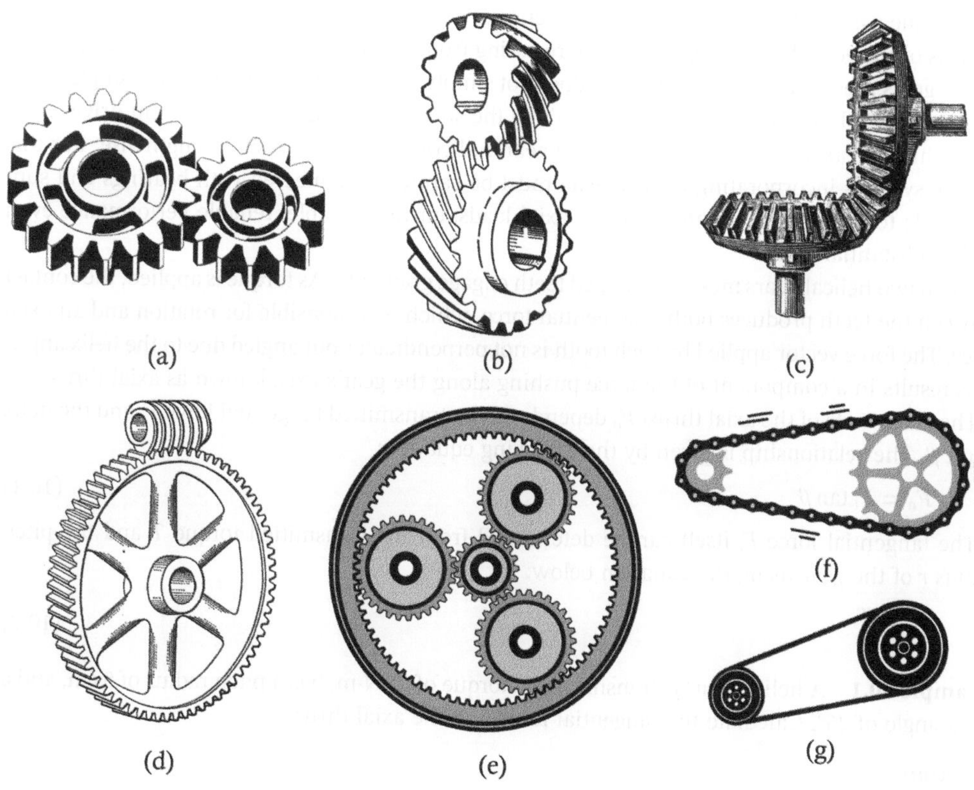

(a) (b) (c) (d) (e) (f) (g)

Figure 10.1 Common rotational drives (a: Spur gears; b: Helical gears; c: Bevel gears; d: Spiral gear; e: Planetary gear; f: Chain; g: Belt and pulley)

10.1.2 Helical Gear

Helical gears are a fundamental type of rotary transmission mechanism, closely related to spur gears but with a significant difference in tooth orientation as shown in Figure 10.1. While spur gears have straight teeth aligned parallel to the axis of rotation, helical gears feature teeth cut at an angle (the helix angle) to the gear axis. This angled design introduces several unique working principles and desirable characteristics in motion transmission.

One notable characteristic of helical gears is the generation of axial thrust. Because the teeth are angled, the force generated during operation has both a tangential (rotational) and an axial (along the shaft) component. This means that when helical gears are used, the shafts and bearings must be designed to accommodate these additional thrust loads.

Helical gears present several advantages over spur gears. The angled teeth provide more gradual and continuous contact between meshing gears. This reduces vibration and noise, especially at higher speeds. At any given moment, several teeth are in contact, distributing the load more evenly and allowing helical gears to transmit greater forces than spur gears of similar size. Helical gears are not limited to parallel shafts; they are also used for angled shaft configurations, increasing their usefulness in complex machinery. The progressive engagement minimizes impact on the teeth, leading to less wear and longer service life (Lynwander, 2019).

Helical gears are widely used in applications requiring high-speed and high-power transmission with smooth and quiet operation. Common uses include automotive transmissions, industrial gearboxes, elevators, and conveyor systems.

A unique aspect of helical gears is the presence of axial thrust, a pushing force generated along the axis of the gear shaft during operation, resulting from the angled orientation of the gear teeth. As the gears mesh and rotate, each tooth does not simply push against its partner in a single plane but instead applies force both around and along the shaft. The tangential component drives rotation, while the axial component pushes the gear and consequently the shaft lengthwise. For this reason, systems incorporating helical gears must be carefully engineered with bearings and supports able to absorb and counteract these axial loads, ensuring stable, reliable performance even under substantial forces.

When two helical gears mesh, the angled teeth engage gradually. As torque is applied, the contact between the teeth produces both a tangential force, which is responsible for rotation and an axial force. The force vector applied by each tooth is not perpendicular but angled due to the helix angle. This results in a component of the force pushing along the gear's axis, known as axial thrust.

The magnitude of the axial thrust F_a depends on the transmitted tangential force F_t and the helix angle β. The relationship is given by the following equation:

$$F_a = F_t \tan\beta \tag{10.1}$$

The tangential force F_t itself can be determined from the transmitted torque T and the pitch radius r of the gear using the equation below:

$$F_t = \frac{T}{r} \tag{10.2}$$

Example 10.1 A helical gear is transmitting a torque of 25 N·m, has a pitch radius of 5 cm, and a helix angle of 25°. Calculate the tangential force and the axial thrust.

Solution

The tangential force is

$$F_t = \frac{T}{r} = \frac{25}{0.05} = 500\ \text{N}$$

The axial thrust generated by this helical gear is

$$F_a = F_t \tan\beta = 500 \times 0.4663 = 233 \text{ N} \therefore$$

10.1.3 Bevel Gears

Bevel gears represent another essential category of rotary transmission mechanisms, uniquely suited for transferring motion between intersecting shafts—most often at a right angle. Unlike spur and helical gears, which transmit rotational force along parallel or, at times, angled axes, bevel gears have conically shaped pitch surfaces and teeth that allow for the redirection of rotational motion.

Typically, a standard bevel gear set consists of two gears mounted on shafts that intersect, usually at 90°. The teeth of bevel gears can be straight or spiraled (spiral bevel gears), with the latter offering smoother, quieter operation and increased load capacity due to gradual tooth engagement—much like the relationship between the helical and spur gears.

Bevel gears are invaluable in applications where direction changes of drive are vital, such as in differential drives in automobiles, right-angle power tools, and various industrial machinery. They offer efficient torque transmission, precise control of direction, and can be engineered for different speed and torque ratios depending on the relative number of teeth.

One consideration with bevel gears is ensuring proper alignment; misalignment can lead to excessive wear and noise. Moreover, because the force is often transferred at angles, bevel gear assemblies must be robustly supported to handle both the radial and axial loads.

For further reading on the design, operation, and applications of helical gears, authoritative resources include *Gear Design and Application* by Raymond J. Drago and *Dudley's Handbook of Practical Gear Design and Manufacture* by Stephen P. Radzevich. Additionally, technical standards such as ISO 1328 and AGMA (American Gear Manufacturers Association) publications offer comprehensive guidelines and specifications for helical gears. These references provide in-depth discussions on gear geometry, force calculations, material selection, and best practices for ensuring efficient and reliable gear performance.

10.1.4 Planetary Gears

Planetary gear systems, also known as epicyclic gear trains, represent one of the most sophisticated and versatile rotary transmission mechanisms in engineering. Unlike simple gear pairs, a planetary gear set comprises three primary components: a central sun gear, multiple planet gears that revolve around the sun, and an outer ring gear with internal teeth. The planet gears are mounted on a carrier that rotates and can itself serve as an output or input shaft.

The unique configuration of planetary gears enables them to achieve a remarkable combination of compactness, high torque density, and versatile speed ratios. By varying which element (sun, planet carrier, or ring) is held stationary, used as input or as output, a single planetary system can provide multiple gear ratios—including both speed reduction and increase and reversal of rotation direction.

One of the advantages of planetary gears is their ability to distribute the load among several planet gears. This load sharing results in exceptional strength, smooth power transmission, and impressive durability, all within a relatively small footprint. Additionally, since the forces are balanced and the gears mesh continuously, planetary systems offer quiet operation and minimal vibration, making them ideal for high-performance applications.

Planetary gear sets are widely employed in automotive automatic transmissions, electric vehicle (EV) drivetrains, industrial machinery, and precision instruments such as robotics and aerospace actuators. Their robust design can handle both high torque and high-speed applications, and their efficiency is prized wherever space and weight constraints are critical.

Despite these advantages, planetary gears require precise manufacturing and assembly to maintain the proper meshing of multiple gears and to ensure even load distribution. Misalignment or excessive wear in any component can compromise performance and longevity.

10.1.5 Belt-Driven Pulleys

Belt-driven pulleys are used for smooth and flexible transmission of rotary motion, especially when shafts are some distance apart. This is common in fans, conveyors, and automotive engines.

Chains looped around toothed wheels efficiently transmit rotary motion, especially where slippage must be minimized (e.g., bicycles, motorcycles, conveyors). Flexible shafts are used to transmit rotary motion around curves or through obstacles, as in some dental or industrial tools.

10.1.6 Rotary Motion Conversion

When rotational speed is modified through a gearbox, the conversion ratio is determined by the number of teeth on the respective gears. When two spur gears are meshed, the output speed is determined by the gear ratio, which is defined as the ratio of the number of teeth on the output (driven) gear to the number of teeth on the input (driving) gear.

$$a = \frac{n_2}{n_1} \tag{10.3}$$

Specifically, if the input gear has n_1 teeth and the output gear has n_2 teeth, then the ratio of the output angular velocity ω_2 to the input angular velocity ω_1 is inversely proportional to the number of teeth.

$$\frac{\omega_2}{\omega_1} = \frac{n_1}{n_2} = \frac{1}{a} \tag{10.4}$$

Thus, the output gear rotates more slowly than the input gear if it has more teeth ($n_2 > n_1$), and more quickly if it has fewer teeth.

Alternatively, when employing a wheel-and-belt mechanism for rotational conversion, the change in speed corresponds to the ratio of the diameters of the wheels involved. In either case, there is no slip between input and output. Therefore, both pulleys make the same tangential displacement.

$$r_1\theta_1 = r_2\theta_2 \Longrightarrow \frac{\theta_2}{\theta_1} = \frac{\omega_2}{\omega_1} = \frac{r_1}{r_2} \tag{10.5}$$

In both techniques, the rotational mechanical power expressed as the product of the angular velocity and torque is transmitted from the input to the output. Therefore, as the angular velocity increases, the torque decreases or vice versa. Suppose that the friction losses are negligible. The relationship between the input and output variables is obtained as below by applying the energy conservation principle:

$$P_1 = P_2 \Longrightarrow \omega_1 T_1 = \omega_2 T_2 \Longrightarrow \frac{T_1}{T_2} = \frac{\omega_2}{\omega_1} = \frac{1}{a} \tag{10.6}$$

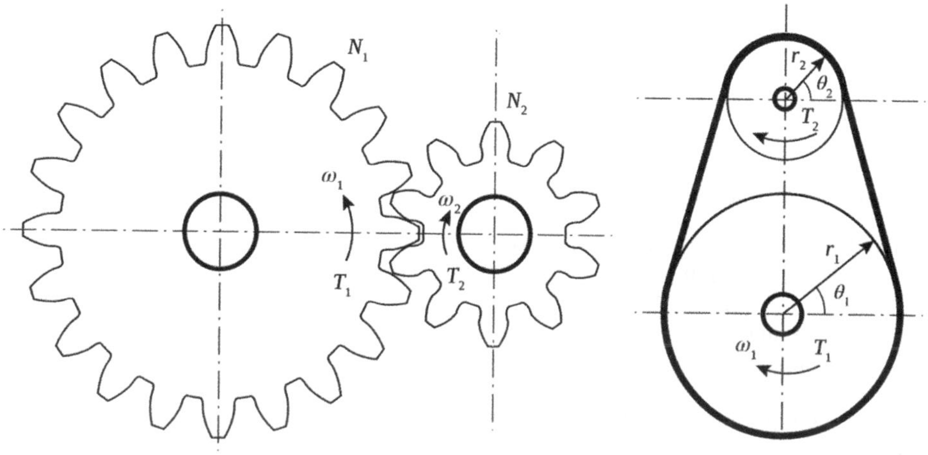

Figure 10.2 Rotary transformation parameters

Let T_L' represents the effective load torque seen from the input shaft, the equivalent work corresponding to the work done by the load torque at the output shaft is

$$\omega_1 T_L' = \omega_2 T_L \Longrightarrow T_L' = \frac{1}{a} T_L \tag{10.7}$$

Similarly, the viscous friction constant β would be seen from the input shaft as β' found by the relationship

$$(\beta' \omega_1)\omega_1 = (\beta \omega_2)\omega_2 \Longrightarrow \beta' = \frac{1}{a^2} \tag{10.8}$$

The effective moment of inertia seen from the input shaft is obtained as

$$\frac{1}{2} J' \omega_1^2 = \frac{1}{2} J \omega_2^2 \Longrightarrow J' = \frac{1}{a^2} J \tag{10.9}$$

These transformed parameters—effective load torque, viscous friction constant, and moment of inertia—are essential for accurately modeling the dynamic response of geared systems. By reflecting the effects of gear ratios on these mechanical quantities, designers can better predict how alterations at the input shaft influence overall system behavior. This understanding becomes particularly valuable when analyzing mechanisms that rely on the interactions between rotational and linear motions, or those that operate under varying load conditions (Figure 10.2).

Example 10.2 A 20-tooth spur gear meshes with a 40-tooth gear. If the output shaft runs at 1,200 RPM and powers a 2 HP load, calculate the input shaft speed and torque.

Solution

The gear ratio is the ratio of the number of teeth on the driven gear to the number of teeth on the driving gear. Gear ratio is $a = 40/20 = 2$.

We obtain the input speed by plugging the given number of teeth and the input speed into equation 10.4.

$$N_1 = N_2 \frac{n_2}{n_1} = 1{,}200 \times 2 = 2{,}400 \text{ RPM}$$

To determine the torque, the given output power must be converted to watt and the rotation speed to radians per second.

$$1 \text{ HP} = 746 \text{ W} \Longrightarrow P_2 = 2 \times 746 = 1{,}492 \text{ W}$$

$$\omega_2 = \frac{2\pi N}{60} = \frac{2\pi \times 1{,}200}{60} = 125.67 \text{ rad/s}$$

$$T_2 = \frac{P}{\omega_2} = \frac{1{,}492}{125.67}$$

To determine the input torque, we apply equation 10.6.

$$T_1 = T_2 \frac{1}{a} = \frac{1{,}492}{125.67} \cdot \frac{1}{2} = 5.94 \text{ N} \cdot \text{m} \therefore$$

10.2 Linear (Translational) Transmission

Many mechatronic systems require conversions between rotary and translational motion, or a combination of different motion types. Rack and pinion systems, lead screw and nut assemblies, and slider–crank mechanisms are commonly utilized motion conversion devices for transforming rotational motion into translational motion, or vice versa.

10.2.1 The Rack and Pinion

The rack and pinion system operates by engaging a circular gear (the pinion) with a flat, toothed component (the rack). As the pinion rotates, its teeth mesh with those of the rack, causing the rack to move in a straight line. Conversely, pushing the rack linearly will cause the pinion to rotate. This direct mechanical linkage enables precise and efficient conversion between rotary and linear motion, making it a preferred choice in steering mechanisms, CNC machines, and various automation setups. The simplicity of the design also allows for smooth, controllable movement with minimal backlash when manufactured to tight tolerances.

10.2.2 Lead Screw and Nut

The lead screw and ball screw mechanisms illustrated in Figure 10.3 are widely used for precise conversion between rotational and translational movement. In a traditional lead screw system, a helical thread is cut along a shaft (the screw), and a nut with matching internal threads travels along it as the screw rotates (Juvinall & Marshek, 2020). By turning the screw, the nut advances or retracts linearly, allowing for accurate positioning. However, standard lead screws can suffer from significant friction due to the sliding contact between the screw and the nut threads, resulting in lower efficiency and more heat generation.

To address this, the ball screw incorporates a series of ball bearings that roll between the screw shaft and the nut. As the screw rotates, the balls circulate within a recirculating track, greatly reducing friction since rolling contact is much more efficient than sliding. This leads to improved efficiency (often exceeding 90%), higher positioning accuracy, and longer service life.

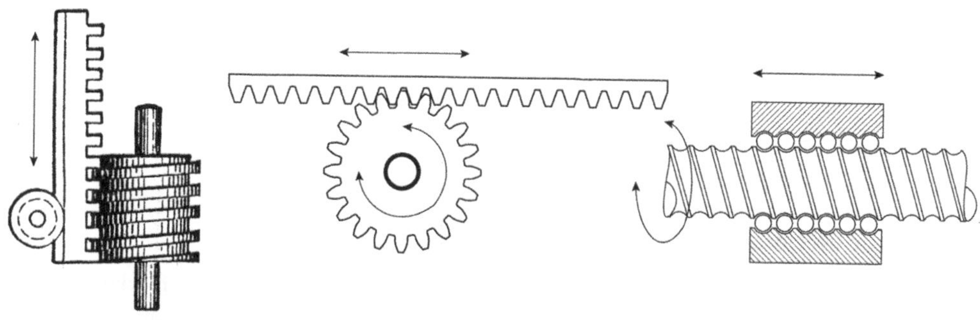

Figure 10.3 Linear motion mechanisms

10.2.3 Ball Screw

Ball screws are thus favored in high-performance applications like CNC machinery, robotics, and precision actuators, where minimal backlash and smooth, efficient motion are critical.

In both lead and ball screw systems, the direction of linear motion depends on the thread's helix and the direction of rotation. The pitch of the thread determines the conversion ratio—the linear distance the nut travels per revolution of the screw. This ratio is crucial for applications requiring precise control of movement (Liu et al., 2021).

10.3 Conversion of Rotary Motion to Translational

Computation of speed, force, and torque transformations is based on the principle of energy conservation and Newton's laws. The kinematic motion conversion factor, or effective gear ratio, of a leadscrew is characterized by its pitch, p rev/in, or rev/mm. Therefore, for a leadscrew, a pitch of p (mm or in per revolution) is the inverse of the distance traveled for one turn of the thread called lead.

$$l = \frac{1}{p} \tag{10.10}$$

The rotational displacement ($\Delta\theta$) at the lead shaft and the translational displacement (Δx) of the nut are related by the following equations:

$$\begin{aligned} \Delta\theta &= 2\pi p \Delta x \Longrightarrow \omega = (2\pi p)v \\ \Delta x &= \frac{1}{2\pi p}\Delta\theta \Longrightarrow v = \frac{1}{2\pi p} \end{aligned} \tag{10.11}$$

where θ is the rotation angle in radians. The translational kinetic energy of a mass at the load side is seen at the rotational side as rotational kinetic energy. Therefore,

$$\frac{1}{2}mv^2 = \frac{1}{2}J\omega^2 \tag{10.12}$$

The equivalent torque seen at the input shaft of the leadscrew due to the load force F at the nut is determined by the equation below:

$$T_{eff} = \frac{1}{2\pi p} F \tag{10.13}$$

10.3.1 Cyclic (Reciprocating) Motion Transmission

Reciprocating motion, often described as cyclic or back-and-forth movement, is fundamental to countless machines and systems in both industrial and everyday settings. Mechanisms that generate or harness reciprocating motion are crucial for transforming rotary motion into linear displacement and vice versa. Their reliability, efficiency, and adaptability make them indispensable in a wide range of mechatronic applications. Common methods to produce reciprocating motion are crank–slider, cam and follower, and four-bar linkage mechanisms.

10.3.2 Slider–Crank

The slider–crank mechanism operates on a fundamental principle of converting rotary motion to linear (reciprocating) motion, or vice versa, through a series of interconnected components. At its core, the mechanism consists of four primary elements: the crank (rotating input), the connecting rod (linkage), the slider (translating output), and the fixed frame.

When the crank rotates around a fixed axis, it drives the connecting rod, which is pinned to both the crank and the slider. As the crank turns, its circular motion is converted to a reciprocating, back-and-forth movement of the slider, which moves along a linear path. The slider's position at any given moment is determined by the angular position of the crank and the lengths of the crank and connecting rod. This geometric relationship enables smooth and continuous conversion between rotational input and linear output.

The slider–crank mechanism is the foundation of many machines where such motion conversion is required—most notably in internal combustion engines, where the piston's linear movement is turned into the rotary motion of the crankshaft, and in reciprocating compressors and pumps. The efficiency and smoothness of the motion depend on precise alignment and minimal friction at the pivots and sliding surfaces.

The slider–crank mechanism is a typical rotary-to-translational motion conversion mechanism used in internal combustion engines to convert the translation of a piston to rotary motion of the crankshaft.

Consider the slider–crank mechanism in Figure 10.4. The distance from the crank axis to the moving component can be determined using the lengths of the two connecting arms.

$$x = r \cdot \cos(\theta) + l \cdot \cos\phi \tag{10.14}$$

Using the trigonometric equalities, we can write

$$\begin{aligned} l \cdot \sin\phi &= r \cdot \sin\theta \\ \cos\phi &= \sqrt{(1 - \sin^2\phi)} = \sqrt{\left(1 - \left(\frac{r}{l}\ \sin\theta\right)^2\right)} \end{aligned} \tag{10.15}$$

The linear displacement is

$$x = r\cos\theta + l\sqrt{1 - \left(\frac{r}{l}\sin\theta\right)^2}$$

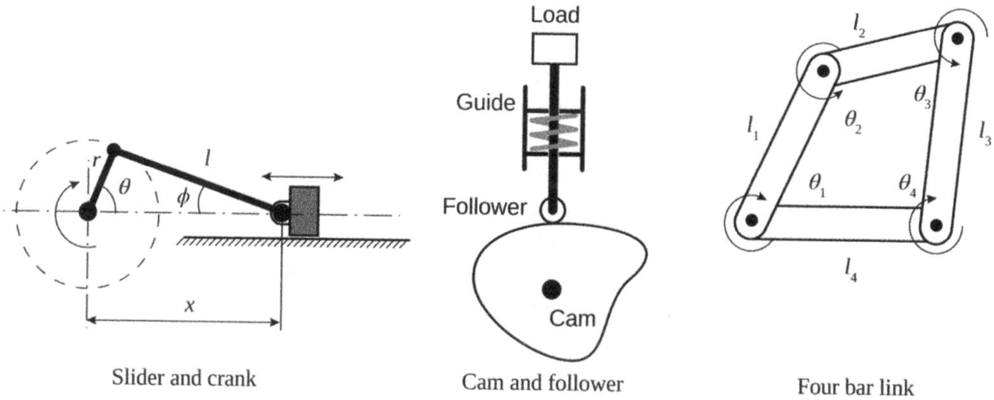

Figure 10.4 Cyclic motion mechanisms

The linear velocity is found by differentiating the linear displacement and substituting ω for $d\theta/dt$.

$$\frac{dx}{dt} = -r\omega\left[\sin\theta + \frac{r}{2l}\frac{\sin 2\theta}{\sqrt{1-\left(\frac{r}{l}\sin\theta\right)^2}}\right] \tag{10.16}$$

10.3.3 Cam and Follower

Cams convert the rotary motion of a shaft into the translational motion of a follower. The relationship between the translational motion and the rotary motion is not a fixed gear ratio, but a nonlinear function. The nonlinear cam function is determined by the machined cam profile. A cam mechanism has three major components: input shaft, cam, and follower.

In the cam-and-follower mechanism, a shaped rotating cam mechanically pushes a follower along a predetermined path, enabling timed, complex reciprocating or oscillatory motion. This arrangement is widely applied in engines and automated machinery for precision timing and actuation.

10.3.4 Four-Bar Linkage

A four-bar linkage is composed of four rigid rods connected in a loop; this mechanism produces complex cyclic movements and is used in robotic arms, folding mechanisms, and various automation systems. The four-bar linkage stands out for its versatility and ability to produce controlled, repeatable motion. A typical four-bar linkage consists of four rigid rods—commonly referred to as links—connected end-to-end in a closed loop by four pivot joints. By anchoring one link as the fixed frame and applying an input motion (usually rotary) to a second link, the linkage translates this movement through the remaining links, resulting in a complex but predictable path for the output link.

Depending on the relative lengths of the links and the input motion, a four-bar linkage can generate a range of outputs, from smooth oscillating arcs to near-linear reciprocation. The path traced by a point on one of the links can be finely tuned by adjusting the proportions of the linkage, making this mechanism highly adaptable for specific tasks. Four-bar linkages are known for their reliability and mechanical simplicity, offering designers a powerful tool to achieve intricate motion patterns with minimal components.

Key Applications of Reciprocating Motion Mechanisms are found in internal combustion engines, compressors, pumps, and textile and sewing machinery.

Reciprocating compressors and pumps utilize pistons or plungers moving back and forth to compress gases or move fluids. These are essential in many industrial applications like refrigeration systems, air conditioning units, hydraulic machinery, and in the oil and gas industry.

Robotic arms often use four-bar linkages and other reciprocating mechanisms to produce repetitive, controlled motion for tasks such as assembly, welding, and packaging. These mechanisms offer a blend of precision and speed, critical for modern manufacturing.

Cyclic motion is harnessed in pick-and-place machines, conveyors, and automated packaging systems for precise, repeated transfer of products and materials in automated packaging and material handling.

Traditional printing presses utilize reciprocating motion to move printing plates and paper back and forth, ensuring even ink distribution and print quality.

Power tools perform reciprocating motion for cutting, sanding, grinding, and polishing in woodworking, metalworking, and jewelry making.

10.4 Characteristic Parameters of Motion Transmission and Conversion

The performance of motion transmission mechanisms is characterized by several parameters. Key parameters are conversion ratio, reversibility, stiffness, backlash, and efficiency. Efficiency is the ratio of the output power to the input power. The efficiency of a real gear ratio is always less than 100%.

Backlash given in units of arc minutes = $1/60°$ in rotary mechanisms, and in inches or mm units in translational mechanisms. The backlash directly affects the positioning accuracy. If the position sensor is connected to the motor, not to the load, it will not be able to measure the positioning error accurately due to backlash. In systems where the backlash is much smaller than the positioning accuracy required, the backlash can be ignored.

Stiffness: The transmission components are not perfectly rigid. They have finite stiffness. The stiffness of the transmission box between the input and output shaft is rated with a torsional or translational stiffness parameter.

Breakaway Friction: This friction torque (or force) is an estimated value and highly dependent function of the lubrication condition of the moving components. This is the minimum torque or force needed at the input shaft to move the mechanism.

Back drivability refers to the property of a transmission mechanism (such as gears, ball screws, or actuators) that allows the output shaft or element to drive the input shaft when an external load is applied.

10.5 Application Examples

10.5.1 Garage Door Opener Design

A typical garage door opener combines multiple motion transmission mechanisms. An electric motor provides rotary motion, and a rack and pinion or screw drive converts this motion into linear movement. Then, a belt or chain transmits rotary motion from the motor to a trolley system to open and close the door. Safety sensors and limit switches integrate with the mechanism to ensure controlled operation.

10.5.2 Robotics

Articulated robotic arms employ servomotors with gear trains for joint rotation. Lead screws or ball screws provide precise linear positioning. Linkages are used for synchronized cyclic or oscillating motion in grippers.

10.5.3 Automated Manufacturing

Advanced manufacturing equipment uses cam-driven mechanisms for repetitive assembly tasks, belt and pulley or chain and sprocket systems for conveying materials, and gearboxes to adjust motor output for specific application needs.

10.6 Selection Criteria for Motion Transmission Mechanisms

When choosing the appropriate mechanism for a mechatronic system, engineers must consider:

- Required motion type (linear, rotary, reciprocating)
- Load and force characteristics
- Speed/torque requirements
- Precision and play (backlash)
- Space and layout constraints
- Cost and manufacturability
- Maintenance and longevity

10.7 Summary

This chapter focuses on motion transmission and conversion mechanisms, which are essential components in mechatronic systems. These mechanisms enable the transfer of motion and force from actuators to the working components of devices, playing a crucial role in the design and operation of systems such as robotic arms, garage doors, and automated manufacturing tools.

- The three fundamental types of motion in mechatronics are:
  - **Translational Motion**: Straight-line movement, such as a piston moving up and down.
  - **Rotary Motion**: Rotational movement, such as a motor shaft turning.
  - **Cyclic/Reciprocating Motion**: Repetitive back-and-forth or oscillatory movement, such as a crank–slider mechanism.
- Various motion transmission mechanisms include the following.
  - **Rotary Transmission**: Gear trains (spur, helical, bevel, and planetary gears) that transmit rotation, change speed or torque, and can reverse direction.
  - **Translational Motion**: Belt-driven pulleys are used for smooth and flexible transmission of rotary motion, especially when shafts are some distance apart. Chains are preferred, especially where slippage must be minimized.

- Common motion conversion mechanisms are:
  - **Rack and Pinion**: Converts rotary motion into linear motion and vice versa.
  - **Lead Screw and Nut**: Used for precise conversion between rotational and translational movement.
  - **Slider–Crank Mechanism**: Converts rotary motion to linear (reciprocating) motion.
  - **Cam and Follower**: Converts rotary motion of a shaft into translational motion of a follower.
  - **Four-Bar Linkage**: Produces complex cyclic movements and is used in robotic arms, folding mechanisms, and various automation systems.

10.8 Check Your Understanding

10.8.1 Self-Assessment Quiz

1 Spur gears have teeth cut at an angle to the gear axis.
- **a.** True
- **b.** False

2 Helical gears generate axial thrust due to their angled teeth.
- **a.** True
- **b.** False

3 A rack-and-pinion system can convert rotary motion into linear motion.
- **a.** True
- **b.** False

4 Planetary gear systems are used for compact, high torque density, and versatile speed ratios.
- **a.** True
- **b.** False

5 A slider–crank mechanism converts linear motion to rotary motion.
- **a.** True
- **b.** False

6 What are the three fundamental types of motion in mechatronics?
- **a.** Translational motion, rotary motion, cyclic/reciprocating motion
- **b.** Linear motion, circular motion, oscillatory motion
- **c.** Translational motion, circular motion, vibrational motion
- **d.** Linear motion, rotary motion, vibrational motion

7 What is the primary function of motion transmission mechanisms in mechatronic systems?
- **a.** To generate electrical energy
- **b.** To transfer mechanical energy with desired characteristics
- **c.** To store mechanical energy
- **d.** To convert electrical energy to mechanical energy

8 Which type of gear has teeth cut at an angle to the gear axis?
  a. Spur gear
  b. Helical gear
  c. Bevel gear
  d. Planetary gear

9 How does a rack-and-pinion system convert motion?
  a. Converts linear motion to rotary motion
  b. Converts rotary motion to linear motion
  c. Converts rotary motion to oscillatory motion
  d. Converts linear motion to oscillatory motion

10 What are the main components of a planetary gear system?
  a. Sun gear, planet gears, ring gear
  b. Spur gear, helical gear, bevel gear
  c. Rack, pinion, nut
  d. Crank, connecting rod, slider

Answers:
1: False; 2: True; 3: True; 4: True; 5: False; 6: a; 7: b; 8: b; 9: b; 10: a.

10.8.2 Short Answer Questions

1 What are the three fundamental types of motion in mechatronics?

2 Describe the primary function of motion transmission mechanisms in mechatronic systems.

3 What is the main difference between spur gears and helical gears?

4 How does a rack-and-pinion system convert motion?

5 What is the purpose of a planetary gear system, and what are its main components?

6 What are the key parameters that characterize the performance of motion transmission mechanisms?

7 Explain the function of a slider–crank mechanism.

8 Describe the role of a cam-and-follower mechanism.

10.8.3 Problems

1 Calculate the axial thrust generated by a helical gear with a helix angle of 30° and a tangential force of 500 N.

2 What is the gear ratio if a 20-tooth spur gear meshes with a 40-tooth gear?

3 A lead screw has a pitch of 5 mm per revolution. Calculate the linear distance traveled by the nut for 10 revolutions of the screw.

4 In a slider–crank mechanism, the crank length is 100 mm, and the connecting rod length is 300 mm. Calculate the linear displacement of the slider when the crank rotates by 90°.

5 In a planetary gear system, the sun gear has 30 teeth, each planet gear has 10 teeth, and the ring gear has 70 teeth. Calculate the speed ratio if the sun gear is the input and the ring gear is stationary.

6 A cam rotates at 600 RPM and has a follower that moves with a maximum displacement of 50 mm. Calculate the frequency of the follower's motion.

7 Design a gear train system for a robotic arm that requires a speed reduction of 5:1. The input motor has a speed of 1,500 RPM and a torque of 10 N m. Calculate the required gear ratios, the number of teeth for each gear, and the output speed and torque.

8 Design a gear train system for a robotic arm that requires a speed reduction of 5:1. The input motor has a speed of 1,500 RPM and a torque of 10 N m. Calculate the required gear ratios, the number of teeth for each gear, and the output speed and torque. Use MATLAB® to simulate the gear train system and plot the speed and torque at each stage of the gear train.

9 Design a helical gear system for an industrial conveyor that requires smooth operation and minimal noise. The system must transmit a tangential force of 800 N with a helix angle of 25°. Calculate the axial thrust and design considerations for the bearings and shafts. Use MATLAB® to simulate the force distribution and plot the axial thrust as a function of the helix angle.

10 Design a lead screw system for a CNC machine that requires precise linear positioning. The lead screw has a pitch of 4 mm per revolution and must move the nut by 200 mm. Calculate the number of revolutions required, the conversion ratio, and the efficiency of the system. Use MATLAB® to simulate the lead screw motion and plot the linear displacement as a function of the number of revolutions.

11 Design a slider–crank mechanism for an internal combustion engine. The crank length is 120 mm, and the connecting rod length is 400 mm. Calculate the linear displacement of the slider when the crank rotates by 60°. Use MATLAB® to simulate the slider–crank mechanism and plot the linear displacement as a function of the crank angle.

12 Design a planetary gear system for an EV drivetrain that requires a speed reduction of 4:1 and can handle high torque. The sun gear has 40 teeth, each planet gear has 20 teeth, and the ring gear has 100 teeth. Calculate the speed ratio, the load distribution among the planet gears, and design considerations for durability and efficiency. Use MATLAB® to simulate the planetary gear system and plot the speed and torque distribution.

13 Design a cam-and-follower mechanism for an automated machine that requires precise timing and actuation. The cam rotates at 600 RPM and has a follower that moves with a maximum

displacement of 50 mm. Calculate the frequency of the follower's motion and design considerations for the cam profile. Use MATLAB® to simulate the cam-and-follower motion and plot the follower displacement as a function of the cam rotation angle.

References

Juvinall, R. C. & Marshek, K. M., 2020. *Fundamentals of Machine Component Design*. 7th ed. Hoboken, NJ: John Wiley.

Liu, C., Zhao, C., Liu, Z. & Wang, S., 2021. Dynamic Analysis of Ball Screw Feed System with the Effects of Excitation Amplitude and Design Parameters. *Applied Sciences*, 11(13), p. 7070.

Lynwander, P., 2019. *Gear Drive Systems: Design and Application*. New York: Marcel Dekker.

11
Analog Control

OBJECTIVES

- Explain the role of automatic control in mechatronic system automation.
- Compare manual (open-loop) and automatic (feedback) control.
- Describe analog and digital control systems.
- Present key performance metrics for control systems.
- Model and analyze analog motion control systems.
- Explain the effects of different control strategies such as proportional, integral, derivative (PID), on system performance.
- Provide practical examples for simulation and analysis of control system behavior.
- Discuss the importance of stability in feedback control systems and introduce classical analysis methods for assessing and ensuring stability.

Automation of mechatronic system involves the combination of mechanical and electronic components, computer science, and automatic control theory. It influences the design, operation, and functionality of engineering systems, such as industrial robots, smart appliances, automotive systems, and manufacturing equipment. In general, automation involves combining hardware and software components to create machines that can sense, make decisions, and execute with limited human involvement.

Automation in mechatronic systems involves increasing the capability of these systems to carry out complex, accurate, and repeatable tasks efficiently and consistently, including adapting to different environmental conditions.

Automatic control is essential for mechatronic system automation, enabling machines to adapt, remain stable, and achieve performance goals. It works by monitoring system output, comparing it to a setpoint, and adjusting inputs to reduce any difference.

11.1 Control Problem

One of the central challenges in mechatronic system automation is the formulation and solution of control problems that arise from the need to maintain optimal system performance in the presence of uncertainty and external disturbances. A typical control problem involves designing a control algorithm that enables a mechatronic device to follow the desired trajectory or maintain a specified output, despite variability in system parameters or environmental influences.

For example, consider a robotic arm tasked with precisely positioning objects on a fast-moving assembly line. Its control system must ensure that the arm's tip reaches the target location within a fraction of a second, compensating for variations in payload weight, joint friction, or sudden vibrations caused by nearby equipment. The control problem, in this case, is to develop a feedback mechanism that dynamically adjusts the actuator commands based on real-time sensor data, thereby minimizing the error between the actual and desired positions.

In practice, control problems frequently demand balancing competing priorities such as speed, accuracy, energy efficiency, and robustness. Designers must select appropriate sensors, actuators, and control strategies to meet the specific requirements of each application. Solutions to control problems are at the heart of successful automation, enabling mechatronic systems to perform reliably in complex, ever-changing environments.

11.1.1 Manual Control

A mechatronic system can operate manually or automatically. Manual control is the simplest form, while automation may use analog or digital controls.

Manual operation of a device, known as "open-loop control," involves a human observing the system and deciding on an action to perform the task. As illustrated in Figure 11.1 (a), a human operator collects information about the state of the system; the brain processes the information and sends signals to the muscles to perform required actions.

The inputs of an "open-loop" system are independent of the outputs; instead, the human operator adjusts the input after observing the result. For example, driving without cruise control requires the driver to monitor speed, stay within lane markings, watch for pedestrians and other vehicles on the road, and adjust speed and direction as needed.

An open-loop system operates based on predetermined instructions, independent of the actual output. Consequently, variations in load or external conditions may result in substantial errors or instability, necessitating continuous human supervision and adjustment to ensure satisfactory performance.

In mechatronic systems where fluctuations of output variables are not critical, open-loop control is acceptable. Ceiling fans, food processors, and inexpensive power tools are examples of open-loop systems. Many mechatronic systems require levels of precision, reliability, and adaptability that open-loop control does not provide; consequently, feedback control is commonly used in advanced automation.

11.1.2 Feedback Control

Automatic control process is realized through feedback loops, where sensors collect information on the actual state of the system and the controller performs corrective actions. Figure 11.1 (b) outlines a feedback control system where actual outputs of the controlled device, equipment, or system are compared in real time to a preset reference or target value. If any difference, which is known as the error, arises between the desired and actual outputs, the controller adjusts the inputs of the system components to minimize this error and drive the system back toward its intended state. For instance, in a temperature-controlled chamber, the system continuously monitors the internal temperature and adjusts the heating or cooling elements to maintain a precise setpoint.

Feedback control is fundamental for achieving stability, precision, and adaptability in modern mechatronic automation, whether in temperature regulation, robotics, automotive systems, or smart household devices. Without feedback control, these systems would either be incapable of achieving their intended functions or require continuous human supervision.

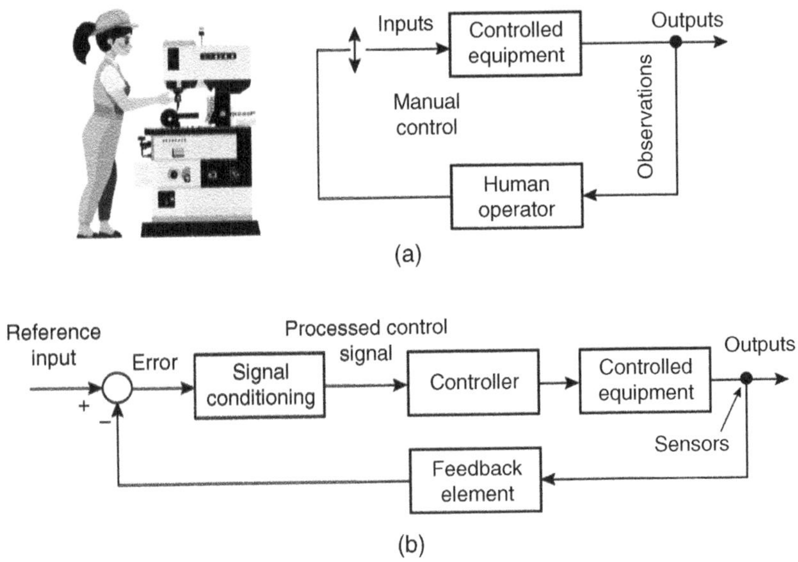

Figure 11.1 Manual and feedback control. (a: Manual (open-loop) control; b): Feedback control)

Mechatronic systems use a variety of feedback control methods such as on–off, continuous, and computer-based feedback control, each suited to specific operational needs and complexities. In a multifunction mechatronic system, multiple feedback loops may be coordinated by a computer or activated in sequence to carry out specific actions. These control strategies enable mechatronic systems to be customized for specific applications, effectively balancing simplicity, responsiveness, and advanced adaptive intelligence.

11.1.2.1 Analog Control

Analog control systems process continuous signals to monitor, command, and regulate the behavior of mechanical and electronic subsystems in real time. These systems involve analog signal conditioning circuits for amplification, noise filtering, and signal processing.

System outputs are continuously measured by sensors. When digital sensors or encoders are used, digital-to-analog converters (DACs) transform digital signals into analog. A feedback element may be needed to scale up and filter sensor signals. An analog comparator circuit compares the feedback signal with the reference inputs and generates an error signal. Error signals are usually further processed using operational amplifier circuits. Proportional, integral, and derivative (PID) control strategies are often implemented to improve responsiveness and accuracy.

As illustrated in Figure 11.2, in an analog-controlled mechatronic system, feedback is handled by comparing a continuous output signal to a reference and generating an error signal, which then drives power converters to activate transducers, manipulators, and motors. Signal processing circuits modify the error signal by simply multiplying with a constant (proportional control), applying analog integration (integral control), or differentiating the error signal (derivative control). The integration of PID control strategies enables enhanced responsiveness and accuracy in system outputs. PID control is a highly effective method widely implemented in conventional motor drives within mechatronic systems.

Analog control is well-suited for applications requiring real-time, smooth adjustments rather than discrete, stepwise changes.

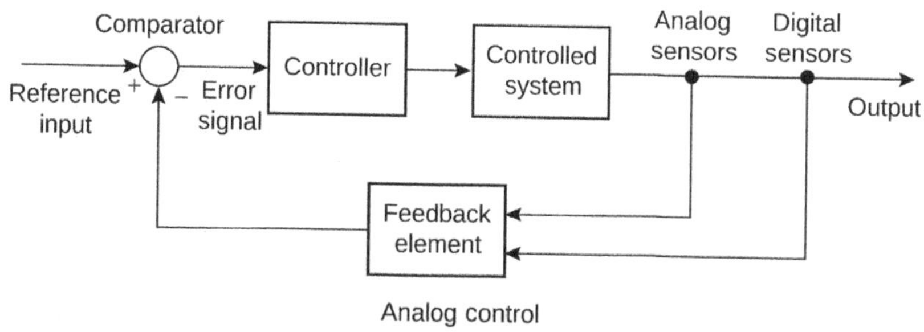

Figure 11.2 Analog control system outline

11.1.2.2 Digital Control

Digital control systems process information in discrete steps, utilizing binary logic and microprocessors. The central component of a digital control system is typically a controller, such as a microcontroller, digital signal processor (DSP), or embedded computer, which receives input data from sensors in the form of digital signals. These systems sample variables such as temperature, position, or speed at regular intervals, converting analog signals into digital values through analog-to-digital converters (ADCs). Figure 11.3 outlines a typical digital control system.

Once digitized, the controller analyzes these values using sophisticated algorithms, ranging from classic PID control to more advanced adaptive or predictive strategies. The controller then determines the appropriate response and sends digital output signals, often via DACs or directly to actuators such as motors, valves, or relays. Because all processing is done mathematically and logically, digital control systems are highly programmable, supporting rapid adjustment or reconfiguration for new tasks.

Computer-based control uses microprocessors and programmable devices to process sensor data, perform control algorithms, and operate actuators with precision and adaptability. This approach supports features including adaptive control, machine learning integration, and communication with other systems.

11.1.2.3 On–Off Control

On–off control operates by switching actuators or output devices between two distinct states—fully on or fully off—based on whether the measured variable crosses a specific threshold. This binary method, although straightforward and reliable, can result in oscillations around the target value and is best suited for processes where exact precision is not critical.

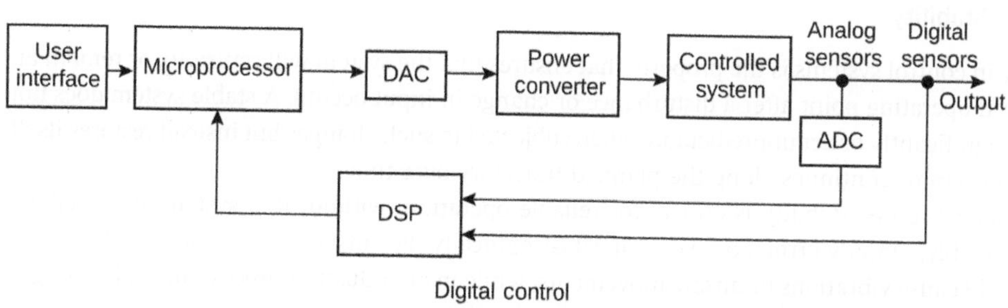

Figure 11.3 Digital control system outline

11.1.2.4 Sequential Control

Sequential control governs systems by following a preset sequence of actions, with each stage initiated by predetermined logic conditions or specific inputs. In this approach, subsystems may employ open-loop, on–off, continuous, or computer-based control methods, depending on the requirements for accuracy and response time. This approach is found in manufacturing automation and household appliances like dishwashers, cloth washers, and microwave ovens, where operations must occur in a precise order to achieve the desired outcomes.

This digital approach offers several advantages. It allows for precise, repeatable, and robust control even in the presence of noise or variations in the environment. Furthermore, digital controllers can implement complex decision-making, self-tuning, and communication with other systems, enabling the development of intelligent, networked mechatronic devices. Digital control has become indispensable in modern automation, robotics, automotive electronics, and smart consumer products.

11.2 Performance Metrics

The performance of a mechatronic control system is evaluated based on criteria such as accuracy, stability, responsiveness, robustness, and adaptability.

11.2.1 Accuracy

Accuracy in automatic control systems refers to the system's ability to achieve and maintain precise control over physical variables such as position, speed, force, temperature, or voltage. It describes how closely the system's output matches the desired reference value or setpoint, even in the presence of disturbances and uncertainties.

A system with high accuracy effectively reduces the discrepancy between the intended objective and the actual outcome. For example, in industrial robotics, accuracy means reaching a target position within a fraction of a millimeter. In smart appliances, it ensures precise thermal regulation for optimal performance and safety. This not only facilitates consistency and repeatability but also reduces the need for manual correction and supervision.

Accuracy is especially critical in mechatronic systems that require precise and repeatable actions. For example, accuracy of a robotic arm means reaching the exact target location. In industrial automation, accuracy of conveyor belts ensures a steady, predefined movement rate. In automated assembly, accurate force control prevents damage to components. In smart appliances, accuracy allows consistent thermal regulation. For electronic circuits, accuracy keeps voltages within safe and functional limits.

11.2.2 Stability

Stability in control systems is the property that ensures that the system will return to, or remain at, its desired operating point after a disturbance or change in input occurs. A stable system does not deviate significantly or act unpredictably when subjected to such changes but instead restores itself to equilibrium or continues along the planned trajectory over time.

In practical terms, stability is critical for reliable operation: without it, a system may oscillate uncontrollably, amplify errors, or even fail catastrophically. For instance, in a robotic arm, instability could cause vibrations or unsafe movements, while in an industrial process, instability might lead to dangerous fluctuations in temperature or pressure.

Mathematically, stability is often analyzed by examining how the system responds to small deviations from its reference point. If all such deviations diminish in time and the system's output settles back to the desired value, the system is considered stable. The stability of mechatronic systems depends on the design of feedback mechanisms, the selection of system parameters, and the implementation of control strategies. These elements are essential for ensuring safe, predictable, and precise system performance.

11.2.3 Responsiveness

Responsiveness in automatic control systems refers to the system's ability to detect and react quickly and appropriately to changes in input signals, external disturbances, or shifts in the environment. It is a measure of how fast the system can adapt its behavior when faced with new commands or unexpected circumstances, ensuring that the controlled process remains on course or reaches the new desired state in minimal time. An ideal system balances rapid reaction with smoothness, avoiding excessive overshoot or oscillation, so that responses to changes are both fast and controlled. In real-world mechatronic applications such as robotic manipulation, automotive control, or industrial automation, high responsiveness enables equipment to handle dynamic tasks, adapt to variable loads, and interact safely with humans or other systems.

11.2.4 Robustness and Adaptability

Robustness is the system's ability to maintain performance in the presence of uncertainties and disturbances like external forces, parameter variations, or unexpected events. Adaptable systems can adjust their internal parameters in real time to optimize performance, while robust systems remain reliable under a wide range of conditions.

Automation algorithms, such as adaptive control or machine learning-based strategies, are increasingly employed in mechatronic systems to enhance their robustness and adaptability.

Robust control is a fundamental concept in automatic control theory, particularly relevant for mechatronic systems operating in unpredictable or variable environments. At its core, robust control refers to the design and implementation of control strategies that maintain a system's desired performance and stability despite uncertainties, disturbances, or inaccuracies in the system model. Real-world systems rarely behave exactly as predicted because there may be unexpected changes in load, sensor noise, wear and tear, or external influences. Robust control ensures the system continues to function reliably.

To achieve robustness, designers must anticipate possible discrepancies between the theoretical model and actual system behavior. This often involves designing controllers that are less sensitive to parameter variations, external disturbances, or incomplete knowledge of the system. Techniques such as H-infinity control, sliding mode control, and adaptive control are commonly used, each providing mechanisms for the system to cope with uncertainty while achieving its goals.

In practice, robust control enables mechatronic systems to operate safely and effectively in a wide range of conditions. For example, a robot working in a factory may encounter uneven surfaces or shifting payloads, while an autonomous vehicle must adapt to changing road conditions and sensor imperfections. Robust control strategies ensure that, even in these variable situations, performance criteria like accuracy, stability, and responsiveness are upheld without extensive manual recalibration.

11.3 Analog Motion Control

Analog motion control systems are used to adjust the velocity or position of a mechanical load continuously in real time. Motors and actuators operating mechanical loads may function in open-loop control or feedback control mode. Feedback motion control systems are known as servomechanisms. In analog motion, sensors continuously monitor the system's actual output, which is compared to a reference input. The controller minimizes any deviation from the desired in real time. Servomechanisms are commonly used in mechatronic systems that require speed or position control.

Permanent Magnet DC (PMDC) motors are widely used in low-power analog servomechanisms, while higher power industrial systems typically employ separately excited DC or asynchronous motors. This text will focus on velocity and position control of PMDC motors.

11.3.1 Open-Loop DC Motor Dynamics

Chapter 6 discusses the structure and operation of DC machines in detail. The PMDC motor is a specific motor type where the magnetic field required for energy conversion is produced by permanent magnets (PMs) mounted on the stator. This section examines the dynamic characteristics of PMDC motors when operating under open-loop conditions.

Figure 11.4 illustrates a PMDC motor operating with a mechanical load. The mechanical load is characterized by an equivalent moment of inertia J, a viscous friction constant β, and torque T_L. The switch turns on to power the armature circuit when the motor is at rest. The resistance and self-inductance of the rotor windings are denoted by R_a and L_a, respectively. e_a represents the velocity voltage induced on the rotor windings. Velocity voltage is a function of the air gap flux φ.

11.3.1.1 State-Space Model of PMDC Motor

KVL applied to the armature loop and Newton's laws to the mechanical elements yields

$$v_a = R_a i_a + L_a \frac{di_a}{dt} + e_a \tag{11.1}$$

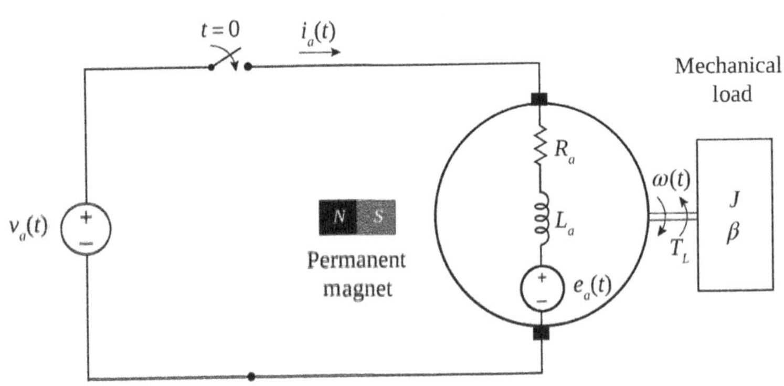

Figure 11.4 Circuit diagram of a PMDC driving a mechanical load

$$T_e = \beta\omega + J\frac{d\omega}{dt} + T_L \tag{11.2}$$

Since the PM provides a constant magnetic field, when the motor operates in the linear region of the core material, induced velocity electromotive force (EMF) is proportional to the angular velocity ω.

$$e_a = k_e\omega \tag{11.3}$$

According to the conservation of energy principle, mechanical power developed in the air gap equals electrical power on the velocity EMF. In the linear region, the motor generates a torque proportional to the armature current.

$$\omega T_e = e_a i_a \Longrightarrow T_e = k_m i_a \tag{11.4}$$

The velocity EMF constant k_e is expressed in volts per radians per second (V·s/rad) and the torque constant in N/A. While their numerical values are identical, they are usually denoted by separate parameters k_e and k_m in equations to match the units.

Field current, armature current, and angular velocity are independent state variables. State-space equations can be organized as below by considering armature voltage and load torque as inputs.

$$\frac{d}{dt}\begin{bmatrix} i_a \\ \omega \end{bmatrix} = \begin{bmatrix} -\frac{R_a}{L_a} & -\frac{k_f}{L_a} \\ \frac{k_m}{J} & -\frac{\beta}{J} \end{bmatrix} \cdot \begin{bmatrix} i_a \\ \omega \end{bmatrix} + \begin{bmatrix} \frac{1}{L_a} & 0 \\ 0 & -\frac{1}{J} \end{bmatrix}\begin{bmatrix} v_a \\ T_L \end{bmatrix} \tag{11.5}$$

Example 11.1 A PMDC motor drives a mechanical load. An armature voltage of 24 V is applied while the motor is initially at rest. Given the motor and load parameters below.

a) Find the initial armature current when the motor is powered on.
b) Using a MATLAB® script, obtain the step response of the motor with the specified load. Plot the time variation of the speed and armature current.

Motor Parameters:

$$V_a = 24\,\text{V}; R_a = 1.7\,\Omega; L_a = 2\,\text{mH}; k_f = 0.0439\,\text{V/A}$$

Load Parameters:

$$T_L = 0.1\,\text{N}\cdot\text{m}; \beta = 0.155 \times 10^{-5}\,\text{N/rad}\cdot\text{s}^{-1}; J = 0.155 \times 10^{-2}\,\text{kg}\cdot\text{m}^2$$

Solution

a) The motor speed is zero when it is energized. Therefore, no velocity voltage is induced on the armature winding. The initial armature current is calculated by applying Ohm's law to the armature resistance.

$$I_{a,0} = \frac{V_a}{R_a} = \frac{24}{1.7} = 14.1\,\text{A}$$

b) The following script computes the step response of the state equations in equation 11.1 and plots the time variation of the motor speed and the corresponding armature current values presented in Figure 11.5. Both the speed variation and current are exponential functions approaching the final value asymptotically.

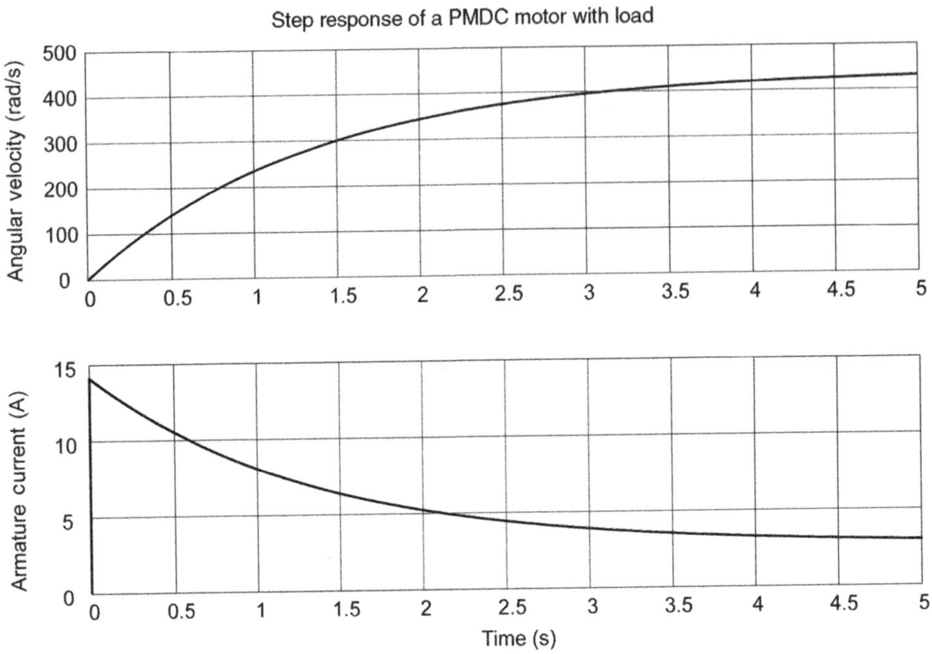

Figure 11.5 Step response of a DC motor starting with a mechanical load

```
% Step response of a PMDC motor with load
clear;
% Motor parameters
Ra = 1.7;          % Armature resistance (Ohm)
La = 2.e-6;        % Armature inductance (H)
Kb = 0.0439;       % Back EMF constant (V/rad/s)
Kt = 0.0439;       % Torque constant (Nm/A)
% Load parameters (referred to rotor)
J = 0.155e-2;      % Total inertia (kg·m^2)
B = 4.545e-5;      % Viscous friction coefficient (N/rad.s^-1)
TL= 0.1;           % Load torque
Va = 24;           % Step armature voltage (Volts)
% Time span
tspan = [0 5];   % Simulation interval
y0 = [0; 0];      % Initial conditions: [current; angular velocity]
% State equations for PMDC motor with load
% State vector x = [ia; w] where:
%   ia = armature current (A)
%   w  = angular velocity (rad/s)
MSE = @(t, x) [
    (1/La)*(Va - Ra*x(1) - Kb*x(2));          % d(ia)/dt
    (1/J)*(Kt*x(1) - B*x(2)-TL)               % d(w)/dt
];
% Solve using ode45
```

```
[t, y] = ode45(MSE, tspan, y0);
% Extract current and speed
ia = y(:, 1);
w  = y(:, 2);
% Plot results
figure;
subplot(2,1,1);
plot(t, w, 'k', 'LineWidth', 1.5);
title('Step response of a PMDC motor with load');
ylabel('Angular Velocity \omega (rad/s)');
ylim([0 500])
grid on;
subplot(2,1,2);
plot(t, ia, 'k', 'LineWidth', 1.5);
ylabel('Armature Current i_a (A)');
ylim([0 15])
xlabel('Time (s)')
grid on;
```

11.3.1.2 Transfer Function of PMDC Motor

Applying the Laplace transform to equation 11.1, assuming all initial conditions are zero, results in the following algebraic equation in the *s*-domain.

$$I_a(s) = \frac{V_a(s) - k_e\Omega(s)}{R_a + sL_a} \tag{11.6}$$

An expression of the angular velocity is obtained by substituting $T_e = k_m i_a$ into the mechanical dynamic equation 11.2 and taking the Laplace transform with initial conditions zero.

$$\Omega(s) = \frac{k_m I_a(s) - T_L}{\beta + Js} \tag{11.7}$$

Figure 11.6 shows a block diagram illustrating the *s*-domain equations above written for the electrical and mechanical aspects of a PMDC or separately excited motor with constant field current in the linear region. This diagram gives a clear picture of the transfer function (TF) of each functional block of the motor. While a DC motor by itself is an open-loop system, the block diagram illustrates a feedback loop resulting from the velocity EMF generated in the rotor windings (Kuo, 1982). The inputs of the block diagram are armature voltage and mechanical load. Depending on the objective of the analysis, either angular velocity or rotation angle is selected as the output. The rotation angle corresponds to the integral of velocity, which is represented by the *s*-domain integration block $1/s$ in the block diagram.

According to the superposition principle, the total impact of two inputs of a linear, time-invariant system is equal to the sum of the individual effects of each input when the other input is zero. In a velocity control system, typically the armature voltage is considered the input and the angular velocity the output.

Considering V_a as input and Ω as output, the TF is derived by substituting equation 11.6 into equation 11.7 and writing the ratio of the output to the input.

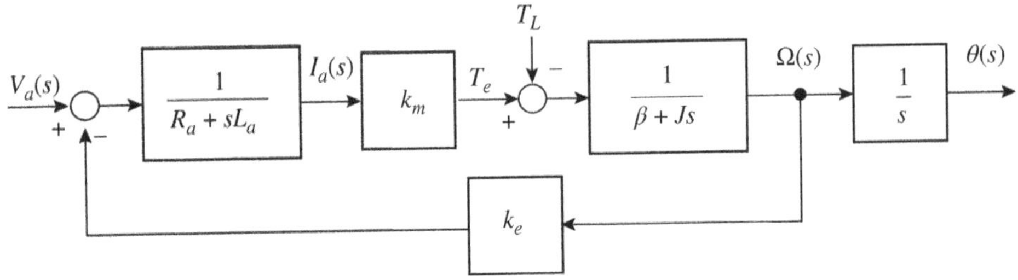

Figure 11.6 Block diagram of a PMDC

$$G(s) = \frac{\Omega(s)}{V_a(s)} = \frac{k_t}{(sL_a + R_a)(sJ + \beta) + k_e k_t} \tag{11.8}$$

This TF is arranged in the following polynomial form using the damping ratio ζ and natural frequency ω_n.

$$G(s) = \frac{k_t}{s^2 + \left(\frac{R_a}{L_a} + \frac{B}{J}\right)s + \frac{k_e k_t}{JL_a}} = K\frac{\omega_n^2}{s^2 + 2\zeta\omega_n s + \omega_n^2}$$
$$\zeta = \frac{1}{2}\left(\frac{R_a}{L_a} + \frac{B}{J}\right); \omega_n^2 = \frac{k_e k_t}{JL_a}; K = \frac{JL_a}{k_e} \tag{11.9}$$

The second-order TF above describes both the electrical and mechanical properties of a PMDC motor. For a typical motor, the damping ratio is greater than unity, which produces an overdamped step response. The electrical time constant L/R is considerably smaller than the mechanical time constant J/β, meaning that electrical transients decrease much faster than mechanical transients. When either only electrical or mechanical dynamics are modeled, the DC motor can be represented by a first-order TF.

Computation of electrical dynamics may be needed for designing electronic converters. In this case, mechanical dynamics are ignored, and the velocity EMF is treated as constant or slowly varying. This results in a first-order response with time constant $\tau_e = L/R$. In the analysis of mechanical response, however, electrical time constant can be ignored, and the armature current can be assumed constant.

After electrical transients have vanished, the armature current remains constant; therefore, $L(di_a/dt) = 0$. Dynamic equations can be written in the following form:

$$V_a(t) = R_a i_a(t) + k_e\omega(t) \tag{11.10}$$

$$k_m i_a(t) = J\frac{d\omega(t)}{dt} + B\omega(t) + T_L(t) \tag{11.11}$$

Laplace transforms of these differential equations with all initial conditions zero result in the following algebraic equations in the s-domain:

$$V_a(s) = R_a I_a(s) + k_e\Omega(s) \tag{11.12}$$
$$k_m I_a(s) = Js\Omega(s) + \beta\Omega(s) + T_L(s) \tag{11.13}$$

The TF is derived by solving the s-domain equations above and dividing the output by the input. When the load torque is zero, the armature current in the s-domain is obtained as

$$I_a(s) = \frac{Js\Omega(s) + \beta\Omega(s)}{k_m} \tag{11.14}$$

The resulting expression is derived through substitution of $I_a(s)$ into equation 11.12 and factorization.

$$V_a(s) = \left[\frac{R_a(Js+\beta)}{k_m} + k_e\right]\Omega(s) \tag{11.15}$$

Armature voltage is considered as input and angular velocity as output; therefore, the TF is obtained as below:

$$\begin{aligned} G_a(s) &= \left.\frac{\Omega(s)}{V_a(s)}\right|_{T_L=0} = \frac{k_m}{R_a(Js+\beta)+k_m k_e} = \frac{K_a}{\tau_m s+1} \\ K_a &= \frac{k_m}{R_a\left(\beta + \frac{k_m k}{R_a\, e}\right)}; \quad \tau_m = \frac{J}{\beta + \frac{k_m k_e}{R_a}} \end{aligned} \tag{11.16}$$

On the other hand, if the armature voltage is zero, armature current is obtained from equation 11.12 as below:

$$I_a(s) = -\frac{k_e \Omega(s)}{R_a}$$

Substitution of $I_a(s)$ into equation 11.13 and factorization results in the TF below, considering load torque as input and angular velocity as output.

$$\begin{aligned} G_T(s) &= \left.\frac{\Omega(s)}{T_L(s)}\right|_{V_a=0} = -\frac{R_a}{R_a(Js+\beta)+\frac{k_e k_m}{R_a}} = -\frac{K_T}{\tau_m s+1} \\ K_T &= \frac{R_a}{\beta + \frac{k_e k_m}{R_a}}; \quad \tau_m = \frac{J}{\beta + \frac{k_m k_e}{R_a}} \end{aligned} \tag{11.17}$$

This is a first-order TF with a negative sign, indicating that increasing load torque reduces angular velocity.

Example 11.2 Given the motor parameters below,

a) Calculate the damping ratio, natural frequency, and poles of the TF.
b) Find the unit-step response using only the pole with the smaller absolute value.
c) Obtain the unit-step response using both poles.
d) Compare the results obtained in (b) and (c).

$$R_a = 1\,\Omega; L_a = 0.5\,\text{mH}; k_m = 0.01\,\text{Nm/A}; k_e = 0.01\,\text{V/(rad/s)}$$

$$J = 0.01\,\text{kg}\cdot\text{m}^2; \beta = 0.001\,\text{Nm}\cdot\text{s/rad}$$

Solution

a) The TF below is derived in

$$G(s) = \frac{k_m}{s^2 + \left(\frac{R_a}{L_a} + \frac{\beta}{J}\right)s + \frac{k_e k_m}{J L_a}}$$

Substituting the given values, we obtain the TF as

$$G(s) = \frac{0.1}{s^2 + 2{,}000.1\,s + 20}$$

The poles of the TF are calculated below:

$$s_1 = \frac{-2{,}000.1 - \sqrt{2{,}000.1^2 - 4 \times 20}}{2} = -2{,}000$$

$$s_2 = \frac{-2{,}000.1 + \sqrt{2{,}000.1^2 - 4 \times 200}}{2} = -0.2$$

b) Since $s_2 \ll s_1$, ignoring s_1 should not affect the response. The initial value of the output is assumed to be $C_0 = 0$. For unit-step input, the steady-state value of the output is $C_s = 1$. The approximate response due to the slower decaying exponent is

$$c(t) = 1 - e^{-0.2t}$$

c) When both poles are considered, the unit-step response is the combination of two decaying exponents.

$$c(t) = 1 + A_1 e^{s_1 t} + A_2 e^{s_2 t}$$

The coefficients A_1 and A_2 are determined based on the initial conditions, which are assumed to be zero.

$$1 + A_1 + A_2 = 0 \Rightarrow A_2 = -(A_1 + 1)$$

$$\frac{d}{dt}(1 + A_1 e^{s_1 t} + A_2 e^{s_2 t}) = s_1 A_1 e^{s_1 t} + s_2 A_2 e^{s_2 t}|_{t=0} = 0$$

$$-2{,}000A_1 + 0.2(A_1 + 1) = 0 \Longrightarrow A_1 = \frac{0.2}{2{,}000.2} = 2.5 \times 10^{-5}$$

$$A_2 = -1$$

$$c(t) = 1 + \left(2.5 \times 10^{-5}\right)e^{-2{,}000t} + e^{-0.1t}$$

d) The results obtained in (b) and (c) clearly show that the TF of the PMDC motor with the given parameters can be considered as first order.

11.3.2 Velocity Control

A velocity-controlled servomechanism regulates the speed of a mechanical system driven by an electric motor. Velocity control methods typically utilize feedback loops that compare measured velocity to a reference signal. Sensors convert the system's output velocity into an electrical signal. The error signal is defined as the difference between the reference input and the sensor output. The controller processes this error signal and activates a motor to adjust the output.

A wide variety of mechatronic applications use velocity control systems. For instance, velocity-controlled servomechanisms maintain precise rotational speeds, ensuring reliable data access in computer and audiovisual disc drives. In automotive engineering, cruise control systems regulate vehicle speed to provide a consistent and comfortable driving experience. Conveyor belts in manufacturing plants use speed control to synchronize product flow and optimize throughput, while industrial mixers depend on regulated motor velocities to achieve uniform blending of materials.

In robotics, speed control facilitates smooth and accurate movement of robotic arms, particularly during assembly or pick-and-place operations. Modern washing machines, food processors, and power tools incorporate velocity control to adjust performance based on user settings or varying load conditions.

11.3.2.1 Proportional Velocity Control

Figure 11.7 shows a simple feedback control system used for speed regulation of a PMDC. A tachometer (T) mounted on the motor shaft produces a voltage proportional to the angular velocity. The pulse-width DC/DC converter supplies the armature circuit with a voltage proportional to the applied error signal. The converter's voltage scaling factor (A) is adjustable to modify the feedback system performance.

Proportional velocity control uses a feedback loop in which the output velocity of the system is continuously compared to a reference signal representing the target velocity. The difference between these values, known as the error signal, is amplified by the PWM converter, which serves as controller. The control voltage proportional to the magnitude of the error is applied to the armature circuit of the motor.

Figure 11.8 shows two block diagrams of the described velocity control system. In (a), internal feedback of the PMDC motor is represented using the block diagram in Figure 11.6. To obtain the TF for $R(s)$ input and $\Omega(s)$ output, the load torque must be set to zero. To find the TF between the load torque and the angular velocity, the reference input should be set to zero. The combined impact of the varying reference input and load torque can be analyzed by applying the superposition principle. In the block diagram (b), the motor is modeled as a first-order system, assuming the electrical time constant is much smaller than the mechanical one and there is no torque load.

TF of the velocity control system can be obtained by simplifying the block diagram. In the forward path of the block diagram, the first-order TF of the PMDC is multiplied by the gain A of the power amplifier (DC/DC converter). In this system, feedback gain is unity. The following TF is obtained for the complete feedback control system:

$$T(s) = \frac{\Omega(s)}{R(s)} \frac{AG(s)}{1 + K_v G(s)} = \frac{AK_a}{\tau_m s + 1 + AK_a} \tag{11.18}$$

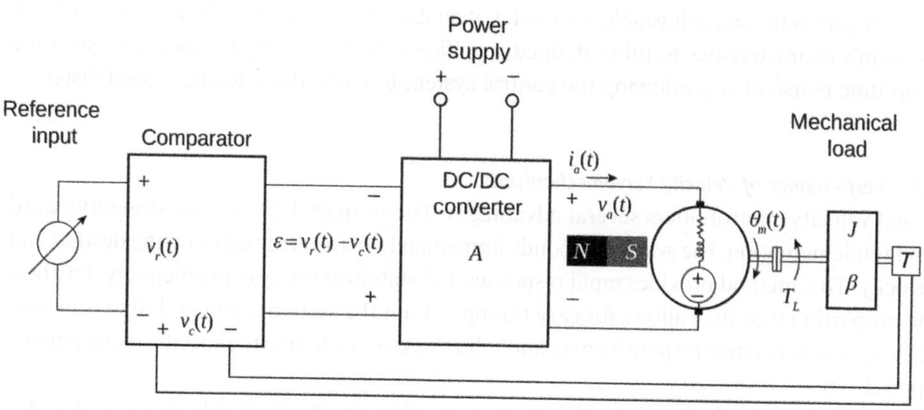

Figure 11.7 An analog speed control system driven by a permanent magnet DC motor

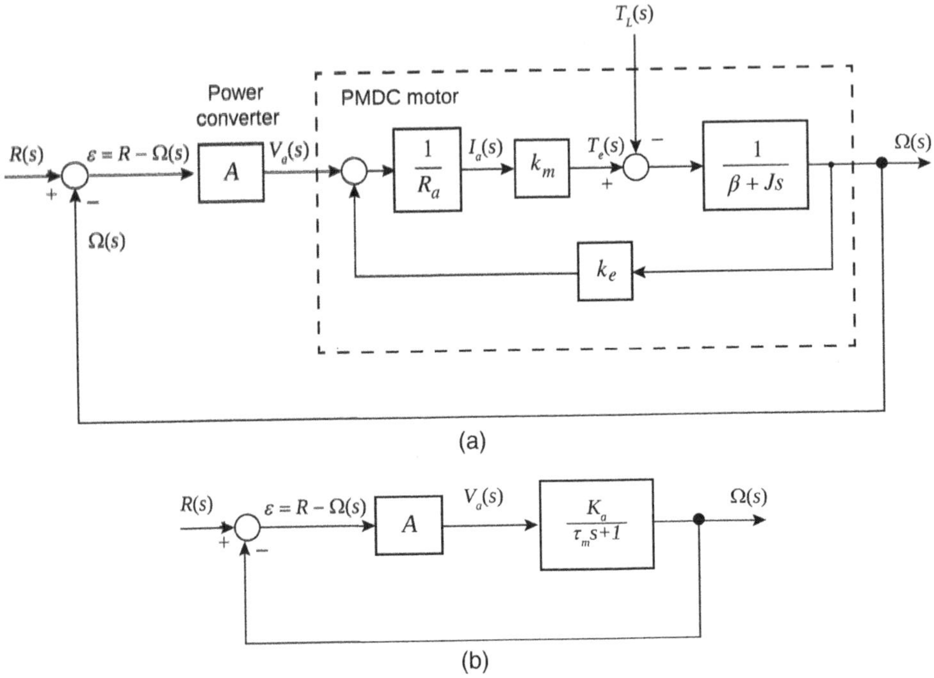

Figure 11.8 Block diagram of a velocity control system (a: PMDC motor block diagram shown explicitely; b: 'PMDC motor represented by a single block).

The resulting TF can be expressed in the standard form below:

$$T(s) = \frac{K_{FB}}{\tau_{FB}s + 1}$$

$$\tau_{CL} = \frac{\tau_m}{1 + AK_a} \tag{11.19}$$

$$K_{CL} = \frac{AK_a}{1 + AK_a}$$

As amplifier gain is the sole adjustable parameter, it enables modification of the feedback velocity control system's characteristics to fulfill defined specifications. Increasing the gain decreases the closed-loop time constant τ_{CL}, allowing the control system to reach the reference speed faster.

11.3.2.1.1 *Performance of Velocity Servomechanisms*

Proportional velocity control offers several advantages. The primary benefit is its straightforward design and implementation; the system responds immediately to the error between the desired and actual velocity. This method provides rapid responses and stabilizes the system efficiently. Furthermore, the proportional control allows for easy tuning of both the system's gain and time constant, enabling designers to balance responsiveness and robustness according to the specific requirements of the application.

However, proportional velocity control has limitations. One significant drawback is the presence of steady-state error. Because the controller only reacts to instantaneous error, it cannot eliminate offset between the target and actual velocity when subjected to constant disturbances or changing loads. Such systems may also be sensitive to parameter variations such as changes in motor characteristics or feedback gains, which can impact performance and stability.

11.3.2.1.2 Steady-State Error

Steady-state error refers to the difference between a system's actual output value after all transients have decayed and its desired target value. The TF between the error signal and the system output can be written as

$$\frac{E(s)}{R(s)} = \frac{R(s) - \Omega(s)}{R(s)} = 1 - \frac{\Omega(s)}{R(s)} \tag{11.20}$$

Substituting the TF given in equation 11.18, we obtain

$$\frac{E(s)}{R(s)} = 1 - \frac{AK_a}{\tau_m s + 1 + AK_a} = \frac{\tau_m s + 1}{\tau_m s + 1 + AK_a} \tag{11.21}$$

Since $T(s) = \dfrac{K_{FB}}{\tau_{FB}s + 1}$, the following error expression can be derived in terms of the time constant τ_{FB} and gain of the closed-loop system K_{FB} for unit-step input.

$$E(s) = \frac{\tau_{FB}s + 1}{\tau_{FB}s + 1 + AK_a} \cdot \frac{1}{s} \tag{11.22}$$

The steady-state value can be determined using the final value theorem of Laplace transform.

$$e_{ss}(t)|_{t \longrightarrow \infty} = \lim_{s \to 0} sE(s) \tag{11.23}$$

$$e_{ss}(t)|_{t \longrightarrow \infty} = \lim_{s \to 0} \left\{ s \frac{\tau_{FB}s + 1}{\tau_{FB}s + 1 + AK_a} \cdot \frac{1}{s} \right\} = \frac{1}{1 + AK_a} \tag{11.24}$$

This steady-state error approaches zero when the gain (A) is increased, indicating the output closely follows the input. Increasing proportional gain enhances system response by reducing the time constant. However, higher gain may also result in undesirable noise amplification.

Example 11.3 Given the motor parameters below, compare the unit-step response of the unity feedback velocity control system for the power amplifier gain values of 1 and 10.

Armature circuit gain: $K_m = 1$; time constant: $\tau_m = 0.2$ s.

Solution

Closed-loop TF is obtained by substituting the given values into equation 11.19.

$$\tau_{CL} = \frac{\tau_m}{1 + AK_m} = \frac{0.2}{1 + A}$$

$$K_{CL} = \frac{AK_m}{1 + AK_m} = \frac{A}{1 + A}$$

$$T(s) = \frac{K_{CL}}{\tau_{CL}s + 1} = \frac{A}{1 + A} \cdot \frac{1}{0.2\,\mathrm{s} + 1}$$

MATLAB® Script to Compute the Step Response of the Given Transfer Function.

```
% Parameters
K_F =1;       % Proportional velocity gain
K_m = 1;      % Motor gain
tau = 0.2;   % Motor time constant (s)
% Plant transfer function: G(s) = K / (tau*s + 1)
s = tf('s');
G_m = K_m / (tau*s + 1);
% Closed-loop transfer function: T(s) = (Kv * G) / (1 + Kv * G)
T = feedback(K_F * G_m, 1);
% Step response
figure;
T1 = feedback(1 * G_m, 1);
T2 = feedback(10 * G_m, 1);
step(T1,T2, 2); % Simulate for 2 seconds
title('Step Response of Proportional Velocity Control
Servomechanism');
xlabel('Time (s)');
ylabel('Angular Velocity (rad/s)');
grid on;
```

Step responses plotted in Figure 11.9 clearly show that for $A = 1$, the steady-state error is 50% and the output reaches its steady-state value in around 0.5 seconds. Increasing the gain to 10 significantly improves the system performance by reducing the response time to 0.1 second and steady-state error to 9%.

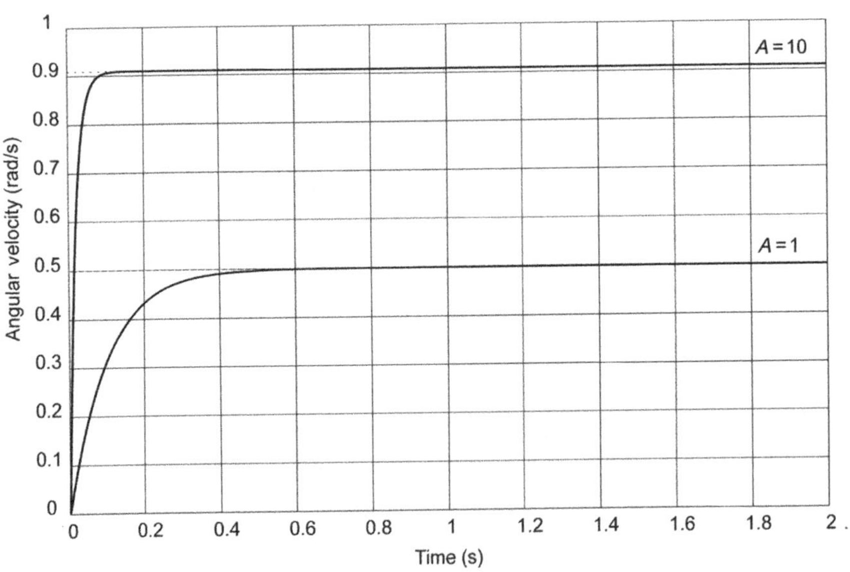

Figure 11.9 Step response of proportional speed control servomechanism

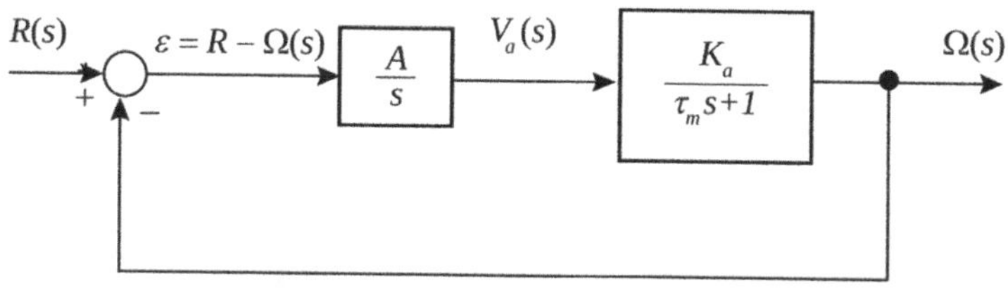

Figure 11.10 A velocity control system with integrator in the feedback loop

11.3.3 Integral Control of Velocity Servomechanism

The steady-state error can be eliminated by including an integrator in the feedback loop as shown in Figure 11.10. The integrator accumulates the error over time, continuously adjusting the control signal until the output matches precisely the reference input. This modification not only improves accuracy but also enhances the system's ability to reject persistent disturbances and compensate for changes in system dynamics.

With the integrator, the closed-loop TF becomes

$$G(s) = \frac{\Omega(s)}{R(s)} = \frac{AK_a}{s(\tau_m s + 1) + AK_a} \tag{11.25}$$

Applying similar operations as in the proportional control, we obtain the following TF between the error signal and the system output:

$$T(s) = \frac{R(s) - \Omega(s)}{R(s)} = 1 - \frac{\Omega(s)}{R(s)} = \frac{s(\tau_m s + 1)}{s(\tau_m s + 1) + AK_a} \tag{11.26}$$

For unit-step input, the error expression can be derived as below:

$$E(s) = T(s) \cdot U(s) \tag{11.27}$$

$$E(s) = \frac{s(\tau_{FB} s + 1)}{s(\tau_{FB} s + 1) + AK_a} \cdot \frac{1}{s} \tag{11.28}$$

The steady-state value can be determined using the final value theorem of Laplace transform given in (10.23).

$$e_{ss}(t)|_{t \to \infty} = \lim_{s \to 0} \left\{ \frac{s^2(\tau_{FB} s + 1)}{s(\tau_{FB} s + 1) + AK_a} \cdot \frac{1}{s} \right\} = 0 \tag{11.29}$$

Evidently, adding an integrator in the feedback loop eliminates the steady-state error. However, integral control causes output oscillations and overshoot, as demonstrated in the following example.

11.3.4 Position Control

The goal of position control servomechanisms is to regulate the final displacement of a mechanical system with precision, speed, and stability. These systems use feedback components that detect the current position and compare it with a reference value, producing an error signal that adjusts the actuator to reach the target. In low-power applications, PMDC motors are cost-effective and reliable. Single- or three-phase induction motors are more suitable for higher power and industrial applications. While stepper motors can be utilized for analog position control, they are particularly advantageous in digital control systems due to their enhanced capabilities. The effectiveness of the

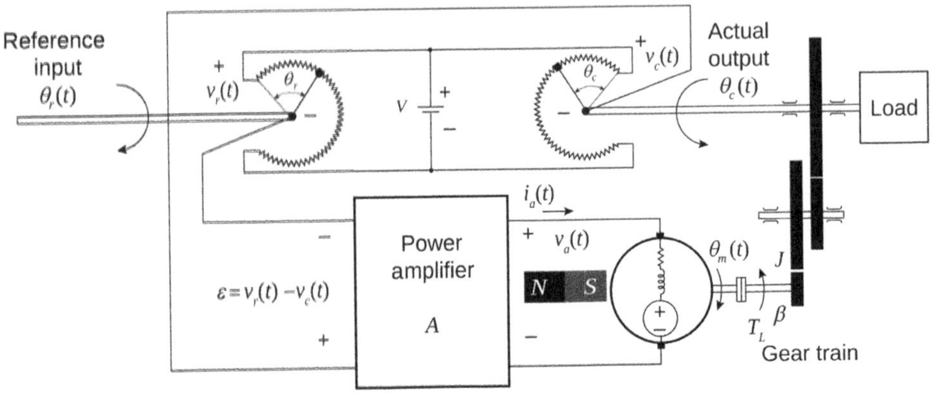

Figure 11.11 Position control servo

servomechanism relies on how well the control strategy can reject disturbances and compensate for uncertainties in load or system dynamics.

Figure 11.11 outlines a simple position control servomechanism with potentiometric feedback. The reference input is set by adjusting the shaft of the left potentiometer, which produces a voltage v_r proportional to the rotation angle. The second potentiometer, linked to the output shaft, produces a voltage v_c proportional to the output position. The circuit configuration shown in the figure acts as a Wheatstone bridge; the potential difference between the center terminals of both potentiometers is proportional to the difference between the reference and output angles.

$$\begin{aligned} &v_r(t) = K_p V\theta_r(t); v_c(t) = K_p V\theta_c(t) \\ &\varepsilon = v_r(t) - v_c(t) = K_p[\theta_r(t) - \theta_c(t)] \end{aligned} \tag{11.30}$$

The gear train reduces the angular velocity of the motor shaft to turn the mechanical load at the desired slower velocity. The moment of inertia J_m and viscous friction constant β_m represent combined mechanical parameters of the load, gear train, and rotor referred to the motor shaft, as previously discussed in Chapter 10.

Under the assumption that electrical transients decay much faster than mechanical transients, the PMDC motor can be represented by a first-order TF. Figure 11.12 shows the block diagram relating the reference input position to the controlled output position without load torque. TF of this system is derived below:

$$\begin{aligned} G(s) &= \frac{\theta_c}{E(s)} = \frac{K_p k_m}{s[R_a(J_m s + \beta_m) + k_m k_e]} \\ T(s) &= \frac{G(s)}{1 + G(s)} = \frac{K_p k_m}{s[R_a(J_m s + \beta_m) + k_m k_e] + K_p k_m} \\ &= \frac{K_p k_m}{s^2 + \left(\dfrac{R_a\beta_m + k_m k_e}{R_a J_m}\right)s + \dfrac{K_p k_m}{R_a J_m}} \end{aligned} \tag{11.31}$$

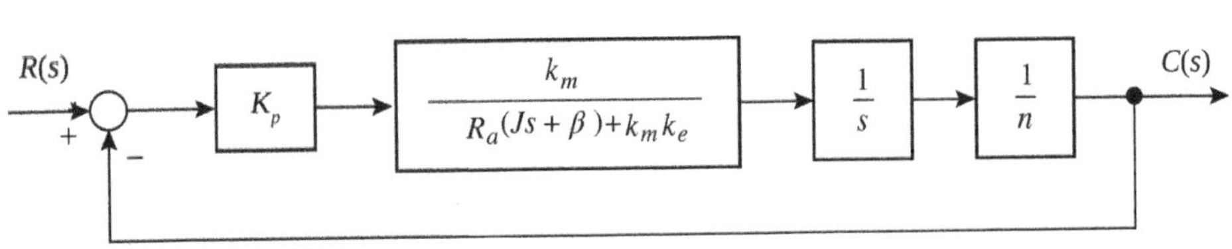

Figure 11.12 Block diagram of a position control servo

The TF of a position control system is second order, which means it has two poles below associated with its time constants.

$$s_{1,2} = -\frac{R_a\beta_m + k_m k_e}{2R_a J_m} \pm \sqrt{\frac{1}{4}\left(\frac{R_a\beta_m + k_m k_e}{R_a J_m}\right)^2 - \frac{K_p k_m}{R_a J_m}} \tag{11.32}$$

In control theory, second-order systems are typically analyzed using parameters such as damping ratio and resonance frequency (Kuo, 1982; Ogata, 2002). The following expressions can be derived for the TF in equation 11.31.

$$\begin{aligned} \omega_n^2 &= \frac{K_p k_m}{R_a J_m} \\ \zeta &= \frac{R_a\beta_m + k_m k_e}{K_p k_m} \end{aligned} \tag{11.33}$$

Resonant angular frequency ω_n is also known as natural frequency and measured in radians per second (rad/s). The damping ratio (ζ) and (ω_n) characterize the dynamic response of a second-order system. The interaction between them influences the type of response, response speed, and the rate at which transients are diminished. TF of a position control system can be expressed in terms of these standard parameters.

$$T(s) = \frac{C(s)}{R(s)} = \frac{K\omega_n^2}{s^2 + 2\zeta\omega_n s + \omega_n^2} \tag{11.34}$$

In the s-domain, the location of the TF's poles in the complex plane corresponds to the system's performance.

11.3.4.1 Position Control System's Unit-Step Response

The response of a position control system to unit-step function depends on the values of the damping ratio ζ and natural frequency ω_n. The system's transient and steady-state performance may be modified to meet the requirements of a particular application by adjusting the damping ratio and natural frequency values. Figure 11.13 illustrates how pole location, damping ratio, and natural frequency are related geometrically.

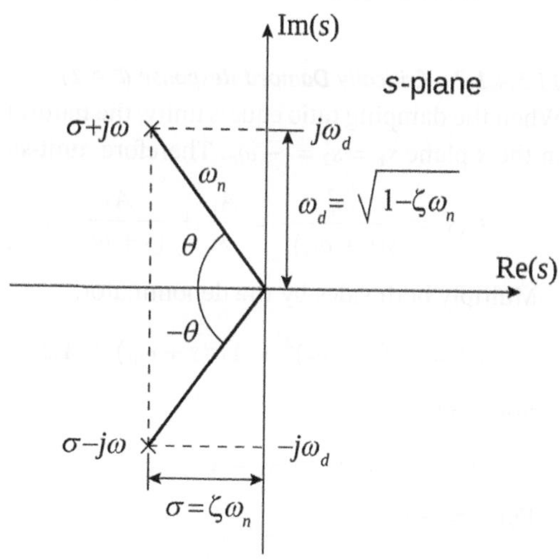

Figure 11.13 Relationship between natural frequency, damping ratio, and the poles

The unit-step response is examined by taking the inverse Laplace transform of the output function for unit-step input.

$$C(s) = \frac{\omega_n^2}{s^2 + 2\zeta\omega_n + \omega_n^2}\cdot\frac{1}{s} \tag{11.35}$$

For a second-order system, three distinct cases can occur based on the values of the damping ratio and natural frequency.

11.3.4.1.1 Overdamped Response ($\zeta > 1$)

In this case, the TF has two distinct negative real poles given by the following formula:

$$\begin{aligned} s_1 &= -\zeta\omega_n + \omega_n\sqrt{\zeta^2 - 1} = -\sigma + \omega_d \\ s_2 &= -\zeta\omega_n - \omega_n\sqrt{\zeta^2 - 1} = -\sigma - \omega_d \end{aligned} \tag{11.36}$$

where $\omega_d = \omega_n\sqrt{\zeta^2 - 1}$ is referred to as damped natural frequency. The output function can be factorized using the poles and separated into partial fractions.

$$C(s) = \frac{\omega_n^2}{s(s - s_1)(s - s_2)} = \frac{A_1}{s} + \frac{A_2}{s - s_1} + \frac{A_3}{s - s_2} \tag{11.37}$$

The residues are calculated as

$$A_1 = s\cdot\left.\frac{\omega_n^2}{(s^2 + 2\zeta\omega_n + \omega_n^2)s}\right|_{s=0} = 1$$

$$A_2 = (s - s_1)\cdot\left.\frac{\omega_n^2}{s(s - s_1)(s - s_2)}\right|_{s=s_1} = -\frac{1}{2\sqrt{\zeta^2 - 1}\left(\zeta - \sqrt{\zeta^2 - 1}\right)}$$

$$A_3 = (s - s_2)\cdot\left.\frac{\omega_n^2}{s(s - s_1)(s - s_2)}\right|_{s=s_2} = \frac{1}{2\sqrt{\zeta^2 - 1}\left(\zeta - \sqrt{\zeta^2 - 1}\right)}$$

Step response of the position control system is the inverse Laplace transform of the TF.

$$c(t) = 1 - A_2 e^{\left(-\zeta\omega_n + \omega_n\sqrt{\zeta^2 - 1}\right)t} + A_3 e^{\left(-\zeta\omega_n - \omega_n\sqrt{\zeta^2 - 1}\right)t} \quad t \geq 0 \tag{11.38}$$

11.3.4.1.2 Critically Damped Response ($\zeta = 1$)

When the damping ratio equals unity, the natural frequency, the TF has a double real negative pole in the s-plane $s_1 = s_2 = -\omega_n$. Therefore, unit-step response in the s-domain becomes

$$C(s) = \frac{\omega_n^2}{s(s + \omega_n)^2} = \frac{A_1}{s} + \frac{A_2}{(s + \omega_n)} + \frac{A_3}{(s + \omega_n)^2} \tag{11.39}$$

Multiply both sides by the denominator:

$$\omega_n^2 = A_1(s + \omega_n)^2 + A_2 s(s + \omega_n) + A_3 s$$

For $s = 0$:

$$\omega_n^2 = A_1(\omega_n)^2 \Rightarrow A_1 = 1$$

For $s = -\omega_n$

$$\omega_n^2 = A_3(-\omega_n) \Rightarrow A_3 = -\omega_n$$

To find A_2, we can differentiate both sides and solve the resulting equation. This results in $A_2 = -1$. Hence, the unit-step response of a position control system is obtained by taking the inverse Laplace transform of the partial fraction expansion of $C(s)$.

$$\mathcal{L}^{-1}\left\{\frac{1}{s} + \frac{1}{s-s_1} + \frac{-\omega_n}{(s-s_1)^2}\right\} = 1 - e^{-\omega_n t}(1 + \omega_n t) \qquad t \geq 0$$

11.3.4.1.3 Underdamped Response ($0 < \zeta < 1$)

In this case, the TF has two complex conjugate poles given by the following formula:

$$\begin{aligned} s_1 &= -\zeta\omega_n - j\omega_n\sqrt{1-\zeta^2} = -\zeta\omega_n - j\omega_d \\ s_2 &= -\zeta\omega_n + j\omega_n\sqrt{1-\zeta^2} = -\zeta\omega_n + j\omega_d \end{aligned} \tag{11.40}$$

The decomposition below makes it easy to find the time domain response using a Laplace transform table.

$$\begin{aligned} C(s) &= \frac{\omega_n^2}{s(s-s_1)(s-s_2)} = \frac{1}{s} - \frac{s+\zeta\omega_n}{(s+\zeta\omega_n)^2+\omega_d^2} - \frac{\zeta\omega_n}{(s+\zeta\omega_n)^2+\omega_d^2} \\ \mathcal{L}^{-1}&\left\{\frac{s+\zeta\omega_n}{(s+\zeta\omega_n)^2+\omega_d^2}\right\} = e^{-\zeta\omega_n t}\cos\omega_d t \\ \mathcal{L}^{-1}&\left\{\frac{\omega_d}{(s+\zeta\omega_n)^2+\omega_d^2}\right\} = e^{-\zeta\omega_n t}\sin\omega_d t \end{aligned} \tag{11.41}$$

Unit-step response can be organized in the following form (Ogata, 2002):

$$\begin{aligned} c(t) &= 1 - e^{-\zeta\omega_n t}\left[\cos\omega_d t + \frac{\zeta}{\sqrt{1-\zeta^2}}\sin\omega_d t\right] \\ &= 1 - \frac{e^{-\zeta\omega_n t}}{\sqrt{1-\zeta^2}}\sin\left[\omega_d t + \tan^{-1}\frac{\sqrt{1-\zeta^2}}{\zeta}\right] \end{aligned} \tag{11.42}$$

Example 11.4 Plot the unit-step response of a second-order system with a natural frequency of 1 rad/s for damping ratios of 0.5, 1, and 2.

Solution

The MATLAB® script below computes the unit-step response using Control Toolbox functions. The function "`tf()`" creates the TF model for the arrays "Num" and "Den" containing the coefficients of the numerator and denominator polynomials. The function "`step()`" calculates and plots the step response of the given system. To calculate the step responses for three values of the damping ratio, the denominator is created in a for loop, and the function is called for each system.

MATLAB® Script to Compute the Unit-Step Response of a Second-Order System

```
% Second-order system parameters
Wn = 1;            % Natural frequency (rad/s)
Z = [0.5 1 2 ];  % Damping ratio
Num = [Wn^2];
```

```
n=size(Z);
for k=1:n(2)
    Zeta=Z(k);
% Define denominator of the transfer function
    Den = [1, 2*Zeta*Wn, Wn^2];
    sys = tf(Num, Den);
    step(sys)
    hold on;
end
```

Figure 11.14 shows the plots obtained from this MATLAB® script. The given values of damping ratio $\zeta = 0.5$, 1, and 2 correspond to underdamped, critically damped, and overdamped responses, respectively, as indicated in the figure.

11.3.4.2 Performance Characteristics of Second-Order Systems

Performance characteristics of mechatronic systems are usually specified in terms of its transient response to a unit-step input. The transient response of a practical mechatronic system often exhibits damped oscillations before reaching steady state. Transient-response characteristics for a control system's unit-step input are typically specified as follows:

- Delay time (t_d) is the time required for the response to reach half the final value for the first time.
- Rise time (t_r) refers to the period during which a system's response changes from a lower percentage to a higher percentage of its final value. For underdamped second-order systems, rise time is typically measured from 0% to 100%. In the case of overdamped systems, rise time is commonly defined as the interval between reaching 10% and 90% of the final value.
- Peak time (t_p) is the moment at which the underdamped step response reaches its first maximum.

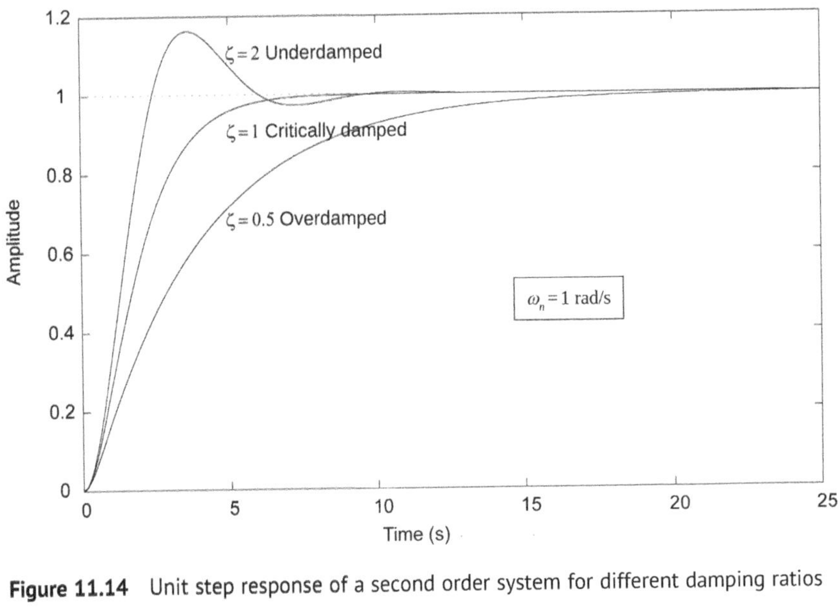

Figure 11.14 Unit step response of a second order system for different damping ratios

- Maximum overshoot indicates how much the peak response exceeds the final steady-state value. Maximum overshoot is usually indicated as the percentage of the final value.
- Settling time is the time required for the response to remain within a specified range around the final value specified by the absolute percentage of the final value (usually 2% or 5%).
- Steady-state error is defined as the difference between the system's output and the desired input once all transient effects have decayed and the system response has settled into a constant value. Steady-state error reflects the system's accuracy in tracking the desired input after the transients have decayed.

System performance depends on its natural frequency and damping ratio, which determine oscillation, settling speed, and overshoot. Adjusting these parameters lets engineers optimize responsiveness and stability by balancing speed, accuracy, and damping for effective control.

11.3.4.2.1 Evaluation of the Rise Time (t_r)

The rise time for a second-order system from 0% to 100% is found by setting the unit-step response in equation 11.42 equal to one.

$$c(t) = 1 - e^{-\zeta\omega_n t}\left[\cos\omega_d t + \frac{\zeta}{\sqrt{1-\zeta^2}}\sin\omega_d t\right] = 1 \tag{11.43}$$

Because $e^{-\zeta\omega_n t} \neq 0$,

$$\cos\omega_d t + \frac{\zeta}{\sqrt{1-\zeta^2}}\sin\omega_d t = 0 \tag{11.44}$$

Therefore, we obtain

$$\frac{\cos\omega_d t}{\sin\omega_d t} = \tan\omega_d t = -\frac{\sqrt{1-\zeta^2}}{\zeta} = -\frac{\omega_d}{\sigma} \tag{11.45}$$

The rise time can be expressed as below:

$$t_r = -\frac{1}{\omega_d}\tan^{-1}\frac{\omega_d}{\sigma} \tag{11.46}$$

The rise time formula is simplified by using the angle θ of the vector corresponding to the complex pole in the s-plane shown in Figure 11.13.

$$t_r = -\frac{\pi-\theta}{\omega_d} \tag{11.47}$$

There is no exact analytical formula for calculating the rise time between arbitrary values of the system response. In general, the system response can be calculated using MATLAB® to determine the times corresponding to the system response values. The interval between these time values gives the rise time.

11.3.4.2.2 Evaluation of the Peak Time (t_p)

The peak time is found by differentiating the step response with respect to time and equating to zero. Differentiating the s-domain expression of the step response simplifies this operation.

$$\mathcal{L}\left\{\frac{dc(t)}{dt}\right\} = sC(s) = \frac{\omega_n^2}{s^2 + 2\zeta\omega_n s + \omega_n^2} \tag{11.48}$$

The inverse Laplace transform of the s-domain differential of the step response results in the following expression for the peak time.

$$\left.\frac{dc(t)}{dt}\right|_{t=t_p} = \sin \omega_d t_p \frac{\omega_n}{\sqrt{1-\zeta^2}} e^{-\zeta\omega_n t_p} = 0 \tag{11.49}$$

This equation yields

$$\sin \omega_d t_p = 0 \Rightarrow \omega_d t_p = 0, \pi, 2\pi \cdots$$

The first peak occurs at the instant

$$t_p = \frac{\pi}{\omega_d} = \frac{1}{2f_d} \tag{11.50}$$

Note that the peak time corresponds to the first half cycle of the linear damped resonance frequency f_d.

11.3.4.2.3 Evaluation of the Maximum Peak (Overshoot M_p)

The maximum overshoot occurs at the peak time t_p found above. Using the time-domain expression of the unit-step response, we obtain

$$\begin{aligned} M_p &= \frac{\pi}{\omega_d} = c(t_p) - 1 \\ &= 1 - e^{-\zeta\omega_n \frac{\pi}{\omega_d}} \left[\cos\pi + \frac{\zeta}{\sqrt{1-\zeta^2}} \sin\pi\right] \end{aligned} \tag{11.51}$$

After simplification, the following formula is derived to calculate the maximum overshoot.

$$M_p = e^{-\left(\frac{\zeta}{\sqrt{1-\zeta^2}}\right)\pi} = e^{-\frac{\sigma\pi}{\omega_d}} \tag{11.52}$$

MATLAB® control toolbox includes the function "stepinfo()" to compute rise time, peak time, maximum overshoot, and settling time for step response of a linear dynamic system illustrated in the following example.

Example 11.5 Given the system parameters $\zeta = 0.5$, $\omega_n = 1$, find the rise time between 10% and 90% for the final value, peak time, overshoot, and settling time of the unit-step response.

Solution

The following MATLAB® script utilizes the Control System Toolbox to perform this analysis. It specifies the TF, calculates the step response, and obtains relevant performance metrics by employing the "stepinfo()" function with customized rise time limits (10%–90%).

MATLAB® Script to Calculate the Step Response Characteristics of the Given System

```
% Percentages of final value between which rise time will be calculated
C1 = 10;                 % Lower limit (10%)
C2 = 90;                 % Upper limit (90%)
```

```
% Define transfer function: H(s) = omega_n^2 / (s^2 + 2*zeta*omega_n*s
+ omega_n^2)
num = [omega_n^2];
den = [1, 2*zeta*omega_n, omega_n^2];
sys = tf(num, den);

% Compute unit step response
[y, t] = step(sys);

% Compute step response characteristics
RT = [C1/100 C2/100]; % Rise time limits as fractions
info = stepinfo(sys, 'RiseTimeLimits', RT);

% Display results
disp('Step Response Characteristics:');
disp(['Rise time (10% to 90%): ', num2str(info.RiseTime), ' s']);
disp(['Peak time: ', num2str(info.PeakTime), ' s']);
disp(['Maximum overshoot: ', num2str(info.Overshoot), ' %']);
disp(['Settling time: ', num2str(info.SettlingTime), ' s']);

% Optional: Plot the step response and annotate
figure;
step(sys);
title('Unit Step Response of Second Order System');
xlabel('Time (s)');
ylabel('Amplitude');
grid on;
```

Figure 11.15 illustrates the step response characteristics on the time plot. The results obtained from this code are presented in Table 11.1.

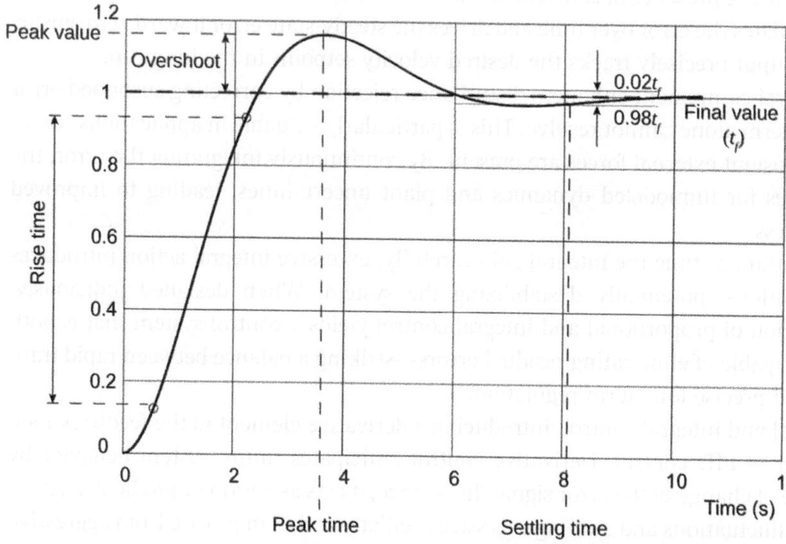

Figure 11.15 Step response characteristics of the second order system with $\zeta = 0.5$, $\omega_n = 1$

Table 11.1 Unit-step response characteristics for second-order system with $\zeta = 0.5$, $\omega_n = 1$

Rise time	1.6390 s
Peak time	3.5920 s
Peak value	1.1629
Maximum overshoot	16.29%
Settling time	8.0759 s

11.4 Proportional, Integral, and Derivative (PID) Control

Both velocity and position control require precise system parameter identification and controller tuning to achieve the desired operation. Factors such as uncertainties in system modeling, nonlinearities, and load variations can influence control performance. Modern analog servomechanisms utilize adaptive methods or feedforward compensation to address these factors and manage disturbances.

Proportional control can be applied to a wide range of mechatronic and industrial applications. One of the primary advantages of proportional velocity control is its simplicity. The control law is straightforward to implement using analog or digital electronics, and the feedback loop can be realized with basic sensors and amplifiers. This simplicity makes proportional control attractive for cost-sensitive or resource-limited applications. It offers relatively fast response to changes in reference input or disturbances. The speed at which the system reaches the new setpoint is a function of the gain settings, motor dynamics, and load characteristics. Proportional velocity control, however, produces a finite steady-state error for a step input, determined by the proportional gain and system parameters. Increasing the gain reduces this error but affects other system dynamics.

Adding an integral element to the feedback control loop offers several key benefits that address some of the inherent limitations of pure proportional control. While proportional action provides immediate response and helps to minimize error, it typically cannot eliminate steady-state error completely, especially in the presence of constant disturbances or system offsets. The integral action systematically accumulates the error over time and drives the steady-state error toward zero, ensuring that the system output precisely tracks the desired velocity setpoint in the long run.

Moreover, the integral component enhances disturbance rejection by correcting sustained error that the proportional term alone cannot resolve. This is particularly valuable in applications where load variations or persistent external forces are present. By continuously integrating the error, the controller compensates for unmodeled dynamics and plant uncertainties, leading to improved accuracy and robustness.

However, it is important to tune the integral gain carefully; excessive integral action introduces overshoot and oscillations, potentially destabilizing the system. When designed judiciously, though, the combination of proportional and integral control yields a control system that is both fast-responding and capable of eliminating residual errors—striking a balance between rapid transient performance and precise long-term regulation.

Beyond proportional and integral control, introducing a derivative element to the feedback loop yields what is known as PID control. Derivative control anticipates future system behavior by responding to the rate of change of the error signal. In essence, it acts as a form of predictive damping, countering rapid fluctuations and curbing excessive oscillations that may arise from aggressive proportional or integral action.

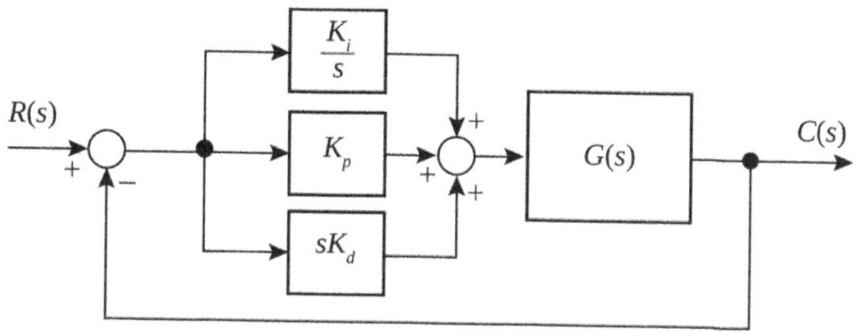

Figure 11.16 Parallel PID controller

The primary benefit of derivative control is its ability to improve transient response—minimizing overshoot and settling time when the system faces sudden changes in setpoint or external disturbances. By generating a corrective action based on how quickly the error is changing, the derivative term tempers sharp responses and helps prevent instability, especially in systems prone to rapid or unpredictable dynamics.

However, implementing derivative control requires careful consideration. Since it amplifies high-frequency signals including noise from sensors, excessive derivative action can actually degrade performance or introduce undesirable jitter. Practical implementations often filter the derivative signal to mitigate the impact of noise.

When properly tuned and combined with proportional and integral control, the derivative component complements the system's ability to achieve fast, accurate, and stable regulation. This synergy enables velocity servomechanisms and other mechatronic systems to deliver robust performance, even in the presence of complex disturbances and changing operating conditions.

Figure 11.16 illustrates the block diagram of a parallel PID controller. For optimal system response, the gain of the PID components must be simultaneously tuned. While manual tuning algorithms are available, MATLAB® Control Systems Toolbox includes functions to create a PID controller and tune its parameters (Mathworks Inc, 2025).

Example 11.6 The closed-loop time constant of the proportional velocity control system described in Example 11.3 is 2 seconds and its closed-loop gain is 0.8. Using MATLAB® Control System Toolbox, design a parallel PID controller and tune the PID parameters to minimize the steady-state error and improve its rise time.

Solution

TF of the described velocity control system as obtained in Example 11.3 is

$$G(s) = \frac{\theta_c}{E(s)} = \frac{K_{CL}}{\tau_{FB}s + 1} = \frac{0.8}{2s + 1}$$

The following MATLAB® script automatically tunes the PID parameters using "pidtune()" function of the MATLAB® Control System Toolbox.

MATLAB® Script for Tuning the PID Parameters.

```
% Define the system transfer function without PID
num = 0.8;
den = [2 1];
G = tf(num, den); % 1/(2s+1)
```

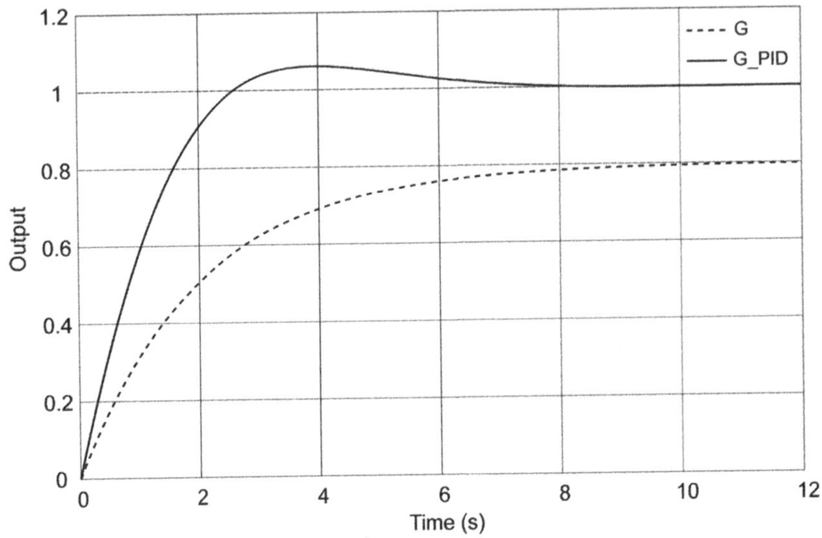

Figure 11.17 Unit step response of the given system without and with PID controller

```
% Automatically tune a PID controller
[C_pid, info] = pidtune(G,'PID');
% Display tuned PID gains
disp('Tuned PID Controller:');
disp(C_pid);
% Unity feedback system with PID controller
G_PID = feedback(C_pid * G, 1);
% Plot step response
figure;
step(G,G_PID);
title('Step Response with Tuned PID Controller');
xlabel('Time (s)');
ylabel('Output');
grid on;
legend
```

The following PID gains are obtained upon executing this code.

$$K_p = 1.7895; \quad K_i = 1.6218; \quad K_d = 0$$

Derivative control gain is zero; therefore, a PID controller is sufficient for this system. Figure 11.17 displays the unit-step response without and with the tuned PID controller. The steady-state error of 20% is eliminated by the PID controller, and the rise time is significantly improved.

11.5 Stability

System stability is an essential consideration in the development of any feedback control strategy, whether proportional, integral, derivative, or their combinations. Stability is the system's ability to return to equilibrium following a disturbance; in practical terms, a stable velocity or position control servomechanism will settle smoothly to the desired setpoint after experiencing a change in load

Table 11.2 Relationship between pole locations and system responses

	System	Pole Locations	Unit Step Response
Stable Systems	First order	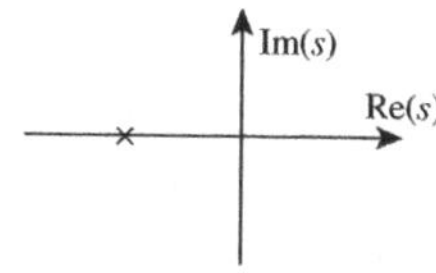	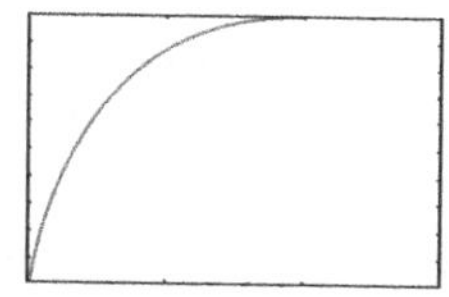
	Second order overdamped	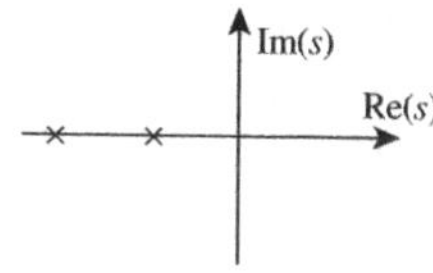	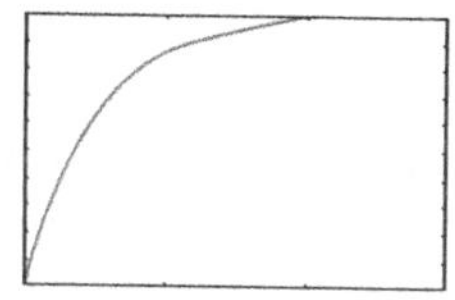
	Second order critically damped	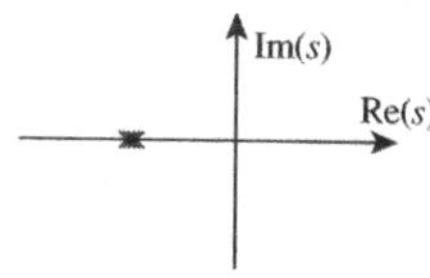	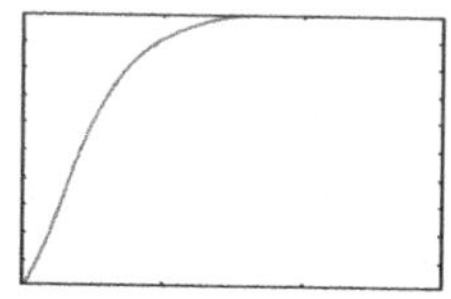
	Second order underdamped	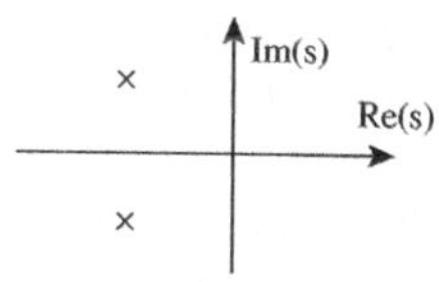	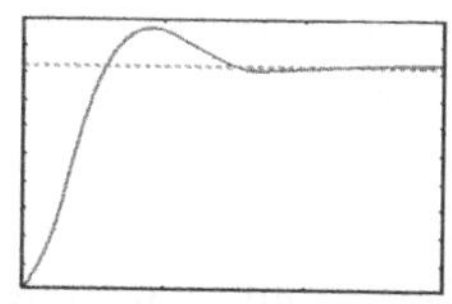
Oscillatory	Undamped oscillation due to poles on the imaginary axis	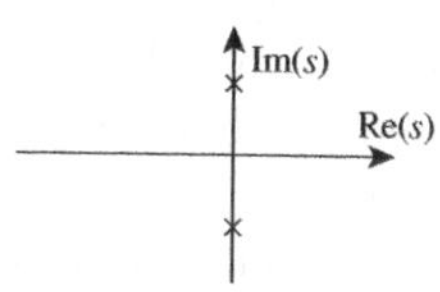	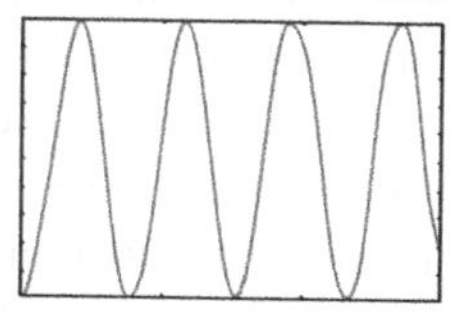
Unstable Systems	Unstable system due to a real pole in the right half-plane	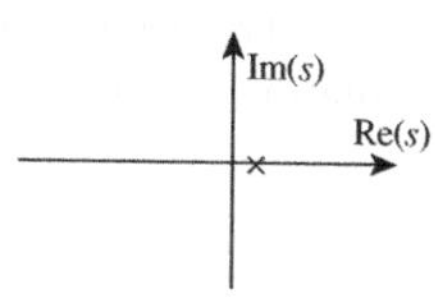	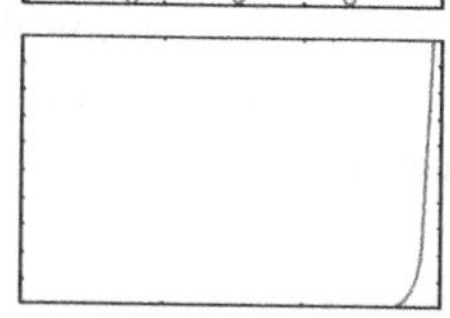
	Unstable system due to complex conjugate poles in the right half-plane	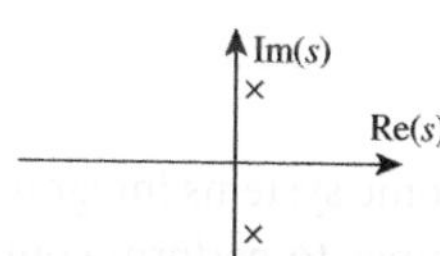	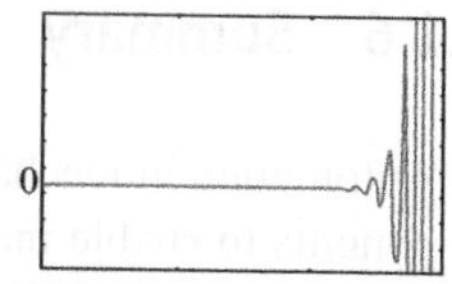

or control input. Classical analysis methods such as Routh–Hurwitz criterion, root locus, Bode plots, and Nyquist criteria are used to assess and design for stability, ensuring that feedback gains and loop parameters are chosen so that oscillations are damped and runaway conditions are avoided.

The location of system poles determines the response characteristics and stability relative to the imaginary axis. Table 11.2 summarizes system responses corresponding to the positions of the system poles on the s-plane.

Absolute stability refers to the condition where a system remains stable for all possible values of its operating parameters within a specified range, regardless of external disturbances or initial conditions. In feedback control, achieving absolute stability means that the system's response always settles to equilibrium—never exhibiting sustained or growing oscillations or diverging to infinity—so long as the underlying assumptions about system structure and feedback gain are respected.

Typically, absolute stability is evaluated based on the location of the system's poles: for continuous-time systems, if all poles of the TF lie strictly in the left half of the complex plane, the system is stable. This guarantees that, for any bounded input, the transient response diminishes over time, and the system ultimately attains its finite steady state.

In practical terms, absolute stability assures that the control strategy maintains consistent function and reliability despite changes in parameters or uncertainties in modeling. Analytical tools such as the Routh–Hurwitz criterion, root locus, and Nyquist plot are employed to rigorously verify absolute stability during the design process. Comprehensive explanations of these analytical methods can be found in standard control theory textbooks, including as Ogata (2002), Kuo (1982), Phillips and Parr (2011), and Nise (2019).

Relative stability is evaluated using metrics such as phase margin, gain margin, and the real part of system poles. For instance, in a continuous-time system, when the poles are located far into the left half of the s-plane, the system returns to equilibrium quickly, with significant damping and minor overshoot. Conversely, if the poles are positioned close to the imaginary axis, the system is considered marginally stable and may experience slower decay, more noticeable oscillations, or instability if there are small changes in gain or system parameters.

Frequency response methods, such as Bode plots and Nyquist diagrams, are commonly used to visualize and assess relative stability. The phase margin indicates how much the system's phase can change before instability occurs, while the gain margin measures how much the system's gain can increase before it crosses into instability. Sufficient margins ensure that the feedback system can withstand unforeseen variations without tipping into oscillatory or divergent behavior.

Relative stability is an important design consideration, particularly for systems exposed to noise, nonlinearities, or time delays. A servomechanism that is designed for stability not only avoids instability but also maintains a margin of safety, supporting reliable performance and operational durability.

11.6 Summary

- Automation in mechatronic systems integrates mechanical, electronic, and computational components to enable machines to perform complex, precise tasks with minimal human intervention. Automatic control plays a critical role in this automation by continuously adjusting system inputs based on output feedback to maintain desired performance despite disturbances or uncertainties.
- The central challenge in mechatronic automation is formulating control algorithms that ensure devices follow desired trajectories or maintain outputs despite variable conditions. Manual control, or open-loop control, relies on human observation and input adjustments but lacks adaptability and precision, making it unsuitable for many mechatronic applications. Feedback control systems overcome these limitations by continuously comparing actual outputs with reference values and adjusting inputs to minimize errors, thereby achieving stability, precision, and adaptability.
- Analog control systems process continuous signals and employ circuits for amplification, filtering, and signal processing. They often implement PID control strategies to enhance

responsiveness and accuracy. These systems are well-suited for real-time smooth adjustments and commonly use PMDC motors for velocity and position control.

- Digital control systems operate on discrete signals using microcontrollers or DSPs. They sample sensor data via ADCs, apply control algorithms such as PID or adaptive control, and output commands through DACs or directly to actuators. Digital control enables programmability, adaptability, and integration with advanced features like machine learning and network communication.
- Other control methods include on–off control, which switches actuators fully on or off based on thresholds, and sequential control, which executes preset action sequences suitable for manufacturing and household appliances.
- Key criteria for evaluating control system performance include:
- **Accuracy**: The ability to maintain outputs close to desired setpoints despite disturbances, essential for tasks requiring precision such as robotic positioning and thermal regulation.
- **Stability**: The property ensuring the system returns to or maintains equilibrium after disturbances, preventing oscillations or failures.
- **Responsiveness**: The speed and appropriateness of the system's reaction to input changes or disturbances, balancing fast adaptation with controlled behavior.
- **Robustness and Adaptability**: The system's capacity to maintain performance amid uncertainties and to adjust parameters dynamically for optimal operation.
- Analog motion control systems regulate velocity or position continuously using feedback servomechanisms. PMDC motors are commonly used in low-power applications, with their dynamics modeled by electrical and mechanical equations. The motor's state-space and TF models describe relationships between armature voltage, current, angular velocity, and load torque. Electrical transients typically decay faster than mechanical ones, allowing simplifications to first-order models for control design.
- Velocity-controlled servomechanisms use feedback loops where the motor speed is compared to a reference, and the error signal drives the motor input. Proportional velocity control amplifies the error to adjust motor voltage, offering simplicity and fast responses but resulting in steady-state errors. Increasing gain reduces error but may amplify noise.
- Adding an integrator in the feedback loop eliminates steady-state error by accumulating error over time but can introduce oscillations and overshoot.
- Position control servomechanisms regulate mechanical displacement precisely by comparing output position to a reference and adjusting actuators accordingly. PMDC motors with potentiometric feedback are typical in low-power applications. Position control systems are modeled as second-order systems characterized by natural frequency and damping ratio, influencing transient response types: overdamped, critically damped, or underdamped.
- The system's step response varies with damping ratio, exhibiting oscillations or smooth convergence. MATLAB® scripts can plot these responses for different damping values, aiding design decisions.
- Transient response metrics include delay time, rise time, peak time, maximum overshoot, settling time, and steady-state error. These are used to quantify and optimize system behavior.
- PID control combines proportional, integral, and derivative actions to improve transient and steady-state performance:
- **Proportional Control**: Simple and fast but leaves steady-state error.
- **Integral Control**: Eliminates steady-state error and improves disturbance rejection but may cause oscillations.

- **Derivative Control**: Predicts error trends to reduce overshoot and oscillations but can amplify noise.
- MATLAB® Control System Toolbox supports automatic PID tuning to minimize steady-state error and improve rise time. An example shows that a PID controller (zero derivative gain) may suffice for some velocity control systems.
- Stability ensures that systems return to equilibrium after disturbances. Absolute stability requires all system poles to lie in the left half of the complex plane, guaranteeing bounded and diminishing responses. Relative stability relates to how far poles are from the imaginary axis, with methods like Routh–Hurwitz, root locus, Bode plots, and Nyquist criteria used for analysis and design. Adequate phase and gain margins provide robustness against parameter variations and noise.

11.7 Check Your Understanding

11.7.1 Self-Assessment Quiz

1 Manual control systems are more adaptable and precise than feedback control systems.
True False

2 Analog control systems process continuous signals and often use PID control strategies to enhance responsiveness and accuracy.
True False

3 Digital control systems cannot integrate advanced features like machine learning and network communication.
True False

4 On–off control is best suited for processes where exact precision is not critical.
True False

5 The damping ratio and natural frequency are parameters that influence the transient response of a second-order system.
True False

6 What is the primary role of automatic control in mechatronic systems?
a. To increase human intervention
b. To continuously adjust system inputs based on output feedback
c. To eliminate the need for sensors
d. To reduce system complexity

7 Which of the following is NOT a key criterion for evaluating control system performance?
a. Accuracy
b. Stability
c. Color
d. Responsiveness

8 In a velocity-controlled servomechanism, what is the purpose of the feedback loop?
   a. To amplify noise
   b. To compare measured velocity to a reference signal
   c. To eliminate the need for actuators
   d. To increase system complexity

9 Which control strategy is used to eliminate steady-state error by accumulating error over time?
   a. Proportional control
   b. Integral control
   c. Derivative control
   d. On–off control

10 What is the effect of adding an integrator in the feedback loop of a velocity control system?
   a. It introduces steady-state error.
   b. It eliminates steady-state error.
   c. It reduces system stability.
   d. It amplifies noise.

Answers:

1: False; 2: True; 3: False; 4: True; 5: True. 6: b; 7: c; 8: b; 9: b; 10: b.

11.7.2 Questions

1 What is the primary role of automatic control in mechatronic systems, and how does it contribute to automation?

2 Compare and contrast manual (open-loop) control with automatic (feedback) control. What are the advantages and disadvantages of each?

3 Describe the main differences between analog and digital control systems. In what scenarios is each type preferred?

4 What are the key performance metrics for evaluating a control system, and why are they important?

5 Explain the importance of stability in feedback control systems. How is stability typically analyzed?

6 How does a feedback loop operate in a velocity-controlled servomechanism? What is the role of the error signal?

7 What are the effects of proportional, integral, and derivative (PID) control strategies on system performance? Provide examples of when each component is most beneficial.

8 Why does proportional control alone result in steady-state error, and how does adding an integrator address this issue?

9 Discuss the trade-offs involved in increasing the proportional gain in a velocity control system.

10 How does the damping ratio affect the transient response of a second-order system? Illustrate with examples of underdamped, critically damped, and overdamped responses.

11 Given a set of motor and load parameters, how would you develop the state-space model for a PMDC motor?

12 What steps would you follow to design a proportional velocity control system for a PMDC motor, and how would you evaluate its performance?

13 How can MATLAB® be used to simulate and analyze the step response of a control system? What insights can be gained from such simulations?

14 Describe how you would design a PID controller for a velocity or position control system and tune its parameters for optimal performance.

15 What are the challenges in achieving robustness and adaptability in mechatronic control systems, and how can modern control algorithms help?

16 Why is it important to consider both absolute and relative stability in control system design? How do phase and gain margins contribute to robust performance?

17 Discuss the practical implications of steady-state error, overshoot, and settling time in real-world mechatronic applications.

11.7.3 Problems

1 A permanent magnet DC motor drives a mechanical load. An armature voltage of 24 V is applied while the motor is initially at rest. Given the motor and load parameters below,

a. Find the initial armature current when the motor is powered on 12 V. Use the following parameters:

b. Find the transfer function of the motor described in Problem 1 with armature voltage as input and shaft speed as output.

c. Compare the electrical and mechanical time constants and reduce the transfer function considering the dominant time constant.

Motor parameters:

- **Armature Resistance (R_a)**: 1.7 Ω
- **Armature Inductance (L_a)**: 2 mH
- **Back EMF Constant (K_b)**: 0.0439 V/rad/s
- **Torque Constant (K_t)**: 0.0439 Nm/A
- **Total Inertia (J)**: 0.155e-2 kg·m^2
- **Viscous Friction Coefficient (B)**: 4.545e-5 N/rad·s
- **Load Torque (T_L)**: 0.1 Nm

2 Given a proportional velocity control servomechanism with the following parameters, compare the steady-state error for the power amplifier gain values of 1 and 10. Use MATLAB® to plot the unit-step response.

- **Armature Circuit Gain (K_m)**: 16.66
- **Time Constant (τ)**: 0.76 s

3 Plot the unit-step response of a second-order system with natural frequency of 1 rad/s for damping ratios of 0.5, 1, and 2. Use MATLAB® to compute the step response and determine the rise time, peak time, maximum overshoot, and settling time.

4 A feedback control system is characterized by undamped oscillation frequency of 3.16 rad/s and damping factor of 0.474.

a. Plot the step response using MATLAB®.
b. Design a PID controller to improve the system's response.
c. Discuss the improvements achieved by PID control.
d. Calculate the poles of the transfer function and analyze the system's stability.

5 Given a DC motor with the parameters below, derive its state-space. Clearly identify the state variables, inputs, and outputs.

- **Armature Resistance (R_a)**: 5 Ω
- **Armature Inductance (L_a)**: 2 mH
- **Back EMF Constant (K_b)**: 0.05 V/rad·s
- **Torque Constant (K_t)**: 0.05 Nm/A
- **Total Inertia (J)**: 0.155e-5 kg·m^2
- **Viscous Friction Coefficient (B)**: 4.545e-5 N/rad·s

11.7.4 Design Assignment

1 Design a position control system for the DC motor using velocity (tachometer) feedback. Specify the block diagram, write the closed-loop transfer function, and discuss how varying the velocity feedback gain affects system damping and transient response.

2 Design a PID controller to improve the performance of the position control system described above. Outline the steps for tuning the PID parameters and explain the expected effects of each component (P, I, D) on system performance.

3 Write a MATLAB® script to simulate the step response of the position control system with and without velocity feedback. Analyze the results in terms of rise time, overshoot, settling time, and steady-state error.

4 For the designed control system, calculate and interpret the following performance metrics:

a. Rise time
b. Peak time
c. Maximum overshoot
d. Settling time
e. Steady-state error

5 Discuss the challenges in achieving robustness and adaptability in your control system design. Suggest modern control algorithms or strategies that could help address these challenges.

References

Kuo, B. C., 1982. *Automatic Control Systems*. 4th ed. Englewood Cliffs, NJ: Prentice Hall.

Mathworks Inc., 2025. *PID Controller*. [Online] Available at: https://www.mathworks.com/help/releases/R2025a/control/ref/pid.html [Accessed 6 9 2025].

Nise, N. S., 2019. *Control Systems Engineering*. 8th ed. Hoboken, NJ: John Wiley & Sons.

Ogata, K., 2002. *Modern Control Engineering*. 4th ed. Upper Saddle River, NJ: Prentice Hall.

Phillips, C. L. & Parr, J. M., 2011. *Feedback Control Systems*. 5th ed. Boston: Prentice Hall.

12

Digital Control

OBJECTIVES

- Explain the differences between analog and digital control systems, and discuss the advantages, challenges, and applications of digital control in mechatronic systems.
- Explain the roles of digital logic, microcontrollers, microprocessors, and programmable logic controllers (PLCs).
- Describe the processes of analog-to-digital (ADC) and digital-to-analog (DAC) conversion, digital signal conditioning.
- Use discrete-time feedback and modeling to design basic mechatronic systems.

In contrast to analog control systems that operate continuously, digital control systems process signals at discrete, periodic time intervals. This fundamental distinction means digital controllers rely on sampling and quantization, allowing them to be programmed, integrated with computer technologies, and capable of complex algorithmic adjustments, while analog controllers respond in real time to continuously varying signals.

Switching from analog to digital systems comes with issues like quantization errors and time delays, but it also offers major advantages, such as greater flexibility, programmability, and easy integration with computer-based technologies. In mechatronics, digital controllers have advanced capabilities such as real-time system monitoring, highly accurate control law adjustments, and the application of complex algorithms for optimizing the overall system performance.

The design of digital control systems typically involves converting analog sensor signals into digital representations using analog-to-digital converters, processing these signals with microcontrollers or digital signal processors, and then generating output signals through digital-to-analog conversion. This workflow allows for implementation of complex control strategies, data logging, and remote supervision.

The simplest digital control system is on–off control. It works by switching devices completely on or off based on a set limit. Logic gates may help make this decision. Although this method is easy and reliable, it can cause the system to bounce around the target level, so it's best for situations where exact accuracy isn't needed.

Sequential control activates multiple subsystems by following a preset sequence of actions, with each stage initiated by predetermined logic conditions or specific inputs. In this approach, subsystems may employ open-loop, on–off, continuous feedback, or computer-based control methods, depending on the requirements for accuracy and response time. Digital logic systems composed of flip-flops, logic gates, and programmable logic circuits (PLCs) are commonly found in sequential control architectures. This technique is applied in mechatronic systems where operations must occur in a precise order to achieve the desired outcomes from manufacturing automation to household appliances. Conveyor belts in assembly lines, material handling robotic arms, bottle filling lines, dishwashers, cloth washers, and microwave ovens are typical examples.

Computer-controlled systems employ general-purpose computers or embedded microcontrollers to process sensor inputs, execute control algorithms, and operate actuators in real time. Figure 12.1 illustrates the outline of a microprocessor-controlled digital system. Unlike fixed-function controllers, computer-controlled systems offer high flexibility, allowing for the implementation of advanced capabilities such as adaptive control, integration with machine learning technologies, and connectivity with other systems or networks through standard communication protocols. Furthermore, the integration of computer control enables remote monitoring and diagnostics, seamless updates to control logic. This method is widely used in industrial automation, robotics, automotive systems, and smart home devices.

Furthermore, digital control facilitates the use of adaptive and predictive algorithms, enhancing both the robustness and responsiveness of mechatronic systems. For example, model predictive control can anticipate future states and optimize actions accordingly, while machine learning methods can refine controller performance based on accumulated data. As industrial and consumer systems grow in complexity, the role of digital control continues to expand, underpinning innovations from autonomous robotics to smart manufacturing (Phillips and Parr, 2011).

In digital control, stability analysis is even more critical. Sampling effects and discretization can introduce delays and alter system dynamics, potentially destabilizing an otherwise well-tuned analog system. Specialized techniques—such as z-domain analysis and discrete-time transfer functions (TFs)—are employed to evaluate and guarantee stability in digital implementations.

Digital control systems offer several advantages allowing for precise, repeatable, and robust control even in the presence of noise or variations in the environment. Furthermore, digital controllers can implement complex decision-making, self-tuning, and communication with other systems, enabling the development of intelligent, networked mechatronic devices. Digital control has become indispensable in modern automation, robotics, automotive electronics, and smart consumer products. The following sections review various levels of digital control techniques, ranging from basic logic circuits to systems based on microcontrollers.

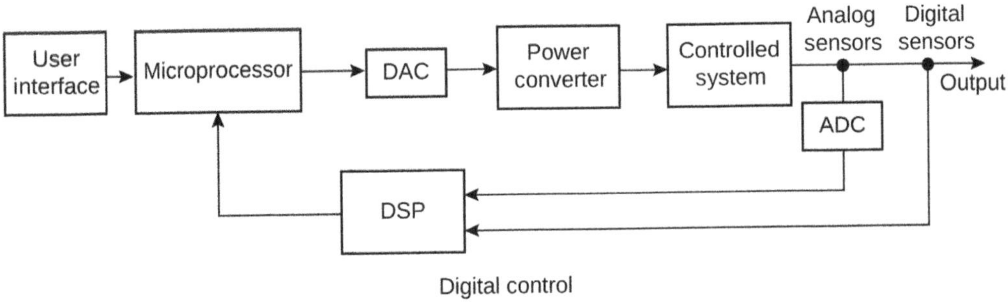

Figure 12.1 Outline of a computer-controlled system

12.1 Common Digital Architectures

Digital logic is the primary component of advanced mechatronics systems, enabling the integration of mechanical, electronic, and computer systems to create complex, automated processes. Logic circuits achieve functions such as signal processing, instrumentation, decision-making, error detection, and communication.

Combinations of logic gates, gate arrays, and flip-flops (refer to Appendix E) can perform various binary functions for encoding and code conversion of digital signals to control electromechanical systems. Stepper motor drivers, for example, consist of flip-flops and logic gates to allow precise rotation of stepper motors in a scanner, printer, or CNC machine. Encoders and code converters transform the pulse trains produced by an optical sensor into usable digital signals. Microcontrollers and microprocessors execute programmed instructions to make real-time decisions in response to changes in the environment. Such functions are essential for complex tasks like robotic motions or adaptive cruise control in vehicles. PLCs use digital logic to automate manufacturing processes.

Moreover, digital logic boards facilitate communication between different components of a mechatronic system. This includes data transfer protocols and communication interfaces that allow various subsystems to work together seamlessly.

Digital systems that include embedded microprocessors enable the implementation of error detection and correction algorithms, which ensure the reliability and safety of mechatronic systems. This is particularly important in applications like medical devices or aerospace systems. User interfaces of modern mechatronic systems have a touch screen and digital displays, which are driven by digital logic circuits and microprocessors.

Machine learning plays a significant role in mechatronics. Machine learning algorithms are used to develop intelligent robots that can learn from their environment, adapt to changing conditions, and make autonomous decisions. Machine learning algorithms can process and interpret data from various sensors, enabling more accurate and efficient operation and optimal control of mechatronic systems. In industrial systems, computer-controlled mechatronic systems can detect faults and predict maintenance needs (The Mechatronics Blog, 2023).

In essence, digital logic coordinates different elements of a mechatronic system to function together, ensuring they achieve precise and reliable operation. This section reviews digital electronic circuits and systems used in mechatronic systems for a variety of logic functions.

12.1.1 Combinational Logic Circuits

If the current states of variables are independent from their previous state, the output of the logic function contains only a combination of the latest states of variables. Logic circuits that perform such logic functions are in the category of combinational logic circuits. Often, the statement of a design problem describes the required conditions for the logic function output to be true or false. In practice, the logic variables correspond to the state of switches or levels of sensors that monitor the physical system components to produce 0 or 1 level binary signals. The design of a combinational logic system starts by listing all possible permutations of the switching variables in a truth table and writing across each row the result of the logic function as either true or false by combining the variables with AND operations. If the variable is one in the combination, it is left as it is; if it is zero, the variable is inverted or complemented. The combinations of variables that produce the 1-state are then joined with OR operations. This design method is referred to as "sum of products" or shortly SOP realization. An alternative method is to combine the terms that make the function false (or 0) using OR operations and join the terms with an AND operation. The term "product of sums"

(POS) refers to this design method. The practical example below shows the design steps for a combinational logic system.

Example 12.1 In an industrial assembly line, a conveyor belt moves the parts and materials. The motor that powers the conveyor belt must start when the material is loaded and if one of two workers is in position and the safety interlock switch is on (closed). Obtain a logic function to operate the combinational logic circuit that will activate the motor's power switch.

Solution

First, we must assign variables for each element of the system. Let's assign A and B for the presence of each worker, M for the existence of material on the belt, and S for the safety interlock switch position. Since there are four input variables, $2^6 = 16$ combinations are possible. Table 12.1 lists all possible combinations of the logic conditions and the corresponding outputs.

To turn on the motor, one of the two workers must be present, material must be present, and the interlock switch must be closed. In the last column of the truth table, 1 indicates the power switch P of the conveyor belt is on and 0 indicates off. The logic function in sum of minterms form is

$$P = \left(\overline{A} \cdot B \cdot M \cdot S\right) + \left(A \cdot \overline{B} \cdot M \cdot S\right) + (A \cdot B \cdot M \cdot S)$$

Implementation of the logic function above requires three 4-input AND gates and one 3-input OR gate. The goal of designing a logic circuit is to meet the requirements by using minimum number of logic gates. This function can be simplified by using the properties of the Boolean algebra listed in Appendix E, Table E.2. Using the distributive property (7), we can combine two of these three

Table 12.1 Truth table for Example 12.1

A–Operator 1	B–Operator 2	M–Material	S–Safety Lock	P–Power Switch
0	0	0	0	0
0	0	0	1	0
0	0	1	0	0
0	0	1	1	0
0	1	0	0	0
0	1	0	1	0
0	1	1	0	0
0	1	1	1	1
1	0	0	0	0
1	0	0	1	0
1	0	1	0	0
1	0	1	1	1
1	1	0	0	0
1	1	0	1	0
1	1	1	0	0
1	1	1	1	1

terms. According to the idempotent law (3), writing one of the terms twice does not change the value of the function.

In addition, since $\overline{A} + A = 1$ according to the complement law (4), the function reduces to

$$P = (\overline{A} \cdot B \cdot M \cdot S) + (A \cdot \overline{B} \cdot M \cdot S) + (A \cdot B \cdot M \cdot S) + (A \cdot B \cdot M \cdot S)$$

$$(\overline{A} \cdot B \cdot M \cdot S) + (A \cdot B \cdot M \cdot S) = (B \cdot M \cdot S) \cdot (\overline{A} + A) = B \cdot M \cdot S$$
$$(A \cdot \overline{B} \cdot M \cdot S) + (A \cdot B \cdot M \cdot S) = (A \cdot M \cdot S) \cdot (\overline{B} + B) = A \cdot M \cdot S$$

$$P = (B \cdot M \cdot S) + (A \cdot M \cdot S) = M \cdot S \cdot (A + B)$$

Karnaugh map method described in Appendix E.3 can be used to simplify this logic function. Each minterm of the logic function obtained from Table 12.1 is marked with 1 on the 4 × 4 K-map in Figure 12.2. The sequence of the rows and columns as 00, 01, 11, and 10 is crucial to obtain the correct answer. Three adjacent cells marked with 1 are circled in groups of two. An overlapping on the K-map is allowed due to the idempotent law (Property 3). The variable in each circled group that changes the state from one cell to the adjacent cell is removed from the minterm according to the complement law (Property 4).

Alternatively, the "simplify()" function provided by the Symbolic Math Toolbox of MATLAB® reduces a logic function to its simplest form shown in Figure 12.2.

MATLAB® Code to Simplify the Logic Function *P*.

```
% Define the symbolic variables
syms A B M S;
% Define the logic function
P = (~A&B&M&S) | (A&~B&M&S) | (A&B&M&S);
% Minimize the logic function
Pmin = simplify(P);
disp(P)
disp(Pmin)
```

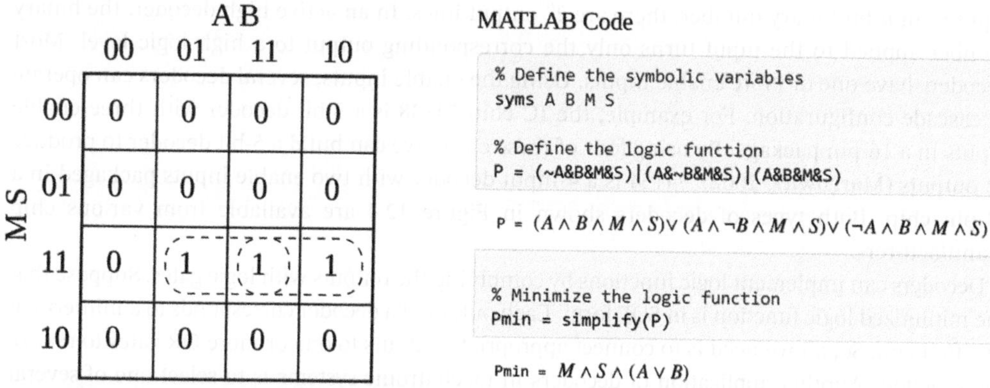

Figure 12.2 Karnaugh map for the logic function derived from the truth table

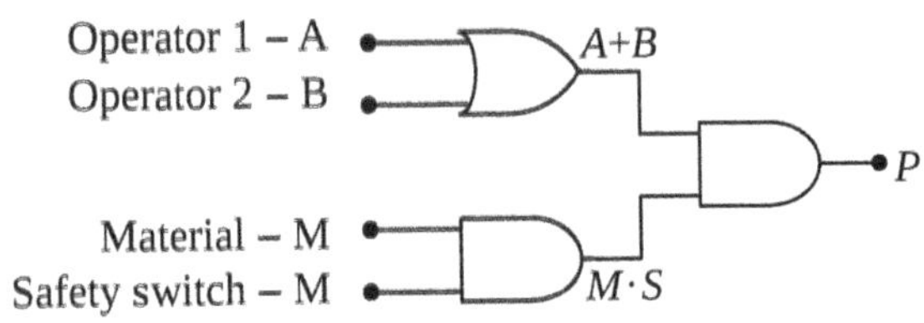

Figure 12.3 Logic circuit for Example 12.1

The MATLAB® code and K-map displayed in Figure 12.2 yield the same simplified function below:

$$P = M \cdot S \cdot (A + B)$$

Logic circuit in Figure 12.3 performs this function using two AND gates and one OR gate instead of three 4-input AND gates combined by one 3-input OR gate. Half of the standard integrated circuit (IC) chip 7408, which contains four 2-input AND gates, and one quarter of the IC chip 7432, which contains four 2-input OR gates, are sufficient to implement this function. Unused gates in these chips can either remain idle or be assigned to other functions within the logic system.

12.1.2 Digital Logic Arrays

In addition to a variety of logic gates, several IC chips are commercially available to build more complex circuits. Often, it is easier and less costly to use these components than building the required circuit with individual gates. Decoders, encoders, multiplexers, and gate arrays are common building blocks of medium-sized logic circuits.

Logic gate arrays are ICs that can be programmed to perform specific logic functions. These arrays consist of multiple logic gates that can be configured to execute complex SOP logic expressions. Unlike fixed-function logic gates, gate arrays can be programmed to perform a wide variety of logic functions, making them versatile for different applications.

12.1.2.1 Decoders

When activated, a decoder elects one of several output lines based on a coded input signal. If the input is an n-bit binary number, there are 2^n output lines. In an active high decoder, the binary number applied to the input turns only the corresponding output to a high logic level. Most decoders have one or more enable inputs. Using the enable inputs, several decoders can operate in cascade configuration. For example, the IC chip 74138 is a 3-bit decoder with three enable inputs in a 16-pin package. By using four of these chips, we can build a 5-bit decoder to produce 32 outputs (Marcowitz, 2002). 74154 is a 4-input decoder with two enable inputs packaged in a 24-pin chip. Both types of decoders shown in Figure 12.4 are available from various chip manufacturers.

Decoders can implement logic functions by combining the outputs with logic gates. Suppose that the minimized logic function is in SOP form. Each output of a decoder corresponds to a minterm of this function. So, all we need is to connect appropriate outputs to one or more OR gates to realize the function. Another application of decoders in mechatronic systems is to select one of several devices, each of which has a unique address. Figure 12.4 (a) shows the schematic diagram of a 2-bit decoder with one EN (enable) input.

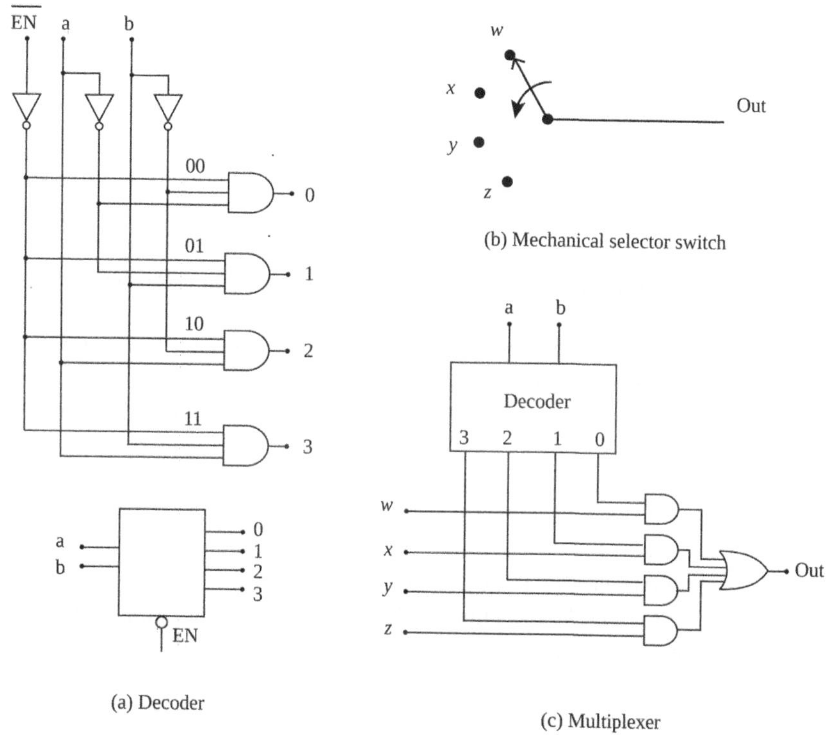

Figure 12.4 Decoder and multiplexer circuits (a: Octal decoder with enable input; b: Switch analogous to 4-input 1-output multiplexer; c: 4-input 1-output solid-state multiplexer)

12.1.2.2 Encoders

An encoder is the inverse of a decoder. An n-bit encoder has 2^n inputs. If only one of the inputs is one and the others are zero, the encoder generates the binary number corresponding to that input. Suppose that several devices are sending a signal to a computer by turning a wire to logic 1 level. The encoder generates the device number, provided that only one of the devices is signaling a 1.

If more than one input signal may occur, a priority encoder must be used to generate the number of the highest priority device with an active input signal. The chip labeled 74147 is a commercial binary coded decimal (BCD) priority encoder taking nine active low input lines and encoding them into four active low outputs.

12.1.2.3 Multiplexers

A multiplexer is a logic switch that passes one of its data inputs defined by the state of the select inputs to the output. A 4-bit multiplexer has two select inputs. If the inputs are labeled w, x, y, z in order, then the output equals input w if the select inputs are 00, x if they are 01, and so on. Figure 12.4 (b) shows a mechanical switch analogy to the multiplexer circuit configured with a decoder and logic gates diagram displayed in (c).

Typical practical applications of multiplexers in computing include data routing, switching between different data sources, memory access, and implementation of complex logic functions. Multiplexers are also used in signal processing to combine multiple signals into one processing unit. In instrumentation and measurement systems, they are used to route multiple sensor signals to a single measurement device. In mechatronic systems, multiplexers are used for sensor data integration and actuator control. If multiple sensors must send data to a single microcontroller or processor, a multiplexer route these signals more efficiently, reducing the required data lines and input pins. In general, multiplexers simplify hardware and reduce the cost of logic systems.

12.1.2.4 Read-Only Memories (ROM)

Read-Only Memory (ROM) is a permanent (nonvolatile) storage of binary information. ROM is a versatile combinational logic device that contains a fixed AND array and a configurable OR array. It retains stored information even when the device is powered off. A typical application of ROM is to store firmware, which provides the basic functions of the device, such as startup processes and system operations.

There are four types of ROM: Mask ROM, Programmable ROM (PROM), Erasable Programmable ROM (EPROM), and Electrically Erasable Programmable ROM (EEPROM).

Mask ROM is programmed during the manufacturing process based on the designer's configuration. The data is hardwired into the chip during production, which makes it highly reliable. Data stored in a Mask ROM cannot be erased or changed after manufacturing. PROM can be programmed once after manufacturing. Special equipment called a PROM programmer is used to write data. A switching element and a fusible link at each intersection during manufacturing of the chip. The manufacturer initializes all ROM cells to 1. If the user needs to store a 0 in a particular cell, blow the fuse at that location by applying an electric impulse. This process is called "programming." Availability of more versatile devices like EPROM, EEPROM, and programmable logic devices (PLDs) made the PROMs obsolete (Rafiqazzam & McNinch, 2019).

EPROM, short for "Erasable Programmable Read-Only Memory," is a type of nonvolatile memory chip that can be programmed and erased multiple times. The data is written using a special device called programmer, and it can be erased by exposing the chip to ultraviolet light.

EEPROM, or "Electrically Erasable Programmable Read-Only Memory," can be programmed and erased electrically. It is often used in modern devices as it allows for easy reprogramming without special equipment. Unlike EPROM, which requires ultraviolet light to erase, EEPROM can be erased and reprogrammed using an electrical charge. This makes it more convenient and versatile for applications where data needs to be frequently updated, such as in microcontrollers and some types of BIOS chips. Flash Memory is a type of EEPROM that allows for quick and efficient programming and erasing of large blocks of data.

12.1.2.5 Programmable Logic Devices

PLDs are IC chips that permit users to program after they are manufactured. They contain arrays of inverters, AND OR gates. They enable rapid and cost-effective development of medium to complex logic circuits that perform logic expressions in POS form. Programmable gate arrays grouped as Programmable Logic Arrays (PLA) and Programmable Array Logic (PAL) has similar layouts. A PLA comprises an array of programmable AND gates in conjunction with programmable OR gates. The PAL includes programmable AND gates and fixed OR gates (Rafiqazzam, 2005).

After the logic function is minimized, the logic gate arrays, input and output buffers, and any required invert/noninvert matrices are configured. Once the connections are defined, the IC is programmed using a technique called "fusing." This involves creating or breaking connections (fusible links) in the logic gate arrays to determine the logic as per the desired application. After programming, the circuit is tested to ensure it functions correctly. Figure 12.5 (a) shows the layout of a PLA programmed to implement three logic functions X, Y, and Z given on the figure. Figure 12.5 (b) is a simplified diagram for the same circuit. The bold dots indicate the gate inputs connected to the appropriate data lines.

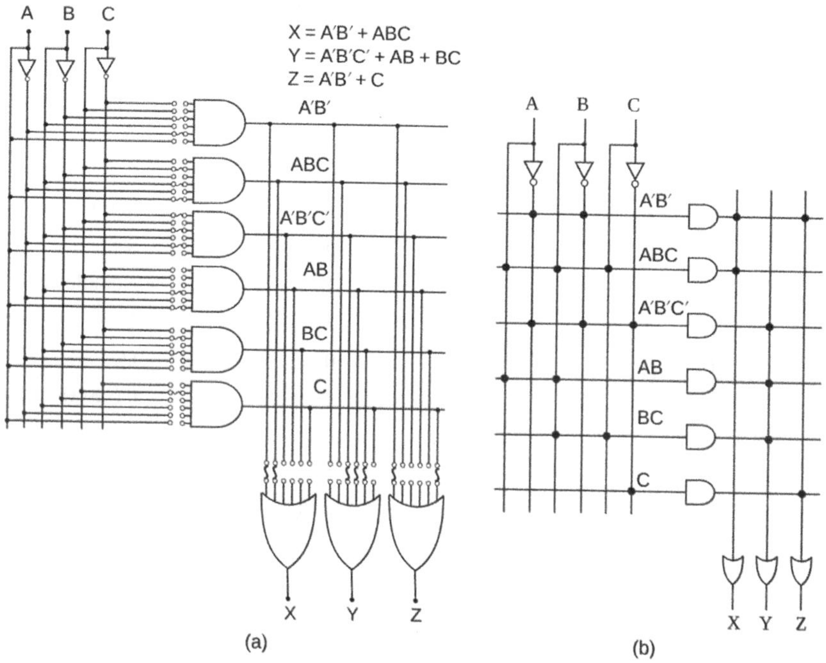

Figure 12.5 A Programmable Logic Array (PLA) configured to implement a logic function (a: Wiring diagram; b: Simplified on-line diagram).

PLDs are versatile circuit elements for custom logic circuits, easy to program, and reprogram using several Hardware Description Languages (HDL), including VHDL, Verilog, Programmable Array Logic Assembler (PASALM), and Advanced Boolean Expression Language (ABEL). PLDs are widely utilized in digital control systems across mechatronics, robotics, and industrial automation applications.

12.1.3 Sequential Logic Circuits and Flip-Flops

Digital systems composed of logic gates are referred to as combinational logic circuits because their outputs are only the function of current inputs. Sequential logic circuits, however, contain memory elements that store past inputs so that the outputs are not only functions of the present inputs but also include prior logic states. Counters, registers, memory units, microcontrollers, and PLCs, which contribute to mechatronic systems, are sequential logic circuits. Microprocessors, microcontrollers, and PLCs are major applications of sequential logic in mechatronics.

The key components of sequential logic circuits are flip-flops, which are basic memory elements capable of storing one bit of data. Flip-flops are constructed from combinational logic gates, but due to the feedback loop, their output depends on the past inputs.

A flip-flop is a bistable logic gate that can remain in one of two stable states until external conditions cause it to change this state. Sequential logic systems employ in general four types of flip-flops outlined in Table 12.2.

SR Flip-Flop (Set-Reset Flip-Flop) has two inputs: Set (S) and Reset (R). When $S = 1$ and $R = 0$, the output is set to 1 ($Q = 1$). When $S = 0$ and $R = 1$, the output is reset to 0 ($Q = 0$). When both S and

Table 12.2 Flip-flop types and properties

Flip-Flop Type	Circuit Symbol	Truth Table		
SR Flip-Flop	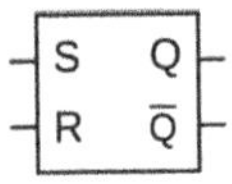	S	R	Q
		0	0	Present state
		0	1	Reset
		1	0	Set
		1	1	Not allowed
JK Flip-Flop	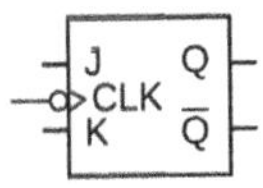	J	K	Q
		0	0	Q_n
		0	1	0 (reset)
		1	0	1 (set)
		1	1	$\overline{Q}_n$ (toggle)
D Flip-Flop	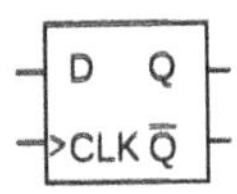	D	CLK	Q
		0	Rising	0
		1	Rising	1
T Flip-Flop	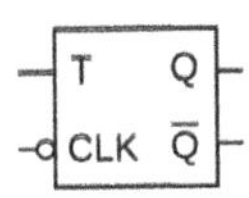	T	CLK	Q_{n+1}
		0	Falling	Q_n
		1	Falling	$\overline{Q}_n$

R inputs are 0, the output retains its previous state. Logic 1 level of both S and R is not allowed and creates an invalid condition.

D Flip-Flop (Data or Delay Flip-Flop) has one data input (D) and a clock input. The output Q follows the input D at the moment of the clock edge (either rising or falling). D Flip-Flop is also known as a data or delay flip-flop; it captures the value of the D input at a specific portion of the clock cycle and outputs it at Q. It is used for synchronization and data storage.

JK Flip-Flop is similar to the SR flip-flop but without the invalid state. It has two inputs: J and K. When $J = 1$ and $K = 0$, the output is set to 1 ($Q = 1$). When $J = 0$ and $K = 1$, the output is reset to 0 ($Q = 0$). When both J and K are 0, the output remains at its previous state. When both J and K are 1, the output toggles (Q changes to the opposite state).

T Flip-Flop is known as toggle Flip-Flop. It has one input, T (toggle), and a clock input. When $T = 1$, the output toggles on each clock pulse. When $T = 0$, the output remains in its current state. It is useful in counters and frequency dividers.

12.1.3.1 Counters

Counters are sequential logic circuits that go through a predetermined sequence of states upon the application of input pulses. They are widely used in digital systems for counting purposes, time measurement, frequency division, and event counting.

Binary counters count in binary numbers, while BCD counters count from 0 to 9 and then reset to 0. Ring counters are a type of counter where only one flip-flop is set at any time, and the "1" bit circulates around the ring. Johnson counter is a special type of ring counter where the inverted output of the last flip-flop is fed back to the input of the first flip-flop.

Counters are driven by clock pulses. Each pulse causes the counter to advance to the next state. A reset input can be used to set the counter back to its initial state (usually 0). An enable input can control whether the counter is allowed to count or not.

Counters are fundamental components in digital electronics, providing essential counting and timing functions in a wide range of applications. Typical counter applications are frequency dividers, event counters, timers, and digital clocks. Counters are used in microcontrollers for various timing and counting functions.

12.1.3.2 Registers

Registers are a type of digital memory used to store binary data. They are composed of flip-flops and can store multiple bits of data. Parallel input-shift register allows data to be shifted in or out one bit at a time. Basic functions of registers are load, shift, clear, and enable.

In a serial-in, serial-out (SISO) register, data is input and output serially. Serial-in, parallel-out (SIPO) register takes data serially and output in parallel such that the data can be read out in parallel from all flip-flops simultaneously. Parallel input-parallel output register allows data to be loaded in parallel and read out in parallel.

There are four types of registers. Data registers store data temporarily for processing. Address register holds the address of a memory location. Status registers contain flags that indicate the status of the processor or the outcome of operations. Accumulator is used to store intermediate results of arithmetic and logic operations.

Registers are essential components in digital systems, providing temporary storage for data and facilitating data manipulation and transfer within the system.

12.2 Programmable Logic Controllers (PLC)

PLCs are specialized modular digital systems designed to monitor inputs from sensors, execute control algorithms, and manage outputs to actuators in real time. Commercial PLCs are designed to operate in harsh industrial environments with high levels of electrical noise, extreme temperatures, humidity, and vibration. A PLC can execute control tasks in real time, ensuring timely responses to input changes and maintaining process control. Their rugged design and ability to withstand harsh environments make them ideal for factory automation, robotics, and complex machinery. PLCs play a central role in industrial systems by providing reliable, flexible, and programmable automation solutions.

In mechatronic systems, PLCs enable precise sequencing, coordination, and synchronization of multiple subsystems, such as motors, valves, and conveyors. Through their integrated digital and analog input/output (I/O) modules, PLCs continuously sample process variables, execute logic operations, and implement control strategies such as on–off, sequential, and even PID control. Additionally, PLCs facilitate communication between devices and allow for easy modification of control logic, which supports rapid reconfiguration and integration with other digital systems in automated production lines.

Most PLC units are modular, allowing for easy expansion and customization with additional I/O modules, communication modules, and specialized function modules. They can be easily reprogrammed and reconfigured for different tasks and processes. Furthermore, they are easy to maintain and highly reliable, which are much desirable properties for continuous industrial operations.

Diagnostic capabilities and modular components simplify troubleshooting and maintenance of PLC systems.

PLCs are typically programmed using ladder logic, a graphical programming language resembling electrical relay logic diagrams. Other programming languages include Function Block Diagram (FBD), Structured Text (ST), Instruction List (IL), and Sequential Function Chart (SFC).

Components of a PLC include CPU, memory, and I/O modules. CPU is responsible for executing control instructions, processing data, and managing communications. Both volatile Random Access Memory (RAM) and nonvolatile ROM/Flash store the control program, data, and configuration settings. PLC modules have interfaces for connecting sensors, actuators, and other field devices. Inputs of a PLC receive signals from sensors, while outputs send control signals to actuators.

Communication ports of a PLC module allow connection to other PLCs, computers, and network devices. Common communication protocols include Ethernet/IP, Modbus, Profibus, and DeviceNet. For more information on industrial communication protocols, refer to Chen (2025) and Flaus (2019).

PLCs are mainly used for manufacturing automation and process control. Building automation systems control assembly lines, robotic arms, conveyor systems, and packaging machines. Process control systems manage industrial processes such as chemical processes, water treatment plants, oil refineries, and food processing. Moreover, PLCs facilitate building automation by controlling HVAC systems, lighting, security systems, and elevators. In transportation systems, PLCs manage traffic lights, railway signaling, and airport baggage handling systems. In power systems, they control power flow in electric distribution, renewable energy systems, and smart grids.

In conclusion, PLCs are essential components in modern industrial automation, providing reliable, flexible, and efficient control solutions for a wide range of applications. Their ability to operate in harsh environments, execute real-time control tasks, and be easily reprogrammed makes them indispensable in manufacturing, process control, and other industrial sectors.

12.3 Microprocessors and Microcontrollers

Microprocessors and microcontrollers are both integral components of computer-controlled mechatronic systems, but they serve different purposes and have distinct characteristics.

A microprocessor is a CPU on a single IC chip that performs arithmetic and logic operations, controls data flow, and executes instructions from memory. Microprocessors typically include the Arithmetic Logic Unit (ALU), registers, and control unit. They require external components like RAM, ROM, I/O ports, and timers to function. Microprocessor architectures are commonly classified as either Complex Instruction Set Computing (CISC) or Reduced Instruction Set Computing (RISC). Intel Core series and AMD Ryzen series are examples of microprocessors.

Microprocessors are suitable for high-performance applications. They are used for running an operating system, executing complex applications, and handling multitasking environments. They are used in general-purpose computing devices such as personal computers, laptops, and servers.

Microcontrollers, on the other hand, are compact IC chips designed to govern a specific operation in an embedded system. They include a CPU, memory (RAM, ROM/Flash), and I/O ports on a single chip. Microcontrollers are used in embedded systems for specific control applications such as home appliances, automotive systems, medical devices, and consumer electronics. Most microcontrollers use RISC architecture for greater efficiency and simplicity.

Microcontrollers are simpler and consume less power, making them ideal for battery-powered and low-power applications. Managing the functions of a mechatronic system, operating digital instrumentation, and controlling robotic functions are among their typical applications. Examples of microcontrollers are Atmel ATmega328P, which is used in Arduino, PIC series, and ARM Cortex-M series. A comprehensive explanation of microcontroller architecture and programming is beyond the scope of this textbook. Interested readers may refer to specialized textbooks (Barrett & Pack, 2022; Sanchez & Canton, 2014).

12.3.1 Widely Utilized Microcontrollers in Mechatronics Applications

Arduino has become a widely used microcontroller platform used in introductory mechatronics projects. As an open-source electronics system, it offers accessible hardware and software designed for straightforward integration and development. It consists of programmable circuit boards (microcontroller-based) and an integrated development environment (IDE) that allows users to write, upload, and execute code to interact with sensors, actuators, and other electronic components. Arduino boards are widely used for prototyping, educational projects, and hobbyist electronics because they simplify embedded system development by providing a straightforward interface for hardware interaction and a large ecosystem of software libraries.

Most microcontrollers use C or C++ for efficient, hardware-level programming. Some advanced MCUs also support Python (MicroPython), which allows for faster prototyping. The Arduino programming language is primarily based on C and C++, providing simplified syntax and a set of built-in functions designed for easy interaction with hardware components. Arduino programs, known as "sketches," are written in the Arduino IDE, which streamlines coding, uploading, and debugging tasks. The language includes libraries for controlling digital and analog I/O pins, communicating with sensors and actuators, and managing timing functions, making it especially suitable for beginners and rapid prototyping in embedded systems. An abridged list of Arduino IDE commands is provided in the Appendix.

Users develop code using functions such as setup() and loop(), which define initialization routines and continuous execution logic, respectively. Its abstraction of hardware details allows developers to focus on high-level logic and device interaction without needing in-depth knowledge of microcontroller registers or low-level configurations. This accessibility has contributed to Arduino's widespread adoption in education, hobbyist projects, and prototyping mechatronic systems.

12.3.2 The Role of Microcontrollers in Mechatronic Systems

Microcontrollers execute control algorithms, process sensor data, and generate actuator commands, thus enabling autonomous system behavior and dynamic decision-making. Beyond their computational function, their programmability, real-time responsiveness, and compactness allow for integration directly into the physical layer of products, supporting direct and distributed digital control.

Microcontrollers play a key role in mechatronics because of the following essential features.

- **Sensor Integration**: Microcontrollers sample, digitize, and pre-process signals from a variety of sensors like temperature, pressure, position, acceleration, optical, and more. Microcontrollers translate real-world phenomena into actionable digital data.
- **Signal Conditioning and Data Processing**: They filter noise, apply calibration, and fuse data streams, thereby supporting high-fidelity control and diagnostics.
- **Actuator Command Generation**: They convert control decisions into electrical signals for actuators, including motors, solenoids, valves, and relays.

- **Real-Time Control**: Microcontrollers are designed for deterministic response, crucial for closed-loop control and applications—industrial robots, medical devices—where timing is paramount.
- **Communication and Networking**: They interface with external systems via UART, SPI, I^2C, CAN, Ethernet, or wireless protocols, allowing distributed control and integration into larger cyber-physical systems.
- **Resource Optimizations**: Their low-power consumption and small form factor enable deployment in embedded, battery-powered, and space-limited environments.

Microcontrollers combine these functions, making it easier to merge mechanical and electrical systems with little additional complexity. This leads to mechatronic products that are both highly integrated and scalable.

12.3.3 Microcontroller Selection and Roles in Mechatronic Systems

Key criteria for selecting MCUs in mechatronic systems include:

- **Processing Power:** Sufficient clock speed and architecture (e.g., ARM Cortex-M0, Cortex-M4, RISC-V) to execute all control and communication tasks within necessary timing windows.
- **Memory Capacity:** Enough RAM and Flash/EEPROM to support sensor buffers, control algorithms, safety code, and future firmware updates.
- **Peripheral Support:** Rich analog/digital peripherals for direct sensor and actuator interfacing—high-resolution ADCs, DACs, PWM timers, communication interfaces (UART, SPI, I2C, CAN).
- **Power Consumption:** Low active and sleep current draw, with advanced power management features, especially for battery-powered mechatronic devices or energy-sensitive industrial automation.
- **Development Ecosystem:** Availability of mature software libraries (CMSIS, Arduino, MicroPython), debugging tools, simulation environments, and community support.
- **Cost, Form Factor, and Reliability:** Physical footprint and integration with the host mechanical design; compliance with industry standards for environmental, safety, and security requirements.

Common MCU families in mechatronics include AVR (Arduino), STM32 series, Microchip PIC, Texas Instruments MSP430, and increasingly, ARM Cortex-M-based chips with integrated connectivity, DSP features, or machine learning capabilities (Table 12.3).

Table 12.3 Comparison of microcontrollers used in mechatronics

Microcontroller	Clock Speed	Memory	Key Interfaces	Power Profile	Application Domains
Arduino (AVR)	16 MHz	2 KB RAM	GPIO, UART, I2C, SPI	Low	Educational, prototyping
STM32F4 (Cortex-M4)	168 MHz	192 KB RAM	GPIO, UART, SPI, I2C, CAN	Low-medium	Robotics, industrial
Raspberry Pi Pico	133 MHz	264 KB RAM	GPIO, UART, SPI, I2C	Low	Signal processing, IoT
TI TDA4VM (SoC)	2 GHz	Up to 16 GB RAM	PCIe, CAN, Ethernet	Variable	AI robotics, Machine vision

In industrial automation and robotics, microcontrollers are selected not just for their raw capabilities but also for their ecosystem—availability of real-time OS support, functional safety certification (e.g., ISO 26262 for automotive), and longevity of supply.

12.4 Digital-to-Analog Conversion

A digital signal represents data as a series of discrete numerical values. Unlike an analog signal, which varies continuously over time, a digital signal is often represented as binary data (0 s and 1 s) that remains constant between fixed sampling steps. Digital signals are widely used in modern mechatronic systems because they are less susceptible to noise and interference, making them more reliable for transmitting data over long distances. They also allow for easy storage, compression, and manipulation of information.

In analog mechatronic systems, electronic circuits process the information as continuous voltages and currents to drive electromechanical elements. However, some sensors and data acquisition boards provide digital data, which must be converted to analog signals to allow analog parts of the system to operate properly. In digital systems, microprocessors and DSP hardware produce digital signals to drive electromechanical devices controlled by analog inputs.

Digital to analog converter (denoted as DAC, D/A, or D2A) converts a digital word stored in its data register in the binary form into an analog value (voltage or current). This property can be used in practical applications such as programmable voltage or current source and a digitally controlled potentiometer.

Digital-to-analog conversion is quite straightforward. Since the decimal equivalent of a binary number is the sum of digits weighed by powers of two, an op-amp circuit in summing amplifier configuration can achieve the conversion if the resistor values are selected properly. Figure 12.6 illustrates an op-amp circuit that performs 4-bit digital to analog conversion and its output as the digital input increases step by step. The switches selecting either 0 (ground) or 1 ($+V_d$) represent

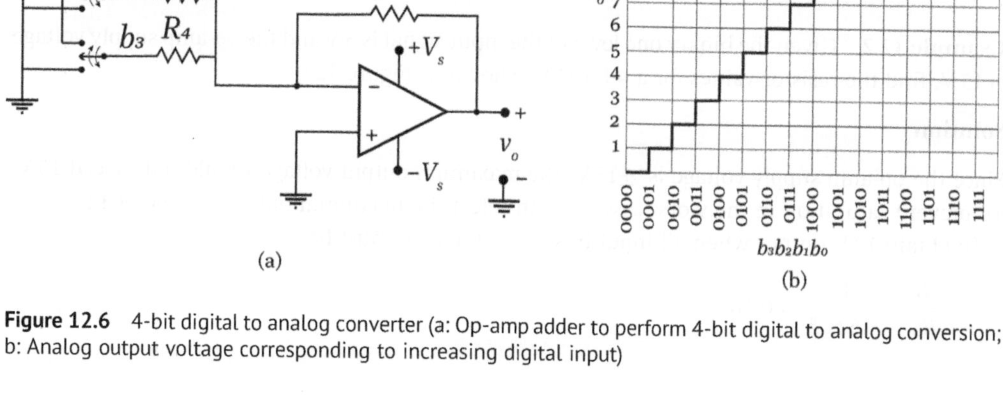

Figure 12.6 4-bit digital to analog converter (a: Op-amp adder to perform 4-bit digital to analog conversion; b: Analog output voltage corresponding to increasing digital input)

the output of the digital that provides the digital signal. Therefore, the binary signal applied to the input of the summing amplifier is a combination of bits at either 0 V or V_d level. In most commercial logic ICs and microprocessors, each bit switch between either 0 or 5 V.

Assuming the supply voltages of the operational amplifier are $\pm V_s$, the maximum output voltage that can be achieved without causing saturation is V_s. Therefore, the minimal change in output voltage happens when the least significant digit transitions from 0 to 1. The output voltage change that corresponds to the least significant digit (LSD) from 0 to 1 specifies the resolution of the DAC unit.

If denote the digits of the 4-bit binary word B and the level 1 voltage of each bit is d, the corresponding output voltage of the summing amplifier will be as follows:

$$v_o = \left(b_3 \cdot 2^3 + b_2 \cdot 2^2 + b_1 \cdot 2^1 + b_0 \cdot 2^0\right)d \tag{12.1}$$

Equation 12.1 gives the output voltage of summing amplifier. In DAC, the input voltage of each bit is either zero or the 1-level voltage of the binary signal. We can write the output voltage of the summing amplifier in Figure 12.7 as

$$v_o = -R_f\left[\frac{b_0}{R_0} + \frac{b_1}{R_1} + \frac{b_2}{R_2} + \frac{b_3}{R_3}\right] \cdot d \tag{12.2}$$

If each resistor is half of the last one ($R_k = R_{k-1}/2$), the output voltage of the op-amp becomes decimal value of the digital number multiplied with the gain $-R_f/R_0$.

$$v_o = -\frac{R_f}{R_0}\left[b_0 + b_1 2^1 + b_2 2^2 + b_3 2^3\right]d \tag{12.3}$$

Feedback resistor must be selected such that the output voltage that corresponds to the highest binary number to be converted does not exceed the supply voltage. The step size or output resolution is the voltage corresponding to the LSD. The maximum output voltage divided by the maximum digital number gives the resolution.

$$\delta = \frac{12}{15} = 0.8 \tag{12.4}$$

The step size causes noise on the analog signal. Increasing the number of digits reduces noise and improves the accuracy of the output signal. In practice, a capacitor or an active filter is typically added to the output to filter out the noise.

Example 12.2 Given the binary one level of the input signal is 5 V and the op-amp supply voltage ± 15 V, find the resistor values for a 4-bit DAC shown in Figure 12.6.

Solution

Since the op-amp supply voltage is ± 15 V, the maximum output voltage should not exceed 15 V. To prevent saturation in the worst case, we can select the maximum output voltage of 12 V.

To obtain 12 V output when all input bits are 1, the gain must be

$$\frac{R_f}{R_0} = \frac{12}{15 \times 5} = 0.16$$

Since input resistors are calculated by successive divisions with 2, let's arbitrarily select $R_0 = 10\,\text{k}\Omega$. Then, the resistor values are

$$R_1 = \frac{10}{2} = 5\,\text{k}\Omega; R_2 = \frac{5}{2} = 2.5\,\text{k}\Omega; R_3 = \frac{2.5}{2} = 1.25\,\text{k}\Omega; R_f = 0.16 \times 10 = 1.6\,\text{k}\Omega$$

The accuracy of DAC depends on the precision and temperature constant of the resistors. Since a variety of DAC chips are available, it is not necessary to construct a DAC circuit using discrete components. The resistors are accurately manufactured using advanced techniques of semiconductor technology and packaged in the same chip. All resistors changing at the same rate by temperature do not affect the conversion accuracy. Manufacturer specifications typically include the resolution, the accuracy, the output range, power supply requirements, and the power dissipation (Rizzoni & Kearns, 2007).

12.5 Analog-to-Digital Conversion

In digital control, the information must flow in numerical form throughout the system until the processor transmits the final control signal to the electromechanical actuators and motors. ADCs convert continuous analog signals into digital. ADCs are essential components of digitally controlled mechatronic systems to send information sensors collect to logic gates or microprocessor inputs.

The first step is sampling the analog signal at regular time intervals and holding this value over the sampling while ADC converts the samples to digital signal.

Sample and Hold Block uses a sampler switch to collect the analog signal magnitude at every sampling interval and stores the sampled voltage in a capacitor. The output signal is continuous in time but discrete in amplitude as illustrated in Figure 12.7.

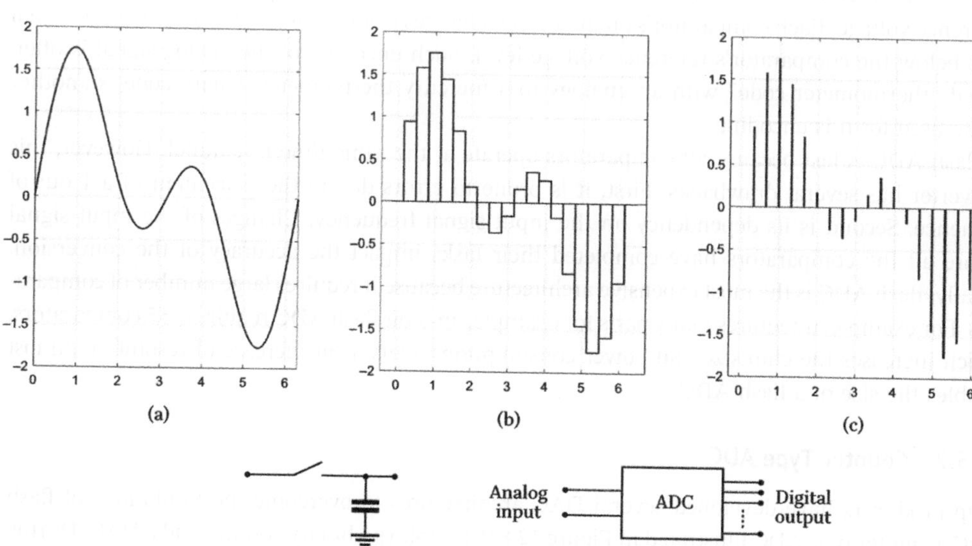

Figure 12.7 Analog to digital conversion process (a: Original signal; b: Sampled signal; c: Quantized signal)

The number of samples per second, which is the inverse of the sampling period, is the sampling frequency. If the sampling frequency is not high enough, a phenomenon called aliasing causes signal distortion. Shannon–Nyquist theorem states that the sampling frequency must be at least twice the highest frequency component of the sampled signal to prevent aliasing. This is equivalent to sampling the signal at least twice over one period of the fastest component.

The sample-and-hold circuit must keep the sampled voltage constant until the quantization block converts its analog value to the corresponding binary number. Therefore, the holding time must be long enough to allow the electronic circuit to convert the analog signal to binary numbers. The accuracy of the conversion depends on the number of bits of the binary output of the ADC. In practice, this is identified as the resolution of the conversion.

Digital output of ADC can be in series or parallel format. In series output, the binary words are transmitted sequentially one after the other to the memory or digital processing unit. In parallel format, several binary words are transmitted at the same time through separate lines. Parallel output is faster but serial output is more economical as it requires fewer pins compared to parallel output. The resolution of the ADC determines the minimum voltage variation required to change the digital output. Quantization error occurs due to the finite resolution, leading to information loss.

12.5.1 Flash ADC

The simplest and fastest analog-to-digital conversion architecture is known as flash or parallel comparator DAC illustrated in Figure 12.8 (a). It consists of a resistive voltage divider, cascaded comparators, and a code converter. For N bit digital signal, the voltage divider with 2^N resistors provides the reference voltage for 2^{N-1} comparators. One of the inputs of each comparator is connected to the analog signal to be converted. The fractions of the reference voltage obtained from the voltage divider supply the other input of the comparators. Each comparator receives the reference voltage corresponding to one least significant bit (LSB) greater than the comparator below it. The output of any one of the comparators is zero if the input voltage is less than the reference voltage connected to it. The output rises to the logic 1 level when the input voltage is equal to or greater than the applied reference voltage. Each comparator's binary output changes from one to zero when the input signal falls below the comparator's reference voltage level. Such encoding of the analog signal is often called "thermometer code" with an analogy to a mercury thermometer. Unary code is another name used for this encoding.

Flash ADC is fast because all comparators operate at the same time in parallel. However, this converter has several drawbacks. First, it is prone to errors due to the placement of a 1 out of sequence. Second is its dependency on the input signal frequency. Changes of the input signal before all the comparators have completed their tasks impact the accuracy of the conversion. Finally, flash ADC is the most expensive architecture because it requires large number of comparators. For example, it requires many parts for example, an 8-bit flash ADC requires 255 comparators, which increases the chip size and power consumption. Every 1-bit increase of resolution almost doubles the size of a flash ADC.

12.5.2 Counter Type ADC

Chip producers have developed several DAC architectures to overcome the limitations of flash ADC. Counter type ADC illustrated in Figure 12.8 (b) employs a binary counter and a DAC. During the conversion process, the counter counts the clock pulses, and a DAC continuously converts the output of the counter to analog signal. A single comparator compares the analog version of the counter output with the applied signal. Counting ends when both signals are equal and the binary

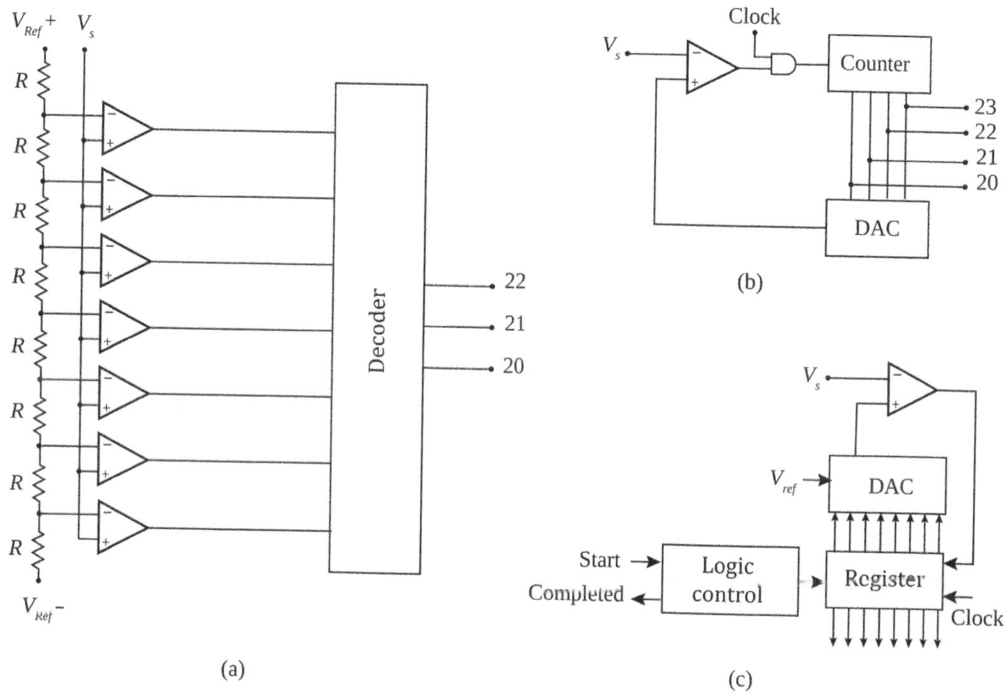

Figure 12.8 Common analog to digital converter architectures (a: Flash (parallel) comparator DAC; b: Counter type ADC; c: Successive approximation DAC (SAR))

output of the counter is sent to the output pins of the DAC. In this type of converter, the maximum conversion time is equal to $2^n - 1$ times the clock pulse duration for n-bit ADC. The conversion time depends on the magnitude of the analog input voltage.

12.5.3 Successive Approximation Converter

Successive approximation converter, also known as successive approximation register (SAR) is one of the most popular commercial ADC architectures. Figure 12.8 (c) shows a block diagram of SAR. A register, DAC, and a single analog comparator are the essential elements. The conversion process is based on iterative comparisons of the analog output of the DAC against the analog input signal. The search algorithm starts with an initial guess by setting the most significant digit in the register to 1 and all other digits to zero. That is, the beginning point is at the center of the search range. The internal DAC converts the digital word at the register output to an analog value, and the comparator compares this analog value with the input signal. If the DAC output is greater than the input signal, the MSB is set to 0; otherwise, it remains 1. The process repeats for the next significant bit. This process continues bit by bit, from MSB to the LSB, refining the digital output value with each step. After all bits have been processed, the final digital value represents the original analog signal.

SAR ADCs are the most common ADCs, having resolutions of 8–16 bits and speeds up to 10 MS/s. SAR type ADCs are cost-efficient. They provide high accuracy conversion. Whereas the speed of the SAR ADC is limited by the rate at which the DAC can settle inside the feedback loop, their conversion time is uniform for any magnitude of the analog signal (Tyler, 2002). SAR ADCs are preferred for applications that require high-speed and accurate conversions.

12.6 Digital Signal Conditioning

Digital signal conditioning refers to the processing steps applied to digital data after analog-to-digital conversion, ensuring that the signal is suitable for subsequent processing or analysis. Common digital conditioning operations include filtering to remove noise, scaling to adjust amplitude, offset correction, and linearization to compensate for sensor or system nonlinearities. These procedures are essential for enhancing signal quality, improving measurement accuracy, and enabling reliable operation of digital systems.

In practical applications, digital signal conditioning can be implemented using algorithms in microcontrollers, digital signal processors, or specialized hardware. Techniques such as moving average filters, finite impulse response (FIR) filters, and error correction codes are frequently used to optimize the signal for digital control, storage, or communication. The flexibility of digital signal conditioning allows for easy adaptation to specific system requirements and facilitates the integration of advanced features like self-calibration and diagnostics.

12.6.1 Digital Amplification

Digital signal amplification refers to increasing the strength or amplitude of a digital signal after it has been converted from analog form. Unlike analog amplification, which directly boosts voltage or current levels, digital amplification is typically achieved through mathematical operations within digital processors, such as multiplying the signal values by a gain factor. This makes it possible to enhance signal clarity and maintain data integrity without introducing the noise or distortion commonly associated with analog amplifiers.

Digital signal scaling is the process of adjusting the amplitude of a digitized signal to fit within a desired range or to match the requirements of subsequent processing stages. This is typically achieved by multiplying each sample in the digital signal by a scaling factor, which can either increase or decrease the signal's overall magnitude. Scaling is important for normalizing data, ensuring compatibility between different systems, and preventing overflow or underflow in digital computations. In digital signal conditioning, scaling is often used alongside filtering, offset correction, and linearization to prepare signals for analysis, storage, or transmission. Because scaling is performed mathematically within digital processors, it preserves signal integrity and enables precise control over the output levels, making it an essential step in many digital control and measurement applications.

Amplification in digital systems is often used in conjunction with other digital signal conditioning techniques, such as filtering and scaling, to optimize signals for further processing, communication, or storage. The precision and repeatability of digital amplification help ensure reliable performance in applications ranging from audio processing to industrial automation.

12.6.2 Digital Filters

Digital filters are essential tools in digital signal conditioning, used to selectively enhance or suppress specific components of a digitized signal. Unlike analog filters, which operate on continuous signals using resistors, capacitors, and inductors, digital filters process discrete-time data using mathematical algorithms implemented in software or hardware. The two main types of digital filters are FIR and infinite impulse response (IIR) filters. FIR filters are known for their inherent stability and linear phase characteristics, while IIR filters are valued for their efficiency and ability to achieve sharp filtering with fewer computations.

Applications of digital filters include noise reduction, signal smoothing, and extracting useful information from complex signals. Because digital filters can be easily adjusted or reconfigured, they offer exceptional flexibility compared to their analog counterparts. Their precision and repeatability make them indispensable in areas such as audio processing, telecommunications, biomedical instrumentation, and industrial automation. Modern digital systems use advanced filtering algorithms to satisfy demanding measurement and control needs.

Digital filters can be customized to target specific frequency bands or eliminate unwanted noise, making them highly adaptable to changing application needs. Furthermore, advances in DSP enable the real-time implementation of complex filtering algorithms that would be impractical with analog circuits. Adaptability and efficient processing are essential in today's control systems to ensure dependable feedback and decision-making.

There are two main types of digital filters used in digital signal conditioning: FIR filters and IIR filters. FIR filters offer stability and linear phase, ideal for tasks needing reliable signal timing. IIR filters are efficient and can deliver sharp filtering with minimal computation. Both types can be customized for targeted frequencies and noise reduction, making them versatile for industrial automation as well as audio and telecommunications.

12.7 Discrete-Time Feedback

Discrete-time feedback systems are used in mechatronics for accurate control of dynamic processes. Unlike continuous feedback systems, discrete-time feedback operates on sampled data, where sensor readings are collected at regular intervals and processed digitally. Figure 12.9 displays a simplified block diagram of a feedback computer-controlled system. In this structure, the input and output are analog signals. As in an analog feedback system, the comparator produces the error signal by comparing the actual output with the desired reference. ADC converts the error into digital signal, and the computer determines the necessary corrective action at each sampling instant. Through digital algorithms, the system can adjust actuator inputs to minimize error, compensate for disturbances, and maintain stability. DAC produces analog signals to drive the motors and actuators in the controlled system through electronic converters.

Common applications of discrete-time feedback in mechatronics include servomotor control, temperature regulation, and position tracking in robotics. Digital controllers, such as PID algorithms, are frequently implemented in discrete-time, leveraging microcontrollers or digital signal processors (Kuo, 1992).

The design of discrete-time feedback systems requires consideration of sampling effects, quantization, and delays, as these factors can affect system performance and stability.

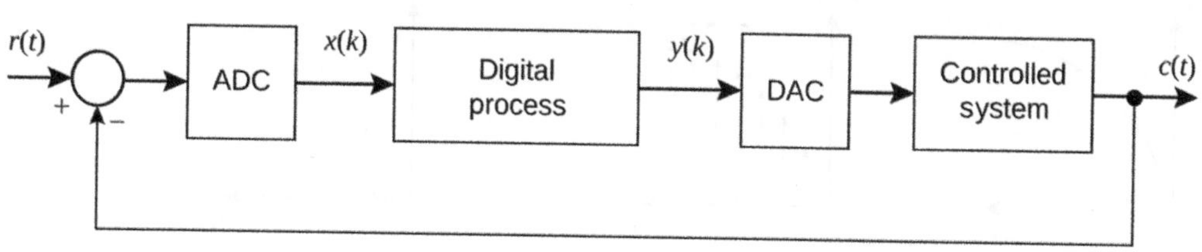

Figure 12.9 Discrete-time unity feedback control system

12.7.1 Modeling Discrete-Time Control Systems

Modeling discrete-time control systems involves representing the system's dynamic behavior using mathematical equations. The most common approaches use difference equations, z-transform techniques, and block diagrams to capture the relationships between system inputs, outputs, and states at distinct-time intervals. These models allow analysis of system stability, transient and steady-state response, and the effects of sampling rates and quantization noise. Accurate modeling is essential for controller design, simulation, and implementation, allowing engineers to predict system performance and optimize parameters for robust operation in real-world digital control applications.

ADCs modify the signal behavior primarily through the process of sampling and holding signals at specified intervals. Figure 12.10 illustrates an ideal sampler alongside a zero-order hold. The ideal switch turns to position 1 periodically at every sampling time t_k and stays at this position during the sampling interval Δt, which determines the sample width. A basic hold element is a capacitor, which charges until its terminal voltage rises to the signal magnitude at the sampling time $v_s(t_k)$ and maintains the same voltage throughout the sampling interval. Practical sample-and-hold circuits are made with fast semiconductor switching elements, which have small rise and fall times compared to the sampling interval Δt.

The sampling rate and nonideal characteristics of the sample-and-hold circuits impact the performance of discrete-time systems. To model discrete-time systems, we use mathematical equations that represent the ideal sample-and-hold element realistically. A common modeling technique is to formulate the sampled data in the time domain using difference equations, then transform this equation to the frequency domain using the z-transform methodology.

12.7.1.1 Difference Equations

The signal flow over the zero-order sample-and-hold (ZOSOH) process is illustrated in Figure 12.10. Linear, time-invariant function $f(t)$ represents the continuous signal. The sampled signal is discrete in time represented by a sequence of delta functions $f^*(t)$. The output of the first-order hold element is a staircase approximation of $f(t)$.

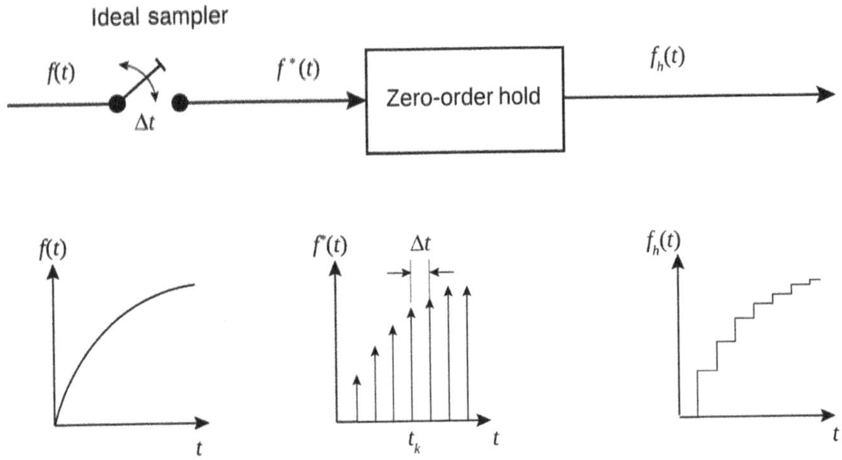

Figure 12.10 Signal flow through a zero-order sample-and-hold (ZOSAH) circuit

The ZOSOH process yields rectangular pulses, which can be considered as a series of delta functions with amplitude of the sampled waveform and area equal to the product of the rectangular pulse width (Nise, 2019). The following mathematical equation corresponds to this waveform:

$$f^*(t) = \sum_{k=-\infty}^{\infty} f(kT)\delta(t-kT) \tag{12.5}$$

Difference equations describe how the output of a system evolves over discrete-time steps based on the current and past inputs and outputs.

12.7.1.2 ZOSAH Model

A single impulse from the sampler results in a step response over the sampling interval. Therefore, the Laplace transform of this single step corresponds to the impulse response of the zero-order holder. We have seen earlier that the Laplace transform of a system's impulse response gives its TF. The Laplace transform of a step that starts at t = 0 and ends at $t = T$ is below:

$$G_h(s) = \frac{1-e^{-Ts}}{s} \tag{12.6}$$

We see that the zero-order hold is an integrator. If the ideal sampled signal $f^*(t)$ is followed by a zero-order hold, the TF is independent of the pulse width (Nise, 2019).

12.7.1.3 The *z*-Transform

The *z*-transform is a mathematical operation that converts a discrete-time signal, represented by a sequence of values at specific time intervals, into its form in the complex frequency domain. Like how the Laplace transform applies to continuous-time signals, the *z*-transform is utilized to examine the behavior of discrete-time systems. By transforming time-domain sequences into the *z*-domain, the *z*-transform facilitates the study of system stability, frequency response, and the design of digital filters.

Laplace transform applied to the difference equation given in equation 12.5 yields

$$F^*(s) = \sum_{k=0}^{\infty} f(kT)e^{-kTs} \tag{12.7}$$

By introducing the variable $z{=}e^{-Ts}$, the following equation defines the *z*-transform:

$$F^*(z) = \sum_{k=0}^{\infty} f(kT)z^{-k} \tag{12.8}$$

While the *z*-transform of a difference equation can be obtained by using this definition, it is more practical to use a table of *z*-transforms for common functions.

To indicate a *z*-transform of a time-domain difference equation and the inverse operation, we will use the following notation:

$$\mathcal{Z}\{f^*(kT)\} = F^*(z) \longleftrightarrow f^*(kT) = \mathcal{Z}^{-1}\{F^*(z)\} \tag{12.9}$$

Appendix A contains a compilation of *z*-transform formulas commonly used in DSP and discrete-time feedback control system analysis.

12.7.1.4 Inverse z-Transform

The inverse z-transform is the mathematical process used to convert a function from the z-domain back to the discrete-time domain. In other words, given a sequence's z-transform $F^*(z)$, the inverse operation reconstructs the original time-domain sequence $f(kT)$ as denoted in equation 12.9. The inverse z-transform is essential for analyzing and interpreting digital signals and systems after they have been represented in the z-domain for easier manipulation and analysis.

There are several methods to compute the inverse z-transform, including power series expansion, partial fraction decomposition, and using lookup tables for common z-transform pairs. These approaches allow us to retrieve the original discrete-time sequence required for system analysis and implementation.

The previous chapter demonstrated that the step response of analog feedback control systems consists of exponential function terms, which yield s-domain expressions that contain $A/(s + a)$. The sampled exponential time function corresponds in the z-domain to $z/(z - e^{aT})$. Therefore, we can expect that a partial fraction expansion should be in the following form:

$$F(z) = \frac{Az}{z - z_1} + \frac{Az}{z - z_2} + \cdots \tag{12.10}$$

We can eliminate the z terms in the numerator by performing the partial fraction expansion of $F(z)/z$, then multiply the result by z to restore the z terms in the numerator, as the following example demonstrates.

Example 12.3 The TF below represents a first-order discrete-time feedback control system. Find the unit-step response.

$$T(z) = \frac{5}{z - 0.8}$$

Solution

The unit-step response is obtained by multiplying the TF by the z-transform of the unit-step function $z/(z - 1)$.

$$F(z) = \frac{5}{z - 0.8} \cdot \frac{z}{z - 1} = \frac{5z}{(z - 0.8)(z - 1)}$$

Dividing by z yields

$$\frac{F(z)}{z} = \frac{5}{(z - 0.8)(z - 1)} = \frac{A}{z - 0.8} + \frac{B}{z - 1}$$

Calculate residues A and B.

$$A(z - 1) + B(z - 0.8) = 5$$
$$(A + B)z - (A + 0.8B) = 5$$
$$A + B = 0$$
$$A + 0.8B = 5$$

By solving this equation, we obtain

$$A = -B$$
$$-B + 0.8B = -25$$

The partial fraction expansion is obtained by substituting the values for A and B and multiplying by z.

$$F(z) = \frac{25z}{z-0.8} - \frac{25z}{z-1}$$

Using the z-transform pairs table in Appendix A, we can find the following inverse z-transform:

$$f(kT) = 25(0.8)^k - 25$$

12.7.1.5 Method for Obtaining Transfer Function in the z-domain

Applying the z-transform to a difference equation involves several systematic steps outlined below:

1) Write the difference equation.
2) Start with the system's difference equation that relates inputs and outputs at discrete-time steps. For example: $y[k] = a_1 \cdot y[k-1] + b_0 \cdot u[k] + b_1 \cdot u[k-1]$.
3) Apply the z-transform.
4) Take the z-transform of both sides of the equation. The z-transform of $y[k]$ is $Y(z)$, and the z-transform of $y[k-1]$ *is* $z^{-1}Y(z)$. Apply similar transformations for each term.
5) Manipulate algebraically.
6) Rearrange the resulting equation in terms of $Y(z)$ and $U(z)$, grouping like terms to isolate the output and input z-domain representations.
7) Express the ratio of output to input, $Y(z)/U(z)$, which yields the system's TF in the z-domain.

This step-by-step application of the z-transform streamlines the analysis and optimization of digital control systems, making it easier to predict system responses and design robust controllers for discrete-time environments.

As in continuous systems, the TF characterizes the system's frequency and stability properties. We can and the inverse operation use the TF to study stability, transient response, and steady-state behavior. The z-domain representation also simplifies controller design and simulation tasks.

12.8 Performance Criteria

When evaluating digital control systems, several key performance metrics are used to assess their effectiveness and reliability. Common criteria include:

- **Stability**: The ability of the control system to maintain controlled variables within acceptable bounds without oscillation or divergence.
- **Accuracy**: The degree to which the system output follows the desired reference or setpoint, often measured by steady-state error.
- **Response Time**: The time it takes for the system to react to changes in input or disturbances, including metrics like rise time and settling time.
- **Overshoot**: The maximum amount the system output exceeds the target value during transient response, which should be minimized in most applications.
- **Robustness**: The system's capacity to maintain performance despite variations in parameters or external disturbances.

- **Noise Immunity**: How well the system rejects or filters out unwanted noise or disturbances in the input or feedback signals.
- **Computational Efficiency**: The speed and resource usage of the control algorithms, which impact real-time performance and hardware requirements.

By analyzing these performance metrics, engineers can optimize digital control systems to achieve desired behavior, ensure reliability, and meet the specific needs of various applications, from industrial automation to consumer electronics.

12.9 Summary

Digital control systems play a crucial role in modern mechatronics by processing signals at discrete-time intervals, offering advantages such as flexibility, programmability, and integration with computer technologies. These systems convert analog inputs into digital data, process them via microcontrollers or processors, and produce analog outputs for actuators, enabling complex control strategies and real-time monitoring.

- Digital control contrasts with analog control by handling signals discretely rather than continuously. The simplest form is on–off control, which switches devices fully on or off based on thresholds but may cause oscillations around target levels. Sequential control activates subsystems in a preset order using logic circuits like flip-flops and PLCs, commonly applied in manufacturing automation and household appliances. Computer-controlled systems utilize general-purpose or embedded microcontrollers for real-time sensor processing, adaptive control, and network connectivity, supporting applications in industrial automation, robotics, and smart devices.
- Advanced digital control incorporates adaptive and predictive algorithms such as model predictive control and machine learning to optimize system performance. Stability analysis in digital control is critical due to sampling and discretization effects, employing z-domain methods and discrete TFs.
- Digital logic circuits form the backbone of mechatronic systems, integrating mechanical, electronic, and computational elements. Logic gates (AND, OR, NOT, NAND, NOR, XOR) implement Boolean operations fundamental to signal processing and control decisions. These gates are realized in ICs and can be combined into complex functions using techniques like SOP or POS.
- Combinational logic circuits produce outputs based solely on current inputs without memory. Their design often starts with truth tables and can be optimized using Karnaugh maps (K-maps) or software tools like MATLAB® for simplification, reducing hardware complexity and cost.
- Medium-complexity logic circuits employ integrated components such as decoders, encoders, multiplexers, and PLDs. Decoders select one output line based on binary inputs, useful for device addressing and logic function implementation. Encoders perform the inverse operation. Multiplexers route one of many inputs to a single output based on select signals, optimizing data routing and sensor integration.
- ROM stores fixed binary data and includes types like Mask ROM, PROM, EPROM, and EEPROM. While Mask ROM is permanently programmed during manufacturing, PROM can be programmed only once while EPROM and EEPROM allow multiple programming cycles with different erasing methods. Flash memory, a type of EEPROM, supports efficient block programming and erasing

- PLAs and PAL devices allow user-defined logic programming post-manufacture, enabling flexible and cost-effective medium to complex logic circuits. These devices are programmable using HDL and are integral to mechatronic control systems.
- Sequential logic circuits incorporate memory elements, making outputs dependent on current and past inputs. Flip-flops are fundamental bistable elements storing one bit of data in SR, JK, D, and T types. Counters and registers, built from flip-flops, perform counting, timing, and data storage functions essential in digital electronics and microcontrollers. Counters may be binary, BCD, ring, or Johnson types, driven by clock pulses with reset and enable functions. Registers store and shift data in various configurations such as SISO or parallel-in/parallel-out.
- PLCs are rugged, modular digital systems designed for real-time industrial control, capable of operating under harsh conditions. They monitor sensor inputs, execute control algorithms (e.g., on–off, sequential, PID), and manage actuators. PLCs support communication via Ethernet/IP, Modbus, and other protocols and are programmed using ladder logic or other graphical/text languages. Their modularity allows easy expansion and maintenance, making them central to automation in manufacturing, process control, building systems, transportation, and power management.
- Microprocessors are CPUs on single chips requiring external memory and peripherals, suitable for complex, multitasking applications like PCs. Microcontrollers integrate CPU, memory, and I/O on one chip, optimized for embedded control tasks in home appliances, automotive, and consumer electronics. Microcontrollers typically consume less power and are used for managing mechatronic functions, digital instrumentation, and robotics.
- DACs transform digital binary words into analog voltages or currents, commonly using summing amplifier circuits with weighted resistors. The resolution depends on bit depth, with higher bits reducing output noise. Practical DACs are often ICs with precise resistor networks.
- ADCs convert continuous analog signals into discrete digital values through sampling and quantization. Sampling frequency must satisfy the Nyquist criterion to avoid aliasing. ADC architectures include flash ADCs (fast but costly), counter-type ADCs (variable conversion time), and SAR ADCs (balance of speed and accuracy).
- Digital signal conditioning involves operations such as amplification, scaling, filtering, offset correction, and linearization to enhance signal quality. Digital amplification is performed mathematically, preserving signal integrity. Digital filters, including FIR and IIR types, selectively modify signal components and are implemented in software or hardware for noise reduction and signal smoothing.
- Discrete-time feedback systems sample sensor data at intervals, process error signals digitally, and output actuator commands via DACs. Digital controllers like PID algorithms operate in discrete time, requiring careful consideration of sampling effects, quantization, and delays for stability and performance.
- Modeling uses difference equations and z-transforms to represent system dynamics in the discrete domain. The ZOSOH circuit maintains sampled signal values during conversion. The z-transform converts time-domain sequences into the complex frequency domain, facilitating stability analysis and controller design. Inverse z-transforms recover time-domain responses from z-domain functions.
- Digital control systems are evaluated on stability, accuracy, response time, overshoot, robustness, noise immunity, and computational efficiency. These metrics guide optimization for reliable and effective control in diverse applications.

12.10 Check Your Understanding

12.10.1 Self-Assessment Quiz

1 Digital control systems process signals continuously, unlike analog systems that handle them at set time intervals.
True False

2 The simplest digital control system is on–off control, which works by switching devices completely on or off based on a set limit.
True False

3 Sequential control activates multiple subsystems by following a preset sequence of actions, with each stage initiated by predetermined logic conditions or specific inputs.
True False

4 Digital logic circuits that include embedded microprocessors enable the implementation of error detection and correction algorithms, which ensure the reliability and safety of mechatronic systems.
True False

5 Flash ADC is the most expensive architecture because it requires many comparators.
True False

6 What is the primary advantage of digital control systems over analog systems?
a. Continuous signal processing
b. Flexibility and programmability
c. Lower cost
d. Simplicity

7 Which of the following is NOT a type of flip-flop used in sequential logic circuits?
a. SR flip-flop
b. D flip-flop
c. JK flip-flop
d. AB flip-flop

8 What is the main function of a multiplexer in digital systems?
a. To convert analog signals to digital signals
b. To pass one of its data inputs to the output based on the state of the select inputs
c. To store binary data
d. To perform arithmetic and logic operations

9 Which type of memory can be programmed and erased multiple times using an electrical charge?
a. Mask ROM
b. PROM
c. EPROM
d. EEPROM

10 What is the primary purpose of a programmable logic controller (PLC) in industrial systems?
   a. To execute control tasks in real time
   b. To store large amounts of data
   c. To perform complex arithmetic operations
   d. Providing high-speed internet connectivity

11 What is the primary function of a flip-flop in digital circuits?
   a. To perform arithmetic operations
   b. To store binary data
   c. To generate clock signals
   d. To convert analog signals to digital

12 Which type of flip-flop has inputs labeled *J* and *K*?
   a. SR flip-flop
   b. D flip-flop
   c. JK flip-flop
   d. T flip-flop

13 Which of the following is a characteristic of a sequential logic circuit?
   a. Outputs depend only on current inputs.
   b. Outputs depend on both current and past inputs.
   c. Outputs are always high.
   d. Outputs are always low.

14 What is the primary use of a counter in digital electronics?
   a. To store data
   b. To perform arithmetic operations
   c. To count events or time intervals
   d. To generate clock signals

15 Which type of register allows data to be shifted in or out one bit at a time?
   a. Parallel input-parallel output register
   b. Serial-in, serial-out register
   c. Serial-in, parallel-out register
   d. Parallel input, serial output register

Answers:

1: False; 2: True; 3: True; 4: True; 5: True; 6: b; 7: d; 8: b; 9: d; 10: a; 11: b; 12: c; 13: b; 14: c; 15: b.

12.10.2 Short Answer Questions

1 What is the primary advantage of digital control systems over analog systems?

2 Describe the simplest form of digital control and its potential drawback.

3 What is the role of sequential control in mechatronic systems?

4 How do computer-controlled systems differ from fixed-function controllers?

5 What are the main types of flip-flops used in sequential logic circuits?

6 Explain the function of a multiplexer in digital systems.

7 What is the purpose of read only memory (ROM) in digital systems?

8 How do programmable logic arrays (PLAs) differ from programmable array logic (PAL)?

9 What is the significance of stability analysis in digital control systems?

10 Describe the process of digital-to-analog conversion and its importance in digital systems.

12.10.3 Essay Questions

1 Discuss the advantages and disadvantages of digital control systems compared to analog control systems.

2 Analyze the functions and applications of registers and counters in digital electronics. How do these components contribute to data storage, manipulation, and timing functions in mechatronic systems, and what are some common uses of counters in digital systems?

3 Analyze the significance of stability analysis in digital control systems and the methods used to ensure stability.

4 Analyze the functions and applications of registers and counters in digital electronics. How do these components contribute to data storage, manipulation, and timing functions in mechatronic systems, and what are some common uses of counters in digital systems?

5 Evaluate the impact of adaptive and predictive algorithms, such as model predictive control and machine learning, on the performance of digital control systems.

6 Discuss the role of programmable logic controllers (PLCs) in industrial automation and their advantages over traditional control methods.

7 Explain the function of digital filters in signal conditioning and their applications in various fields.

8 Analyze the importance of discrete-time feedback systems in mechatronics and the challenges associated with their design and implementation.

9 Describe the components and operation of an automated conveyor system, highlighting the integration of sensors, actuators, controllers, and safety systems.

12.10.4 Problems

1 Given a 4-bit digital-to-analog converter (DAC) with a reference voltage of ($V_{ref} = 5$ V):
a. Calculate the output voltage for the digital input (1010_2).
b. Determine the resolution of the DAC in volts.

2 An analog signal is sampled at a rate of 1 kHz and converted to a 10-bit digital signal using a successive approximation register (SAR) ADC:
a. What is the quantization step size if the input range is 0–10 V?
b. Calculate the digital output for an analog input of 3.75 V.

3 Consider a logic circuit with three inputs (A, B), and (C):
a. Write the Boolean expression for the output of a circuit that uses AND, OR, and NOT gates to implement the function ($F = (A \cdot B) + (\bar{C})$).
b. Simplify the Boolean expression using Boolean algebra rules.

4 A JK flip-flop has inputs ($J = 1$) and ($K = 0$):
a. Determine the next state of the flip-flop if the current state is ($Q = 0$).
b. Explain the behavior of the flip-flop when both (J) and (K) are set to 1.

5 Design a discrete-time system to model the behavior of a simple RC (resistor-capacitor) circuit.
a. Derive the difference equation for the discrete-time system using the backward Euler method with a sampling period (T_s).
b. Implement the difference equation in MATLAB® and simulate the system response for a step input.

References

Barrett, S. & Pack, D. J., 2022. *Microcontrollers Fundamentals for Engineers and Scientists*. Amsterdam: Springer.

Chen, M., 2025. *PLC Communication Protocols Explained: Comparing Modbus, Profinet, and EtherNet/IP*. [Online] Available at: https://rabwellplc.com/blogs/news/compare-plc-protocols-for-automation [Accessed 26 11 2025].

Flaus, J.-M., 2019. Architecture and Communication in an Industrial Control System. In: *Cybersecurity of Industrial Systems*. Hoboken, NJ: Wiley, pp. 25–55.

Kuo, B. C., 1992. *Digital Control Systems*. 2nd ed. Ft. Worth: Saunders College Publishing.

Marcowitz, A. B., 2002. *Introduction to Logic Design*. Boston, MA: McGraw Hill.

Nise, N. S., 2019. *Control Systems Engineering*. 8th ed. Hoboken, NJ: John Wiley & Sons.

Phillips, C. L. & Parr, J. M., 2011. *Feedback Control Systems*. 5th ed. Boston: Prentice Hall.

Rafiqazzam, M., 2005. *Flash Memory: A type of EEPROM that Allows for Quick and Efficient Programming and Erasing of Large Blocks of Data*. 5th ed. Hoboken, NJ: Wiley-Interscience.

Rafiqazzam, M. & McNinch, S. A., 2019. *Digital Logic with Verilog and FPGA-Based Design*. Hoboken, NJ: John Wiley & Sons.

Rizzoni, G. & Kearns, J., 2007. *Fundamentals of Electrical Engineering*. 6th ed. New York: McGraw-Hill.

Sanchez, J. & Canton, M. P., 2014. *Microcontrollers.* Boca Raton, FL: CRC Press.
The Mechatronics Blog, 2023. *How Machine Learning is Used in Mechatronics Engineering.* [Online] Available at: https://www.themechatronicsblog.com/2023/05/%20how-machine-learning-is-used-in-mechatronics-engineering.html?form=MG0AV3 [Accessed 31 12 2024].
Tyler, M., 2002. A/D and D/A. In: R. H. Bishop, ed. *The Mechatronics Handbook.* Boca Raton, London, New York, Washington DC: CRC Press, pp. 46.1–46.4.

13

Design Considerations

OBJECTIVES
• Review the integrated design process for mechatronic systems. • Explain the principles and methodology used in mechatronic system design. • Discuss fault detection and diagnostics techniques. • Address reliability and safety issues. • Describe standards commonly used in mechatronic design.

Mechatronic systems have changed modern engineering by combining mechanical, electrical, and software technologies. This interdisciplinary approach helps industries such as manufacturing and healthcare create systems that are more reliable, functional, and intelligent. Designing mechatronic systems allows for strong, flexible, and high-performance solutions that work better together than each part on its own (Ogata, 2002).

This chapter reviews how engineers design mechatronic systems by combining mechanical, electrical, electronic, and software parts. It covers the main ideas, methods, and technologies used in the field. The focus is on how these parts work together, as well as important topics like role of advanced control techniques, system modeling, simulation, new types of sensors and actuators, fault detection, and communication standards. The chapter also discusses reliability and safety issues.

13.1 Integrated Design Process

Mechatronic systems combine mechanical, electronic, and computational components to achieve specific tasks. Their integrated design enables greater precision, agility, and intelligence than each part alone. Applications range from simple household appliances to autonomous vehicles and industrial robots.

Mechatronic system design uses a teamwork approach called concurrent engineering, instead of the old sequential design methods. This means mechanical, electronic, and software parts are developed at the same time after breaking down the required functions. By working together and checking progress often, problems between parts are found and fixed early. The main goal is to make the whole system meet the prescribed requirements, focusing on reliability, flexibility, and cost, rather than just making individual parts work properly. This approach helps deliver products faster and reduces mistakes. Using feedback during design allows engineers to quickly make changes as needed. Advanced tools for product life cycle management (PLM) also make it easier for teams to work together, test system behavior, and find problems early before building real prototypes (Morales-Cruz et al., 2021).

The core principles of mechatronic design emphasize an integrated systems approach, as detailed below:

- Concurrent development of hardware and software components to ensure seamless interoperability.
- Iterative integration and refinement to identify cross-domain issues early, reducing the risk of costly post-production modifications.
- Functional partitioning to assign responsibilities to the domain best suited for implementation (mechanical, electrical, or software).
- Modularity and scalability to support future system upgrades and easy maintenance.
- Design for testability and reliability, enabling built-in fault detection, self-diagnostics, and robust error detection.

These principles promote innovation, accelerate development, and enhance adaptability to evolving requirements and technological advancements (Janschek, 2014).

The integration process generally consists of multiple phases, which are thoroughly detailed in numerous publications on engineering design and product development (Ford & Coulston, 2008; Haik, 2003; Ulrich & Eppinger, 2012).

1) **Requirements Analysis**: Define system-level functional and performance criteria, considering real-time response, safety, and cost constraints. Stakeholder interviews, surveys and questionnaires, usage cases and scenarios, workshops, and focus group meetings are typical tools to identify engineering requirements.
2) **Creating Testable and Verifiable Specifications**: State measurable requirements, use clear language, and develop methods for testing and validation.
3) **System Architecture Design**: Map out the hardware and software components, specifying microcontroller platforms, communication networks, data flows, and control loops.
4) **Component Selection**: Choose sensors, actuators, embedded processors, and interface electronics based on the system needs and integration compatibility. Quality, cost, compatibility, and market availability are key criteria for component selection.
5) **Prototyping and Simulation**: Use tools such as MATLAB/Simulink to simulate component interactions, validate control laws, and identify integration issues before hardware deployment.
6) **Implementation**: Integrate and validate hardware and software modules, ensuring that sensor–actuator chains are synchronized and meet timing requirements.
7) **Testing and Verification**: Employ unit, integration, and environmental tests to ensure robustness, followed by system-level validation under operational scenarios. Hardware-in-the-loop (HIL), software-in-the-loop (SIL), and model-in-the-loop (MIL) are common techniques that can be implemented to validate system performance.
8) **Managing Discrepancies Between Specifications and Actual System Performance**: To resolve discrepancies, it is essential to identify their root cause, implement corrective actions, and establish a change control process to ensure that any changes to specifications are thoroughly evaluated and approved.

13.1.1 Mechanical Design Principles

Mechanical design in mechatronic systems typically involves the definition of moving elements, housing, frames, and energy-transfer interfaces. The voltage, mechanical forces, inertia, and torque play a critical role in determining dimensioning and robustness. Choices of whether metals, polymers, or composite materials would be used are guided by criteria such as weight, strength, fatigue resistance, manufacturability, cost, and environmental factors such as thermal expansion and corrosion resistance.

Modern tools like finite element analysis (FEA) assist engineers in verifying load pathways, identifying stress concentrations, and optimizing the geometry for performance and manufacturability. This simulation-driven design philosophy minimizes over-engineering and supports lightweighting, crucial in automotive and aerospace systems.

13.1.1.1 Kinematics and Dynamics

The kinematic and dynamic modeling of mechanical elements forms the foundation of mechatronic motion control and precision. Dynamic equations governing rigid and flexible bodies are derived using methods such as Newton's laws. Accurately capturing multi-body dynamics is essential for applications like robotic manipulators or high-speed pick-and-place mechanisms, where inertial effects, damping, and friction impact control and positioning accuracy.

13.1.1.2 Design for Manufacturability and Maintenance

Mechanical structures are designed with attention to their performance during use, as well as factors related to manufacturing, assembly, and maintenance. Features such as standardized fasteners, accessible joints, and self-aligning interfaces greatly shape operational costs and downtime. Increasingly, design for additive manufacturing influences mechanical design choices, enabling topology-optimized structures unattainable through conventional means.

13.1.2 Electrical and Electronic Design

Design of electrical parts and electronic circuits involves electromechanical systems, power converters, signal processing, and conditioning boards.

13.1.2.1 Electromechanical Energy Conversion

The electrical subsystem in mechatronic systems generally includes motors, actuators, power electronics, signal conditioning circuits, and embedded processors. The selection of motors—such as DC, stepper, brushless DC, servo, and piezoelectric actuators—is based on parameters such as torque, speed profiles, energy efficiency, and controllability. For applications requiring precision, actuators with high dynamic response and minimal backlash are often utilized.

13.1.2.2 Electronics

Electronic hardware converts controller signals into drive currents and processes sensor feedback, influencing system response and stability. Many modern mechatronic designs also incorporate energy harvesting and management circuits for use in portable and autonomous systems

13.1.2.3 Signal Processing and Conditioning

Sensors generate analog or digital signals that often require amplification, filtering, isolation, or modulation. Analog-to-digital converters (ADCs), operational amplifiers, isolation devices, and wireless transceivers are integral to modern mechatronic circuit design. The fidelity of signal processing directly affects measurement accuracy and the granularity of control (de Silva, 2016).

13.1.3 Software

The software architecture in mechatronic systems generally spans three layers:

- Device-level firmware (microcontroller/FPGA code for sensor data acquisition, motor drive, and I/O)
- Control logic (real-time algorithms for feedback and supervisory control)
- User interface (UI) and networking applications (HMI, remote monitoring, cloud interconnect)

Modularity and code reusability are critical in firmware development, ensuring that basic drivers and libraries can support multiple system configurations and product variants.

Robust software integration relies on proven engineering methodologies such as

- **Model-Based Design**: Specification of algorithms and logic in high-level modeling environments (e.g., MATLAB/Simulink), generating tested code that streamlines validation and deployment.
- **Version Control and Continuous Integration**: Systematic source code management and automated test pipelines to detect integration issues early.
- **Software in the Loop and Hardware in the Loop Testing**: Deploying code models in simulated or real hardware environments to verify functionality and timing constraints prior to physical integration.

The software in mechatronic systems is usually organized into three main layers:

- **Device-Level Firmware**: Code running on microcontrollers or FPGAs to read sensors, control motors, and handle basic input/output.
- **Control Logic**: Real-time software that makes decisions based on feedback and supervises system behavior.
- **User Interface and Networking**: Software for displays, remote monitoring, and connecting to networks or the cloud.

Good firmware is modular and reusable, which means basic drivers and libraries can be used in different versions of a product or system.

Reliable software integration uses engineering best practices like:

- **Model-Based Design**: Creating algorithms in tools like MATLAB/Simulink, then generating code that is easier to test and deploy.
- **Version Control and Continuous Integration**: Tracking changes in code and using automated tests to catch problems early.
- **Software in the Loop (SIL) and Hardware in the Loop (HIL) Testing**: Running code in simulated or real hardware environments to check that it works correctly before final integration.

13.2 Embedded System Control Strategies

An embedded system is a specialized computing system that is designed to perform dedicated functions or tasks within a larger mechanical or electrical system. Unlike general-purpose computers, embedded systems are typically built into the devices they control and operate with real-time constraints, often using microcontrollers, digital signal processors, or FPGAs. These systems are optimized for efficiency, reliability, and responsiveness, making them essential for applications like industrial automation, automotive electronics, consumer devices, and medical equipment.

Embedded microcontrollers, digital signal processors, and field-programmable gate arrays (FPGAs) constitute the computational heart of electronic subsystems. These platforms execute real-time control algorithms, handle communication tasks, and enable UIs. Tight constraints on latency, timing determinism, and energy use drive the choice of hardware and real-time operating systems (RTOSs) for high-performance applications such as industrial robots and automotive systems (Schraft et al., 2000).

Control strategies regulate the behavior of a mechatronic system to achieve the desired performance—tracking, stability, rapid response, and disturbance rejection. Control may be centralized or distributed across multiple processors.

Proportional-Integral-Derivative (PID) control dominates due to its simplicity, robustness, and ease of implementation in industrial systems. The PID algorithm adjusts the actuator input based on the error between setpoint and measurement, integrating proportional (P), integral (I), and derivative (D) terms to correct steady-state errors and respond quickly to disturbances.

For systems with rapidly changing dynamics, nonlinearity, or multiple interacting degrees of freedom, the following strategies are necessary:

- **State-Space Control**: Models plant behavior in vector form, supporting multiple input and output variables (MIMO systems).
- **Robust Control**: Designs controllers tolerant of modeling uncertainty and external disturbances.
- **Adaptive Control**: Adjusts controller parameters in real time as plant dynamics evolve (e.g., due to changing load or wear).
- **Optimal Control and Model Predictive Control (MPC)**: Predict future behaviors based on a plant model, optimizing performance over a moving horizon subject to constraints.

Modern controllers achieve real-time performance using embedded microcontrollers, Digital Signal Processing (DSPs) or FPGA capable of executing control loops in sub-millisecond time frames. Software must be carefully designed to meet timing deadlines and prioritize task scheduling, particularly in safety-critical applications.

An FPGA is a type of integrated circuit that can be configured by the user after manufacturing to perform specific logic functions. FPGAs are widely used in embedded systems for tasks requiring high-speed parallel processing, real-time control, and flexibility, such as signal processing and custom hardware acceleration. Their reconfigurability allows engineers to tailor hardware architectures to challenging applications like industrial automation, robotics, and automotive electronics.

Loop architectures of typical control systems are listed in Table 13.1. Closed-loop control predominates in mechatronics, but cascade and feedforward architectures offer enhanced precision and disturbance rejection, critical in high-performance machines.

Table 13.1 Control loop architectures

Control Type	Description	Example Application
Open loop	No feedback: output not measured	Fan, simple pump
Closed loop	Output measured, errors corrected	Temperature control
Cascade	Two nested loops for improved response	Motion control
Feedforward	Anticipatory correction	Motor speed control

13.2.1 Control Functions of Embedded Systems in Mechatronics

Embedded systems serve as the central nervous system for nearly all mechatronic devices. Their main control functions include:

- **Real-Time Data Acquisition**: Continuous collection and preprocessing of sensor data (position, velocity, force, temperature, etc.), often requiring precise synchronization and digital filtering.
- **Signal Conditioning**: Amplification, scaling, filtering, and analog-to-digital conversion of raw sensor signals before control logic processing.
- **Control Algorithm Execution**: Running real-time feedback and feedforward control laws—PID, fuzzy logic, sliding mode, MPC, adaptive control—that govern actuation based on system state and desired outcomes.
- **Decision-Making**: Implementing higher-level logic for task sequencing, fault detection, supervisory safety (e.g., watchdog timers), and complex behaviors driven by artificial intelligence (AI) or rule-based systems.
- **User Interface and HMI**: Managing user interactions through either graphical interfaces, tactile controls, or networked dashboards for remote monitoring and diagnostics.
- **Data Logging and Communication**: Storing operational data and interacting with other networked devices or cloud systems for predictive maintenance, analytics, or optimization.

13.2.2 Hierarchical and Modular Integration

Mechatronic systems are inherently complex, incorporating multiple mechanical and electronic subsystems. Traditionally, control architecture in this domain has followed either hierarchical or modular models. In hierarchical integration, embedded controllers are layered—lower-level controllers handle real-time signal processing and actuation, while supervisory layers perform decision-making and system coordination. Modular integration, gaining traction with the rise of distributed intelligence, organizes embedded systems into interoperable modules, each responsible for specific tasks (such as motion control or sensor fusion), communicating over standardized protocols like CAN bus or Ethernet.

A key feature of modern mechatronic integration is the seamless blending of hardware components (sensors, actuators, microcontrollers) with embedded firmware and application-layer software. For example, in automotive "drive-by-wire" systems, the steering, throttle, and braking functions are governed by specialized embedded controllers connected via real-time buses, orchestrating precise coordination for safety and comfort.

Mechatronic systems are complicated because they mix mechanical and electronic parts. Their control setups are usually either hierarchical or modular. In a hierarchical system, controllers are arranged in layers: basic controllers handle fast tasks like moving parts, while higher-level controllers make decisions and coordinate all functions. In a modular system, different modules (like those for motion control or sensor fusion) oversee specific jobs and talk to each other through standard connections like CAN bus or Ethernet.

Today's mechatronic systems smoothly combine hardware (sensors, motors, controllers) with software. For example, in modern cars, "drive-by-wire" systems use special controllers for steering, gas, and brakes. These are linked by fast data lines to make sure the car operates safely and comfortably.

13.2.3 Real-Time Operating Systems for Mechatronic Control

Mechatronic systems often require deterministic scheduling and concurrency that exceeds the capabilities of simple infinite loop or polling architecture. An RTOS provides preemptive multitasking, hardware abstraction, intertask messaging, and timing guarantees.

Mechatronic systems need reliable scheduling and the ability to run several tasks at once, which cannot be achieved by simple loops. An RTOS helps by managing multiple tasks, making hardware easier to use, letting tasks communicate, and ensuring actions happen on time. Critical properties of RTOS for mechatronics are determinism, multitasking, resource management, and peripheral handling.

Determinism means the system always responds to events in a predictable way and within a set amount of time. Multitasking and scheduling allow several jobs to run at once, with rules like rate monotonic scheduling (RMS) or earliest deadline first (EDF) deciding, which task goes next. Resource management makes sure memory and processor time are shared fairly, so nothing gets stuck or ignored. An RTOS also manages drivers for things like network connections (TCP/IP, CAN, USB) and helps make the system easier to update or move to new hardware. Popular RTOS options in mechatronics are FreeRTOS (free and common in industry), Zephyr, and paid choices like VxWorks, QNX, or TI-RTOS, depending on needs for licensing, safety, and support.

Advantages realized with RTOS include:

- **Improved Modularity**: Allows adding new features (diagnostics, UI, networking) without compromising critical control loops.
- **Debugging Support**: Built-in profiling and monitoring tools to analyze timing, memory use, and stack overflows.
- **Reduced Development Time**: Eliminates the need to develop custom basic software, freeing engineers for application logic.

An RTOS becomes essential when task complexity increases (typically as software size exceeds 64 KB or several interrupt sources need real-time management), and safety or reliability certification is required.

13.2.4 Automation Enhancements Enabled by Embedded Systems

Embedded systems have revolutionized automation, enabling machines, smart factories, vehicles, and consumer devices to operate autonomously, efficiently, and adaptively.

- **Manufacturing Automation**: PLCs and embedded MCUs orchestrate assembly lines, welding, milling, and quality inspection, employing sensor feedback to dynamically tune process variables (temperature, position, pressure) and optimize throughput and energy use.
- **Industrial Robotics**: Articulated arms with embedded controllers execute intricate tasks (pick-and-place, welding) with repeatable microsecond-scale precision, adapting in real time to environmental changes (e.g., variable part positions or force feedback).
- **Logistics and AGVs**: Autonomous guided vehicles (AGVs) navigate factory floors using sensor fusion and embedded trajectory planning, reducing human labor and allowing for 24/7 operation.
- **Consumer and Medical Devices**: Smart home thermostats, wearables, and surgical robots leverage embedded control for responsive, user-driven operation and closed-loop feedback.

Industry 4.0 advances (integration of IoT, cloud, and big data analytics) are propelled by embedded systems, which serve as the "edge nodes" connecting the physical world to digital management and optimization platforms.

13.2.5 Sensor and Actuator Interfacing in Embedded Mechatronic Systems

Robust interfacing between microcontrollers and sensors/actuators is foundational for mechatronic system performance. Sensors (e.g., temperature, pressure, position encoders, accelerometers, vision modules) and actuators (e.g., motors, solenoids, valves) often present challenges stemming from disparate electrical and data standards.

For accurate feedback and precise control, it is crucial to tightly integrate sensors and actuators along with suitable signal conditioning methods like amplification and filtering. When working in compact setups or areas with strong EMI, factors such as signal-to-noise ratio, electromagnetic compatibility, and electrical isolation should be carefully addressed (Nise, 2019).

- **Signal Conditioning**: Analog sensor outputs frequently require amplification, voltage translation, filtering, and impedance matching before digitization by the MCU's ADCs.
- **Digital Interfacing**: Increasingly, sensors offer digital outputs (I2C, SPI, CAN, UART), simplifying CPU-side connections but requiring careful protocol selection to balance speed, noise immunity, and scalability.
- **Actuator Drivers**: MCUs, typically incapable of sourcing required power, interface through MOSFET-based drivers, H-bridges for DC/stepper motors, or specialized servo amplifiers. Isolation (optocouplers) is used to protect processor logic from high-power domains.

Best practices in sensor/actuator integration include:

- Utilize ready-to-use interface libraries or middleware that abstract protocol details like Hardware Abstraction Layer (HAL) for STM32, or Arduino libraries, which provide a standardized interface between application code and the underlying hardware components, such as microcontrollers, sensors, and actuators. They simplify cross-platform migration and reduce development time.
- Carefully manage power supplies, separate analog and digital grounds, voltage levels, and provide robust control of inductive kickback from motors.
- Implement interrupts, direct memory access (DMA), or real-time polling strategies to balance responsiveness and computational load.
- Provide diagnostics in the interface code to detect issues such as wiring faults, communication errors, or device malfunctions.

Integration is further complicated in distributed systems, where network reliability, signal latency, and node synchronization must be managed.

13.2.6 Precision and Accuracy Improvements with Embedded Control

Quantitative performance—in terms of accuracy, repeatability, settling time, and error minimization—is fundamentally enhanced by embedded digital control:

- **High-Resolution Sensor Sampling**: Using high-bit ADCs and sophisticated digital filtering, MCUs extract fine detail from physical processes, supporting closed-loop control in high-precision applications (e.g., CNC machining, medical imaging, microfluidics).

- **Advanced Control Algorithms**: Algorithms such as MPC, sliding mode, and adaptive control—enabled by powerful MCUs—offer robustness to model uncertainties and external disturbances, reducing steady-state error and improving transient response.
- **Calibration and Diagnostics**: Embedded systems facilitate online calibration, auto-tuning, and self-testing, compensating for sensor drift, nonlinearity, and mechanical wear.

For example, in industrial robotics such as precision soldering or semiconductor packaging, embedded controllers utilizing FPGAs and high-speed signal processing achieve positioning precision down to microns, integrating vision feedback and AI-based trajectory adjustment to maintain yield (PUSR, 2025).

13.2.7 Efficiency and Energy Optimization in Mechatronic Systems

Energy efficiency is critical for both large industrial facilities and mechatronic devices powered by batteries. Embedded controllers achieve energy savings through dynamic power management techniques such as dynamic voltage and frequency scaling (DVFS), power gating, and selective wake–sleep cycles, which help reduce consumption during varying operational loads or idle periods. RTOS and smart firmware further optimize task scheduling, taking advantage of low-power modes and minimizing unnecessary computation or sensor polling.

Hardware selection also plays a significant role, with low-power microcontroller units (MCUs), integrated sensor-processor modules, and power-efficient communication protocols like BLE, Zigbee, and CAN FD, contributing to reduced energy usage. Energy harvesting and regeneration methods, including piezoelectric, thermal, or solar harvesting, supplement battery power and extend the autonomous operation of devices used in applications such as remote monitoring and mobile robotics.

Predictive and condition-based maintenance, enabled by embedded analytics, allows early detection of potential failures, thereby optimizing maintenance schedules, lowering downtime, and preventing catastrophic energy loss. Together, these energy-aware embedded mechatronic strategies support sustainable automation, reduce the total cost of ownership, and promote environmentally responsible operation across sectors such as smart factories, healthcare, and smart cities.

13.3 Communication Networks and Middleware in Mechatronic Systems

Modern mechatronic systems depend heavily on robust communication networks to facilitate real-time data exchange among distributed sensors, actuators, and controllers. These networks must meet stringent requirements for low latency, high reliability, strong noise immunity, and deterministic data delivery. Table 13.2 highlights the networking protocols most frequently adopted in these scenarios.

13.3.1 Distributed Collaboration and Middleware

When multiple processors or nodes work together within a mechatronic system, such as in distributed sensor–actuator networks, middleware manages communication, synchronization, and data exchange, streamlining system integration. Industry standards like the Robot Operating System (ROS) and domain-specific protocols, including Controller Area Network (CAN) and Ethernet

Table 13.2 Key networking protocols

Protocol	Application Domain	Key Features
CAN (controller area network)	Automotive, industrial robotics	Real-time, robust, multi-master, error detection
EtherCAT	Industrial automation	Real-time Ethernet, high bandwidth, deterministic
Modbus, PROFIBUS	Process automation	Legacy support, wide adoption
RS-232/RS-485	Point-to-point, simple I/O	Simplicity, low speed
Wireless protocols	IoT, AGV	Flexibility, decreasing latency (Wi-Fi 6, 5G)

for Control Automation Technology (EtherCAT), enable straightforward, plug-and-play interfacing. This approach accelerates development and reduces time-to-market for new products.

CAN is a powerful and reliable communication protocol commonly used for real-time data exchange in cars and industrial machines (UPLATZ, 2025). It offers built-in error detection and handling mechanisms and supports data rates up to 1 Mbps for short lengths, with newer versions like CAN FD offering even higher speeds. On the other hand, the robustness of CAN comes with increased complexity in both hardware and software setup. Implementing CAN is more costly due to the need for CAN transceivers and controllers (Eureka, 2025).

EtherCAT is a fast, Ethernet-based protocol made for industrial automation. Both allow quick and predictable communication between controllers, sensors, and actuators by passing data efficiently as it moves through each device, which reduces delays and increases speed. EtherCAT can connect devices in different layouts—like line, tree, or star shapes—and is especially useful in applications that need precise timing, such as robotics and motion control. Its speed, flexibility, and strong error checking make EtherCAT a popular choice for modern systems that require both reliability and performance (Wang et al., 2021).

13.3.2 Key Networking Protocols

Because mechatronic systems generate larger data volumes and require greater connectivity, real-time Ethernet standards and wireless networking are increasingly being adopted. These technologies support advanced applications, including flexible production lines and autonomous vehicles.

The CAN bus protocol is extensively used for communication in automotive and industrial robotics applications. Its popularity stems from its cost-effectiveness and strong noise resistance. In addition, CAN's event-driven architecture, robust error management, and scalability make it well suited for supporting modular mechatronic system architectures.

Modbus is a widely used communication protocol in industrial automation and mechatronic systems. Originally developed in the late 1970s, Modbus enables data exchange between electronic devices, such as programmable logic controllers (PLCs), sensors, and actuators, over serial lines or Ethernet networks. It is valued for its simplicity, reliability, and openness, making it easy to integrate devices from different manufacturers within a distributed system (Chen, 2025).

In mechatronic applications, Modbus facilitates real-time monitoring and control by supporting master–slave and client–server architectures. Its straightforward messaging structure and broad industry adoption help ensure compatibility and efficient communication, particularly in scenarios where interoperability and robust data transfer are essential.

RS-232 and RS-485 are serial communication standards commonly used in industrial and mechatronic systems for connecting devices, such as sensors, actuators, and controllers. RS-232 is a point-to-point protocol, ideal for short-distance communication between two devices, and known for its simplicity and wide adoption in legacy equipment. In contrast, RS-485 supports multi-drop configurations, allowing multiple devices to share the same communication line over longer distances and in electrically noisy environments.

Both protocols facilitate reliable data exchange in real-time monitoring and control applications. RS-485's differential signaling provides robust noise immunity, making it well suited for harsh industrial settings, while RS-232 remains popular for straightforward device-to-device connections and configuration tasks. Their compatibility and ease of integration make them valuable tools for achieving interoperability in distributed mechatronic systems.

Wireless communication protocols are increasingly vital in modern mechatronic systems, supporting flexible layouts and remote monitoring where wired connections are impractical. Commonly adopted wireless standards include Wi-Fi (IEEE 802.11), Bluetooth, Zigbee, and emerging industrial wireless protocols like WirelessHART and ISA100.11a. These protocols enable reliable, low-latency data exchange between sensors, actuators, and controllers, even across large facilities or in mobile robotic platforms.

Wireless solutions contribute to system scalability, ease of reconfiguration, and cost reduction by eliminating the need for extensive cabling. Security features such as encryption and authentication are integral to these protocols, ensuring safe operation in environments where data integrity and confidentiality are paramount. As industrial IoT and smart factory initiatives expand, robust wireless networking becomes essential for seamless integration and real-time control in distributed mechatronic systems.

13.3.3 Design Verification and Validation and Reliability Assessment

Verification ensures the system design meets technical specifications, while validation checks that the system fulfills its intended operational requirements. Common approaches include using simulation to test system models in both normal and challenging conditions, conducting HIL testing to validate controllers with real-time simulations, and performing physical prototyping to assess accuracy, tolerance, and durability.

Reliability engineering within mechatronic systems involves several key methodologies. Accelerated life testing involves subjecting products to stress beyond typical operational loads to identify potential failure mechanisms. Fault tree analysis and failure modes and effects analysis (FMEA) are used to systematically identify possible failure points.

Fault tree analysis is a deductive, top-down approach used to identify the root causes of system failures by visually mapping potential faults within a system. It starts with a specific undesirable event, such as a system malfunction, and breaks it down into contributing factors using logical gates (AND, OR) to represent how combinations of faults can lead to the top event. This method helps engineers systematically analyze complex systems, understand interdependencies among components, and prioritize corrective actions to enhance reliability and safety.

FMEA is a structured, systematic approach used to identify potential failure modes within a system, subsystem, or process; analyze their causes and effects; and prioritize actions to mitigate associated risks. During FMEA, each component or process step is examined for ways it might fail, considering both the likelihood and impact of each failure. This proactive method helps engineers address possible weaknesses early in the design phase, improving overall reliability and safety of mechatronic systems.

Statistical quality control focuses on monitoring production and operational data to detect anomalies. A comprehensive verification and validation process is essential, particularly in sectors where system failures may pose risks to safety or result in substantial damage, such as automotive, aerospace, and healthcare industries.

13.4 Fault Detection and Diagnostics

Faults in mechatronic systems are unexpected problems that cause the system not to work as designed. These problems can happen suddenly, develop slowly, or become permanent. Faults can come from any part of the system or from how different parts interact, including mechanical, electrical, or software components (Hu et al., 2025; Lee, 2025).

Common sensor faults include bias (constant error in readings), drift (gradual change over time), saturation (sensor hits its maximum or minimum value), noise (random errors in readings), stuck-at (sensor gets stuck at one value), crosstalk (interference from other signals), and total failure (sensor stops working).

Common actuator faults are reduced effectiveness (lower power), jamming (immobilized), stiction (difficulty initiating movement), floating (inability to maintain position), and locking (complete inability to move).

Mechanical faults include worn bearings, unbalanced parts, misaligned components, broken gears, fatigue, looseness, or cracks in the structure.

Software and control faults can be logic mistakes, timing problems, software crashes, lost communications, or cybersecurity issues.

Sometimes, several faults happen at once or one problem causes another, leading to more serious failure across different parts of the system. To maintain high performance, early detection and accurate diagnosis are essential, enabling timely maintenance, system reconfiguration, or fail-safe operation to mitigate catastrophic failure and economic loss. Table 13.3 provides a summary of typical faults in complex mechatronic systems.

Table 13.3 Common faults in mechatronic systems

Fault Category	Description	Example Effects
Sensor bias	Systematic offset from true value	Incorrect feedback, degraded control
Sensor drift	Gradual deviation over time	Loss of calibration, late detection
Actuator jamming/lock-in-place	Unable to follow control commands	Loss of maneuverability/stability
Loss of effectiveness (actuator)	Reduced output or responsiveness	Sluggish, inaccurate actuation
Mechanical wear	Fatigue, surface degradation, misalignment	Vibration, excessive noise, failure propagation
Software/control error	Logical, runtime, or timing fault	Unexpected shutdown, improper sequence

13.4.1 Definitions and Objectives of Fault Detection and Diagnosis

- Fault Detection is the process of identifying the presence of an abnormal condition (fault or failure) in a system based on the deviation of observed behavior from expected norms.
- Fault Isolation/Diagnosis involves determining the location, type, cause, and often the severity of the detected fault.
- Fault Identification may refer to ascertaining the fault's magnitude or progression (e.g., developing severity or remaining useful life [RUL]).
- Fault Management refers to the procedures implemented in response to fault detection, such as issuing alarms, reconfiguring system operations, or transitioning to alternative operational modes through fault-tolerant control.

The main goals of fault detection and diagnosis within mechatronic systems are as follows:

- Ensuring operational safety and compliance with standards.
- Minimizing unplanned downtime and maintenance costs.
- Enhancing system reliability, lifespan, and overall productivity.
- Supporting condition-based or predictive maintenance strategies.

The main goals of fault detection and diagnosis within mechatronic systems are to enhance system reliability and ensure operational safety and compliance with standards. Effective fault detection minimizes unplanned downtime and maintenance costs, lifespan, and overall productivity.

13.4.2 Fault Diagnosis Approaches

Traditional fault detection methods can be grouped as model-based, hardware-based, and data-driven approaches as summarized below.

13.4.2.1 Model-Based Diagnostic Approaches

Model-based diagnostics use mathematical models to estimate the system response when it is functioning normally without any faults. Possible faults can be detected by comparing the actual signals from the system to those predicted by the model.

Residuals, which represent the deviation of the expected signal from actual, are generated through parity equations, observers, state estimators, or parameter estimation. Residuals are processed using statistical tests (RMS, thresholding) or machine learning (classification or clustering) to declare faults, possibly identifying faulty components and their severity.

Parity equations compare measurements to detect inconsistencies that could indicate faults. When actual data deviates from expected relationships beyond a preset threshold, it signals abnormal system behavior. These equations are most effective with multiple sensors, allowing the formation of residuals sensitive to specific faults. Signature matrices then map patterns of residual violations to precise fault sources (Isermann, 2006).

Kalman filter serves as an observer or state estimator that predicts the internal states and outputs of a mechatronics system based on a mathematical model. During normal operation, the filter continuously compares its estimated values to actual sensor measurements. When faults occur, such as sensor degradation or actuator failure, discrepancies between the predicted and measured signals increase beyond expected tolerances (Gertler, 1998).

Parameter estimation is based on continuously monitoring specific physical parameters of a system, such as resistance, capacitance, inductance, stiffness, or damping. In normal operation, these parameters remain within expected ranges as determined by a mathematical model of the system.

Online estimation algorithms, like recursive least squares (RLS) or support vector regression, track these parameters in real time. If a parameter begins to drift or change abnormally due to wear, damage, or degradation, it signals a potential fault. By comparing the estimated values to their nominal or baseline figures, developing faults can be identified early, particularly in mechanical subsystems like gearboxes and bearings, and take preventive action before more severe failures occur (Gertler, 1998; Isermann, 2006).

13.4.2.2 Hardware-Based Diagnostics

Hardware-based diagnostics approach refers to schemes that employ physical redundancy, such as installing multiple sensors or actuators to provide backup in the event of component failure. This is widely used in safety-critical applications, such as aerospace, where triplicated or even quadruplicated sensor arrays and voting logic are employed.

While robust, hardware redundancy increases cost, system weight, and complexity. This approach is not always feasible, especially in weight or space-constrained systems.

13.4.2.3 Data-Driven Diagnostics

Data-driven diagnostics map measured signals, often high-dimensional time series, to system health or fault states through classification or anomaly detection. Key steps of this approach are:

1) **Data Collection**: Extensive acquisition of healthy and faulty system data, often using sensors for vibration, current, temperature, and position.
2) **Preprocessing and Feature Extraction**: Signal processing to remove noise, augment data, and transform signals into suitable representations.
3) **Model Training**: Using classical statistical tests like hypothesis testing, regression, and machine learning.
4) **Pattern Recognition/Fault Classification**: Outputting healthy/faulty states, fault types, or even fault severity/RUL.

The following classical machine learning techniques are frequently used for data-driven fault detection:

Support Vector Machines (SVMs) are supervised machine learning algorithms used for classification and regression tasks. They work by finding the optimal hyperplane that separates different classes in the data, making them particularly effective for distinguishing between linearly or nonlinearly separable faults in applications like vibration-based fault classification. They are highly effective for classifying linearly or nonlinearly separable faults; popular for vibration-based fault classification in bearings and motors.

K-Nearest Neighbors (KNN) is a simple, nonparametric classification algorithm that relies on the intuition that similar data points are likely to belong to the same class, making it effective for certain fault scenarios when sufficient labeled examples are available. KNN assigns a data point to the majority class among its closest neighbors in the feature space, referring to an abstract, multidimensional space where each dimension represents a distinct characteristic, measurement, or feature extracted from the data. In machine learning and fault detection, data points are mapped into this space based on their features, allowing algorithms to analyze similarities, differences, and patterns for classification or anomaly detection. For example, in vibration-based fault classification, features might include statistical measures or frequency components derived from sensor signals, and the feature space organizes these into a structure that algorithms like SVM and KNN can use to distinguish between healthy and faulty states.

Convolutional Neural Networks (CNNs) are a type of deep learning model especially well suited for analyzing spatial or temporal patterns in data, such as images or time-series signals. They automatically extract features from raw data, which has led to high accuracy in fault classification tasks like monitoring bearings or gearboxes.

13.4.2.4 Hybrid Diagnostics

Hybrid approaches are combination of fault detection and diagnostics approaches summarized above. They are gaining popularity for complex scenarios where analytical modeling is not feasible for the whole system but feasible for major subsystems, complemented by machine learning for others.

13.4.2.5 Emerging Diagnostic Technologies

Digital twin technology refers to the creation of a high-fidelity, real-time, virtual models of a physical system, process, or asset. This digital representation continuously receives real-time data from sensors and other sources, enabling it to mirror the state, behavior, and performance of its physical counterpart. In fault detection and diagnostics, digital twins are used to simulate, monitor, and predict the health and operation of mechatronic systems, supporting predictive maintenance and operational optimization by providing insights that are difficult to obtain through physical monitoring alone. Digital twins are continuously updated with live operating data from the plant environment.

This method allows for simulation-based diagnosis and what-if analysis to simulate failures. By combining models with real-time data, it supports predictive analytics for monitoring conditions, estimating outcomes, and scheduling maintenance proactively and efficiently.

Digital twins are often used in settings like semiconductor lithography systems, wind turbines, and particularly industrial robot arms and manipulators where it is impractical to build fully redundant physical systems or to install sensors at every site.

Fault detection and diagnostics methods in mechatronic systems are rapidly evolving and combine knowledge from different fields. Traditionally, experts use detailed physical models to understand and analyze faults, which is especially important in areas like aerospace and automotive where safety and accuracy matter. However, newer data-driven and machine learning methods are becoming more important because they can adapt and scale to different systems and work well for real-time diagnostics. By combining both approaches and using technologies like digital twins, industries are improving predictive maintenance and the way people and machines work together. Table 13.4 compares the strengths and drawbacks of principal approaches used for fault detection and diagnostics.

Table 13.4 Comparison of fault detection and diagnostic approaches

Approach	Pros	Cons	Typical Use
Model-based	Physically interpretable, supports root cause analysis	Requires detailed models; sensitive to uncertainties	Aerospace, automotive, precision systems
Data-driven	Learns complex/nonlinear patterns, less expert input	Requires large datasets; less interpretable (often)	Production lines, mass manufacturing
Hybrid (model + data)	Leverages strengths of both; adaptable	Integration complexity; overhead in model management	High-value/critical systems, robotics

13.5 Standards Commonly Used in Mechatronic Design

Standards and guidelines are essential in developing mechatronic products because they provide a shared basis for ensuring quality, safety, and compatibility. They help products comply with both regulatory standards and industry guidelines, support effective communication between diverse teams, and encourage the use of best practices during every stage of design and production. Adhering to these standards not only lowers risks but also streamlines validation, enhances documentation, and increases the reliability and efficiency of mechatronic systems. These factors are essential benefits for industries that are highly regulated and of significant value.

Additionally, standards make it easier for various engineering fields such as mechanical, electrical, and software to work together under a single development framework. By streamlining integration, standards help speed up both deployment and acceptance in the market, since products that meet well-known standards tend to gain trust and widespread adoption worldwide.

As mechatronic products have advanced, valuable knowledge has led to the creation of industry guidelines that support efficient design in this field. These standards outline key steps and strategies for developing mechatronic solutions that are both practical and cost-effective. Detailed documentation and thorough validation, backed by traceable requirements and risk assessments, ensure compliance with regulations and simplify deployment, especially in industries with strict oversight (Vasic & Lazarevic, 2008).

13.5.1 Safety Standards

Safety engineering involves the processes of hazard identification, risk assessment, transitioning to safe states, and the use of watchdog timers that are typically incorporated during the development and mitigation. Principles such as fail-safe design, redundancy in hardware and software, transitioning to safe states, and the use of watchdog timers are typically incorporated during the development phase. The following standards ensure the protection of users and operators from mechanical and electrical hazards:

- **ISO 26262**: Functional safety for automotive systems.
- **IEC 61508**: Functional safety of electrical/electronic/programmable electronic systems.
- **ISO 13849**: Safety for machinery-related control systems.
- **IEC 60204-1**: Safety of machinery—Electrical equipment of machines. Covers wiring, protection devices, and control circuits for industrial machinery.
- **ISO 12100**: *Safety of machinery—General principles for design*. Provides a framework for risk assessment and reduction during design.
- **ISO 13849-1**: Safety-related parts of control systems. Focuses on the reliability of control systems used in safety functions.

13.5.2 Electromagnetic Compatibility

Standards designed to prevent interference between mechatronic systems and electronic devices include:

- **IEC 61000-6-2**: Immunity for industrial environments. Ensures systems can withstand electromagnetic disturbances.
- **IEC 61000-6-4**: Emission standard for industrial environments. Limits electromagnetic emissions from mechatronic devices.

13.5.3 Environmental Standards

The purpose of the following standards is to promote sustainability and compliance with environmental regulations:

- **RoHS (Restriction of Hazardous Substances)**: Limits the use of specific hazardous materials in electrical/electronic equipment.
- **WEEE (Waste Electrical and Electronic Equipment Directive)**: Governs the disposal and recycling of electronic systems.
- **ISO 14001 (Environmental management systems)**: Encourages eco-friendly design and manufacturing practices.

13.5.4 Functional and Design Standards

These standards guide the integration of mechanical, electrical, and software components:

- **VDI/VDE 2206**: Design methodology for mechatronic systems. A German guideline offering a structured approach to mechatronic product development, including modeling, simulation, and validation.
- **ISO/IEC 15288**: System life cycle processes. Defines processes for system design, development, and maintenance.

13.6 Mechatronic Design Workflow and Standards

The mechatronic design workflow ensures systematic integration of multidisciplinary engineering efforts, from the initial requirements phase through conceptualization, detailed design, implementation, and verification. Several key regulations and standards apply to mechatronic design and product development. While the regulations and standards vary depending on the industry and the specific application, referring to these standards throughout the product development process enhances quality, safety, and regulatory compliance. Furthermore, standards and guidelines support efficient collaboration among engineering teams. Table 13.5 shows how various stages of mechatronics product development correspond to applicable standards.

13.7 Example: Design Concept for an Automated Warehouse Conveyor System

13.7.1 Background

Automated conveyor systems form the backbone of modern industrial mechatronics, meeting the ever-escalating demands for efficiency, flexibility, quality, and traceability in manufacturing, warehousing, and distribution sectors. These systems not only transfer items from one point to another but also operate with a high level of autonomy, precision, and adaptability. Achieving this sophistication requires the seamless integration of mechanical components (such as belts, rollers, actuators), electrical hardware (sensors, motors, controllers), digital control logic and PLCs, human machine interface (HMI), and supervisory control and data acquisition (SCADA) systems. Essential operations, including part detection, conveyor movement, accurate stopping at designated

Table 13.5 Mechatronic design workflow and relevant standards

Design Phase	Focus Area	Relevant Standards
1. Requirements and planning	System life cycle, documentation	ISO/IEC 15288 (system life cycle), VDI/VDE 2206 (mechatronic design methodology)
2. Conceptual design	Risk analysis, safety planning	ISO 12100 (safety principles), ISO 13849-1 (control system safety)
3. Detailed design	Electrical, mechanical, software	IEC 60204-1 (electrical safety), IEC 61000-6-2/6-4 (EMC), ISO 25119 (Ag machinery)
4. Integration	Interoperability, interfaces	ISO 10303 (STEP for data exchange), OPC UA (industrial communication)
5. Testing and validation	Functional safety, performance	ISO 26262 (automotive functional safety), IEC 61508 (general functional safety)
6. Deployment	Environmental compliance	RoHS, WEEE, ISO 14001 (environmental management)
7. Maintenance and updates	Life cycle support, documentation	ISO/IEC 15288 (life cycle), ISO 9001 (quality management)

positions, and robotic arm activation, are executed in coordinated sequences regulated by control algorithms and real-time feedback systems.

13.7.2 Problem

Design an automated conveyor system for high-speed sorting with dynamic routing in warehouse distribution.

Include the following in the design report:

1) Provide a conceptual and technical overview of automated conveyor systems.
2) Develop a block diagram.
3) Explain principal components—sensors, actuators, and controllers.
4) Discuss possible control logic and sequential operation of key functions.
5) Recommend safety functionalities, HMI/SCADA integration, energy management solutions, and predictive maintenance capabilities.

13.7.3 Design Report

13.7.3.1 Overview of the Proposed Design Concept

Automated conveyor systems integrate mechanical, electrical, and computational components to automate the transport, sorting, accumulation, and delivery of materials along predefined routes in manufacturing, logistics, or processing facilities. Modern conveyor systems use integrated electronics and software for closed-loop feedback and coordinated operations, unlike older manual or semi-automated versions.

Conveyor automation is common due to its efficiency, lower error rates, scalability, reduced labor costs, higher throughput, and better safety. Various conveyor types such as belt, chain, roller, slat, bucket, pneumatic, and overhead are adaptable for different sizes of materials.

Modern systems employ distributed smart sensors, high-efficiency motors, servo-driven actuators, and robust industrial controllers (PLC/IPC) that collectively support advanced features like variable speed, intelligent routing, real-time diagnostics, and energy optimization. Sophisticated communication protocols and integrated safety architecture make such systems enable real-time data analytics, and seamless HMI/SCADA UIs.

13.7.3.2 Block Diagram for Conveyor Control

The proposed system includes the following elements:

- **Input and Signaling**: The operator interacts with the system through HMI/SCADA units, start/stop commands, operator settings, logic triggers from equipment.
- **Sensors**: Devices providing real-time feedback about process variables (e.g., object presence, speed, position, weight, faults).
- **Controller**: Usually a PLC or industrial PC, processing sensor inputs, running control logic, and generating actuator commands.
- **Actuators**: Components effecting physical change—motors, robotic arms, diverters, stoppers—based on controller output.
- **HMI** and **SCADA** monitoring/warning system.
- **Safety Circuits and Interlocks**: Emergency stops, guard interlocks, safety relays, ensuring compliance with industry safety and standards.
- **Communication Links**: Wiring/networks transmitting signals (digital/analog) and data (Ethernet, fieldbus, wireless, etc.).

Figure 13.1 shows a conceptual block diagram of the automated conveyor system for high-speed sorting with dynamic routing in warehouse distribution. Principal functions of the system components are described in Table 13.6.

13.7.3.3 Process Flow

Operator initiates command (start/stop) via HMI/SCADA or physical switches. Sensors detect item presence and send feedback to the PLC. PLC processes logic and actuates conveyor motors and robotic arm as per the sequence.

13.7.3.4 Description of Automated Conveyor System Components

1) **Sensors**: Sensors collect data from the conveyor system, providing the real-time process with necessary information for intelligent, adaptive, and safe control. Sensor selection and placement are pivotal for system responsiveness, fault prediction, object traceability, and overall automation reliability. Primary sensor types and their roles are explained below:

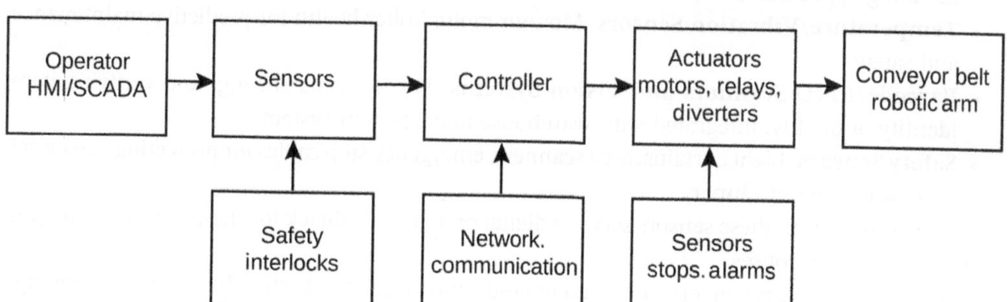

Figure 13.1 Conceptual block diagram for the conveyor belt system

Table 13.6 Key components of the automated conveyor system

Component	Principal Function	Example Devices/Variants
Sensors	Detect presence, position, speed, weight, faults, safety status	Photoelectric, proximity, ultrasonic, limit, encoders, load cells, barcodes, vision, misalignment
Actuators	Convert control signals to mechanical motion/action	Motors (AC/DC/servo), pneumatic/hydraulic cylinders, solenoids, brakes, stoppers, diverters, robotic arms
Controller	Execute control logic, process inputs/outputs, coordinate sequence	PLC (Siemens S7, Allen Bradley, Delta), IPC, microcontroller, distributed I/O modules
HMI/SCADA	Provide operator interface, visualization, diagnostics, alarms	Touch screens, panels, software dashboards, mobile HMI, SCADA servers
Safety devices	Ensure safety and compliance, intervene on faults/emergency	E-stop switches, guarded interlocks, light curtains, relays, safety PLCs
Communication	Link components, provide real-time data exchange	Modbus, Profinet, EtherCAT, Ethernet/IP, fieldbus, wireless, OPC UA
Power equipment	Supply energy and protection	High-efficiency drives, inverters, VFDs, panels, sensors for monitoring
Process stations	Places for special action (inspection, picking, packing)	Quality station, robotic hand-off, sorter, weighing station

- **Photoelectric Sensors**: Use light beams to detect presence/absence or counting of objects on the belt.
- **Proximity Sensors**: Ultrasonic noncontact detection where physical touch is undesirable.
- **Limit Switches**: Physical contact-based; used for end-of-travel detection, safety guard interlocking, and confirming positions of mechanical components such as diverters.
- **Encoders**: Rotary or linear encoders measure conveyor or robotic arm positions, speeds, and direction, enabling position control, synchronization, and variable speed regulation.
- **Load Cells**: Weigh items in motion; critical for batching, overload protection, and process control.
- **Belt Misalignment/Tear Sensors**: Early warning for belt drift/damage, crucial in bulk handling applications to prevent downtime.
- **Temperature/Vibration Sensors**: Monitor motor/roller health for predictive maintenance and safety.
- **Barcode/RFID Scanners and Vision Systems**: For tracking, sorting, and verifying item identity or quality, integrated with warehouse management system.
- **Safety Sensors**: Light curtains, area scanners, emergency stop cables for protecting personnel around moving machinery.

The outputs from these sensors serve as digital or analog feedback for the controller to trigger precise, rapid responses.

2) **Actuators**: Actuators convert control commands into mechanical action like moving, stopping, diverting, or manipulating items on the conveyor line. Essential actuators to be used in the system are explained below:

- **Motors**: The following motor types will be used as primary source of motion:
  - DC Motors to adjust the speed, allow rapid reversals, and precise control.
  - Stepper/Servomotors for accurate position control in robotic picking and diverting packages.
- **Brakes (Friction, Electromagnetic, Hydraulic)**: Safely halt the conveyor belts, hold items in place during stops, and decelerate loads to avoid product damage.
- **Robotic Arms/End-Effectors**: Perform pick-and-place, or orientation adjustment.

The actuators' efficiency, type, and control method (on–off, open-loop, or feedback control) will be selected for each actuator based on the speed accuracy, and energy consumption.

3) **Controller**: An external PC executes special software to operate system functions, supporting complex high-throughput lines, extensive data and logging needs, real-time vision processing, and integration with SCADA systems. The PC communicates with sensors and actuators through serial bus. Local controllers for each actuator/motor include on–off or PID elements.

13.7.3.5 Sequence of Operations

1) **Operational steps**
   - *System Initialization*
     - Operator starts the system via remote panel.
     - Safety interlocks are checked; all E-stops and guards must be reset.
     - Robotic arms are homed.
   - *Part Detection*
     - Start-of-line photoelectric sensor signals the approach of a part/package.
     - Controller starts conveyor motor.
     - Downstream proximity sensors confirm part presence at the predefined stations for inspection, labeling, or picking.
   - *Conveyor Movement and Position Control*
     - Encoder feedback maintains or adjusts speed for accumulation, sorting, or batch operations.
     - Parts are stopped at specific stations via diverters when required.
   - *Station-Specific Operations*
     - Robotic arm is triggered by the controller upon part arrival confirmation.
     - Arm picks the part from conveyor and places it into a bin or onto another conveyor.
     - Sensors confirm successful pick/place event. Alarms if missing.
   - *Exit/Hand-Off*
     - After operation completion, conveyor resumes.
     - Additional weight or barcode sensors may verify completion.
     - System can pause for jam detection, misalignment, or missed pick-up events (with alarms).
   - *Fault Handling and Emergency Stop*
     - If a sensor or actuator detects a fault (overload, jam, misalignment), controller stops relevant sections and indicates error via lights and buzzers.
     - The safety system allows reset only by authorized personnel after diagnostics.
   - *Maintenance/Predictive Checks*
     - Diagnostic or maintenance mode can be entered for sensor/actuator health checks, with noninterfering slow motion or manual control as required.
2) **Human-Machine Interfaces (HMI) and SCADA Integration**: Sophisticated HMI panels and SCADA systems are essential in modern conveyor automation for real-time system

visibility, diagnostics, and control. HMI/SCADA implementations range from panel-mounted screens to PC screens. Such systems offer the following functions and benefits:

- **Visualization**: Real-time display of conveyor operation—active motors, sensor status, alarms, productivity metrics—using graphical dashboards.
- **Manual Override**: Operator can start/stop belt segments, reset faults, or trigger station-specific operations.
- **Parameter Tuning**: Adjust speed setpoints, timing intervals, batching logic, or actuator force settings.
- **Diagnostics**: Immediate fault identification, guided troubleshooting, and alarm history.
- **Remote Access**: Web- and mobile-compatible interfaces support remote monitoring and control across plant networks.
- **Data Logging**: Complete operation histories to support process optimization, quality analytics, and maintenance planning.

3) **Safety Systems and Interlocks**: Safety, regulatory compliance, and risk reduction are cornerstones of conveyor automation. Safety integration must meet both regulatory requirements (OSHA, ANSI, EN/ISO 13849, IEC 62061) and industry best practices. Key safety devices used in the designed system will include:
   - **Emergency Stops (E-Stop)**: Strategically placed, directly cuts power to motors and control circuits.
   - **Guard Interlocks**: Prevent system activation when gates or guards are open. They send stop commands and prevent restarting until safe state is confirmed.
   - **Safety Relays**: Provide reliable, fail-safe signal processing, and meet required PL (performance level) for the application.
   - **Lockout/Tagout (LOTO)**: Ensures complete de-energization during maintenance.
   - **Light Curtains/Scanners**: Create protected zones; halt operation when penetrated.
   - **Belt Guarding**: Protects personnel against nip points, moving parts, overhanging loads.
   - **Safe Torque Off (STO)**: Integrated with motor/drive electronics for instantaneous de-energization.
   - **Warning Alarms**: Audible/visual signals for status, startup, abnormal conditions.

13.8 Summary

Mechatronic system design integrates mechanical, electrical, electronic, and software engineering to develop intelligent, high-performance systems. This interdisciplinary approach emphasizes concurrent engineering, system-level optimization, and advanced modeling and control techniques to meet evolving industrial demands.

- **Interdisciplinary Integration**: Mechatronic systems combine mechanical, electrical, electronic, and software components synergistically to enhance agility, precision, and intelligence across applications from household appliances to autonomous vehicles.
- **Concurrent Engineering Approach**: Design involves parallel development of subsystems with iterative integration to detect conflicts early, focusing on system-level performance, reliability, cost, and flexibility rather than isolated components.
- **Mechanical Design Principles**: Mechanical elements are dimensioned considering forces, inertia, and material properties, supported by FEA for optimized geometry, manufacturability, and lightweighting. Maintenance and additive manufacturing influence design choices.

- **Electrical and Electronic Design**: **Selection** of motors and actuators depends on torque, speed, and controllability; signal conditioning ensures accurate sensor data processing; embedded processors execute control and communication tasks.
- **Advanced Control Strategies**: Beyond PID control, methods like state-space, robust, adaptive, and MPC address nonlinearities and dynamic system behaviors, implemented on microcontrollers, DSPs, or FPGAs for real-time performance.
- **Software Architecture and Methodologies**: Firmware, control logic, and UI layers are developed modularly using model-based design, version control, continuous integration, and HIL testing to ensure reliability and reusability.
- **Communication Protocols**: Robust, low-latency networks like CAN, EtherCAT, Modbus, and wireless standards enable real-time data exchange among distributed components, supporting modular and scalable systems.
- **Modeling, Simulation, and Verification**: Multiphysics modeling tools and simulation platforms facilitate virtual prototyping, control design, and system validation, reducing reliance on physical prototypes and enhancing robustness through HIL testing.
- Fault Detection and Diagnosis include model-based, hardware-based, and data-driven, and hybrid approaches.
- **Standards and Safety**: Compliance with international standards ensures functional safety, with design practices incorporating hazard analysis, redundancy, fail-safe mechanisms, and thorough documentation for regulated industries.

13.9 Check Your Understanding

13.9.1 Self-Assessment Quiz

1 Mechatronic systems integrate mechanical, electrical, electronic, and software engineering to develop intelligent systems.
True False

2 The V-model is used in mechatronic system design to ensure traceability from requirements through validation.
True False

3 PID control is the only control strategy used in mechatronic systems.
True False

4 Digital twins are used for continuous performance monitoring and predictive analytics in mechatronic systems.
True False

5 Which of the following is NOT a principle of mechatronic design?
a. Concurrent development of hardware and software components
b. Iterative integration and refinement
c. Functional partitioning
d. Sequential design approach

6 Which control strategy is known for its simplicity and robustness in industrial systems?
   a. State-space control
   b. Robust control
   c. PID control
   d. Adaptive control

7 What is the primary purpose of using finite element analysis (FEA) in mechanical design?
   a. To optimize the geometry for performance and manufacturability
   b. To design electrical circuits
   c. To develop software algorithms
   d. To manage communication protocols

8 Which communication protocol is commonly used in automotive and industrial robotics for real-time, robust communication?
   a. EtherCAT
   b. CAN (Controller Area Network)
   c. Modbus
   d. RS-232/RS-485

9 What is the role of embedded microcontrollers in mechatronic systems?
   a. To execute real-time control algorithms
   b. To design mechanical components
   c. To manage energy harvesting
   d. To perform finite element analysis

Answers:

1: True; 2: True; 3: False; 4: True; 5: d; 6: c; 7: a; 8: b; 9: a.

13.9.2 Questions

1 Discuss the interdisciplinary nature of mechatronic system design. How do mechanical, electrical, electronic, and software engineering components integrate to create intelligent systems? Provide examples of applications where this integration is crucial.

2 Explain the principles of concurrent engineering in mechatronic system design. How does this approach differ from conventional sequential design methods, and what are the benefits of concurrent engineering in developing high-performance systems?

3 Analyze the role of finite element analysis (FEA) in mechanical design within mechatronic systems. How does FEA contribute to optimizing geometry for performance and manufacturability, and what are its advantages in modern engineering applications?

4 Evaluate the various control strategies used in mechatronic systems, including PID control, state-space control, robust control, adaptive control, and model predictive control. Discuss the scenarios in which each strategy is most effective and the challenges associated with their implementation.

5 Describe the importance of communication protocols in mechatronic systems. Compare and contrast protocols such as CAN, EtherCAT, Modbus, and wireless standards in terms of their application domains, key features, and suitability for real-time, reliable communication.

6 Examine the role of embedded systems and real-time control in mechatronic systems. How do embedded microcontrollers, digital signal processors (DSPs), and field-programmable gate arrays (FPGAs) contribute to executing real-time control algorithms, and what are the key considerations in their design and implementation?

7 Discuss the significance of system-level methodologies such as the V-model in mechatronic system design. How do these methodologies ensure traceability from requirements through validation, and what are the advantages of using digital twins and product life cycle management (PLM) tools in this context?

8 Explore the emerging trends in mechatronic systems, including artificial intelligence (AI), digitalization, open-source platforms, and sustainability. How are these trends shaping the future of mechatronic systems, and what are the potential benefits and challenges associated with their adoption?

9 Compare and contrast model-based, hardware-based, and data-driven approaches to fault detection and diagnosis in mechatronic systems. What are the strengths and limitations of each method, and in what scenarios are they most effectively applied?

10 Discuss the role of residuals and observers (such as the Kalman filter) in model-based fault diagnosis. How do these techniques help identify faults, and what challenges might arise when implementing them in complex mechatronic systems?

11 Explain the importance of hardware redundancy in fault detection for safety-critical mechatronic applications. What are the trade-offs involved in using physical redundancy, and how does it impact system design and reliability?

12 Analyze the use of machine learning techniques (such as SVM, KNN, and CNN) in data-driven fault diagnosis. What are the key steps in developing a data-driven diagnostic system, and how do these methods improve fault detection accuracy?

13 Evaluate the benefits and challenges of hybrid fault diagnosis approaches that combine model-based and data-driven methods. Why are hybrid approaches increasingly popular in modern mechatronic systems?

14 Describe the concept of digital twin technology in fault detection and diagnostics. How does the integration of real-time data and simulation-based diagnosis enhance predictive maintenance and operational optimization in mechatronic systems?

15 Discuss the objectives of fault detection and diagnosis in mechatronic systems. How do these objectives contribute to operational safety, reliability, and cost-effectiveness?

16 Examine the typical faults encountered in mechatronic systems, such as sensor bias, actuator jamming, mechanical wear, and software errors. How can early detection and accurate diagnosis mitigate the impact of these faults on system performance?

References

Chen, M., 2025. *PLC Communication Protocols Explained: Comparing Modbus, Profinet, and EtherNet/IP.* [Online] Available at: https://rabwellplc.com/blogs/news/compare-plc-protocols-for-automation [Accessed 26 10 2025].

de Silva, C. W., 2016. *Sensors and Actuators - Control System Instrumentation.* 2nd ed. Boca Raton: CRC Press.

Eureka, 2025. *Digital Communication Protocols Compared: I^2C vs SPI vs CAN for Embedded Sensors.* [Online] Available at: https://eureka.patsnap.com/article/digital-communication-protocols-compared-ic-vs-spi-vs-can-for-embedded-sensors [Accessed 29 10 2025].

Ford, R. M. & Coulston, C. S., 2008. *Design for Electrical and Computer Engineers: Theory, Concepts, and Practice.* Boston: McGraw-Hill.

Gertler, J. J., 1998. *Fault Detection and Diagnosis in Engineering Systems.* New York, NY: Marcel Dekker.

Haik, Y., 2003. *Engineering Design Process.* Pacific Grove, CA: Thomson-Brooks/Cole.

Hu, J. et al., 2025. *Research and Prospects of Digital Twin-Based Fault Diagnosis of Electric Machines.* [Online] Available at: https://www.mdpi.com/1424-8220/25/8/2625 [Accessed 29 10 2025].

Isermann, R., 2006. Fault Detection with Parity Equations. In: *Fault-Diagnosis Systems.* Berlin, Heidelberg: Springer.

Janschek, K., 2014. *Mechatronic Systems Design.* Heidelberg: Springer.

Lee, S., 2025. *Mastering Robotic Fault Detection.* [Online] Available at: https://www.numberanalytics.com/blog/ultimate-guide-robotic-fault-detection-autonomous-robotics [Accessed 29 10 2025].

Morales-Cruz, C., Ceccarelli, M. & Portilla-Flores, E. A., 2021. *An Innovative Optimization Design Procedure for Mechatronic Systems with a Multi-Criteria Formulation.* [Online] Available at: https://doi.org/10.3390/app11198900 [Accessed 16 9 2925].

Nise, N. S., 2019. *Control Systems Engineering.* 8th ed. Hoboken, NJ: John Wiley & Sons.

Ogata, K., 2002. *Modern Control Engineering.* 4th ed. Upper Saddle River, NJ: Prentice Hall.

PUSR, 2025. *The Application of Embedded Industrial Computers in Industrial Robot Control Systems.* [Online] Available at: https://www.pusr.com/blog/The-Application-of-Embedded-Industrial-Computers-in-Industrial-Robot-Control-Systems.html [Accessed 18 10 2025].

Schraft, R. D., Hagele, M., Heni, M. & Seid, R., 2000. Mechatronic System Techniques for Robots for Service Application Systems. In: C. T. Leondes, ed. *Mechatronic Systems Techniques and Applications.* London: CRC Press.

Ulrich, K. T. & Eppinger, S. D., 2012. *Product Design and Development.* 6th ed. New York, NY: McGraw-Hill Education.

UPLATZ, 2025. *Automotive Embedded Systems & Applications: A Deep Dive into Smart Vehicle Technology.* [Online] Available at: https://uplatz.com/blog/automotive-embedded-systems-applications-a-deep-dive-into-smart-vehicle-technology/ [Accessed 29 10 2025].

Vasic, V. S. & Lazarevic, M. P., 2008. Standard Industrial Guideline for Mechatronic Product Design. *FME Transactions*, 36(3), pp. 103–108.

Wang, H. et al., 2021. Real-Time Control Systems with Applications in Mechatronics. In: Y. Tian & D. Levy, eds. *Handbook of Real-Time Computing.* Singapore: Springer Nature, pp. 1–36.

Appendix A

Integral Transforms

A.1 Continuous-Time Signals and Systems

A.1.1 Laplace Transform

Laplace transform is an integral transform method that converts a time-domain function $f(t)$ into a complex frequency-domain function $F(s)$, simplifying the analysis of linear systems and differential equations. If $f(t)$ is a function defined for $t \geq 0$, its Laplace transform $F(s)$ is given by

$$F(s) = \mathcal{L}\{f(t)\} = \int_{0}^{\infty} e^{-st} f(t)\, dt$$

Laplace transform simplifies solution of ordinary differential equations by converting differentiation and integration into algebraic operations. Table A.1 lists the essential properties of Laplace Transform.

A.1.2 Fourier Transform

The Fourier transform is an integral transform method that converts a time-domain signal into its frequency-domain representation. For a continuous-time signal $f(t)$, the Fourier transform $F(\omega)$ is defined as

$$F(\omega) = \mathcal{F}\{f(t)\} = \int_{-\infty}^{\infty} f(t) \cdot e^{-j\omega t}\, dt$$

where

- ω: Angular frequency (radians/second).
- j: Imaginary unit.
- $F(\omega)$: Frequency-domain representation of $f(t)$.

Fourier transform as a spectral decomposition: It breaks a signal into its constituent sine and cosine waves. Important properties of Fourier transform are shown in Table A.2.

Table A.3 presents a combined list of commonly used Laplace and Fourier transforms.

Table A.1 Properties of Laplace transform

Linearity	$\mathcal{L}\{af(t)+bg(t)\}=aF(s)+bG(s)$
Time shifting (delay)	$\mathcal{L}\{f(t-t_0)u(t-t_0)\}=e^{-st_0}F(s)$
Frequency shifting	$\mathcal{L}\{e^{at}f(t)\}=F(s-a)$
Time scaling	$\mathcal{L}\{f(at)\}=\frac{1}{\lvert a\rvert}F\left(\frac{s}{a}\right)$
Differentiation in time-domain	$\mathcal{L}\left\{f^{(n)}(t)\right\}=s^nF(s)-\sum_{k=0}^{n-1}s^{n-1-k}f^{(k)}(0)$
Integration in time–domain	$\mathcal{L}\left\{\int_0^t f(\tau)d\tau\right\}=\frac{1}{s}F(s)$
Convolution	$\mathcal{L}\{f(t)*g(t)\}=F(s)\cdot G(s)$
Initial value	$f(0^+)=\lim_{s\to\infty}sF(s)$
Final value	$\lim_{t\to\infty}f(t)=\lim_{s\to 0}sF(s)$

Table A.2 Properties of Fourier transform

Linearity	$\mathcal{F}\{af(t)+bg(t)\}=aF(\omega)+bG(\omega)$
Time shifting (delay)	$\mathcal{F}\{f(t-t_0)\}=e^{-j\omega t_0}F(\omega)$
Frequency shifting	$\mathcal{F}\{e^{j\omega_0 t}f(t)\}=F(\omega-\omega_0)$
Time scaling	$\mathcal{F}\{f(at)\}=\frac{1}{\lvert a\rvert}F\left(\frac{\omega}{a}\right)$
Differentiation in time-domain	$\mathcal{F}\left\{\frac{d^nf(t)}{dt^n}\right\}=(j\omega)^nF(\omega)$
Integration in time-domain	$\mathcal{F}\left\{\int_{-\infty}^t f(\tau)d\tau\right\}=\frac{F(\omega)}{j\omega}+\pi F(0)\delta(\omega)$
Convolution	$\mathcal{F}\{f(t)*g(t)\}=F(\omega)\cdot G(\omega)$
Multiplication	$\mathcal{F}\{f(t)\cdot g(t)\}=\frac{1}{2\pi}F(\omega)*G(\omega)$
Duality	$\mathcal{F}\{F(t)\}=2\pi f(-\omega)$

Table A.3 Common Laplace and Fourier transforms

Time Domain $f(t)$	**Laplace Transform** $F(s)=\mathcal{L}\{f(t)\}$	**Fourier Transform** $F(j\omega)=\mathcal{F}\{f(t)\}$
Unit impulse: $\delta(t)$	1	1
Unit step: $u(t)$	$\frac{1}{s}$	$2\pi\delta(\omega)$
Unit ramp: t	$\frac{1}{s^2}$	Not defined
Exponential: $e^{-at}u(t)$	$\frac{1}{s+a}$	$\frac{1}{a+j\omega}$
Sine: $\sin(\omega t)u(t)$	$\frac{\omega}{s^2+\omega^2}$	$\pi[\delta(\omega-\omega_0)-\delta(\omega+\omega_0)]$
Cosine: $\cos(\omega t)u(t)$	$\frac{s}{s2+\omega 2}$	$\pi[\delta(\omega-\omega_0)+\delta(\omega+\omega_0)]$

A.2 Discrete-Time Signals and Systems: z-Transform

z-Transform is a mathematical tool used to analyze discrete-time signals and systems by converting time-domain sequences into a complex frequency-domain representation. It is the discrete-time counterpart of the Laplace transform. For a discrete-time signal $x[k]$, the unilateral z-transform is defined as

$$F(z) = \mathcal{Z}\{f[k]\} = \sum_{k=0}^{\infty} f[k] \cdot z^{-k}$$

where

- z is a complex variable: $z = re^{\{j\omega\}}$.
- $f[k]$ is the signal in the time-domain.
- $F(z)$ is the transformed signal in the z-domain.

z-Transform converts difference equations into algebraic equations and allows analysis of discrete time system behavior in terms of poles and zeros. z-transform is commonly used in digital filter design and stability analysis. z-transform is frequently used in digital signal processing, discrete-time control systems, digital filter design, and digital system stability analysis. Properties of z-transform are listed in Table A.4. Table A.5 presents a compilation of commonly used z-transform formulas.

Table A.4 Properties of z-transform

Linearity	$\mathcal{L}\{af(t) + bg(t)\} = aF(s) + bG(s)$
Time shifting (delay)	$\mathcal{L}\{f(t-t_0)u(t-t_0)\} = e^{-st_0}F(s)$
Frequency shifting	$\mathcal{L}\{e^{at}f(t)\} = F(s-a)$
Differentiation in time-domain	$\mathcal{L}\{df(t)/dt\} = sF(s) - f(0)$
Integration in time–domain	$\mathcal{L}\left\{\int_0^t f(\tau)d\tau\right\} = \frac{F(s)}{s}$
Convolution	$\mathcal{L}\{f(t)*g(t)\} = F(s) \cdot G(s)$
Initial value	$f(0^+) = \lim_{s\to\infty} sF(s)$
Final value	$\lim_{t\to\infty} f(t) = \lim_{s\to 0} sF(s)$
Multiplication by t^n	$\mathcal{L}\{t^n f(t)\} = (-1)^n \frac{d^n F(s)}{ds^n}$

Table A.5 Common z-transforms

$f(t)$	$f(kT)$	$F(z)$
$u(t)$	$u(kT)$	$\frac{z}{z-1}$
t	kT	$\frac{Tz}{(z-1)^2}$
t^n	$(kT)^n$	$\lim_{n\to 0} (-1)^n \frac{d}{da^n}\left[\frac{z}{z-e^{-aT}}\right]$
e^{-at}	e^{-akT}	$\frac{z}{z-e^{-aT}}$
$t^n e^{-at}$	$(kT)^n e^{-akT}$	$(-1)^n \frac{d}{da^n}\left[\frac{z}{z-e^{-aT}}\right]$
$\sin\omega t$	$\sin\omega kT$	$\frac{z \sin \omega T}{z^2 - 2z\cos\omega T + 1}$
$\cos\omega t$	$\cos\omega kT$	$\frac{z(z - \cos\omega T)}{z^2 - 2z\cos\omega T + 1}$

Appendix B

Common Fourier Series Expansions

A periodic function $f(t)$ can be expressed as a sum of sine and cosine functions:

$$f(t) = a_0 + \sum_{n=1}^{\infty}[a_n \cos(n\omega_0 t) + b_n \sin(n\omega_0 t)]$$

where

- $\omega_0 = \frac{2\pi}{T}$ is the fundamental angular frequency.
- $a_0 = \frac{1}{T}\int_0^T f(t)\,dt$ is the average (DC) component.
- $a_n = \frac{2}{T}\int_0^T f(t)\cos(n\omega_0 t)\,dt$.
- $b_n = \frac{2}{T}\int_0^T f(t)\sin(n\omega_0 t)\,dt$.

Alternatively, Fourier series can be expressed in the complex form as follows:

$$f(t) = \sum_{n=-\infty}^{\infty} c_n e^{jn\omega_0 t}$$

$$c_n = \frac{1}{T}\int_0^T f(t)e^{-jn\omega_0 t}\,dt$$

Fourier Series Coefficients for Periodic Functions Commonly Used in Mechatronics

Function	Definition Over One Period	Fourier Series Coefficients
Square wave	$f(t) = \begin{cases} +V_m & 0 < \omega t < \pi \\ -V_m & \pi < \omega t < 2\pi \end{cases}$	$a_0 = 0; a_n = 0; b_n = \dfrac{4V_m}{n\pi}$ $n = 1, 3, 5 \ldots \infty$
Pulse train with pulse width D	$f(t) = \begin{cases} +V_m & 0 < t < DT \\ 0 & DT < t < T \end{cases}$	$a_0 = V_m D$ $a_n = \dfrac{V_m}{n\pi}\sin(n2\pi D)$ $b_n = \dfrac{4V_m}{n\pi}[1 - \cos(n2\pi D]$ $n = 1, 2, 3, \ldots \infty$
Half-wave rectified sinusoid	$f(t) = \begin{cases} V_m \sin(\omega t) & 0 < \omega t < \pi \\ 0 & \pi < \omega t < 2\pi \end{cases}$	$a_0 = \dfrac{V_m}{\pi}$ $a_n = \dfrac{2V_m}{(n^2 - 1)\pi}$ $b_n = \dfrac{V_m}{2}$ $n = 2, 4, 6 \ldots \infty$
Full-wave sinusoid rectified (single phase)	$f(t) = \lvert\sin(\omega t)\rvert$	$a_0 = \dfrac{2V_m}{\pi}$ $a_n = \dfrac{4V_m}{\pi} \cdot \dfrac{1}{n^2 - 1}$ $b_n = 0$ $n = 1, 3, 5 \ldots \infty$
Full-wave sinusoid rectified (three phase)		$a_0 = \dfrac{3V_{m,L-L}}{\pi}$ $a_n = \dfrac{6V_{m,L-L}}{\pi(n^2 - 1)}$ $b_n = 0$ $n = 6, 12, 18 \ldots \infty$

Appendix C

Symbols, Units, and Conversions

C.1 Metric Prefixes

Prefix	Symbol	Multiplier	Power of Ten
tera	T	1,000,000,000,000	10^{12}
giga	G	1,000,000,000	10^{9}
mega	M	1,000,000	10^{6}
kilo	k	1,000	10^{3}
hecto	h	100	10^{2}
deca	da	10	10
none	—	1	1
deci	d	0.1	10^{-1}
centi	c	0.01	10^{-2}
milli	m	0.001	10^{-3}
micro	μ	0.000001	10^{-6}
nano	n	0.000000001	10^{-9}
pico	p	0.000000000001	10^{-12}

C.2 Derived Units Frequently Used in Mechatronics

Quantity	Math. Definition	Metric Unit (SI)	Imperial Unit (IS)
Velocity	$\frac{dx}{dt}$	Meter per second (m/s) 1 m/s = 3.28 ft/s Kilometer per hour (km/h) 1 km/h=	Feet per second (ft/s) 1 ft/s = 0.3048 m/s Miles per hour (mph) 1 mph=
Acceleration	$\frac{d^2x}{dt^2}$	m/s^2 $1\ m/s^2 = 3.28\ ft/s^2$	ft/s^2 $1\ ft/s^2 = 0.3048\ m/s^2$
Force	$F = ma$	Newton (N) 1 N = 9.81 kg·m	Pound-force (lbf) lbf = 32.174 lb·ft

Quantity	Math. Definition	Metric Unit (SI)	Imperial Unit (IS)
Torque	$T = rF$	Newton-meter (N·m)	foot-pound force (ft·lb or lb·ft)
Work/Energy	$W = \vec{F}\vec{x}$	Joule (J) 1 J = 1 N·m 1 J = 0.73756 ft·lb 1 kJ = 0.948 BTU	foot-pound (ft·lb) 1 ft·lb = 1.3558 J British Thermal Unit (BTU) 1 BTU = 1,055.06 J
Power	$P = \frac{dW}{dt}$	Watt (W) 1 W = 1 J/s 1 kW = 1.3596 HP	Horsepower (HP) 1 HP = 0.735 kW 1 HP = 0.745 $kW_{electric}$ 1 HP = 2,509.6 BTU/h
Pressure	$p = \frac{F}{A}$	Pascal (Pa) 1 Pa = 1 N/m^2 1 bar = 100,000 Pa 1 Pa = 0.000145038 psi	pound per square inch (psi) 1 psi = 6,894.76 Pa

Appendix D

Analog Circuit Analysis Methods

The goal of circuit analysis is to find voltages and currents needed for practical purposes such as calculation of power, energy, or evaluation of performance characteristics of the given circuit. Several circuit analysis methods exist. The combination of Kirchhoff's and Ohm's laws is sufficient to find voltages and currents in many practical circuits. In more complicated circuits, the number of unknowns in equations reaches larger numbers that make numerical solutions challenging. Electrical engineers have developed several network analysis methods to minimize the number of equations to solve simultaneously. The designer must choose the proper method depending on the problem. The two commonly used methods summarized in this section are based on application of Kirchhoff's current and voltage laws with Ohm's law. We will discuss methods to analyze transient behavior of circuits that involve energy-storing elements in a later chapter.

In mechatronics problems, it is convenient to model complex systems using electric circuit analogy. Mechatronic systems contain electronic circuits, magnetic elements, motors, mechanical, hydraulic, and pneumatic parts. Replacing each element with their electrical analog and building an equivalent electric circuit enables the designer to implement well-established network analysis methods or electric circuit simulation software like PSPICE® or Multisim®.

D.1 Nodal Analysis

The goal of the nodal analysis is to write a set of algebraic equations where the independent variables are unknown node voltages referred to as a common reference (ground) node. To obtain nodal equations, you can follow the steps listed as follows:

1) Select one reference node: Reference node is often marked with a ground symbol since it defines the zero potential for node voltages. Usually, the reference node has most elements tied to it. All other nodes are referenced to this node.
2) Label the remaining node voltages. If a node is not connected to a voltage source, then its voltage is an independent variable.
3) Select reference directions for branch voltages and currents. Reference directions are arbitrary, but since we will use the passive notation, the current arrow enters the element from the terminal marked with a positive sign. After solving the circuit, if the current value happens to be negative, we will understand that the actual direction is opposite of the selected reference.
4) Apply KCL at each node labeled as an independent variable. It is usually convenient to add all currents entering and subtracting all currents leaving the node.

5) Express branch currents in terms of the adjacent node voltages using Ohm's law. To simplify the upcoming arithmetic operations, it is better using the conductance (rather than the resistance) and writing Ohm's law as $I = GV$.
6) Arrange the equations such that all unknown node voltages are on the left side and all known source currents or voltages are on the right side.

In a circuit with n nodes with independent voltages, this yields to $n-1$ equations with $n-1$ unknowns. If there are more than two equations, it is more advantageous to use the matrix form to enable the use of linear algebra methods, MATLAB®, or other software.

Example D.1 Write the nodal equations for the circuit shown in Figure D.1. Upper case voltages are node voltages with respect to the ground. Variables in the lower case represent voltages and currents of each element.

Solution

1) The circuit has five nodes, and the node voltages are numbered according to the node numbers. Bottom node is selected as the ground, and each node voltage is defined with respect to this reference potential. This selection eliminates one of the nodal equations because the node voltage V_1 is known since the voltage source is connected between the ground and node 1.
2) The voltage and current references are shown on the circuit diagram. We are going to number the node voltages with the corresponding node numbers and the branch currents with the number of the resistors. When writing the KCL equations, the signs of the currents coming into a node will be positive and the signs of the currents leaving a node will be negative.
3) Apply KCL to all nodes other than the ground.

   Node 1: The voltage source is connected between N1 and ground. Hence, the voltage of this node is known as V_S. There is no need to write an equation for this node.

   Node 2: $i_1 - i_2 - i_5 = 0$

   Node 3: $i_2 - i_3 - I_4 = 0$

   Node 4: $i_3 + i_5 - i_6 = 0$
4) Write Ohm's law in the form $i = GV$, where $G = 1/R$ for each resistor to obtain branch currents in terms of the branch voltages. For the voltage across each element, substitute the potential

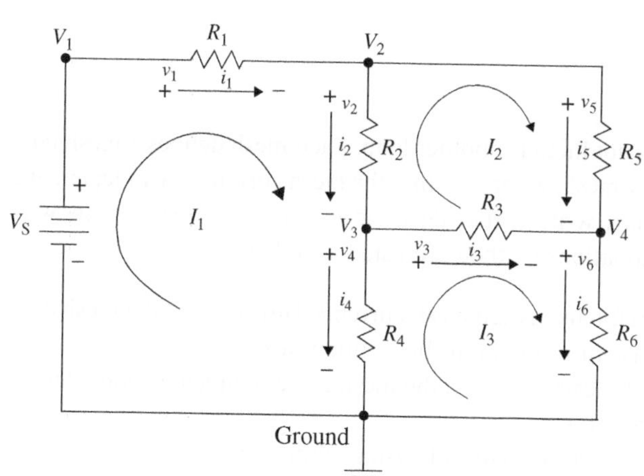

Figure D.1 Circuit diagram for Examples D.1 and D.2

difference $V_k - V_l$, where V_k is on the positive side of the voltage reference (arrow tail) and V_l is on the negative side of the voltage reference (arrowhead).

$$i_1 = G_1(V_1 - V_2) = G_1(V_s - V_2)$$
$$i_2 = G_2(V_2 - V_3)$$
$$i_3 = G_3(V_3 - V_4)$$
$$i_4 = G_4(V_3 - 0) = G_4V_3$$
$$i_5 = G_5(V_2 - V_4)$$
$$i_6 = G_6(V_4 - 0) = G_6V_4$$

5) Substitute branch currents obtained in step 4 into KCL equations in step 3.

$$G_1(V_s - V_2) - G_2(V_2 - V_3) - G_5(V_2 - V_4) = 0$$
$$G_2(V_2 - V_3) - G_3(V_3 - V_4) - G_4V_3 = 0$$
$$G_3(V_3 - V_4) + G_5(V_2 - V_4) - G_6V_4 = 0$$

Arrange the equations in a standard form of linear equation set.

$$(G_1 + G_2 + G_5)V_2 - G_2V_3 - G_5V_4 = G_1V_s$$
$$-G_2V_2 + (G_2 + G_3 + G_4)V_3 - G_3V_4 = 0$$
$$-G_5V_2 - G_3V_3 + (G_3 + G_5 + G_6)V_4 = 0$$

Nodal equations in matrix form:

$$\begin{bmatrix} (G_1 + G_2 + G_5) & -G_2 & -G_5 \\ -G_2 & (G_2 + G_3 + G_4) & -G_3 \\ -G_5 & -G_3 & (G_3 + G_5 + G_6) \end{bmatrix} \cdot \begin{bmatrix} V_2 \\ V_3 \\ V_4 \end{bmatrix} = \begin{bmatrix} G_1V_s \\ 0 \\ 0 \end{bmatrix}$$

D.2 Mesh Analysis

By definition, a mesh is a loop that does not include another loop. Each mesh defines a mesh current. By including a new element in every mesh, we ensure that the mesh currents are independent variables. The goal of the mesh analysis is to write a set of algebraic equations where the unknowns are mesh currents. The steps of mesh equations method are listed as follows:

1) Identify all possible meshes in the circuit and assign mesh currents. Directions of the mesh currents are arbitrary but must be consistent throughout the solution process.
2) Apply Kirchhoff's voltage law (KVL) for each mesh. In the mesh current direction, add all voltage increases and subtract all voltage drops.
3) Express branch voltages in terms of the branch currents using Ohm's law.
4) Arrange the equations such that all unknown mesh currents are on the left side and all known source currents or voltages are on the right side.

In a circuit with n meshes and m current sources, mesh analysis method results in $n - m$ independent equations. The unknown mesh currents are the $n - m$ independent variables. If there are more than two equations, it is more advantageous to use the matrix form to enable the use of linear algebra methods, MATLAB®, or other software.

Example D.2 Write the mesh equations for the circuit shown in Figure D.1.

Solution

1) The circuit has three meshes as shown in the figure.
2) KVL equations for all meshes are
   Mesh 1: $+v_1 + v_2 + v_4 = V_s$
   Mesh 2: $-v_2 - v_3 + v_5 = 0$
   Mesh 3: $v_3 - v_4 + v_6 = 0$
3) Use Ohm's law to express branch voltages in terms of the branch currents.

$$v_1 = R_1 i_1 = R_1 I_1$$

$$v_2 = R_2 i_2 = R_2(I_1 - I_2)$$

$$v_3 = R_3 i_3 = R_3(I_3 - I_2)$$

$$v_4 = R_4 i_4 = R_4(I_1 - I_3)$$

$$v_5 = R_5 i_5 = R_5 I_2$$

$$v_6 = R_6 i_6 = R_6 I_3$$

4) Substitute branch voltages obtained in step 3 into KVL equations in step 2.

$$R_1 I_1 + R_2(I_1 - I_2) + R_4(I_1 - I_3) = V_s$$

$$-R_2(I_1 - I_2) - R_3(I_3 - I_2) + R_5 I_2 = 0$$

$$R_3(I_3 - I_2) - R_4(I_1 - I_3) + R_5 I_2 = 0$$

Arrange the equations in a standard form of linear equation set:

$$(R_1 + R_2 + R_4)I_1 - R_2 I_2 - R_4 I_3 = V_s$$

$$-R_2 I_1 + (R_2 + R_3 + R_5)I_2 - R_3 I_3 = 0$$

$$-R_4 I_1 + -R_3 I_2 + (R_3 + R_4 + R_5)I_3 = 0$$

Mesh equations in matrix form:

$$\begin{bmatrix} (R_1 + R_2 + R_4) & -R_2 & -R_4 \\ -R_2 & (R_2 + R_3 + R_5) & -R_3 \\ -R_4 & -R_3 & (R_3 + R_4 + R_5) \end{bmatrix} \cdot \begin{bmatrix} I_1 \\ I_3 \\ V_4 \end{bmatrix} = \begin{bmatrix} V_s \\ 0 \\ 0 \end{bmatrix}$$

Appendix E

Binary Numbers and Logic Functions

E.1 Binary Numbers

Binary numbers correspond to the physical quantities that change between two values defined as zero and one like a switch or a light source turned on or off, a magnetic field in one direction or the other, and a voltage at a specified high or low value are some examples. In most digital systems, the output of a two-state (bistable) logic element called flip-flops indicates the value of the binary number. One voltage level that a flip-flop generates corresponds to logic one, and the absence or low level of a voltage corresponds to binary zero. Information presented as one or zero is one binary digit, which is shortened as bit. Since each bit is either 0 or 1, the binary number system is more convenient than the decimal system for electronic devices, where the state of a switch can represent a binary digit or one bit of information.

Integers are normally written using a positional number system where each digit represents the coefficient of a power series. An n-digit number N is the sum of the coefficients multiplied by the powers of the base r.

$$N = a_{n-1}r^{n-1} + a_{n-2}r^{n-2} + \cdots + a_2r^2 + a_1r + a_0 \qquad \text{(E.1)}$$
$$0 \leq a_i \leq r$$

For example, the decimal number 273 is $2 \times 10^2 + 7 \times 10 + 3$. Obviously, the powers of the base number are omitted in natural language. The binary number system is based on the powers of 2, instead of powers of 10. In the natural binary code, each bit represents a power of 2, starting from 2^0 on the rightmost digit.

To convert a binary number to its decimal equivalent, each binary digit from right to left is simply multiplied by the corresponding power of 2. A decimal number can be converted to binary by performing repetitive divisions by 2 and keeping track of the quotient and the remainder. The remainder obtained from division of the given number by 2 gives the most significant digit. The quotient from the previous division is the new number and the remainder is the next digit. The divisions are repeated until the quotient is zero. The binary equivalent of the given number is the sequence of digits read in reverse order, that is, from the last remainder to the first.

Example E.1 Find the binary equivalent of 25 and convert the result back to decimal for verification.

Solution

$$\frac{25}{2} = 12 \longrightarrow r = \mathbf{1};\ \frac{12}{2} = 6 \longrightarrow r = \mathbf{0};\ \frac{6}{2} = 3 \longrightarrow r = \mathbf{0};\ \frac{3}{2} = 1 \longrightarrow r = \mathbf{1};\ \frac{1}{2} = 0 \longrightarrow\ r = \mathbf{1}$$

The remainder from right to left gives the binary number 11001 corresponding to the decimal number 25.

The binary number 11001 is converted to decimal below:

$$(11001)_{\text{base } 2} = 1 \times 2^4 + 1 \times 2^3 + 0 \times 2^2 + 0 \times 2^1 + 1 \times 2^0 = 16 + 8 + 1 = 25$$

In information technology, the term "word" refers to a fixed number of bits that the arithmetic and logic unit (ALU) of a microprocessor manipulates. Eight bits form a byte, which is the fundamental building block of data in computers. Each byte corresponds to one character. Decimal multiples of byte (kilo, mega, giga, and tera) specify the size of computer files and capacity of hardware.

Since n bits form 2^{n-1} combinations, 4-bit information can represent more than one decimal digit, which ranges from 0 to 9. In computer science, 16 combinations of a 4-bit are frequently used as hexadecimal digit. A hexadecimal digit represents decimal numbers from 0 to 9, and the first six letters of the alphabets A, B, C, D, E, and F. Octal numbers with radix 8 are also used in information technology. In binary coded decimal (BCD) coding system, the first 10 combinations of 4 bits represent decimal digits from 0 to 9 and the rest of the combinations are ignored. BCD is often used in logic integrated circuits to form decimal digits with 4 bits. In digital communication and control systems, a particular coding system known as Grey code is often used to minimize errors in analog-to-digital conversion. Table E.1 shows binary codes frequently used in digital systems to represent decimal numbers from 0 to 15.

Table E.1 Comparison of common numerical codes

Decimal	Hexadecimal	Octal	Binary	BCD	Grey Code
0	0	0	0000	0000	0000
1	1	1	0001	0001	0001
2	2	2	0010	0011	0011
3	3	3	0011	0011	0010
4	4	4	0100	0100	0110
5	5	5	0101	0101	0111
6	6	6	0110	0110	0101
7	7	7	0111	0111	0100
8	8	10	1000	1000	1100
9	9	11	1001	1001	1101
10	A	12	1010	0001 0000	1111
11	B	13	1011	0001 0001	1110
12	C	14	1100	0001 0010	1010
13	D	15	1101	0001 0011	1011
14	E	16	1110	0001 0100	1001
15	F	17	1111	0001 0101	1000

E.2 Logic Functions

A logic function is a mathematical representation of logic expressions, which include binary variables and Boolean operations. Binary variables can take one of two possible values, either 0 or 1, which correspond to the two states expressed as false or true. In electric circuits, switching elements naturally represent binary numbers. Depending on the definition, a logic one may correspond to the on (or closed) position and binary zero may correspond to the off (or open) position of a switch.

Logic functions result in a single output value expressed as true or false according to the binary variables and Boolean operations. Three basic operations of Boolean algebra are AND, OR, and NOT. In Boolean algebra, usually a dot (·) represents the AND operation and a plus sign (+) represents the OR operation. The symbol (∧) is also used for AND operation and the symbol (∨) for OR operation. The terms product and sum of the conventional arithmetic are often associated with the binary AND operation and OR operation. A bar above a variable or apostrophe (') indicates that the number is complemented. Table E.2 displays fundamental properties of Boolean algebra. Each property is expressed in two equations that are dual to each other. The first eight properties are the basic laws of Boolean algebra.

The last property, known as DeMorgan's law, converts the complement of the AND operation into the sum of the variables complemented, and the complement of the sum operation into the product of the variables complemented. De Morgan's law is an example of duality in binary operations. Duality means that a binary expression remains valid if we swap AND (∧) with OR (∨) and swap 0 with 1 and vice versa.

E.3 Minimization of a Logic Function Using a Karnaugh Map

Reducing the number of ICs will obviously reduce the material cost and the power requirement. In Boolean algebra, a minterm is a product (AND) term that includes all the variables of the function, either in the actual or complemented form. The number of minterms may be reduced by applying the properties listed in Table E.2. In particular, the associativity (7) and complement (4) properties are effective in reducing the logic functions.

Table E.2 Fundamental properties of Boolean algebra

1	Identity law	$A+0=A$	$A\cdot 1=1$
2	Null law	$A+1=1$	$A\cdot 0=0$
3	Idempotent law	$A+A=A$	$A\cdot A=A$
4	Complement law	$A+\overline{A}=1$	$A\cdot\overline{A}=0$
5	Commutativity law	$A+B=B+A$	$A\cdot B=B\cdot A$
6	Associativity law	$(A+B)+C=A+(B+C)$	$(A\cdot B)\cdot C=A\cdot(B\cdot C)$
7	Distributivity law	$A\cdot(B+C)=(A\cdot B)+(A\cdot C)$	$A+(B\cdot C)=(A+B)\cdot(A+C)$
8	Absorption law	$A+(A\cdot B)=A$	$A\cdot(A+B)=A$
9	DeMorgan's law	$\overline{A\cdot B}=\overline{A}+\overline{B}$	$\overline{A+B}=\overline{A}\cdot\overline{B}$

One of the most effective methods to minimize logic functions is using Karnaugh maps. A Karnaugh map (K-map) is a graphical representation that helps minimize complex logic expressions, making it easier to design and implement digital circuits.

A K-map is a different type of the truth table. If there are three logic variables, $2^3 = 8$ possibilities are represented in either 2×4 or 4×2 table. In the case of four variables, the 16 permutations are mapped on a 4×4 table. In a K-map, each cell represents one of these minterms. When simplifying a Boolean expression, we group adjacent 2, 4, or 8 cells (minterms) that have a value of 1. The K-map has a spherical property; the top row is adjacent to the bottom row, the left column is adjacent to the right column, and the corner cells are also adjacent. After forming the largest groups of cells, the logic term corresponding to each group is written by removing the variables that alter the state from one cell to the next. K-maps provide a visual method to simplify Boolean expressions without the need for complex algebraic manipulations.

E.4 Logic Gates

While series and parallel connections of mechanical switches can reflect the input–output relationships of these logic operations, electronic components known as logic gates are widely available in IC chips. Table E.3 shows the circuit symbols and truth tables of the frequently used logic gates.

Table E.3 Logic gates

Boolean Operation	Circuit Symbol	Mathematical Expression	A	B	Output
AND	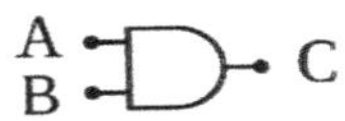	$C = A \cdot B$	0	0	0
			0	1	0
			1	0	0
			1	1	1
OR		$C = A + B$	0	0	0
			0	1	1
			1	0	1
			1	1	1
NOT	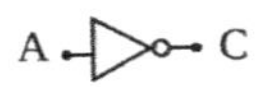	$C = \overline{A}$	0	–	1
			1	–	0
NAND	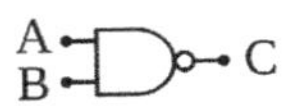	$C = \overline{A \cdot B}$	0	0	1
			0	1	1
			1	0	1
			1	1	0
NOR	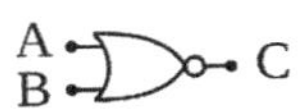	$C = \overline{A + B}$	0	0	1
			0	1	0
			1	0	0
			1	1	0
XOR		$C = A \oplus B$	0	0	0
			0	1	1
			1	0	1
			1	1	0

Appendix F

Arduino Programming

Arduino programming enables users to write code that communicates with and controls a variety of hardware components through an Arduino microcontroller. Programs are developed in a simplified C/C++ language and uploaded using the Arduino IDE, which revolves around two main functions: `setup()` and `loop()`. Common IDE commands include pin configuration, digital/analog I/O, timing, and serial communication.

The simple code below to blink a built-in LED shows the core structure of Arduino code.

- `setup()`: Runs once at startup. Used to initialize settings.

```
void setup() {
  pinMode(LED_BUILTIN, OUTPUT); // Set built-in LED pin as output
}
```

- `loop()`: Runs repeatedly. Contains the main logic.

```
void loop() {
  digitalWrite(LED_BUILTIN, HIGH); // Turn LED on
  delay(1000);                     // Wait 1 second
  digitalWrite(LED_BUILTIN, LOW); // Turn LED off
  delay(1000);                    // Wait 1 second
}
```

Essential operations include configuring pins, reading inputs, and sending outputs to devices like sensors, LEDs, and motors. Frequently used Arduino IDE Commands are listed as follows (ArduinoDocs, 2025).

F.1 Common Arduino Commands

Pin Configuration

pinMode(pin, mode)—Sets pin as INPUT, OUTPUT, or INPUT_PULLUP
digitalWrite(pin, value)—Sets digital pin HIGH or LOW
digitalRead(pin)—Reads digital pin value

Analog I/O

analogRead(pin)—Reads analog value (0–1023)
analogWrite(pin, value)—Writes PWM value (0–255)

Timing Functions

delay(ms)—Pauses for milliseconds
millis()—Returns time since program started
delayMicroseconds(us)—Microsecond-level delay

Serial Communication

Serial.begin(baudRate)—Starts serial communication
Serial.print(data)—Prints data to serial monitor
Serial.println(data)—Prints data with newline

Math and Logic

map(value, fromLow, fromHigh, toLow, toHigh)—Re-maps a number
constrain(val, min, max)—Limits value to a range
random(min, max)—Generates random number

Control Flow

if, else, for, while, switch—Standard C/C++ control structures
break, continue—Loop control
Advanced Commands
attachInterrupt()—Handles external interrupts
tone(pin, frequency)—Generates sound
pulseIn(pin, value)—Measures pulse duration
pinMode(pin, mode): Sets a pin as either an input or output.
digitalWrite(pin, value): Writes a HIGH or LOW value to a digital pin.
digitalRead(pin): Reads the value from a digital pin.
analogRead(pin): Reads the value from an analog pin.
analogWrite(pin, value): Writes a PWM value to a pin.
delay(ms): Pauses the program for a specified number of milliseconds.
millis(): Returns the number of milliseconds since the Arduino began running the current program.
Serial.begin(baudrate): Sets up serial communication at the specified baud rate.
Serial.print(data): Prints data to the serial port, useful for debugging.
Serial.println(data): Prints data to the serial port followed by a new line.
attachInterrupt(digitalPinToInterrupt(pin), ISR, mode): Attaches an interrupt to a pin for asynchronous event handling.
noInterrupts(): Disables interrupts.
interrupts(): Re-enables interrupts.
map(value, fromLow, fromHigh, toLow, toHigh): Re-maps a number from one range to another.
constrain(value, min, max): Constrains a value to lie between a minimum and maximum.

F.2 Arduino–MATLAB® Interaction

Arduino boards can be programmed and controlled using MATLAB® by installing the MATLAB Support Package for Arduino. This allows you to write high-level MATLAB® code to interact with Arduino hardware without using the Arduino IDE or C/C++. Two workflows for this interaction are summarized below.

1) **Interactive Control via MATLAB**: This method is ideal for prototyping, data logging, and algorithm testing. Arduino stays connected to your PC. MATLAB sends commands and receives data in real time. Key features of this operation mode include using read/write digital and analog pins, controlling motors, LEDs, and sensors and use of MATLAB plotting and analysis tools. The basic script below illustrates the interactive control for blinking a built-in LED on an Arduino board.

```
a = arduino();                  % Auto-detect board
writeDigitalPin(a, 'D13', 1); % Turn on LED on pin 13
pause(1);
writeDigitalPin(a, 'D13', 0); % Turn off LED
```

2) **Standalone Deployment via Simulink**: This mode of interaction is well suited for real-time control systems and embedded applications. The control algorithm, developed as a Simulink model, can be deployed directly to an Arduino device, enabling the code to execute independently on the board. The Simulink environment provides configurable blocks for interfacing with sensors, actuators, and logic; supports real-time simulation and tuning; and facilitates code generation for independent operation (Mathworks, 2025).

References

ArduinoDocs, 2025. *Language Reference.* [Online] Available at: https://docs.arduino.cc/language-reference/ [Accessed 25 10 2025].

Mathworks, 2025. *Arduino Programming.* [Online] Available at: https://www.mathworks.com/discovery/arduino-programming-matlab-simulink.html [Accessed 27 10 2025].

Index

Printed and bound by CPI Group (UK) Ltd, Croydon, CR0 4YY
21/09/2026
14965823-0002